Soil Microbiology & Biochemistry

Soil Microbiology & Biochemistry

by :

GHULAM HASSAN DAR

Professor and Head
Division of Plant Pathology
S.K. University of Agricultural Sciences & Technology of Kashmir
SKUAST-K, Shalimar, Srinagar (J&K)

New India Publishing Agency
Pitam Pura, New Delhi-110 088

Published by
Sumit Pal Jain *for*
New India Publishing Agency
101, Vikas Surya Plaza, CU Block, L.S.C. Mkt.,
Pitam Pura, New Delhi- 110 088, (India)
Phone: 011-27341717, Fax: 011-27341616
E-mail: newindiapublishingagency@gmail.com
Web: www.bookfactoryindia.com

ISBN : 978-93-80235-13-4

Typeset at: Typographiya # 9811482328
Printed at: Jai Bharat Printing Press, Delhi

Foreword

The processes that operate in the soil-plant-microbe continuum has emerged as a fascinating branch of agricultural science. The Mother Nature has created soil through continuous weathering process/disintegration of rocks over million years. This has also resulted in the decomposition of litters of vegetation residues, producing as a consequence a complete porous mass of inorganic and organic substances as well as micro-flora and fauna. Organic carbon and its complex compounds act as a source of energy, making the soil as a life-sustaining medium. The microorganisms harbouring in soil play a dynamic role. Some are responsible for fixing nitrogen, solubilizing phosphorus and decomposing the organic biomass through bacteria, actinomycetes, moulds, fungi etc. The dynamic activity enriches the soil which, in turn, makes the essential nutrients available to the plants. There are, however, some microbes that are detrimental to the living system as well.

The discipline of soil microbiology has been amazing, diverse and exciting and has provided a link for many much sought biological processes, occurring in nature, which are of greater significance to the mankind. It holds great promise in restructuring agriculture, forestry and fermentation industries. The information generated during past two centuries have transformed the concept of microbes from descriptive and purely academic nature to a truly interdisciplinary applied science. The soil microbiology is still in the process of revealing new dimensions of microbial potential and applications with thrust on bioengineering and biotechnological avenues. It has greatly helped in developing new models for sustainable organic agriculture. With the growing appreciation of microbial contributions in agriculture, horticulture and forestry through taking care of wear and tear of soil and maintaining better soil health, many microbial processes have been found indispensable to man's survival on earth. The significant role of soil biological components in soil development, organic matter recycling, nutrient transformations, biological nitrogen fixation, sequestering of pollutants, disease control and production of various chemicals involved in plant growth promotion, etc. or as biofertilizers, biopesticides, biofuel and bioenergy source have become the spheres of greater interest.

The book **'Soil Microbiology and Biochemistry'** provides comprehensive information on basic and applied aspects of soil microbes and has been designed as per the requirements of undergraduate and postgraduate students. The book has revealed many facets of microbial diversity and activities, which have direct and indirect bearing on plant growth. The chapters in the book are well-organized with the objective of familiarizing the students and researchers with the basic fundamental concepts of the field. The flow of information in the book appears spontaneous and well-connected. It is written in lucid style and comprehensive manner, supported by appropriate illustrations and tables. At the end of each chapter selected references are given for further reading by the interested readers. The glossary describes the terms frequently used in the field.

Dr. G.H. Dar, author of this book deserves compliments for the painstaking work in writing this book which would fill information gap that exists in this specialized field. This will severe as a text book for graduate & post-graduate students and scientists.

Prof. ANWAR ALAM

Vice Chancellor
S.K. University of Agricultural
Sciences & Technology of Kashmir
Shalimar, Srinagar (J&K)

Preface

Soil is a basic vital component of agriculture. This natural resource has not been fully explored nor exploited with respect to its live components, their enormous biochemical activities and soil processes. The huge hidden world present in soil has yet remained a great mystery although several folds of diversity, versatility, function and biochemical activities of its various components have been explored. These revelations have helped the mankind in improving its standard of living and cultivation practices.

Soil microbiology involves wide ranging multidisciplinary fields and in this book information on the fundamental and applied aspects of soil microbiology have been covered in simple comprehensive manner. The book has primarily been aimed at to adequately introduce **Soil Microbiology and Biochemistry** with thrust on understanding the various microbial processes occurring in soil. The book is expected to be useful to undergraduate and postgraduate students, teachers and researchers dealing with agriculture, horticulture and forestry in general and agricultural microbiology, soil science and environmental sciences in particular.

The issues related to soil fertility, microbial ecology and environment, diversity and biochemical activities, soil health, biological nitrogen fixation, phosphorus solubilisation, nutrient cycling, mineral transformations, waste recycling, pesticide-microbe interaction, biofertilizers/ biopesticides, soil protection, plant-soil microbe-continuum and soil biotechnology are some of the aspects dealt comprehensively in this book. Though good knowledge has been accumulated on various processes of mineral transformations occurring in soil, there is urgent need to intensify the efforts to integrate these processes so as to harness maximum dividends from the knowledge generated. These processes have wide role in achieving sustainable crop growth and soil fertility. The book is designed to provide an insight about the current trends and concepts in the field of soil microbiology supported with tables, diagrams and glossary of the terms frequently used in the field. Those desirous to have deep insight in the subject may consult the selected references given in the end of each chapter.

I express my sincere thanks to all the colleagues in the S.K. University of Agricultural Sciences & Technology of Kashmir, Shalimar, Srinagar, J&K (SKUAST-K) who one way or the other assisted me in completing this book. I am especially grateful to Dr. Javid A. Wani, Head, Division of Soil Sciences and Dr. M.A. Beig, Dr. Nissar A. Qazi, Dr. Vinay Sagar, Dr. M.D. Shah, Dr. T.A. Shah and Mr. N.A. Ganai of the Division of Plant Pathology, SKUAST-K who critically evaluated the manuscript and gave valuable suggestions to improve the manuscript. I am grateful to all Indian expatriate experts serving in the Department of Plant Sciences, Alemaya University, Ethiopia during 1999-2001 whose encouragement prompted my inspiration to proceed ahead with this assignment. Thanks are due to my wife Mrs. Shamima Jahan for endurance and patience and my children, Arbaz Hassan and Andleeb Hassan, for the preparation of neat and beautiful diagrams.

Although extra-ordinary care has been taken in presenting the information, yet unintentional mistakes can not be overruled. The errors and omissions are, therefore, expected. I shall be highly obliged if the reader's provide me their perception in the form of constructive comments and suggestions so that the quality of book can be improved further.

GH. HASSAN DAR

Contents

Foreword .... *v*
Preface .... *vi*

Chapter-1
Soil Microbiology - Origin, History and Diversification 1
1.1 Origin of Life and Soil Microbes 3
1.2 Origin of Soil 7
1.3 Major Historical Milestones in Soil Microbiology 8
1.4 Diversification in Soil Microbiology 15
1.5 Microorganisms and Soil Health 17
Selected References 20

Chapter-2
Soil Biota - Structure and Function 23
2.1 Plant Roots and Their Contribution to Soil C Sink 24
2.2 Soil Microbial Biomass and Their Nutrient Dependence .. 24
2.3 Viruses 29
2.4 Bacteria 30
2.5 Actinomycetes 35
2.6 Cyanobacteria 38
2.7 Fungi 42
2.8 Algae 51
2.9 Lichens 55
2.10 Soil Fauna 56

2.11 Protozoa 56
2.12 Nematodes 58
2.13 Arthropods 59
2.14 Oligochaetes 62
2.15 Molluscs 64
Selected References 65

Chapter-3
Microbe - Soil Environment Interactions 69
3.1 Distribution of Soil Biota 70
3.2 Soil Structure and Microbial Habitat 74
3.3 Soil Water 77
3.4 Soil Aeration 80
3.5 Soil Redox Potential 84
3.6 Soil Reaction 87
3.7 Soil Temperature 89
3.8 Solar Radiation 92
3.9 soil Nutrients 92
Selected References 95

Chapter-4
Microbial Exploration — Methods and Analysis 99
4.1 Microscopy 99
4.2 Soil Sampling 101
4.3 Culture Methods 101
4.4 Microscopic Examinations 103
4.5 Estimation of Microbial Diversity 106
4.6 Biovolume and Biomass Estimations 108
4.7 Soil Enzymes and Their Measurement 112
4.8 Molecular Methods in Soil Microbiology 115
Selected References 118

Chapter-5
Carbon Cycle and Soil Organic Matter Transformation 121
5.1 Carbon Cycle and Global Changes 121
5.2 Primary Production and Carbon Turnover 122
5.3 Forms of Organic Carbon 125
5.4 Incorporation of Organic Matter in Soil 134
5.5 Factors Affecting the Soil Organic Matter Decomposition 139

5.6 Humus and Humic Substances - The Stable Soil Organic Fraction 141
5.7 Dynamics of Organic Matter Decomposition 152
5.8 Microbial Growth Rates in Soil 153
5.9 Reaction Kinetics in Organic Matter Transformation 157
5.10 Modelling in Organic Matter Transformations 163
Selected References 169

Chapter-6
Transformation of Nitrogen in Soil 173
6.1 Nitrogen Cycle in Nature 174
6.2 Organic Nitrogen in Soil 177
6.3 Kinetics of Nitrogen Mineralization 186
6.4 Nitrification 187
6.5 Denitrification 193
6.6 Plant uptake of Inorganic Nitrogen 205
6.7 Nitrogen Losses Through Non-Biological Ways 205
6.8 Biological Nitrogen Fixation 207
Selected References 237

Chapter-7
Transformation of Phosphorus in Soil 243
7.1 Phosphorus Cycle in Nature 244
7.2 Forms of Phosphorus in Soil 246
7.3 Solubilization of Phosphates by Microbes 249
7.4 Immobilization of Phosphorus 250
7.5 Mineralization of Organic Phosphorus 251
Selected References 253

Chapter-8
Sulphur Transformation in Soil 255
8.1 Sulphur Cycle 256
8.2 Forms of Sulphur 257
8.3 Mineralization of Sulphur Compounds 259
8.4 Assimilation of Sulphur Compounds 260
8.5 Oxidation of Inorganic Sulphur Compounds in Soil 262
8.6 Reduction of Inorganic Sulhpur Compounds in Soil 267

8.7 Environmental Consequences of SO_4 Reduction 269
8.8 Volatilization/ Leaching of Sulphur From Soil 270
Selected References 270

Chapter-9
Microbial Transformation of Metals 273
9.1 Transformation of Iron 274
9.2 Transformation of Copper 278
9.3 Transformation of Manganese 280
9.4 Transformation of Potassium 282
9.5 Transformation of Zinc 283
9.6 Transformation of Mercury 284
9.7 Transformation of Arsenic 286
9.8 Transformation of Selenium 287
9.9 Transformation of Other Metals 288
9.10 Bioleaching 292
Selected References 293

Chapter-10
Mycorrhizae - Plant Relationships 295
10.1 Mycorrhizal Morphology... 296
10.2 Mycorrhizosphere 305
10.3 Physiology of Mycorrhizae 306
10.4 Functions of Mycorrhizae 309
10.5 Factors Affecting Mycorrhizal Development.................... 318
10.6 Plant Response to Mycorrhizal Inoculations 321
10.7 Mycorrhizal Inoculants for Plant Use 322
Selected References 326

Chapter-11
Soil Microbial Interactions with Plants 329
11.1 Types of Interactions 329
11.2 Spermosphere 334
11.3 Rhizosphere 336
11.4 Plant Growth Promoting Rhizosphere Microorganisms (PGPR) 343
11.5 Deleterious Rhizosphere Microorganisms (DRMO) 344

11.6 Root-pathogen Interaction 345
11.7 Phyllosphere 347
Selected References 350

Chapter-12
Soil Microbial Products — Phytohormones, Phytotoxins and Antibiotics 355
12.1 Phytohormones 356
12.2 Phytotoxins 360
12.3 Antibiotics 361
12.4 Mycotoxicoses 366
12.5 Hydrogen Sulphide Production 370
Selected References 370

Chapter-13
Organic Waste Recycling 371
13.1 Agricultural Uses 372
13.2 Bioenergy Production 388
13.3 Aquacultural Reuse 391
13.4 Land Application 392
Selected References 392

Chapter-14
Pesticide — Microbe Interactions 395
14.1 Effect on Non-target Microorganisms 399
14.2 Persistence, Accumulation and Biomagnification 402
14.3 Tolerance or Resistance Development in Pests 407
14.4 Biodegradation of Pesticides in Soil 407
14.5 Genetics of Pesticide Biodegradation 414
14.6 Bioavailability of Pesticide / Pollutants 415
Selected References 416

Chapter-15
Biofertilizers and Biopesticides — Production and Application 419
15.1 Biofertilizers 419
15.2 Emerging Needs 422

15.3 Intensification of Biological Nitrogen Fixation in Soil 422
15.4 Nitrogen Fixing Biofertilizers 423
15.5 Phosphate Solubilizing and Mobilising Biofertilizers 431
15.6 Mycorrhizal Inoculants 432
15.7 Compost Inoculants 433
15.8 Biopesticides 434
15.9 Constraints in Biofertilizer and Biopesticide Use 448
Selected References 450

Chapter-16
Soil Biotechnology - The Emerging Trends 453
16.1 Biotechnology in Improving The Crop Productivity 454
16.2 Strain Improvement in Microbial Symbionts 457
16.3 Plant Disease Control 457
16.4 Gene Cloning for Insecticide Resistence/ Toxin Production 458
16.5 Phyllosphere Protection 459
16.6 Microbial Inocula for Plant Growth Regulation 459
16.7 Gem for Detoxification of Contaminants/ Bioremediation 460
16.8 Biofuel and Bioenergy Generation 460
16.9 Risks Involved in Use of Modified Organisms and Biosafety of Soil Ecosystem 461
Selected References 463

Glossary 465

Index 497

Chapter 1

Soil Microbiology — Origin, History and Diversification

"The role of infinitely small is infinitely large"
(Louis Pasteur)

Definition : Soil microbiology is the study of microorganisms living in soil which complete whole or some part of their life cycle in soil. They include both beneficial as well as deleterious microorganisms but as a specialized branch main thrust is laid on their metabolic activities and the role they play in energy flow, cycling of nutrients associated with the primary productivity and soil ecosystem development. Besides, it focuses on environmental impact, whether favourable or unfavourable, on soil microorganisms and the various soil biochemical processes mediated by various microbes. A clear understanding of the interactions between soil microbial communities and elemental soil components is imperative to comprehend total ecosystem function. Atlas and Bartha have defined the soil microbiology as *'the study of organisms that live in soil; their metabolic activity; their roles in energy flow; their roles in nutrient cycling'*. As per Tate soil microbiology involves *'the study of soil ecosystem which is the product of intricate interactions between physical and chemical matrix of highly variable composition and living communities of all form'*. The Soil Science Society of America has defined soil microbiology as *'the branch of soil science concerned* *with soil inhabiting microorganisms, their functions, and activities.'* Since soil microbiology also involves the study of various biochemical transformations/ reactions occurring in soil, so it has indispensable role in understanding soil biochemistry, rather the two disciplines are inseparable.

Diversity : One kg of soil, on an average, weighs 2200 t ha^{-1} furrow slice and possesses about 2.7 t nitrogen, of which about 35% is in microbial biomass form. The number of microbes kg^{-1} soil is quite enormous (around 17 billion organisms kg^{-1}), diverse in nature and function. The organisms, excluding plants and large animal, involved in various soil metabolic processes are categorized into four major groups on the basis of their body size, *viz.*, microflora, microfauna, mesofauna and macrofauna (Table 1.1). Microflora is the group having smallest living organisms.

Table 1.1: Classification of soil organisms on the basis of their body size

Group	Body width	Organisms
i. Microflora	<10 μm	Bacteria, fungi, algae, actinomycetes and viruses
ii. Microfauna	10-100 μm	Protozoa and nematodes
iii. Mesofauna	0.1-0.2 mm	Termites, collembolan and springtails
iv. Macrofauna	0.2-20 mm	Mites, earthworms, spiders, moles, prairie dogs, millipedes, centipedes etc.

These soil organisms influence the soil properties and functioning of larger ecosystem. Every microbe has special qualities, which enable it to survive under stressful environments. Soil microbes generally show high degree of adaptability and versatility, and always work in consortia. So in real sense soil microbes act as translators in recycling process by converting chemicals to the forms which either plants or animals can use. The primary role in soil metabolic activities is that of microflora especially fungi, actinomycetes and bacteria which through their biochemical processes bring major organic matter transformations and help in maintaining the soil health. In comparison to these, micro- and meso-fauna lack such wide ranging enzyme activities.

Significance : Recent scientific developments and improvements in methodologies have revolutionized the concept of soil microbiology and biochemistry. Tailoring of an organism to perform Herculean tasks has now become a reality. The increasing role of soil microbes as detoxifiers of toxic compounds, recalcitrant, hydrocarbon and other compounds; as biological deterrent against pests and diseases; as a cheap source for antibiotic/vaccine production; as biofertilizers and modifiers of soil structure and as the main agents in biogeochemical cycling and humus/ organic matter turnover has now widely been acknowledged. Since microbes in soil have their own micro-habitats so their intricate interactions with biotic and abiotic components have helped to understand soil ecology and environment in a better way.

1.1 ORIGIN OF LIFE AND SOIL MICROBES

The component of earth bearing life, known as 'biosphere', is 22.5 km in thickness and comprises of hydrosphere (water component), lithosphere (a heterogeneous solid layer) and atmosphere (a gaseous mantle). The earth is differentiated into outer crust, middle mantle and central core. The crust is outer 8 to 40 km thick divergent solid layer made of granite. The upper mantle is solid, in contact with crust, and lower mantle is semi-molten. The crust and upper mantle constitutes lithosphere. The core is inner homogenous oceanic region formed of basalt. The present form of earth has come into existence after a series of chemical and biological evolutions in chronological order.

1.1.1 CHEMICAL EVOLUTION OF LIFE

The earth long back about 5 billion years was a molten magma (lava) without any oxygen or ozone shield. The hot gaseous nebula, with the passage of time, as a result of cooling and condensation formed a core of heavy metals (iron and nickel), mantle of moderately heavy elements and crust of light elements like carbon, oxygen, hydrogen, nitrogen, sulphur, phosphorus, potassium, etc. The atoms of early atmospheric elements (H, O, C, N and S) combined to form molecules like N_2, H_2, CO, NH_3, CO_2, H_2S, etc. in hidden crevices at rock-sea water interfaces. These molecules were deemed as the key elements to initiate precursors of life (progenote) which probably took 3 billion years. Since the emergence from 'primordial soup' (sea) the organic base of life took many strides and turns. Because of unusual bonding properties of carbon, continuous colliding of atoms with new and more complex combination and UV rays, the reaction process continued till some unstabilized molecules (probably both straight and ring chain types) became stabilized. It seems that stabilized pyrites, carbon monoxide (CO) and methyl groups combined as precursors of life to form activated acetic acid, a crucial chemical for the synthesis of other organic compounds. Clay minerals were formed before the evolution of life. The clay minerals seem to have served as hot spots for the concentration of organic monomers like formic acid, acetic acid, methylamine, etc.

Current research favours the hypothesis that clays or rock surfaces were more likely the sites for polymer formation rather than free solution. Clay crystals in fact show great variations and capability to evolve into new crystal structures. Clays consist of alternating layers of inorganic ions and water molecules. Their highly ordered lattice structure can attract organic molecules and promote chemical reactions between them, especially when clays contain abundant metallic ions. Through adsorption and accumulation of simple organic molecules, clays seem to have catalyzed their polymerization. Benzene and phenols polymerize spontaneously on Fe (III) and Cu (II) clays. Organic compounds such as amino acids may have acted as primitive catalysts of clay-crystal growth and altered the clay structure. At the same time the positive and negative charges within stacked sheets of clay or mica seem to have acted as simple catalysts to promote linking of amino acid into polypeptide called 'protenoids'. Later, pyrimidine compounds combined with phosphorus and sugars to produce nucleotides. The clay also seems to be the site for prototype nucleic acids or polynucleotides (early RNA's) formation from bases.

L-amino acids and D-sugars are capable to bind certain clays while their enantiomers do not. All macromolecules of living cells on earth contain only L-amino acid or D-sugars. Hence, the first living form on the earth seems to be free living RNA, so presumed to be primordial genome, which may have arisen from simple amino acid (probably lysine) capable of self duplication, replication and mutation. Some RNA might have served as templates for protein manufacture and finally DNA copies of RNA templates may have arisen, thereby yielding the extremely stable DNA genetic material common to most organisms on earth. Thus the chemical evolution was followed by the appearance of cellular life and its subsequent biological evolution, which dates back 3.8 to 4.2 billion years.

1.1.2 BIOLOGICAL EVOLUTION

Viruses can be considered as a link between non-living and living matter because of their possession of some properties of living things and many others of non-living. Thus viruses probably fill the hollow vacuum between these two transition stages. The evidences have led us to believe that microorganisms were first of the multitude of biological forms to appear on this planet. Whatever the mechanism of the origin of cell, it seems that first cell did not have organized structure such as nucleus. The first cell in all probability was an anaerobe, capable to survive in the absence of free oxygen or heterotrophs capable of consuming acetate, formate, amino acids, polypeptides, nucleotides, etc. that had formed in nature. The archaebacteria are the microorganisms that appears to be the remnant of primeval life that transformed prebiotic environment by using a variety of compounds as carbon and energy source before the evolution of complex life process. Archaebacteria have the potential to thrive extreme aquatic and terrestrial habitat (anaerobic, hyper-saline, hyper-pressure, ultrathermophilic, etc.). Such microbes have been active in water samples at 300°C or in polar ice-capes. Somehow mutations in some of the archean cells probably may have resulted in shift to autotrophic mode.

It has been proposed that chemosynthetic microorganisms, especially sulphur reducing and iron oxidizing bacteria (utilizing S and Fe deposits of primitive sea), were pioneer colonizers of primitive earth. Geological evidences seem to support this view. Rod shaped bacteria like organisms have been encountered on the fossils in border regions of South Africa and Swaziland, which date back 3.2 billion years. The extremely ancient rocks (stromatolites; layered stones), about 3.5 billion years old, found in Western Australia also contain a large quantity of organic carbon in 'organically preserved' microfossils, biogenic in nature, which resemble the present day microbes and include unicellular coccoid organisms and multi-cellular filamentous forms.

Changes in life followed a chronological order before communities have evolved to colonize the heterogeneous soil environment, penetrate fine pores, and adapt under extreme/stress conditions (Fig 1.1). Photosynthetic microbes and eukaryotes followed with the changes in adaptability and their capacity to synthesize their own food. Photosynthetic cyanobacteria have evolved 3.3 billion years ago. They evolved O_2, which changed the atmosphere of earth from reducing to oxidizing state. Ammonia (NH_3) was converted to nitrogen (N_2), and water (H_2O) and methane to carbon dioxide and water. Some atmospheric oxygen was converted by sunlight to ozone ($3O_2 \rightarrow 2O_3$), which began to get collected in a high-altitude layer (ozone layer) and formed a shield around the atmosphere. The build up of O_2 in atmosphere has been attributed to the photosynthetic splitting of water by primitive cyanobacteria. With the start of oxygenic photosynthesis by cyanobacteria, S^{2-} was oxidized to SO_4 and Fe^{2+} to Fe^{3+}. Later microbes participated in the mobilization of iron (in granite) and its re-precipitation while solubilization of iron seems to have been promoted by organic acids. This was followed by the evolution of aerobic cells, capable of living in presence of oxygen and even exploiting it through very efficient metabolic reactions. Cellular respiration also gave rise to carbon cycle from the land to atmosphere and back.

In contrast to the long existence of unicellular prokaryotes on earth, the first microscopic unicellular eukaryotes are thought to have arisen around 1.7 billion years ago. The step from prokaryote to first eukaryote was one of the big evolutionary transitions. The interesting hypothesis gaining acceptance is that large and more complex cells evolved partly as a result of certain prokaryotes taking up resilience inside other cell. Oxygen began to accumulate in the atmosphere about 2.5 billion years ago as a result of photosynthetic activity of cyanobacteria. Those prokaryotes that were able to use O_2 in ATP increased. Some of these evolved to aerobic bacteria and others became symbionts within larger cells and evolved into mitochondria, as evidenced by their many similarities with chemical and genomic components of bacteria. Mitochondria are most likely derived from lineage of purple non-sulphur bacteria in which capacity for photosynthesis has been lost.

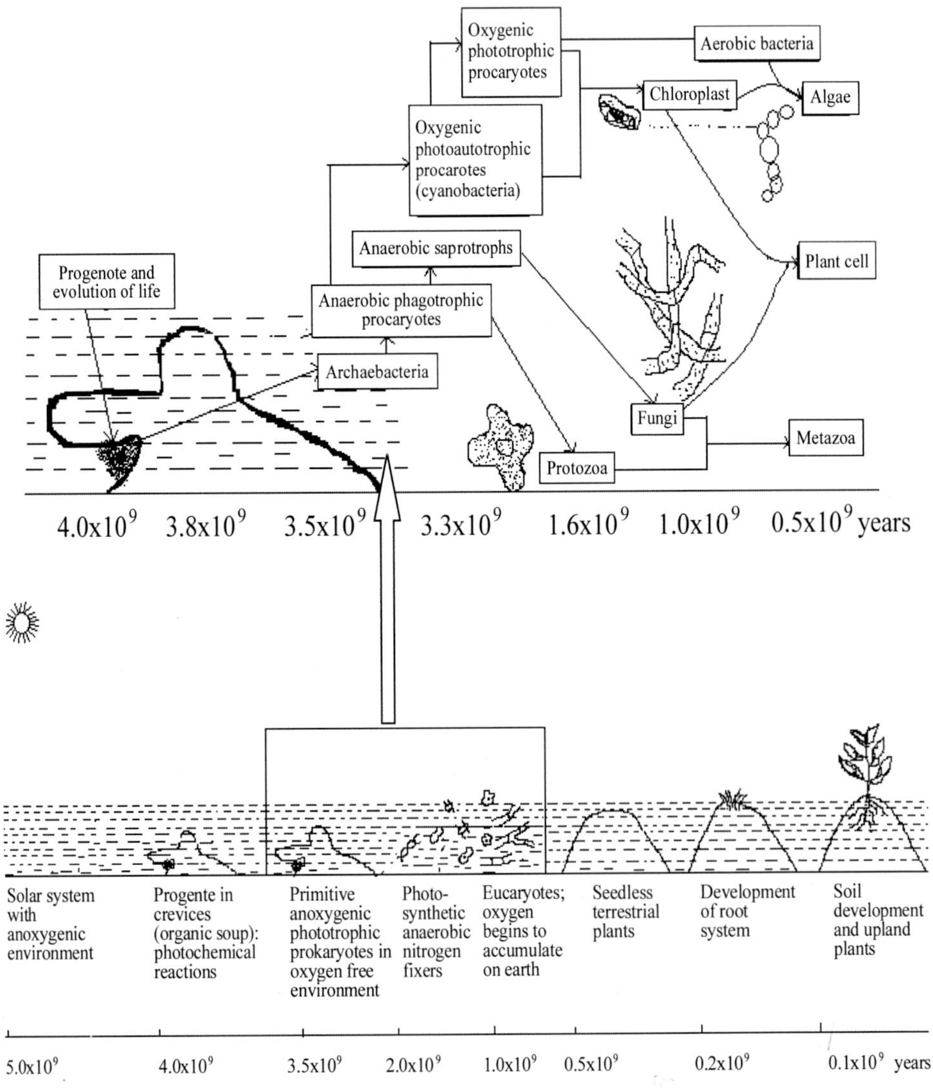

Fig. 1.1: Chronological events of chemical and biological evolution of life

Eukaryotic diversification seems to have occurred around 1.0 billion years ago. Somewhat 0.5 to 1.0 billion years back during 'Precambrian era' phytoplanktons appeared and oxygen on the earth began to accumulate. In the meanwhile ozone layer slowly became visible thereby shielding the evolved life from harmful effects of ultra violet rays. Seedless lowland plants then appeared on bare rocks in sea during Devonian period and in Carboniferous era root system in plants seem to have developed. This was followed by upland soil development in Jurassic era (200 million years back). The organic matter accumulation and organic soil formation as well as angiosperm plants seem to date back 50-100 million years during 'Cretaceous period'. Meanwhile mammalian evolution began and *Homeo-sapiens* started modern agriculture somewhat 20 millions years back.

1.2 ORIGIN OF SOIL

Soil is considered live owing to the presence of diverse living cells in its organic-mineral matrix which always keeps on changing its composition and number with time, location and extrinsic and intrinsic factors (see Chapters 2 and 3). The soil, we see around, were created slowly over many hundreds of years and are still its formation. Rocky parent materials of land initially were subject to chemical and physical weathering (breakdown and erosion by wind, rain, freezing and thawing, glacial ice, etc., thereby changing the underlying minerals into smaller pieces, etc.). Later microorganisms and primitive land plants contributed to the breakdown of rocks as they colonized cracks and depressions. This initiated pedon (soil) development as a result of net product of chemical, biological and climatic factors interacting with a particular parent material over time. Close observation of a soil reveals that it has a three dimensional body which is heterogeneous in nature. Vertical cut of a pedon gives us the overall profile of a soil. A soil profile may reveal many distinct horizons (Fig. 1.2; Table 1.2).

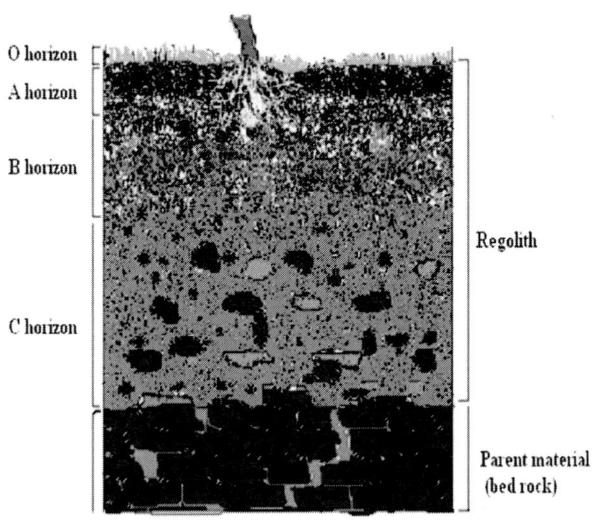

Fig 1.2: Soil profile: Avertical view

Table 1.2: Horizons in a typical soil

Horizons	Characteristics features
O-Horizon	The surface layer (0-2 cm or more) is usually rich in plant and animal residues, dominated by organic matter and exemplified by forest litter layer etc. The zone lies above the mineral soil.
A-Horizon	This topmost mineral layer of soil is relatively thin silicacious layer below O-horizon (several cm) which has partially decomposed and living organic matter (humus) intimately associated with soil mineral matter.
B-Horizon	Below O and A layer lies a mixture of precipitated humus and small pieces of mineral layer (0-2 cm). These are dominated by carbonates, silica, gypsum alone or in combination, with sesquioxides, silicate clays etc. also being present.
C-Horizon	This is the zone of least weathering. Here greater accumulation of calcium and magnesium carbonates occurs so this portion has high bulk density. This layer lies above bed rock and possess small pieces of mineral components with little or no organic matter so is least affected by soil genesis.
R-layer	This is the underlying parent material and has consolidated bed rocks such as granite, lime stone, basalt, etc.

A soil horizon is a layer of soil approximately parallel to the land surface and differs from adjacent in physical, chemical and biological characteristics such as structure, texture, pH, consistence, colour, biotic population, organic matter content, etc. Mature soils have clear-cut distinct horizons while young and immature soils possess merely a few horizons only.

1.3 MAJOR HISTORICAL MILESTONES IN SOIL MICROBIOLOGY

i) Early Developments

Unlike soil science whose origin can be traced back to Roman and Aryan times, soil microbiology emerged as a distinct branch of soil science and microbiology in 1838 after French agricultural chemist J.B. Boussingault revealed that legumes are able to obtain nitrogen from soil when grown in unheated soil. He also suggested that microbial organic matter decomposition results in the accumulation of nitrates and is, therefore, linked with soil fertility. Fifty years later a Dutch scientist M.W. Beijerinck isolated bacteria from the nodules of legume roots. In between this era, a fundamental base for medical and general microbiology had been established due to the persistent efforts of Robert Koch and Louis Pasteur who enunciated bases of germ theory of disease and fermentation. So Koch and Pasteur are considered as fathers of bacteriology and microbiology, respectively.

The beneficial effects of leguminous plants are mentioned in early Roman literature of more than 2800 years ago. The rotation of cereals with legumes in agricultural fields enrich the soil was known to Romans but its scientific basis was unknown. Chinese literature *Chin Min Tao-Shu* (Art of feeding the people), written in 540 AD, describes the beneficial effects and techniques of water fern (*Azolla*) use in rice cultivation. The possible presence of microorganisms in soil is said to be suggested by a Roman namely Collumella in his treatise *Husbandry* in 60 BC. The treatise describes several agricultural practices which were in vogue in Roman Empire for several generations.

ii) Advent of Microscope

The simple lens was developed in 1267 by Roger Bacon after long persistent efforts. Zaccharis Jansen, a Dutch national was pioneer in the field of microscopy and is said to have made the first compound lens system in 1550. Based on circumstantial evidences, it is presumed that Athanasius Kircher (1601-1680) and Barret (1620-1671) were first to have the opportunity to observe microbes. Kircher reported microbial presence in marshy lands in the vicinity of homes in 1671. Robert Hooke of Netherlands in 1664 using elegantly ornate microscope described fruiting structures of moulds. His book *Micrographia* was published in 1665 which described his observations. The book may be considered to be the first textbook on microbiology. However, it was Antony van Leeuwenhoek, a Dutch linen draper (1632-1723), who in 1676 with his crude microscope examined various body fluids, infusions of natural products, beer, wine etc. and by his careful manipulations was able to observe organisms as small as bacteria. He drew first authentic drawings of microbes, the then called "animalcules".

iii) Biogenesis v/s Abiogenesis

Earlier men believed that living organisms generate spontaneously from non-living matter. This belief, known as 'doctrine of abiogenesis', was proposed by Anaximander (641-611 BC) and Aristotle (384-322 BC). Van Helmont (1577-1644) and John Needham (1745) were amongst the strong proponents of this theory. However, Francesco Redi (1665) by covering/uncovering boiled rotting meat, Lezzaro Spalanzani (1775) through sealed flask experiment, Schwann (1836) by simple heat experiment and Cagniard-Latour (1836) and Schulze (1836) by passing sulphuric acid and caustic potash treated air into boiled broth in flask accumulated good evidences disproving spontaneous generation theory. Schroeder and Dusch (1850) allowed the air to pass into boiled broth through cotton that helped to filter microorganisms, and broth failed to show any life. Controversy persisted till strong experimental evidences from Louis Pasteur (1829-1859) settled the debate of spontaneous generation once for all. In a simple experiment he allowed untreated air to pass through long goose-neck into a flask containing heated broth. The microorganisms failed to develop. John Tyndall (1820-1893) proved that air around contained fine dust particles carrying microorganisms.

iv) Fermentation and Putrefaction

Schawann (1837) and Cagnaird Latour (1838) discovered yeast cells and maintained that it caused fermentation whereas Berselius (1939), a Swedish chemist, and Liebig, a German chemist, strongly defended mechanistic explanation of fermentation. In 1862 Louis Pasteur (1822-1895) demonstrated beyond doubt that fermentation and putrefaction were the products of microbial action. Theodor Schwann, C. Cagniard-Latour and F. Kutzing in 1836-1839 studied putrefaction and fermentative activities by microorganisms. The investigations on lactic acid fermentation from sugars were made by Burton and Tremy in 1841, whereas that of butyric acid production under anaerobic conditions were made by Pelouze and Gelis in 1844. Mitscherlich in 1843 stressed on the role of microorganisms in fermentation and putrefaction of plant cells and carbohydrate decomposition. Pasteur in 1857 proved that microorganisms are the agents that bring about chemical changes. Pasteur explained the part played by microbes in decomposition of urea and other nitrogenous substances. He also explained that anaerobiosis and fermentation of wines and beers are because of microbial activities and gave the methods of preservation as well. Heat treatment killed most of the organisms lead to the development of pasteurization principle. His contributions of identifying pebrine disease of silk worms and vaccination production are significant ones.

v) Germ theory and Koch's Postulates

Earlier Greeks suggested that small invisible worms cause disease. Work on human and animal diseases were in advance as compared to soil borne plant diseases. M.J. Berkeley in 1845 established that potato blight disease, cause of havoc in Ireland, was incited by a soil fungus. In meanwhile, the basic principles for authenticity of disease/pathogen, well known as Koch's postulates, were established by German physician Robert Koch (1843-1910). These postulates are summarized as under:

- A specific microorganism must always be present in every case of the disease;
- The specific organism must be isolated and grown in pure culture;

- The specific disease must be reproduced when inoculated into a susceptible healthy host;
- The microorganism must be recoverable once again from the host into which it has been inoculated.

However, postulates do not hold true in case of physiological diseases and those diseases where microbes involved are obligate and uncultivable parasites. Koch worked on the mode of transmission of animal diseases such as anthrax and cholera. Later his postulates were found applicable to plant pathogens as well.

vi) Pure Culture Isolations

Koch, Pasteur and their contemporaries established the basic concepts and laboratory techniques for culturing microbes, methods of disinfection and sterilization. Joseph Lister, an English surgeon, gave the techniques of preparation of pure bacterial cultures by serial dilutions in 1878. These concepts and methods laid foundation for later studies on microorganisms and thus ushered, in real sense, the beginning of microbiology as a discipline of science. Earlier microbiology was medical-oriented. The techniques for studying soil microorganisms were initially provided by Conn (1918) and later Rossi (1927) and Cholodney (1930) gave contact slide method. Fred, Allen and Baldwin (1932) worked on nodule bacteria and gave various techniques for their studies.

vii) Microbial Transformations

Russian microbiologists Sergei N. Winogradsky (1856-1953) and Martinus W. Beijerinck (1851-1931) are considered as harbingers in soil bacteriology who discovered microbial transformation of N and S, autotrophic mode of bacterial life and nitrogen fixation by nodule bacteria. They also showed that in contrast to plants and animals, soil microorganisms exhibit extraordinary wide range of physiological diversity. W.B. Leishman (1858) and M. S. Woronin (1866) demonstrated that root nodules in legumes were formed as a result of infection by a specific group of bacteria. Conversion of ammonia, present in sewage sludge, into nitrates is because of microbial activities was demonstrated by Th. Schloesing and Alferd Muntz in 1877. They added sewage sludge effluents to a long tube filled with sterile sand and $CaCO_3$. After 20 days, NH_4^+ had disappeared and NO_3^- was present. Heating or addition of antiseptic agent stopped the transformation. However, process was reinitiated by the addition of small quantity of garden soil. Boussingault in 1878 suggested that decomposition of organic matter by microorganisms results in the accumulation of nitrates and associated increase in soil fertility.

The beneficial effects of legumes in improving the soil fertility were long known. Boussingault and Hellrigel in 1880 for the first time initiated the work on biological nitrogen fixation. In 1885 Nobbe and Hiltner developed laboratory grown rhizobial culture in laboratory under the name "Nitragen" for the first time. H. Hellriegel and H. Wilfarth in 1886 suggested that bacteria in root nodules of legumes fix atmospheric nitrogen and make it available to plant root. In this maiden attempt 17 different inoculants for important leguminous crops were produced. They also demonstrated that unlike legume plants, the growth of rhizobia in non-legumes is directly related to the amount of N supplied. Meanwhile Th. Schloesing and M. Laurent (1892) demonstrated that the weight of nitrogen adsorbed by bacteria from air was approximately equal to the gain by plant and soil. However, it was M.W. Beijerinck who successfully isolated bacteria growing in root tissues of leguminous plants and forming nodules therein. These root nodule bacteria were then called *Bacillus radicicola* and later renamed as *Rhizobium*. He also isolated asymbiotic N_2 fixing *Azotobacter chroococcum, Azospirillum* and other important soil bacteria. S. N. Winogradsky (1890-1901) obtained pure cultures of bacteria (*Clostridium*) responsible for nitrification and nitrogen fixation and demonstrated the role of bacteria in nitrification. Winogradsky made fundamental observation on the role of microorganisms in performing chemical changes involving S, Fe and other compounds. He found several physiologically distinct groups among bacteria; each characterized by the ability to use a particular inorganic energy source e.g. sulphur bacteria oxidize inorganic sulphur compounds and nitrifying bacteria oxidize inorganic nitrogen compounds.

Winogradsky and Beijerinck (1893-1901) discovered that certain soil microorganisms play vital role in fixation of atmospheric nitrogen to the form that can be used as nutrient by plants. They also developed a new enrichment culture technique for the isolation and study of various physiological groups of microorganisms existing in nature. R. Warington in 1878 demonstrated that nitrification in soil was a microbial process and occurred in two stages involving two groups of microorganisms, one group oxidized NH_4^+ to NO_2^- and second one converted NO_2^- to NO_3^-. However, he did not isolate the organisms responsible for conversions. Omeliansky (1902) isolated anaerobic soil bacteria responsible for the degradation of cellulose. The ammonification of organic N in soil is microbe-mediated process was discovered by J. G. Lipman and P.E. Brown in 1903. They also worked on sulphur oxidizing bacteria. The publication of book *Principles of Soil Microbiology* by Salmon A. Walkman in 1927 was a classic contribution in soil microbiology.

Van Neil (1931) worked on chemoautotrophic bacteria and bacterial photosynthesis. J. Barker (1945) discovered methane bacteria and anaerobic fermentation. R. L. Starkey (1945) discovered iron-transforming bacteria. N. C. Acharya (1948) standardized the methods of biogas production and compost making from organic manure and crop wastes.

viii) Rhizosphere Studies

Some microbes live in close intimacy with the root system. They were discovered in early 19th century and the term rhizosphere microbe was coined for such organisms by L. Hiltner (1904). The presence of plant roots in soil provides nourishment to organic matter and stimulates the growth and multiplication of organisms, was explained by R. L. Starkey (1929). Clarke (1949) suggested that root surface (rhizosphere) itself be used in studying the rhizosphere phenomenon, since it may be more sensitive index of the specific qualitative effects of root on soil organic matter. The other prominent workers in rhizosphere microbiology are Lockhead (1940) and A. D. Rovira (1956). In 1909, John Russell and H. B. Hutchinson studied soil protozoa and found that protozoa control bacterial population and their activities in soil. They also held protozoa responsible for soil sickness because of their pathogenic nature. Pringscheim, Robbins and Bristol contributed towards soil algae.

ix) Mycorrhizal Studies

Robert Hartig illustrated the mycorrhizal fine roots of pine in 1840. However, he did not recognize the fungal component as a separate entity. The fungal cells associated with orchids were described and illustrated by S. Reissek in 1847. In 1881 F. Kamaienshi mentioned about the fungal layers surrounding certain trees and postulated that nutrient may be taken up by plants through this fungal layer. A.B. Frank (1885) coined the term "mykorrhizen" for those fungi living in symbiotic association with plant roots. He characterized sheath forming fungus on tree roots as ectotropisch (ectotrophic) and those without compact external sheath but penetrating root cells as endotropisch (endotrophic). Rayner and Melin (1921-1927) focused attention towards their beneficial role. J. W. Gerdemann, D. H. Marx, J. Trappe and Bowen gave detailed accounts about ectomycorrhiza whereas Barbara Mosse was a pioneer in VAM studies.

x) Antibiotic Production

British physician Alexander Flemming in 1929 discovered penicillin, an antibiotic isolated from *Penicillium notatum,* through a chance finding while examining bacterial culture which was destroyed by the growth of a fungus.

In 1943 S. A. Walkman discovered streptomycin from actinomycetous bacteria unfolding the significance of soil actinomycetes as antagonistic microorganisms. Thus search started for the isolation of organisms having antibiotic/antibiosis potential.

Garrett in 1926 established a school in UK on soil fungi and their ecological classification, whereas Alexander (1961) laid bases for "School of Soil Microbiology" at Cornell University with main thrust on pesticide degradation. Ruinen (1956) observed microbes living on plant leaf surfaces and gave phyllosphere concept.

xi) Biological Nitrogen Fixation (BNF)

Cornahan (1960) discovered biological nitrogen fixing (BNF) potential in cell free extract of *Clostridium pasteurianum*. Burris and Wilson (1952) used N isotopes in quantification of BNF whereas Vincent and Bergersen (1958-60) worked on symbiotic nitrogen fixation. P.K. De (1939) suggested that fertility in tropical fields is due to the activity of nitrogen fixing blue green algae now classified as cyanobacteria. Fogg and Stewart (1950-70) established the role of blue green algae (cyanobacteria) in BNF, especially in wetlands. Krasilnikov and Mishustin (1962) first used Azotobacterin (*Azobacteria*) and Phosphobacterin (*Bacillus megaterium* var. *phosphaticum*) in improving soil fertility (in Russia). Hardy (1968) gave acetylene reduction assay technique for BNF estimation. Callahan, Torry and their associates (1978) found BNF activity in non-legume roots especially by an actinomycete, *Frankia*, which lived as an endophyte.

J. Doberiener from Brazil in the meanwhile found BNF in grass roots by *Azotobacter paspali* and named it "associative symbiosis". Doberiener and J. M. Day (1975) revealed high N_2 inputs and extensive growth in Brazilian grass *Paspalum notatum* in association with *Azotobacter paspali*. In 1976, Doberiener demonstrated the significance of *Azospirillum* as an associated symbiont in grasses in fixing the atmospheric N_2 and coined associative symbiosis as 'rhizobiocoenosis' to designate the intimate association between *Azospirillum* and plant roots. Dommergues and his associates were first to notice nitrogen fixation in stem nodules of *Sesbania rostrata*. Cocking attempted to induce nodulation on roots of rice, wheat, rape etc. by inoculating stem-nodulating bacterial specie *Azorhizobium caulinodans*. R. J. Swaby (1976) developed a preparation of rock phosphate, sulphur and *Thiobacillus* known as "biospur" for enhancing phosphorus solubilization in soils. Gerrestser and Mulder discovered that phosphate solubilization is also microbe mediated and further, molybdenum plays role in N_2 fixation.

xii) Molecular Biology

With the advancement in molecular biology after the establishment of gene structure by J. D. Watson and F. H. C. Crick (1953) followed a series of discoveries. Ochoa and Kornberg (1965) demonstrated the method of RNA and DNA synthesis using *Azotobacter* and *Escherichia,* respectively, as tools in their studies. In mid 1970's, new and powerful set of techniques known as 'recombinant DNA technology' helped the soil microbiology to advance tremendously. There is a great scope for the exploitation of beneficial microorganisms in near future for the welfare of mankind.

1.4 DIVERSIFICATION IN SOIL MICROBIOLOGY

The last twenty five years have witnessed tremendous advances in soil microbiology with its diversified applications in agriculture and allied fields. More information are percolating on C, N, P, S and organic matter turnover/ transformations, soil enzymes, microbe-plant inter-relationships, molecular studies, microbial participation in soil structure-development and plant protection, soil ecosystem diversity and protection, degradation of man-made recalcitrant, pollutants and pesticides, genetic engineering, transformation kinetics etc. Soil microbiology is no longer confined to bulk measurements but is, nowadays, studied in a wider perspective, describing its implications/ applications in soil formation, serving as driving force for interactive reactions in soil, pore space colonizer or as a small micro-site machinery helping to develop soil aggregates and moulding the organic matter in such a manner so as to improve soil fertility and crop productivity. There is a general concept that more the number of microorganisms in a soil, more productive it will be; and more their nature is diverse, more will be the functional diversity in a soil.

Microorganisms have become the vehicles of choice in molecular biology for understanding biochemical processes at molecular and genetic level. Much of the current understanding on macro-organisms, including plant and mammalian genetics, has come from research with microbes. New microbial inoculants, genetically modified microorganisms and nano-biofertilizers have been developed but there is further need for their thorough scanning, characterization and upgradation with respect to their role in ecosystem development, environmental impact assessment and field application. Noted soil scientist Dr. Rattan Lal sums up it as "understanding of basic soil processes requires strong collaborative programs between soil scientists and biologists, climatologists, hydrologists and geologists. Developing technological innovations to meet societal needs require collaboration with engineers, bio-nano-/info-technologists, economists and social scientists. Soil scientists have a major challenge in addressing a wide range of global issues and societal needs".

Microbiologists nowadays participate in empirical approaches to agriculture productivity, range and forest management. Isotope techniques (^{15}N, ^{14}C, ^{31}P) have made quantification of various N, P and S processes in relation to C cycles possible. About 125 years ago soil microbiology was in infancy. The strides made since then in this branch of science has contributed to better understanding of the complex hidden world in soil that literally covers the earth 25 times in biomass than total mass of animal life. Now a number of new factors influencing the soil microbiology have come into fore which include:

a) Bioinformatics: Present scenario calls for computer application for data processing, modelling, information exchange and automated instrumentation. The information technology has provided better connectivity and made data readily available to the scientists even working in remote areas of the world. There is a strong urge to utilize new innovative methods in soil microbiology based on nanotechnology, biotechnology and information technology;

b) Application in whole-system management techniques: These include zero-till, alternate cropping, whole tree harvest, preserving the organic matter reserves, etc.

c) Genetic engineering: Soil microbes and their biochemical reactions etc. are closely linked in biotechnology. Biotechnology has found numerous applications in soil, especially with respect to microbiological assessment of soil quality, phyto-remediation and detoxification of polluted/contaminated soils, waste recycling and in environmental safety. Recombinant biotechnology and transgenic plants have strong applications in the alleviation of biotic and abiotic stresses in soil systems;

c) Biosensor and bioprocessor use: Realizing that soil microbiological processes are affected by environmental changes and, in turn, influence major environmental problems (e.g. sewage sludge decomposition, reducing or enhancing pollutants like nitrates in soil, scavenging recalcitrants, oil spills, etc.), the biosensor use and bio-processing has gained great importance;

d) Conservation of microbial diversity: Enormous biodiversity, observed in soil, are conserved for taping natural resources through bioconversions, biodegradation, bioremediation, antibiotic production, etc.;

e) Biodynamic farming: Biodynamic farming is a new approach for more efficient agriculture and forest management system with thrust on organic recycling rather than showing complete dependence on inorganic farming thus viewing the whole farm as a living system. Development of criteria to objectively assess the real value of soil organic matter is desired to understand the societal value of carbon. In addition to cropland soils, the feasibility of trading carbon sequestered in forest, rangeland, and urban soils (turfs, lawns, golf courses etc.) need thorough studies.

f) Maintenance and restoration of basic soil quality: Soil health determines the future produce, so its maintenance and quality restoration has received more attention in recent years. Innovative technologies need to be adapted/tailored to site-specific situations in consideration of both biophysical and socio-economic factors.

1.5 MICROORGANISMS AND SOIL HEALTH

Soil is a renewable natural resource. The maintenance and restoration of basic quality of soil is primarily a social objective. Soil may be viewed like a living tissue with soil solution as life-line like blood, transporting scavenger propagules, nutrients and waste products. Below soil surface is a hidden microbial world of incredibly diverse nature that is significant to a soil scientist, plant pathologists, entomologist or agronomist. The underground environment of a plant is as important for plant health as the aboveground part. However, meager attention has been focused to the world under our feet. Harmonious friendly organisms normalize the soil health and put the disease causing pathogens or toxicity creating organisms to stressed conditions/defense, thus extend the life span of a soil. A range of organisms involved in various transformations affect the soil and plant by selectively stimulating the growth and/or activity of one or more organisms in soil.

Microbes and soil productivity : During crop production many chemical, physical and biological factors influence crop and soil productivity. *Soil productivity is the capability of a soil to produce a specified plant or sequence of plants under a specified management system.* If climatic conditions remain conducive, then poor crop growth, if any, is mostly attributed to soil factors. These include nutrient deficiency, disease and pest problems, toxicity, poor management practices, etc. Such a situation warrants immediate remedial measures. 'Soil sickness' arises as a result of limitation of growth factors influencing the crop growth and yield. Microbes in soil may be both beneficial and deleterious, thus dominance of beneficial microbes is considered as a sign of improving 'soil health' while dominance of those inducing diseases or toxicities etc. are viewed as the signs of 'soil sickness'.

Microbial processes and soil health : In soil, microbes play pivotal role in the transformation of nutrients and help in overcoming some of the stress conditions. The end products of their activities are oxides of elements, which are taken up by the plants. Soil formation and development, apart from chemical and physical processes, involves biological weathering of rocks and minerals during which nutrient cations and anions are released infusing new vigour to soil health. Microbes excrete and produce enzymes that act as solubilizing agents. The soluble products serve as food for the growth and energy of these microorganisms. Microbial processes in soil result not only in decomposition of organic matters but also in the synthesis of new organic compounds whose nature depend on the prevailing conditions. Most of these synthesized products are absorbed directly by crop plants with stimulatory effect. The antibiosis/antibiotic activities of some organisms help in overcoming the diseases.

Soils can be evaluated in terms of rates at which carbon dioxide is released or nitrates/phosphates and sulphates are produced in them on incubation under optimum moisture and temperature conditions for a standard period of time. The efficiency of any suggested remedy for low productivity can also be estimated in laboratory in a few weeks by measuring the effects of the treatment on microbial processes like NO_3 accumulation, CO_2 evolution, phosphate solubilization, etc. Frequent use of *Aspergillus niger, Cuninghamella* sp. and *Azotobacter* sp. is made in testing soils on agar plates for assessing the deficiencies of elements such as Mo, Zn, etc. in soil. The extent of colony growth directly relates to the quantities of these elements available for use. Thus examination of soil through microbial eyes provides clues for not only deficiencies or soil sickness in a very short growth period and but also helps in diagnosing the soil health problems.

Soil resilience : Microbes in soil are always in a state of constant competition for space, energy and growth. Therefore, understanding microbial ecology (refers to the behavior and significance of microorganisms in natural environment) has far reaching impact on soil health management. Durable life in soil is attained only if biotic and abiotic components, humus/organic matter, including microbial biomass, minerals, air and water, are in dynamic equilibrium and capable of self-regulating natural processes in ecosystem by complementing and supplementing each other. This buffering capacity of a soil in respect of physical, chemical and biological impacts in relation to the biological quality of soil is often referred to as 'soil resilience'. *Resilience can be* *considered as the capacity of soil microbial biomass to maintain its structure, diversity and function*. This is closely related to the concept of conservation of biodiversity. In well-established soils microbial population reach very stable equilibrium. A naturally 'fertile soil' is the one in which soil organisms release inorganic nutrients from organic reserves at a rate sufficient to sustain plant growth as well as replenish it at ease from its own resources.

Soil fertility : Soil fertility serves as an index of soil health as it describes the quality of a soil to provide proper nutrients in balanced manner and in proper quantity for the growth of specified plants when temperature and other factors are favourable. It is now increasingly becoming clear that organic amendments have less adverse consequences as compared to inorganic ones. The ill-health problems in soil can be rectified to a greater extent by building up a strong organic matter base.

Microbial role in maintaining soil quality : Contamination of soil by heavy metal, pesticides and other recalcitrant elements is of great human concern. The present day organic or biological agriculture, therefore, aspires for a system where the maintenance of soil fertility and control of pests and diseases are achieved by the enhancement of natural processes and cycles with only moderate input of energy and resources, while maintaining reasonable productivity. The number, diversity and distribution of microorganisms reflect the overall soil productivity or its health. The soil health is influenced by over-cultivation, deforestation, overgrazing, unskilled irrigation, faulty management, population pressure, organic matter depletion, etc. leading to soil susceptibility to wind and water erosion *vis-à-vis* decrease in productivity. Sewage sludge and animal slurry introduce human pathogens such as *Vibrio cholera* (cholera), *Salmonella typhii* (typhoid), *S.paratyphii* (paratyphoid), etc. into the soil and pose serious threat to soil water quality.

Microbial applications in soil have beneficial socio-economic implications. For example, the use of nitrogen fixers reduces exclusive dependence on inorganic chemicals and offsets the possibility of any soil health problem caused due to pollution. It is now increasingly becoming clear that soil microbiology and plant pathology are sister concerns because both deal with microorganisms in terrestrial environment. Further, pathogens and symbionts in soil share the same host, influencing each other's biochemical activities as well as host root physiology. The knowledge about harmful and beneficial effects of microorganisms in agricultural processes can be manipulated in better way to enhance crop yields and farm animal productivity.

Selected References

Alexander, M. 1977. *Introduction to Soil Microbiology* (2nd edn.), John Wiley & sons, New York, USA.

Atlas, R.M. and Bartha, R. 1981. *Microbial Ecology: Fundamentals and Applications.* Addison-Wesley, Reading, Massachuestts, USA.

Atlas, R.M. and Bartha, R. 1993. *Microbial Ecology: Fundamentals and Applications* (3rd edn.). Benjamin- Cummings, Redwood City, USA.

Black, C.A. 1968. *Soil Plant Relationships.* John Wiley & sons, New York, USA.

Borneman, J., Skroch, P.W., O'Sullivan, K.M., Palus, J.A., Rumjanck, N.G., Jansen, J.L., Nienhuis, J. and Triplett, E.W. 1996. Molecular microbial diversity of an agricultural soil in Wisconsin. *Appl. Environ. Microbiol.,* 62: 1935-1943.

Brocks, T.D. (ed.) 1959. *Milestones in Microbiology.* Prentice-Hall, Englewood Cliff, New Jersey, USA.

Brocks, T.D., Smith, D.W. and Madigan, M.T. 1984. *Biology of Microorganisms* (4th edn.) Prentice-Hall, Englewood Cliff, New Jersey, USA.

Burns R.G. and Slater, J.H. (eds.).1982. *Experimental Microbial Ecology,* Blackwell Scientific Publication, Oxford, UK.

Cairns-Smith, A.G. 1985. *Seven Clues to the Origin of Life.* Cambridge Press, Cambridge, UK.

Campbell, R.E. 1977. *Microbial Ecology.* Blackwell Scientific Publications, Oxford, UK.

Crick, F. 1981. *Life, Its Origin and Nature.* Simon and Schuster, USA.

Ellwood, D.C., Hedger, J.N., Latham, M.J., Lynch, J.M. and Slatert, J.H. (eds.). 1980. *Contemporary Microbial Ecology.* Academic Press, London, UK.

Foth, H.D. 1990. *Fundamentals of Soil Sciences* (8th edn.), John Wiley and sons. New York, USA.

Foth H.D. and Ellis, B.G. 1997. *Soil Fertility* (2nd edn.). Lewis Publishers, CRC Press, Boca Ratoon, Florida, USA.

Gray, T.R.G. and Williams, S.T. 1971. *Soil Microorganisms.* Oliver & Boyd, Edinburg, U.K.

Haldane, J.B.S. 1954. The Origin of Life. *New Biol.,* 16: 12-27.

Hawker, L.E. and Linton, A.H. (eds.). 1979. *Microorganisms: Function, Form and Environment* (2nd edn.). Edward Arnold, London, UK.

Lechevalier, H.A. and Solotorovsky, M. 1965. *Three Centuries of Microbiology.* McGraw Hill, New York, USA.

Lynch, J.M. 1983. *Soil Biotechnology: Microbiological Factors in Crop Productivity.* Blackwell Scientific Publications, Oxford, UK.

Metting Jr. F.B. (ed.) 1993. *Soil Microbial Ecology.* Marcel Dekker Inc., New York, USA.

Nisbet, E. 2000. The realms of Archaean life. *Nature,* 405 (6787): 625-626.

Norman, A.G. 1946. Recent advances in soil microbiology. *Soil Sci. Soc. Am. Proc.,* 11: 4-15.

Oparin, I.A. 1953. *The Origin of Life.* Dover Publication Inc., New York, USA.

Oparin, I.A. 1961. *Life: Its Nature, Origin and Development.* Oliver and Boyd, Edinburgh, UK.

Paul, E.A. and Clark, F.E. 1989. *Soil Microbiology and Biochemistry.* Academic Press, New York, USA.

Pirie, N.W. 1958. A discussion on anomalous aspects of biochemistry of possible significance in discussing the origin and distribution of life. *Proc. Royal Soc., B* 171:1-89.

Rangaswami, G. and Bagyaraj, D.J. 1998. *Agricultural Microbiology* (2nd edn.) Prentice-Hall of India Pvt. Ltd., New Delhi, India.

Roberts, D. McL., Sharp, P., Alderson, G. and Collins, M. (eds.) 1996. *Evolution of Microbial Life: Society for General Microbiology 54th Symposium.* Cambridge University Press, Cambridge, UK.

Russel, J.R.1962. *Soil Conditions and Plant Growth.* Longmans Green & Co., London, UK.

Rutten, M.G.1971. *The Origin of Life by Natural Causes.* Elsevier Publishing Co., Amsterdam, Holland.

Schidlowski, M. 1988. A 3,800-million-year isotopic record of life from carbon in sedimentary-rocks. *Nature,* 333 (6171):313-318.

SSSA, 1998. *Glossary of Soil Science Terms.* Soil Science Society of America, Wisconsin, Madison, USA.

Stanier, R.Y., Adelberg, E.A. and Ingraham, J.L. 1976. *General Microbiology.* (4th edn.). Collier MacMillan, London, UK.

Subba Rao, N.S. 1982. *Advances in Agricultural Microbiology.* Oxford & IBH Publishing Co., New Delhi, India.

Subba Rao, N.S. 1999. *Soil Microbiology.* Oxford & IBH Publishing Co., New Delhi, India.

Swift, M.J., Heal, O.W. and Anderson, J.M. 1979. *Decomposition in Terrestrial Ecosystems.* Blackwell Scientific Publ., Oxford, UK.

Tate, R.L. III. 2000. *Soil Microbiology.* John Wiley & sons, New York, USA.

Tortora, G.J., Funke, B.R. and Case, C.L. 1998. *Microbiology: An Introduction* (6th edn.), The Benjamin-Cummings, Redwood, USA.

Troeh, F.R. and Thompson, L.M. 1993. *Soil and Soil Fertility* (5th edn.). Oxford University Press, New York, USA.

Wainwright, M. 2003. A microbiologist looks at panspermia. *Astrophy. Space Sci.,* 285:563-570.

Waksman, S.A. 1961. *Soil Microbiology.* John Wiley & sons, New York, USA.

Wedberg, S.E. 1966. *Introduction to Microbiology.* Reinhold publishing Co., New York, USA.

Woese, C.R. 1987. Bacterial evolution. *Microbiol. Rev.,* 51: 221-271.

Chapter 2

Soil Biota — Structure and Function

The soil is a graveyard in a wider sense and yet it is very main spring of new life. If any object can demonstrate the possibility of resurrection, it is the soil upon which we walk. Equally wonderful is that it is not an inert mass but the scene of endless preparations and changes not only brought by chemical actions but by living organisms.

(J.W. Wrighton and J.C. Newsham)

The plant roots, soil animals and microorganisms make up soil biological communities, accelerate biochemical activities through intra- and extra-cellular enzymes and interact in such a manner that biogeochemical cycling on this earth continues uninterrupted. Without intermeshing vital processes of soil microflora, microfauna and associated mesofauna, the soil would have become a repository of plant residues and animal remains with no facility for recycling the vital nutrients for plant growth, thus halting the whole process of geo-chemical recycling in nature. Though living component rarely makes up more than 4% of the total soil organic carbon, it plays vital role in geochemical recycling, soil development and ecosystem stability.

2.1 PLANT ROOTS AND THEIR CONTRIBUTION TO SOIL C SINK

Plant roots are incredibly labile and variable component of soil biomass and usually occupy 1-5% volume of surface soil depending upon the nature of vegetation and soil conditions. Roots often contribute 20-30% of total soil respiration and rest respiration comes from the decomposition of organic matter. Plants pump fixed C from atmosphere into the soil through roots in the form of root exudates. Conservative estimates have revealed that roots release 10-40% of the total carbon fixed by plants into soil ecosystem in the form of root exudates or as respired CO_2. On an average, roots release 50-100 mg organic matter g^{-1} root day^{-1}. The decomposition of these organic substrates and their subsequent uptake provides a ready sink for this carbon, thereby keeps the carbon around roots within the acceptable limits of plant growth. Most roots growing in soil bear a cluster of root hair of 0.1-1.0 mm length numbering 100-1000 cm^{-3} root. The hairs appear a few millimeters behind the root tips and increase the effective volume of soil being exploited by roots for water and nutrients.

Roots are constantly exposed to myriads of organisms and varied physico-chemical environments in soil. They serve as an important substrate for microbial activities once sloughed off from the main root. They persist only for a few days and are then easily decomposed by soil microorganisms. In some tree genera, root hairs become suberized and lignified and persist as non-functional brown hair for years together. Although roots comprize a minor constituent of living biota, their passage and distribution through soil may have a greater influence on soil processes than what their pool size suggests. Rhizosphere generally experiences greater rates of mineral weathering than the bulk soil. Microbial population and quality differ in rhizosphere and major biotic interactions do occur in root zone. In arable soils the mass of roots left after crop harvest is usually 15-40% to that of the above-ground crop which weighs 25 to 85 t ha^{-1}. This induces alteration and succession in rhizosphere microflora.

2.2 SOIL MICROBIAL BIOMASS AND THEIR NUTRIENT DEPENDENCE

Soil microbial biomass is defined as '*living part of organic matter, excluding plant roots and soil animal larger than approximately* 5×10^3 *μm*'. The free living components of soil biota are bacteria, moulds, fungi, algae, protozoa and soil fauna. In addition, there are viruses which grow within living hosts. Each component has morphologically and physiologically wide range species. The value of microbial biomass reflects the organic matter content in soil. The microbial biomass in soil is the driving force of terrestrial ecosystems as it controls the rate of turnover and mineralization of organic substrates (Table 2.1). Soil biomass is a large, mostly dormant, population with great diversity and ability to survive hard times.

Table 2.1: Approximate biomass range, C:N ratio and proportion of decomposer biomass of major soil biotic components in soils

Components	Biomass ($t\ ha^{-1}$)	Number (g^{-1} soil)	C:N ratio	Approx. decomposer biomass (%)
Plant roots	10.0-40.0	-	15:1 to 100:1	-
Bacteria	1.0-2.0	10^7-10^9	5:1 to 8:1	50.5
Actinomycetes	0.5-2.0	10^6-10^8	5:1 to 8:1	4.2
Fungi	2.0-5.0	10^5-10^6	8:1 to 20:1	38.9
Algae	0-0.5	10^4-10^5	5:1 to 9:1	-
Protozoa	0-0.5	10^4-10^5	5:1 to 10:1	1.9
Nematodes	0-0.2	10 -10^2	5:1 to 10:1	0.3
Mites	0-0.2	1-10	-	0.8
Earthworm	0-2.5	-	5:1 to 10:1	2.0
Other soil animals	0-0.5	-	5:1 to 10:1	1.2
Viruses	Negligible	-	5:1 to 6:1	-

Nutritional grouping of soil microbes : Microorganisms require nutrients to supply energy for their physiological processes (catabolism) and for building up of their cell constituents (anabolism). Both energy source and C source are useful for describing the basic physiological differences among soil microorganisms. On these basis microbes are categorized into the various groups as given in Table 2.2 and the relationship between various types of soil organisms in soil-plant ecosystem is shown in Fig. 2.1.

i) Autotrophs: Also known as lithotrophs, autotrophs require only simple nutrients and obtain their required carbon for synthesis by assimilation of atmospheric carbon dioxide. Among the autotrophs, photoautotrophs convert the energy of sunlight into the energy of chemical bonds. These include algae, cyanobacteria and green sulphur bacteria. Algae and cyanobacteria have photosynthetic capability using chlorophyll in a way related to those of higher plants and release oxygen from the photolysis of water under adequate oxygen supply. Other photosynthetic bacteria like purple and green bacteria do not produce oxygen but depend on substrates such as H_2S, therefore their activities are restricted to water logged conditions like paddy or bog soils, etc. Chemoautotrophs use diverse energy sources e.g. NH_3^+, NO_3^-, NO_2^-, H_2, Fe^{2+}, S^{2-} and $S_2O_3^{2-}$ and oxidize these reduced inorganic molecules. The energy and electrons gained is used to reduce CO_2 thus providing carbon skeleton for cell synthetic activities as well as energy for biochemical reactions. Chemoautotrophs are restricted to a few types of bacteria such as *Nitrosomonas, Nitrobacter, Thiobacillus, Ferrobacillus,* etc. and are of greater significance in crop production and biogeochemical cycling of some essential nutrients.

Table 2.2: Microbial classification on the basis of energy and carbon source utilization

Group	Energy source	Carbon source	Microorganisms involved
Autotrophic			
i. Photoautotrophic	Solar radiation	CO_2	Algae, cyanobacteria, photosynthetic bacteria
ii. Chemoautotrophic	Organic/inorganic chemicals	CO_2	Specialized bacteria {such as N, S, Fe oxidizers}
Heterotrophic			
i. Photoheterotrophic	Solar radiation	Organic substrate	Purple sulphur bacteria
ii. Chemoheterotrophic	Organic chemicals	Organic substrate	Most bacteria, moulds, fungi, protozoa

ii) Heterotrophs: Also known as organotrophs, heterotrophs show wide range of nutritional requirement from simple to complex substrates. There are a few photosynthetic organisms that require compounds other than CO_2 as a carbon source. These are known as photoheterotrophs. The vast majority of soil microbial species are chemoheterotrophs (organotrophs), which require organic compounds as a source of carbon. Chemoheterotrophs are important in soil organic matter turnover. Although energy generation is an essential life process, soil microorganisms show a wide range of adaptation from aerobic or microaerophilic to anaerobic conditions and vice versa.

Autochthons and zymogens : Most of the soils represent a nutritionally poor (oligotrophic) ecosystem and some scattered pockets of hot spots. Oligotrophic microorganisms show relatively high growth rate at low concentration of substrate than at higher concentrations. However, soil microbes and their activities are concentrated in these niche or hot spots like rhizosphere, organic matter substrate, rich nutrient sites, water-filled pores etc. Winogradsky in 1925 classified soil microorganisms into two groups viz., autochthonous and zymogenous whose characteristics are as under:

Group	Characteristics
Autochthonous	i. Represent indigenous population and are almost uniform in soils.
	ii. Grow slowly and derive energy from native soil organic matter. The examples are *Arthrobacter, Nocardia,* etc.
Zymogenous	i. Organisms are fermentative ones, which show rapid growth and burst of activities when fresh organic matter is added to soil.
	ii. Population in soil under normal conditions is low.
	iii. Require external source of energy like cellulose, proteins etc. The examples are *Pseudomonas, Bacillus,* etc.

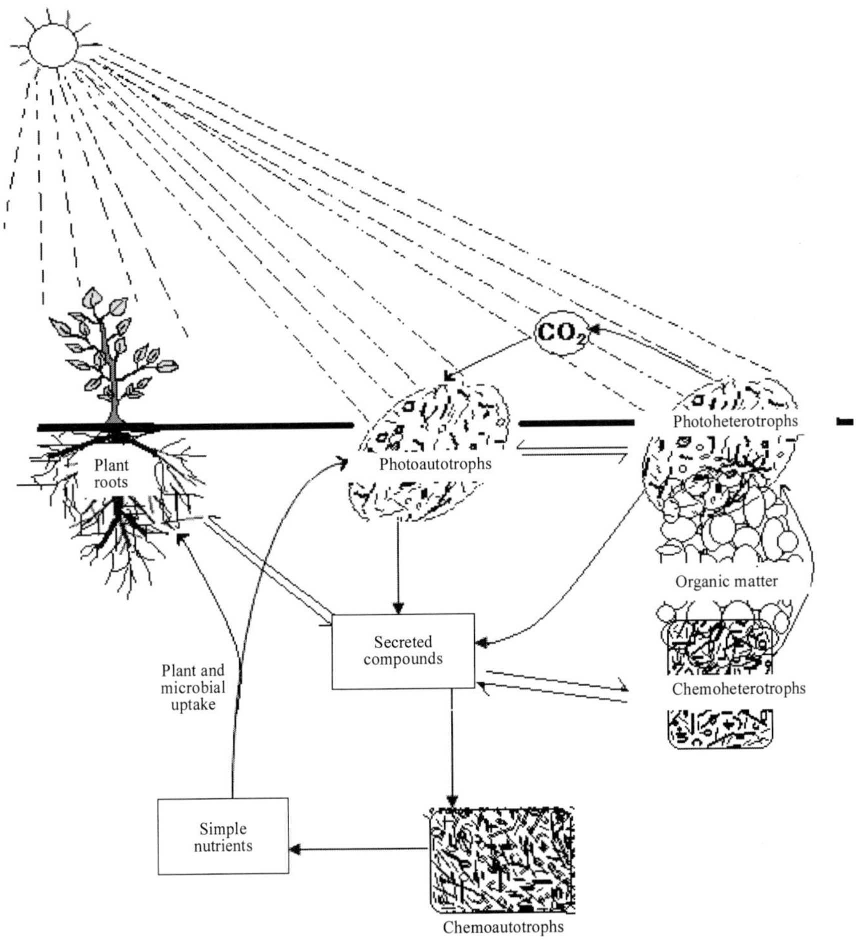

Fig 2.1: Relationship between various types of organisms in soil-plant ecosystem

Primary producers and consumers : The activities of soil flora and fauna are intimately related in a food web. At the top are primary producers (plants, algae and photosynthetic bacteria), which use energy from sun, carbon from atmosphere and water to make organic molecules and cell organelles. Hardly 5% of primary production by green plants is consumed by animals and rest 95% is ultimately decomposed by microorganisms. The organic matter produced by primary producers is used by other microorganisms, which are called primary consumers and include herbivores and detritivores. The bodies of primary consumers serve as food for the predators and parasites in soil and they are called secondary consumers (such as mites, earthworm, fungi, bacteria, protozoa, etc.). Even they may fall prey to other carnivores called tertiary consumers (such as centipedes, ants, spiders, etc.). They are, sometimes, also referred to as ultimate decomposers.

r and K selected organisms : Ecologically, soil microorganisms are categorized into 'r' and 'K' selections, because there is sucession of interacting populations of microorganisms with change in time, substrate and space. The 'r' selected species include the pioneer species, which exploit new or disturbed habitat and are typical of the early stage of ecological succession. 'K' selected species are later stage colonizers in ecological succession. The major characteristics of 'r' and 'K' selected species are given in Table 2.3.

Table 2.3: Characteristics of 'r' and 'K' selected organisms

S. No.	'r' selected species	'K' selected species
1.	These are redural species; opportunistic microorganisms capable to exploit temporary habitats;	Found in stable habitats;
2.	Reproduction rate rapid; short life span but have high specific growth rate;	Reproduction rate slow; long life span and low specific growth rate;
3.	Regeneration time short; organisms show intrinsic growth rate;	Regeneration time long so a few off-springs only; asymptotic growth curve;
4.	Good dispersal but poor competitors;	Poor dispersal but good competitors;
5.	Population densities fluctuate; biomass in spores or other resting spores;	Population densities are more stable; biomass in growing vegetative cells; live at more crowded densities;
6.	Capture primary nutrient source for survival and have good saprophytic ability;	Respond slowly to food but compete well for space or for a limited resource;
7.	Low resistance or susceptible to density-dependent mortality (starvation, predation, toxins etc.);	High resistance to density-dependent mortality;
8.	Low resistance to density-independent mortality (i.e. external factors like moisture, temperature etc.)	Variable to high resistance to density-independent mortality;

i) 'r' selected organisms: These show intrinsic growth rate or in other words they have characteristic ability to quickly exploit and exhaust ephemeral resource of habitat they colonize. They are primary exploiters of freshly added substrates but generally show little competitive ability. They live in restricted and uncrowded environment and exceed the carrying capacity of environment if remain too long.

ii) 'K' selected organisms: These are adapted to live under more stable habitat than 'r' selected species. They have greater competitive ability and operate at the levels close to the carrying capacity of environment. Most soil organisms growing on flushes of substrate are usually 'r' selected whereas rhizosphere organisms and plant pathogens may be either 'r' or 'K' selected ones.

2.3 VIRUSES

The viruses are non-cellular infectious entities. They are obligate parasites and possess either RNA or DNA molecules surrounded by a protein coat (Fig 2.2). Viral particles are metabolically inert and do not carry any respiratory or biosynthetic function. The soil viruses are 0.02-0.30 μm in size. Viruses being small-sized are able to enter soil pores that exclude all other soil organisms. Viruses are adsorbed by organic residues, humus and clay particles. Their persistence in soil is increased by moist conditions. Many plants, insects and human viruses, found in soil, persist for several years. However, soil does not serve as a natural habitat for viruses as they, in most cases, do not remain viable and infective for a long period there. Therefore, soil viruses are mostly confined to their hosts and in most cases cause pathogenic problems. Some plant viruses may survive in soil upto 9 years even in the absence of their hosts. Nuclear polyhydrosis virus of cabbage looper (*Trichoplusia* sp.) although surviving for only I month on leaves persists in soil upto 4 years. Viruses do not reproduce independently but multiply in host cells and induce the host to produce necessary viral components. The host cell is usually damaged and often destroyed as a result of viral infection.

There are viruses which parasitize bacteria, actinomycetes, fungi, algae, etc. so are accordingly named as bacteriophages, actinophages, mycophages or cyanophages, etc. The ability of viral particles, pathogenic to man and animal, to survive in soil and move into water tables is now a major concern in public health. Human viruses such as picornaviruses are introduced into soil through effluents or sludge from sewage. Owing to their capability to serve as vector for genetic transfer, viruses have proved to be important vehicles in genetic engineering in wide diversity of cells. Little is known about the field ecology of viruses that infect soil organisms except that these persist in soil as dormant units and retain parasitic capability. Soil conditions such as moisture and the type of absorbing surface e.g. clay minerals, etc. are likely to be important in affecting the survival and movement of viruses in soil.

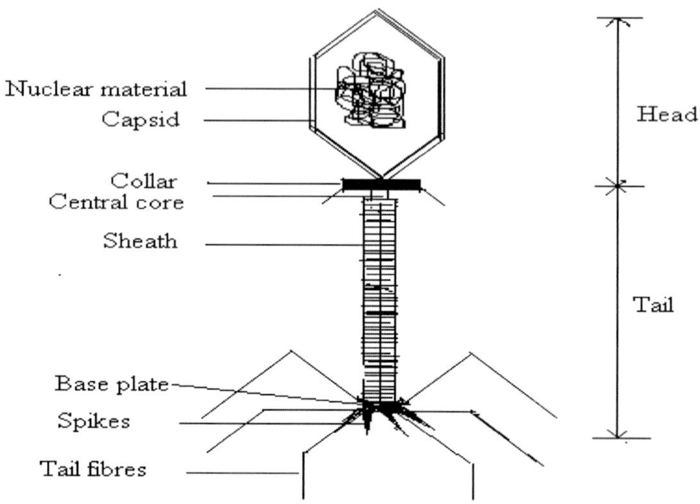

Fig 2.2: The structure of a bacteriophage - A typical virus

2.4 BACTERIA

Bacteria along with the viruses belong to group prokaryotes. The prokaryotes differ from eukaryotes structurally as well as in their some biochemical constituents (Table 2.4).

Diversity and distribution in soil : A typical bacterial cell weighs 10^{-12} g and has a volume of 1 μm^3. In soil, bacteria are most numerous of the microorganisms present, and form about 25% of the total microbial biomass (range 10-40%) with viable populations estimated up to 2 x 10^8 cells g^{-1} soil (range 10^6 to 10^9). The bacteria are nutritionally and physiologically most diverse and dominant group of microorganisms in soil but their biomass is generally exceeded by fungal biomass in soil. The diversity of bacterial community in soil as revealed by using DNA reassociation analysis technique is as much as 170 fold greater than revealed by the collection of bacterial isolates from soil. About 4,000-7,000 different genomic equivalents are estimated to be present in a gram of soil. Bacteria are present in all types of soils but their number/population decreases as soil depth increases, especially being maximum in A horizon of soil. A single soil sample possesses over 4,000 genetically diverse bacterial species but hardly 1% are culturable species.

Table 2.4: Difference between prokaryotes and eukaryotes

S. No.	Important features	Prokaryotes	Eukaryotes
1.	Cell size	Small, 1-10 μm; a few >50 μm	10-100 μm; some > 1 mm
2.	Membrane-bound organelles	None to few (e.g. chlorobium vesicles)	Many (nucleus, mitochondria, lysosomes, chlorophyll, etc.)
3.	Endoplasmic reticulum	-	+
4.	Mitochondria	-	+
5.	Lysosomes	-	+
6.	Golgi bodies	-	+
7.	Nuclear membrane	-	+
8.	Nuclear material (DNA)	Free in cytoplasm	Bound with nuclear membrane
9.	Chromosome	Single, naked, circular	Multiple linear (associated with histone)
10.	Genetic material transfer	Unidirectional (if present), conjugation, transduction and transformation	Exchange of genetic material through meiosis, biparental, independent assortment, crossing over
11.	DNA	Little or none	Often very high
12.	Transcription and translation	Coupled	Uncoupled
13.	Asexual reproduction	Binary fission	Mitosis
14.	Sexual reproduction	No gametes	Meiosis predominates in multicellular eukaryotes
15.	Introns (splitting genes)	-	Common, especially in higher forms
16.	Plasmids	+	Only in yeasts
17.	Cytoskeleton/endocytosis and exocytosis	-	+
18.	Chromosomal attachment for cell division	Plasma membrane	Tubulin spindle
19.	Plasma membrane	+, without sterols (except in mycoplasma)	+, with sterols
20.	Capsules	Frequently present	-
21.	Cell wall	Complex structured with peptide-glycan layer, protein and lipids	Absent or composed of chitin or cellulose
22.	Muramic acid in cell wall	+	-
23.	Membrane lipid glycerol-phosphates	L-isomer	D-isomer
24.	Cellular differentiation	Sporulation, heterocyst, bacteroids	Extensive in multicellular forms
25.	Sensitivity to antibiotics:		
	-Anisomycin	+	-
	-Chloramphenicol & Kanamycin	+	-

Contd...

Table 2.4. Contd...

26. EF_2 sensitivity to diphtheria toxin	-	+
27. Ribosomes (sub-unit size)	70 S type (30 S, 50 S)	80 S type (40 S, 60 S)
-Smaller r RNA size	16 S	18 S
-Larger r RNA size	23 S	28 S
28. CCUCC binding site at 3′ end of small r RNA	+	-
29. tRNA initiating amino acid	Formyl-methionine	Methionine
30. Photosynthesis	Associated with membrane systems and vesicles in cytoplasm	Chloroplast present in algae and plants
31. Microbial arrangement	Unicellular or colonial	Multicellular

+ = Present; - = Absent

Structure : Morphologically bacteria are unicellular (Fig. 2.3) and show great variation in size. Bacterial size range 0.5-1.0 x 1.0-2.0 µm. They may be rod (bacillus), spherical (coccus), helical (spirillum), curved (vibrio) or pleomorphic in shape. The cells may remain together in groups or clusters or form long chains. Some bacteria produce resistant endospores within their cells during unfavourable conditions. Motility is through flagella which may be monotrichous (one polar flagella), lophotrichous (tuft of flagella at one polar end only), amphitrichous (tuft of flagella on both sides) or peritrichous (several flagella all around the cell). Small cocci and short rods predominate in most of the soils. The structure of a typical soil bacterium is shown in Fig. 2.3.

The complete characterization of a soil in terms of genera and species of all bacterial communities present is yet impracticable despite the improvements in isolation techniques. Out of 170 bacterial genes listed in the 7th edition of *Bergey's Manual of Determinative Bacteriology*, 97 genera contained the species that may be considered soil bacteria. The bacteria involved in major biochemical transformations in soil probably constitute <10% of the total bacterial population. The function of remaining population is unknown. The most common genera found in soil are presented in Table 2.5. *Arthrobacter, Pseudomonas, Bacillus* and *Streptomyces* dominate bacterial population and as per the traditional plate count estimates account about 40-80% of the soil bacterial population.

Ecology and function : Most bacterial cells are absorbed on soil particles (clay) by a variety of mechanisms mostly by simple ion exchange. The bacteria and clay, both negatively charged, are held together via ionic bridges involving polyvalent cations. The soil bacteria show numerous morphological and physiological adaptations that enable them to resist adverse conditions e.g. presence of mucilaginous capsule surrounding the cell, slowing down of metabolic activities under nutrient starvation, etc. Although soil temperature and moisture do influence their populations, bacteria can withstand extreme climates like arctic (sub-zero temperate zone) and desert conditions. Bacterial activity is greatest at 20-40°C. Ordinary soil temperature extremes seldom kill bacteria but, of course, suppress or suspend their activities temporarily. Bacterial biomass in soil is also influenced by soil pH, aeration, farming practices, fertilizers, organic matter additions, pesticides, etc. Under anaerobic conditions bacteria dominate because fungi, algae and actinomycetes do not grow well in absence of oxygen. High calcium concentration, pH range of 6.0 to 8.0 and moisture potential of –10 to –100 bar are conducive for most aerobic bacteria. Some bacteria can exceptionally grow even at pH as low as 2.0 or at higher temperatures (>100°C).

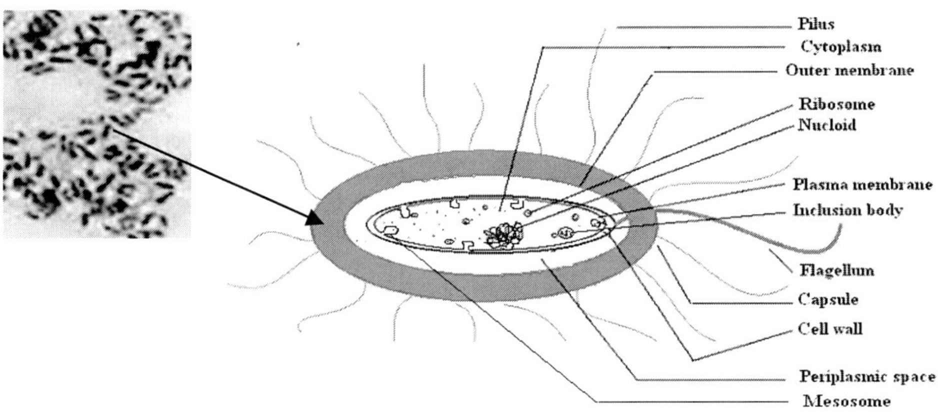

Fig. 2.3: The structure of a typical bacterial cell

Bacterial cells tend to occur in small soil pore spaces thereby not only acquire water regime as per their requirement but are also protected from grazing by protozoa. Inside aggregates of soil mostly G-ve bacteria and actinomycetes dominate whereas inter-aggregate populations consist mainly of G+ve bacteria and other soil microbes like moulds, fungi, protozoa, algae, etc. This distribution is one of their survival strategies. The composition of bacterial population can often be used as an indicator of physicochemical conditions of soil. The presence of actively growing bacteria like *Clostridium*, for example, is indicative of anaerobic conditions either in bulk soil or at/within microsites.

Soil bacteria are heterotrophic or autotrophic in their metabolism, the former are involved in the decomposition of organic matter. Specialized bacterial communities along with fungi, etc. degrade macromolecules into micro-molecules, which serve as substrate for much wider array of bacteria and fungi. Soil bacteria possess a broad range of enzymatic capability so digest from naturally occurring materials to even xenobiotic and recalcitrant compounds.

Table 2.5: Most common bacterial genera (excluding actinomycetes and cyanobacteria) present in soils

Genus	Characteristics	Remarks
Arthrobacter	Pleomorphic, G variable, freely motile, use a variety of substrates, obligate aerobes in soil, slow grower.	10 to 40% of soil bacterial biomass;
Streptomycetes	G+ve, oxidative organotrophs, produce well-developed compact branched mycelia, reproduction by aerial spores and mycelial fragmentation, intolerant to acidity, some cause plant diseases and galls, and many produce antibiotics.	5-20% of soil bacterial biomass);
Pseudomonas	G-ve, straight or curved rod, mostly aerobic, oxidative organotrophs, a few facultative lithotrophs, use H_2or CO_2 as energy source; some cause plant diseases, some produce siderophores, fluorescence, pigments, etc.	5-20% of soil bacterial biomass);
Bacillus	G+ve to variable, rods, motile, some produce endospores; species either organotrophs or facultative lithotrophs. One specie *B. polymyxa* N_2 fixer; some produce antibiotics and lytic enzymes (e.g. *B. thuringiensis*); others are pathogens; tolerate wide temperatures, pH and salts.	1-10% of soil bacterial biomass);
Clostridium	G+ve, motile, anaerobic, rods, sporulating, some species fix N_2; some are pathogenic.	-
Xanthomonas	G-ve, aerobic rods, mostly organotrophic and pathogenic.	-
Azotobacter	G-ve, aerobic rods, organotrophs, fix N_2 symbiotically	-
Azomonas	G-ve, fix N_2 asymbiotically; organotrophs	-
Beijerinkia	G-ve, fix N_2 asymbiotically; organotrophs	-
Azospirillum	G-ve, fix N_2 asymbiotically; organotrophs	-
Rhizobium	G-ve, fix N_2 asymbiotically; organotrophs	-
Bradyrhizobium	G-ve, fix N_2 symbiotically in root nodules of legumes	-
Agrobacterium	G-ve, induces galls or hypertrophy	-
Notrosomonas	G-ve, chemolithotrophs; nitrifiers	-
Nitrobacter	G-ve, chemolithotrophs, nitrifiers	-
Lactobacillus	G+ve, fermentative organotroph associated with plant herbage.	-
Enterobacter	Fermentative; common in sewage sludge and feces, some found on plants and in soil.	-
Corynebactrium	G+ve, non-sporulating rods.	-
Mycobactrium	G+ve, non-sporulating rods.	-
Flavobactrium	G-ve, rods.	-
Achromobacter	G-ve, rods.	-
Aerobacter	G-ve, rods.	-
Micrococcus	G+ve, spherical, aerobic, immotile.	-
Sarcina	G+ve, spherical.	-
Cytophaga	Gliding bacteria (glide as rapidly as 10 μm min^{-1} on moist surfaces); cellulolytic.	-

Arthrobacter species, for instance, are major catabolizers of heterocyclic compounds and hydrocarbon. Many soil bacteria possess capability to oxidize or reduce several elements like N, S, Fe, Mn, etc. Nitrifiers through biochemical oxidation oxidize ammonia to more mobile nitrate form of nutrition. Similarly, sulphur oxidation by bacteria yields sulphate ions. Some specialized bacteria carryout non-symbiotic and symbiotic nitrogen fixation. In soil there are some predacious bacteria like *Bdellovibrio, Myxococcus, Polyangium, Cytophaga, Sporocytophaga, Achrangium, Chondrococcus,* etc. which feed on some common soil bacterial species by secretion of extracellular enzymes that dissolve/lyse the bacteria and then absorb their liquid cell contents.

2.5 ACTINOMYCETES

Structure : Actinomycetes are G+ve aerobic bacteria which produce mycelium-like structures, typically 15-20 μm long and 0.5-2.0 μm in diameter. Actinomycetes belong to class Schizomycetes of order Actinomycetales. Because of their resemblance with fungi in many characters such as filamentous nature, branching pattern and conidia formation the actinmomycetes are also called 'ray fungi' (Fig. 2.4). However, size, spore and many other characters of actinomycetes are similar to bacteria. Actinomycete colonies on agar plates appear slowly, show low consistency and stick firmly to agar surface (Plate 2.1). The cell wall of actinomycetes has no chitin or cellulose. The growth of soil actinomycetes occurs from the hyphal apices with regular branching occurring behind the main apices of leading hyphae. The vegetative mycelium is lichinoid and usually exists below the surface of agar medium.

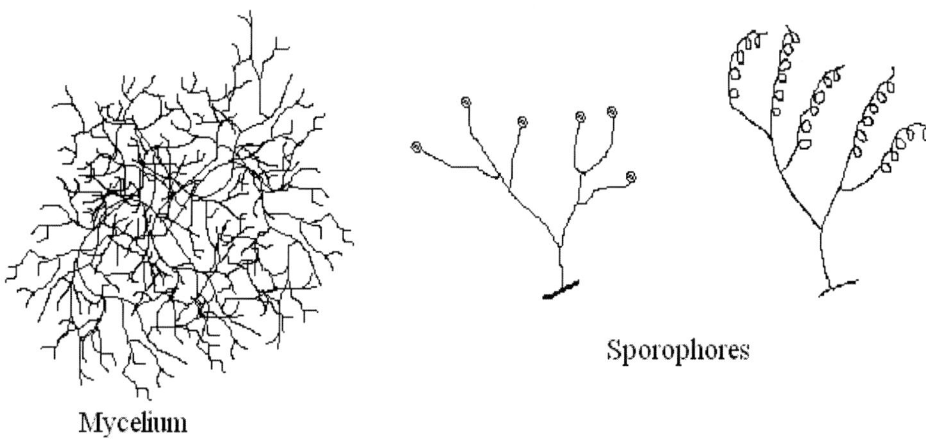

Fig. 2.4: The mycelial structure and sporophores of soil actinomycetes

Cell division is bacteria-like in most cases. Actinomycetes produce true conidia or chlamydospores. The filamentous bodies of actinomycetes break into rod-shaped smaller fragments called 'arthrospores', each of which is capable of growing independently. The conidia or spores may be spherical, oval or cylindrical in shape, borne on aerial sporophores. On germination, spores form germ tube from which primary hyphae are produced. This gives rise to a network of thin mycelium. The type of aerial mycelium varies with the genus and species of actinomycetes. Spores are formed within hyphal wall. The hyphal contents divide simultaneously to form septate dense particles, which are transformed into chains. There are seven groups of actinomycetes classified on the basis of their hyphal nature, mycelial characteristics, sporulation, fragmentation, etc. (Table 2.6).

Plate 2.1: Fried egg appearance of actinomycetes on Petri-plate

Ecology and function : Streptomyces is the most common genera present in soil and forms about 70% of actinomycete population. Actinomycetes are more common in warm than cool soils, and under normal soil conditions their percentage in total microbial population increases with depth. Actinomycetes are largely intolerant of acidity although a few streptomycetous species are active below pH 5.0. Most conducive pH and temperature range for the growth of actinomycetes is 6.5-8.5 and 25-30°C, respectively. Actinomycetes have the ability to withstand desiccation though streptomycetes require moist conditions for growth. The actinomycetes which are capable to tolerate high temperatures play an important role in active compost piles whose temperature can reach 50°C. The thermotolerant actinomycetes include genera *Pseudonocardia, Thermomonospora, Thermopolyspora, Thermoactinomyces,* etc.

Table 2.6: The characteristic features of various groups of actinomycetes

Group	Characteristic features	Common species
Streptomycetaceae	Mycelium abundant, remain intact, aerial mycelium also present; no fragmentation of hyphae; long chains of conidia (5-10);	*Streptomyces, Agromyces, Rothia, Arachnia*
Nocardiaceae	Mycelium abundant mostly fragment to form cocoid or elongated elements, hyphae small rounded structure, conidia very short; reproduction by fragmentation, aerobes;	*Nocardia, Pseudonocardia, Rhodococcus*
Micromonosporaceae	Mycelium remains intact, no fragmentation of hyphae into small segments; conidia (spores) found singly or in pairs or in clusters (short chains);	*Micromonospora, Walksmania (Microbispora), Micropolyspora, Thermoactinomyces*
Actinoplanaceae	True mycelium produced, no hyphal fragmentation; spores formed inside sporangia;	*Streptosporangium, Atinoplana, Planobispora*
Dermatophilaceae	Mycelium filamentous, divide transversely and in at least two longitudinal planes to form a mass of motile cocoid elements; aerial mycelium absent; fragmentation of hyphae leads to the formation of small rounded motile structures;	*Geodermatophilus, Dermatophilus*
Frankiaceae	Nitrogen fixing actinomycetes, true mycelium present, microaerophilic, no hyphal fragmentation; inhabit many non-legume plants;	*Frankia*
Actinomycetaceae	True mycelium absent, branching filaments present, anaerobic; rod, coccoid or coryneform; no hyphal fragmentation;	*Actinomyces*

Most actinomycetes are free-living saprophytes and decompose carbonaceous substances including recalcitrant molecules like chitin, cellulose, hemicellulose, phospholipids, keratin, lignin, etc. under alkaline conditions. When fresh addition of organic matter is made to soil, actinomycetes are much less competitive than bacteria and fungi. Only when resistant materials remain actinomycetes show good competitive ability and prominence. The actinomycetes are also known to aid in the development of water stable soil structure by secreting non-water soluble gummy substances. In forest ecosystems, much of the nitrogen supply depends on certain actinomycetes that are capable of fixing atmospheric nitrogen to ammonium-N that is then available to higher plants. Some actinomycetes produce vitamins, growth substances and antibiotics as secondary metabolites. More than 2,000 antibiotics have been obtained from actinomycetes of which 30 are most widely used in medicine and a few in agriculture and animal production. Some of the most common antibiotics are streptomycin (from *Streptomyces grieseus*), aureomycin (from *S. nouresii*), chloramphenicol (from *S. venezuele*), tetracyclin (from *S. aureofasciens*), neomycin (from *S. fradiae*), erythromycin (from *S. erythreus*), etc. *Frankia* group of actinomycetes develop nitrogen fixing associations with non-leguminous plants. A few actinomycete species cause plant diseases such as *Streptomyces scabies* and *S. ipomeae* inciting scab in potato and sweet potato under alkaline/neutral soil conditions. Some actinomycetes produce a volatile secondary metabolite terpene derivatives (e.g. trans-1,10-dimethyl-trans-9-decalol) called 'geosmin' which give soil its earthy odour especially soon after the rainfall.

2.6 CYANOBACTERIA

Structure: The cyanobacteria are G–ve procaryotes containing chlorophyll *a* and phycobiliprotein pigment capable of performing oxygenic photosynthesis and fix atmospheric nitrogen. Cyanobacteria exist in unicellular, colonial or filamentous form and some even show simple parenchymatous organization (Fig. 2.5). Flagellate cells never occur at any stage in their life cycle. In cyanobacteria filaments (excluding sheath), if present, are known as trichomes. The filaments, in some cases, may be branched which may be genuine (true) or false. The structural part of cell wall consists of murein (peptidoglycan) outside which is a lipopolysaccharide layer. Some cells or filaments may be embedded in a mucilaginous (hydrated polysaccharides) sheath. Cells are generally blue green to violet in colour but sometimes red or green depending on the type of phycobiliprotein present. The green colour of chlorophyll is masked by blue accessory pigments phycocyanin and allophycocyanin, and red accessory pigment phycoerythrin. The phycocyanin granules (composed of arginine-asparagine polymers), cyanophycean starch (reserve polysaccharide α-1,4 glucan granules) and highly polymerized polyphosphate granules serve as the main storage products in cyanobacteria. The cyanobacteria often exhibit phototrophic gliding movement. They have ability to live under the conditions of low oxygen and survive at high temperatures. The cyanobacteria can control their buoyancy via gas vacoules.

The cyanobacteria unlike algae lack sexual reproduction and asexual reproduction by ciliated bodies (zoospores). In many unicellular and colonial forms reproduction is through cell division. The unicellular cyanobacteria show binary or multiple fission or serial release of apical cells as exospores. The young dwarf daughter cells are known as 'baecytes or nanocytes', if they do not have sufficient time to grow to their full size before next division. Other types of asexual reproduction include endospore production.

Filamentous cyanobacteria reproduce through random fragmentation of their filaments known as 'hormogonia formation' as well as trichome breakage so these organisms show great structural and genetic diversity and possess high differentiation. The bacterial parasexual processes of transformation and conjugation, however, may bring about genetic recombination.

Some filamentous cyanobacteria form oversized resting spores called 'akinetes' which are large thick walled cells developed from vegetative cells. Akinetes enable cyanobacteria to survive unfavourable and extreme environmental conditions. They contain large amounts of cyanophycin, polypeptides and cyanophycean starch but not polyphosphates. Cyanobacteria contain fatty acids which have composition different from other bacteria. β-carotene has been found in all the strains investigated so far. Some cyanobacteria possess heterocysts which are well-differentiated specialized cells having transparent cell contents and thick wall. The heterocyst containing cyanobacteria are able to fix atmospheric nitrogen under aerobic conditions whereas non-heterocystous ones do so under anaerobic conditions only.

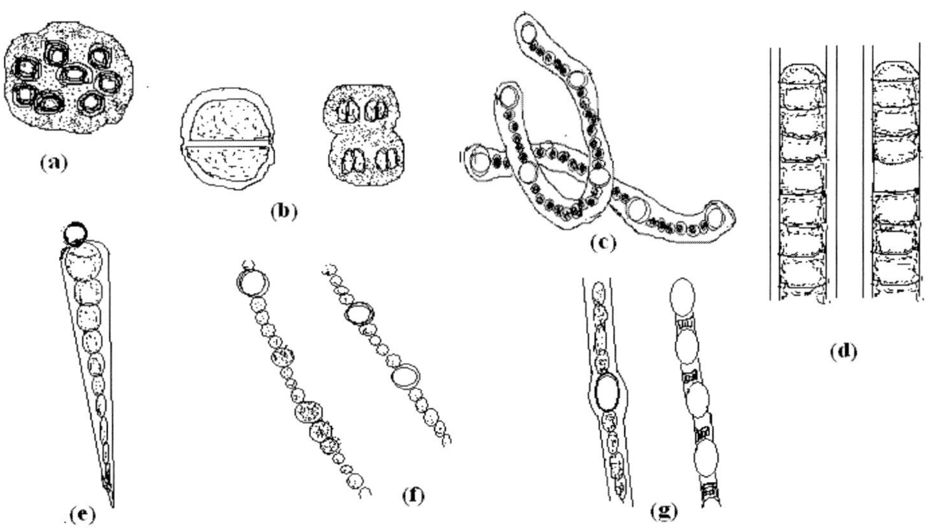

Fig. 2.5: Some cyanobacteria frequently found in soil a) *Gleoecapsa;* b) *Croococcus;* c) *Nostoc;* d) *Lyngbya*; e) *Rivularia;* f) *Anabaena;* g) *Aulosira*

The photosynthetic pigments are located on thylakoids, which lie free in cytoplasm. Thylakoids contain chlorophyll *a* but chlorophyll *b* and *c* are absent. The carboxysomes present contain primary enzyme ribulose 1,5 bisphosphate carboxylase-oxygenase for photosynthetic CO_2 fixation. The mechanism of photosynthesis is identical to that of photosynthetic eucaryotes. Assimilation of CO_2 is through Calvin cycle with the formation and deposition of glycogen as an intracellular reserve material. Other pigments present are phcobilisomes, zeaxanthin, echinenone, cantheraxanthin, myxoxanthin, oscillaxanthin, β-cryptoxanthin, isocryptoxanthin and mutachrome.

Taking into consideration the subdivisions given in the recent bacteriological treatment of blue green bacteria given by Waterbury in 1989, Waterbury and Rippka in 1989 and Hoek and his associates in 1995 have classified cyanobacteria into five major orders on the basis of their cell organization, reproduction and presence or absence of trichome, sheath, heterocyst, akinete etc. (Table 2.7).

Table 2.7: Important characteristics of agriculturally important orders of cyanobacteria

Characters	Orders				
	Chroococcales	**Pleurocapsales**	**Oscillatoriales**	**Nostocales**	**Stignometales**
Cell form and structure	Mostly unicellular, some form clusters held together by mucilaginous sheath;	Unicellular, some form colonies of irregular aggregates bearing short branched or unbranched filament, often united into pseudoparenchy ma;	Filamentous, false branching, trichomes always composed of vegetative cells only;	Filamentous, false branching occur but no true branching;	Filamentous, often multi-seriate (with several rows of cells), true branching present;
Reproduction	Binary fission; formation of nanocyte or through budding; sporogenous;	Binary and multiple fission (baecyte or nanocyte); formation of endospore; sporogenous;	Random trichome breakage; formation of hormogonia; or by akinete germination; division in one plane asporogenous	Formation of hormogonia; trichome breakage; or by akinete germination; division in one plane only; asporo-genous;	Formation of hormogonia; trichome breakage; or by akinete germination; division in more than one plane; asporogenous
Heterocyst	Absent	Absent	Absent	Present; inter-calary or terminal;	Present; terminal or lateral;
Akinete	Absent	Absent	Absent	Present	Present
Common genus	*Gloeocapsa* *Chroococcus* *Aphanotheca* *Microcystis* *Gloeotheca*	*Pleurocapsa* *Dermcarpa* *Cyanocystis* *Myxosarcina*	*Oscillatoria* *Lyngbya* *Plectonema* *Phormidium* *Calothrix*	*Nostoc* *Anabaena* *Rivularia* *Nodularia* *Cylindrospermum*	*Stigonema* *Fischerella* *Hapalosiphon* *Westelliopsis*

Ecology and function : The cyanobacteria at present have about 150 genera and 2,000 species. They are ubiquitous in distribution and found in most diverse habitat from fresh water, damp soil to even extreme and inhospitable places as glaciers, deserts, deep sea and hot springs. Cyanobacteria are the only organisms able to perform oxygenic photosynthesis as well as fix atmospheric nitrogen. Since they do not show dependence on preformed organic matter so can colonize even bare rocks surfaces and initiate soil formation. They can grow in presence of light in a mineral medium, using CO_2 as a carbon source and N_2 as nitrogen source. The nitrogen fixing representatives are conspicuous in the environments where combined nitrogen is a limiting nutrient, notably in tropical and sub-tropical soils, which are often nitrogen poor. Certain thermophilic cyanobacteria grow abundantly in neutral or alkaline hot springs where they are the predominant members of the photosynthetic population. In deserts, unicellular cyanobacteria grow in microfissures just below the surface of rocks where small moisture is trapped and sufficient light penetrates to permit photosynthesis. The cyanobacterial biomass value of 2-62 g m^{-2} rock from desert area in Middle East and North America has been reported. The cyanobacteria have the ability to use even low light intensities effectively, so can thrive below the surface, deep in epilimnion. Cyanobacteria are important net primary producers contributing about 16% of the annual net productivity.

The cyanobacteria can occur in soil even at considerable depths and under suboptimal conditions, congregate at the surface and become exposed to direct sunlight which can bring about photo-oxidation of photosynthetic pigment and ultimately the death and disintegration of organism. In lakes polluted with organic sewage or nutrient salts like phosphates, certain bacteria often form enormous bloom. This eutrophication or algal bloom is a symptom of carbon and nutrient access leading to the depletion of oxygen and creation of anaerobic conditions, especially in polluted water bodies. The growth of cyanobacteria is stimulated with organic substances; perhaps lowered oxygen concentration in water brought about by heterotrophic bacterial breakdown stimulates cyanobacterial photosynthesis and nitrogen fixation. Some cyanobacteria live as symbionts with or even within other plants. They occur in many lichens, roots of *Cycas*, leaf cavities of water fern *Azolla*, liverworts and angiosperms, in unicellular fungus *Geosiphon*, etc. About 8% lichens do have cyanobacteria as their symbiotic partner. Such associations are primary colonizers and initiators of soil formation. Cyanobacteria are estimated to contribute about 40 kg bound N ha^{-1} yr^{-1} to soil. Water fern *Azolla* harbouring symbiotic *Anabaena* in leaf cavities can add 120-310 kg N ha^{-1} yr^{-1}.

2.7 FUNGI

Diversity in soil fungi : Fungi constitute a very large diverse group of organisms found virtually in every ecological niche. Conservative estimates have revealed that world-wide 1.5 million species of fungi are present, of which only 90,000 species have been described. Fungi are next to bacteria in abundance in soil. The fungi generally dominate soil biomass (70% by weight). More than 700 species belonging to 170 genera have been isolated from soil with a small number of genera such as *Penicillium* and *Aspergillus* accounting for more than half of the species.

General structure of fungi : Fungi are eukaryotes, devoid of chlorophyll and generally have filamentous mycelium with uni-, bi- or multi-nucleate hyphae. The hypae may be septate or aseptate, usually 0.5-10 μm in diameter. Cells in septate hypae are separated by a cross wall or septum. Hypha grows characteristically at tips and can usually produce lateral branches in the older parts. The cell wall, which surrounds the fungal somatic structures, is a rigid structure composed of chitin or cellulose generally having layers of semi-crystalline chitin microfibrils that are embedded in an amorphous matrix of α-glucan. Occasionally, both chitin and cellulose may occur simultaneously. In addition, cell wall contains proteins, glycoproteins, lipids and other polysaccharides like chitosan, mannan, glucan or galactose, etc. Mycelium sometimes forms organized stranded or coiled structures like rhizomorphs. Some fungi are unicellular such as yeasts, which multiply through binary fission or budding. Fungi have well developed nuclear material and reproduce generally sexually and asexually. Asexual reproduction occurs through fragmentation of soma, fusion of somatic cells, production of spores like conidia, sporangiospores, zoospores etc. In sexual reproduction fertilization occurs through planogametic copulation, gametangial contact, gametangial copulation, spermatization and somatogamy. In fungi, there is generally a cycle of haploid and diploid structures, corresponding to gametophyte and sporophyte in green plants. Diploid phase begins with karyogamy after plasmogamy and ends with meiosis. In majority of fungi, there is no distinct alteration of generation.

Fungi are obligate aerobes and mostly heterotrophic in metabolism but some species like yeasts are capable of living in low O_2 tension. Fungi are biochemically less versatile than bacteria. Currently fungi are considered monophyletic group where all members within a phylum are descendents of one ancestor. Fungi have four accepted phyla which are distinguished from each other on the basis of difference in vegetative thallus and reproductive structures as well as their antogeny. These are Chytridiomycota, Zygomycota (both lower fungi), Ascomycota and Basidiomycota. Deuteromycota, as per the *Dictionary of Fungi* (9th edn.) is not accepted as formal taxonomic category because they are not monophylectic unit but fungi with either lost sexual phase or which are anamorp of other pylla. There are some additional related phyla collectively called 'moulds', which show some morphological similarities and have absorptive nutritional mechanism but are phylogenetically distant from true fungi. Without going into the controversies regarding fungal classification, the major classes of fungi along with their spore and thallus characteristic is given in Table 2.8.

2.7.1 MOULDS

Slime moulds : Slime moulds, also called 'protistan fungi' because of their protistan nature, are amoeboid organisms that lack cell wall in their assimilative stages. They feed by phagotrophy (engulfment) as well as by absorption of soluble nutrients. They form cell-walled spores and supporting structures during sporulation. Slime moulds include Plasmodiophoromycota, Acrasiomycota and Myxomycota.

<u>Plasmodiophoromycota:</u> Plasmodiophoromycota has organisms which are protozoa-like intracellular obligate endo-parasites of algae, fungi and plants. They are multinucleated having unwalled protoplast within host cells and cause hypertrophy and hyperplasia of surrounding host tissue. The somatic phase is a plasmodium. The swarm cells/zoospores are anteriorly biflagellate. The flagella are unequal. Organisms reproduce by cysts formed within host cells, which on germination produce anteriorly biflagellate zoospore that swim to roots and actively penetrate them. This group contains the important plant pathogen *Plasmodiophora brassicae*, which causes clubroot disease of cabbage and related plants.

<u>Acrasiomycota :</u> Acrasiomycota comprise of soil dwelling 'cellular slime moulds' mostly found in moist soils, dung and decaying materials. They are uni-nucleate haploid amoebae for most of the life cycle and feed by engulfing bacteria. The cells on starvation aggregate into a characteristic multicellular mass called plasmodium or pseudoplasmodium or myxamoebae or grex or slug by simple migration of a population of amoebae to a central location. Pseudoplasmodium behaves like a motile communal unit but no cell division does occur in it. Plasmodium later on forms sporocarp that bears asexual spores. Externally, plasmodium is differentiated into foot, stalk and sporangium. During sporulation some amoebae form cellular stalk of sorocarp whereas rest are held together by exocellular mucoid secretion. Ascarsiomycota produce flagellate sporangia. *Dictyostelium* is an example of Ascarsiomycota.

Table 2.8: Major groups of fungi and their spore and thallus characteristics

Group	Class/ form-class	Spore		Thallus
		Sexual	Asexual	
Slime moulds	Plasmodiophoromycota (endoparasitic)	-	Zoospores/ sporosori	Plasmodium
	Acrasiomycota (Acrasid cellular amoeboid)	Amoebae	Soprcarp spores	Pseudoplasmodium
	Dictyosteliomycota (Dictyostelid cellular amoeboid)	Amoebae	Soprcarp spores	Pseudoplasmodium
	Myxomycota (true cellular swarm cell/amoeboid)	Sporongiospores	None	Plasmodium
Water moulds	Oomycota	Oospore	Biflagellate zoospores	Non-septate mycelium or chytrid like
	Hypochytridiomycota	-	Uniflagellate zoospores	Non-septate mycelium or chytrid like
Flagellate fungi	Chytrydiomycota	Zoospores (meiospores)	Zoospores (mitospores)	Non-septate mycelium or chytrid like
Sugar fungi	Zygomycota	Zygospores/spores	Sporangiospores	Non-septate mycelium
Higher fungi (true fungi)	Ascomycota - Hemiacsomycetidae (yeast, true fungi)	Ascospores	Yeast buds	Yeasts or septate mycelium
	- Euacsomycetidae (true fungi)	Ascospores	Conidia	Septate mycelium
	- Loculoascomycetidae (true fungi)	Ascospores	Conidia	Septate mycelium
	Basidiomycota - Teliomycetidae (true fungi)	Teliospores & basidiospores	Aeciospores or urediniospores	Septate mycelium &/or yeast
	- Eubasidiomycetidae (true fungi)	Basdiospores	Conidia or none	Septate mycelium &/or yeast
Higher fungi (imperfect)	Deuteromycota - Hypomycetidae	-	Exposed conidia on conidiophores or on synnemata & sporodochia	
	- Coelomycetidae	-	Conidia enclosed within pycnidia/ acervuli	
	- Blastomycetidae	-	No differentiated conidia/asexual yeasts	

Myxomycota : Myxomycota are 'acellular true slime moulds'. They can migrate in large masses through crevices of the substrate and engulf food particles like bacteria and spores of many fungi, so are phagotrophic in nutrition. They, however, are fungus-like in their reproductive structures and spores. Myxomycota form uni-nucleate haploid myxamoebae (for their vegetative state only) or swarm cells (on germination of their spores). Myxamoebae and swarm cells are inter-convertible. Myxomycota show plasmodial acellular creeping locomotion. During vegetative phase, these fungi are found as naked multinucleate mass of protoplasm called plasmodium. Flagella are often unequal, attached posteriorly. In some cases, myxamoebae form flagellate swarm cells or may fuse in pairs to form a zygote, which later on develops into a giant amoebae thallus called 'plasmodium'. The plasmodia are coloured, mostly yellow due to the presence of anthracene pigments. Plasmodia may either fuse or break into smaller plasmodia. During reproductive phase they develop spore-bearing structures called sporangia (sporocarp) or plasmodiocarp which are characteristically covered with a thin membrane called peridium. The spores germinate under favourable conditions especially in organic substrates, form naked body and multiply rapidly. Fruiting bodies are similar to those of cellular slime moulds. Myxomycota are found abundantly in decaying vegetation or dung and often attain 50 cm or more size. The genus *Physarum* is largest genus having more than 100 species.

Water moulds : Water moulds (Division Chromista) are flagellate fungi of phylla Oomycota, Hypochytridiomycota and Labrinthulomycota. They are normally found in water and soil. They have several non-mycelial coenocytic (multi-nucleate thallus) forms (Fig. 2.6). The simplest holocarpic chytrids are sac-like multinucleate cell with no appendage or branches while eucarpic chytrids have rhizoids attached to thallus body, which may be simple or branched. Rhizoids are filamentous anchoring organs, which lack nuclei. Most chytrid-like fungi show a determinant limited growth habit but others form a mycelium-like thallus, known as rhizomycelium, of unlimited extent that consist of a series of multinucleate centres connected by rhizoids. Many water moulds form true mycelium that lack septa but may form rhizoids. The thallus body of chytrid may entirely (holocarpic) be converted to sporangium that produces motile zoospores e.g. *Allomyces, Achlyes, Saprolignia,* etc.

Oomycota : Oomycota have cellulose containing cell wall and a diploid vegetative state. They have mycelial and chytrid-like thali which are non-septate and holo- or eu-carpic. They produce zoosporangia asexually and oogonium and antheridium as sexual organs. They reproduce asexually by bi-flagellate motile zoospores and sexually produce thick walled oospores. Zoospores are of two types i.e. primary zoospores (both flagella anteriorly placed) and secondary zoospores (flagella laterally inserted). Many species are highly destructive to higher plants, algae, insect larvae, etc. and some are saprobic on plant and insect debris. *Perenospora, Pythium* and *Phytophthora* are prominent oomycetes causing many diseases like downy mildew, damping off, rot, blight, etc.

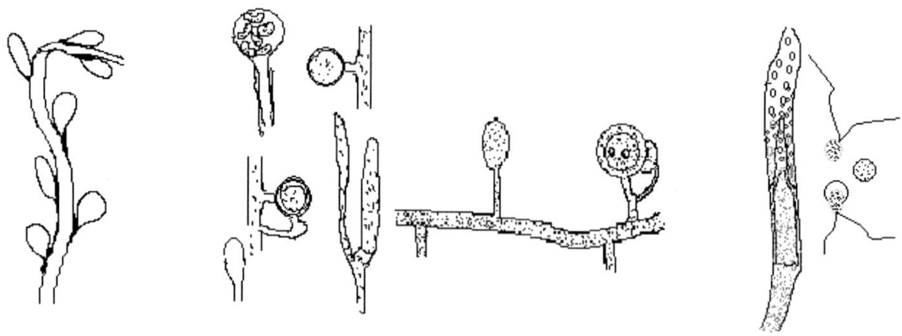

Allomyces ***Pythium*** ***Saprolegnia***

Fig. 2.6: Some soil water moulds

Hypochytridiomycota : Hypochytridiomycota possess single flagellate zoospores and show dominance of haploid vegetative state. All possess eucarpic monocentric thallus that has both cellulose and chitin in cell wall. They may be parasitic or saprobic in nutrition. *Rhizodiomyces* is an example of this group.

2.7.2 TRUE FUNGI

The true fungi have chytrid, yeast or mycelial thalli, mostly being mycelial. They have chitin and non-cellulosic glucans and distinctive sterols in their cell wall. The true fungi are broadly grouped into lower and higher fungi.

Lower fungi : Lower fungi are primitive ones and most prevalent in aquatic or soil water systems. Their vegetative mycelium is non-septate and complete septa are only found in reproductive structures. Lower fungi include sugar and flagellate fungi (Fig. 2.7).

Chytridiomycota (flagellate fungi) : Chytridiomycota have posteriorly uniflagellate zoospores with whip-lash like flagella. The spores and gametes are produced within differentiated segments of mycelium or thallus called zoosporangia and gametangia, respectively. Thallus is coenocytic. Sexual reproduction involves fusion of gametes in pairs. They are saprobic or parasitic on algae, animal and plants. For instance, *Synchytrium, Allomyces* and *Physoderma* cause diseases in plants while *Batrachochytrium, Coelmomyces* and *Catenaria* parasitize soil animal.

Zygomycota (sugar fungi) : Zygomycota are called sugar fungi as many specie of this group are involved in the fermentation of diverse carbohydrate substrates. The fungi of this group possess coenocytic multi-nucleate mycelium. Mycelium is usually well-branched, hyaline and thin-walled. The characteristic asexual non-motile spore bearing structure is called 'sporangium'. Sporangia may be terminal or intercallary. The fungi produce non-motile asexual sporangiospores in sporangium as well as sexual thick-walled resting spore called zygospores. Sexual reproduction is by means of zygospores produced by the fusion of two similar gametangia which are formed as terminal multinucleated segments on specialized branches of mycelium. Some fungi of this category produce single-spored sporangia which are sometimes called 'conidia' because of resemblance to asxual spore of Deuteromycetes. The group is mostly saprobic, some are plant pathogens and others myco-parasitic. The soil fungi belonging to this group include *Mucor, Cumminghamella, Zygorhynchus, Rhizopus, Mycotypha,* which are of commercial importance in alcohol and organic acid production. Some fungi in this group form obligate symbiotic association with roots of higher plants (see chapter 10 and 11). These are called endomycorrhizae such as *Glomus, Gigaspora, Endogone, Acaulospora,* etc.

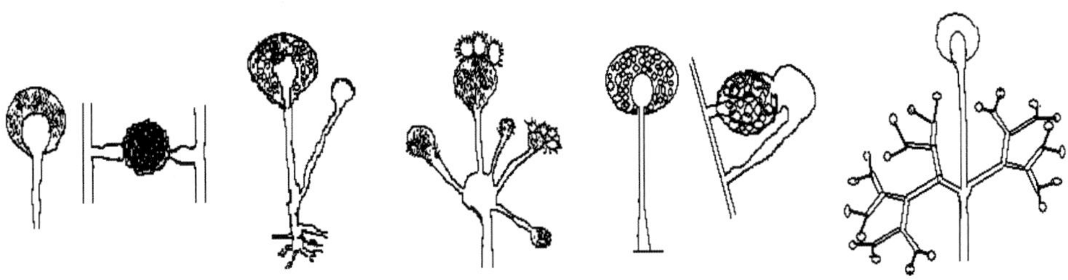

Fig 2.7: Some important lower fungi

Higher fungi : Higher fungi possess septate mycelium and include pylla Ascomycota and Basidiomycota and fungi imperfecti. In contrast to lower fungi, higher fungi are adapted to grow on solid surfaces of organic substrates.

Ascomycota : Also known as 'sac fungi', Ascomycota are the largest fungal group with over 40,000 species identified to date. About half of these species exist as symbionts with photosynthetic microorganisms in lichen. Ascomycota include septate fungi as well as unicellular yeasts. They produce sexual ascospores in sac-shaped cells called 'asci' which are contained in open or closed fruiting bodies called "ascoma". Sexual fusion is by gametangia or through somatic hyphae. Sexual reproduction is characterized by ascospore formation inside ascus. Each ascus contains 2-8 ascospores. The asexual reproductive structures are thick-walled chlamydospores or sclerotia. However, asexual spores, such as conidia, are mainly formed through which fungus propagates. Conidia are produced on conidiophores (Fig. 2.8).

Ascomycetes include ascus forming yeasts, morels and truffles, cup fungi, etc. Some are phyto-pathogenic (powdery mildew) and others are saprophytic in soil.

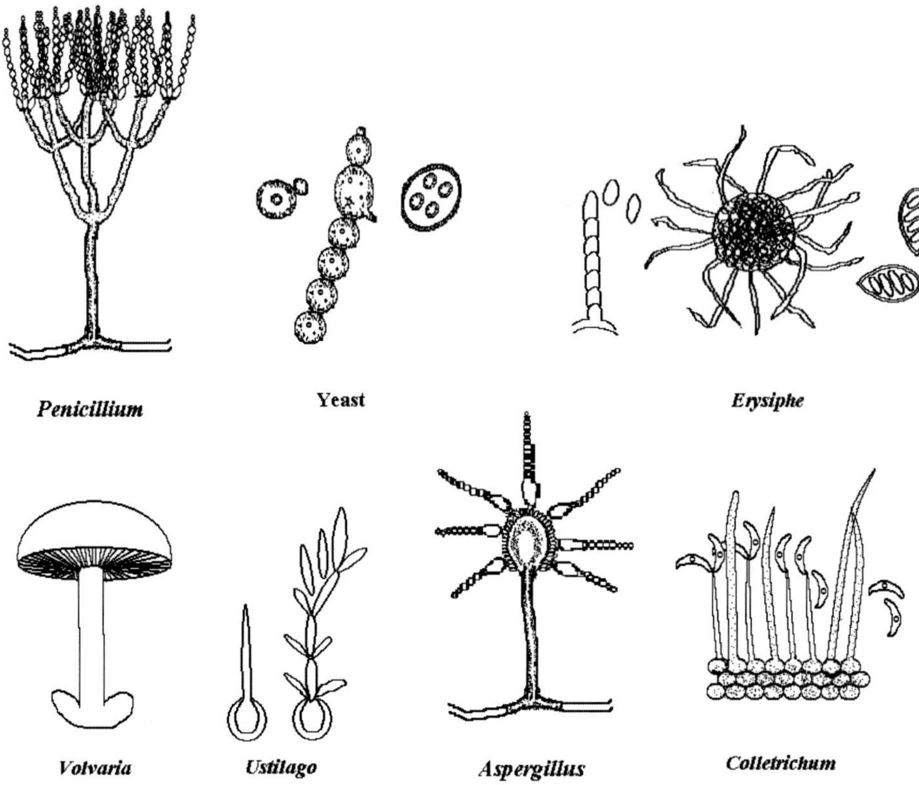

Fig. 2.8: Some important higher fungi of soil

Euascomycetidae : Euascomycetous fungi form asci with ascohymenial/ ontogeny within mycelial structures called ascoma that differentiates as part of sexual cycle into cliestothecia, perithecia or apothecia. The representative soil organisms are *Neurospora, Giberella, Sclerotinia, Claviceps, Chaetomium* etc.

Loculo-ascomycetidae : Loculo-ascomycetidae produce asci in pseudothecia formed within a stroma in preformed cavities with ascolocular ontogeny, a dense black-pigmented tissue. The representative soil organisms are *Pleospora, Elosinoe,* etc.

Hemi-ascomycetidae : Also called yeasts, hemi-ascomycetidae lack mycelial fruiting bodies. Asci are naked formed by the differentiation of yeast cells or hyphal segments, with no protective sterile tissues associated with them. The examples are *Saccharomyces, Schizosaccharomyces, Hansenula,* etc.

Basidiomycota : Also known as 'club fungi', Basidiomycota constitute nearly 25% of the described species of fungi. They are characterized by the formation of spores on well differentiated specialized club-shaped hyphal structures called 'basidium' which are usually the sites of nuclear fission and meiosis. Basidium is analogous to ascus but differs in having sexual spores called basidiospores which are borne externally on it. Dicaryotic phase is predominant phase in the life cycle. Except rusts, most basidiomyceteous fungi possess a dolipore septum. Basidiomycota includes heterobasidiomycetous, homobasidiomycetous and eubasidiomycetous fungi. Heterobasidiomycetous fungi have septate basidium, deeply divided or consisting of a teleutospore, which germinates into a promycelium. Basidiospores are usually capable of germinating by repetition as in *Melampsora, Ustilago, Tilletia,* etc. Homobasiomycetous fungi have aseptate basidium; basidiospores usually germinate by germ tube e.g. *Agaricus, Scleroderma, Clavaria, Polyporus, Boletus, Paxillus, Russula,* etc. Eubasidiomycetous fungi have fleshy basidiomata like mushrooms, bracket fungi, puff balls, jelly fungi, etc. e.g. *Coprinus, Schizophyllum,* etc. Basidiomycetous fungi rarely produce asexual spores. Basidiomycota include mycorrhizae, mushrooms, phytopathogens and vigorous decomposers of wood. The decomposition of most recalcitrant polymers like lignin and cellulose are the main activities of basidiomycetous fungi.

Deuteromycota (fungi imperfecti) : Deuteromycota have septate hyphae and reproduce asexually mainly through conidia (Fig. 2.9). Deuteromycota has numerous diverse fungi composed of asexual stage of fungi that are related to the members of Ascomycota and Basidiomycota, most of them have affinities to ascomycota. Their sexual state is unknown or absent. Whenever their respective sexual stage is discovered, they are reclassified accordingly. The classes of phyllum Deuteromycota are distinguished by the structure conidiomatain or on which conidia and conidiogenous hyphae (conidiophores) are borne. Reproduction in Deuteromycota is by either conidia or odia or budding or by hypal fragmentation.

Deuteromycota include hypomycetous, coleomycetous and blastomycetous fungi. Hypomycetes produce exposed conidia which are scattered loosely over the surface of mycelium or on the surface of differentiated hyphal aggregates (conidiopores) called 'synnemata (coremia)' and sporodochia. The examples of this group are *Aspergillus, Fusarium, Penicillium,* etc. Coleomycetes produce conidia within spherical or flask-shaped mycelial structures (pycnidia or pycniotium) or underneath host and fungal tissues (acervuli) e.g. *Phoma, Botryodiplodia, Gleosporium* etc. Blastomycetes are asexual yeasts that bud or even fail to produce conidia and reproduce by hyphal fragmentation. The true mycelium in Blastomycetes is either lacking or not well-developed. Majority of deuteromycetous fungi are saprobic in nature, though some are phyto-, animal- or myco-parasitic.

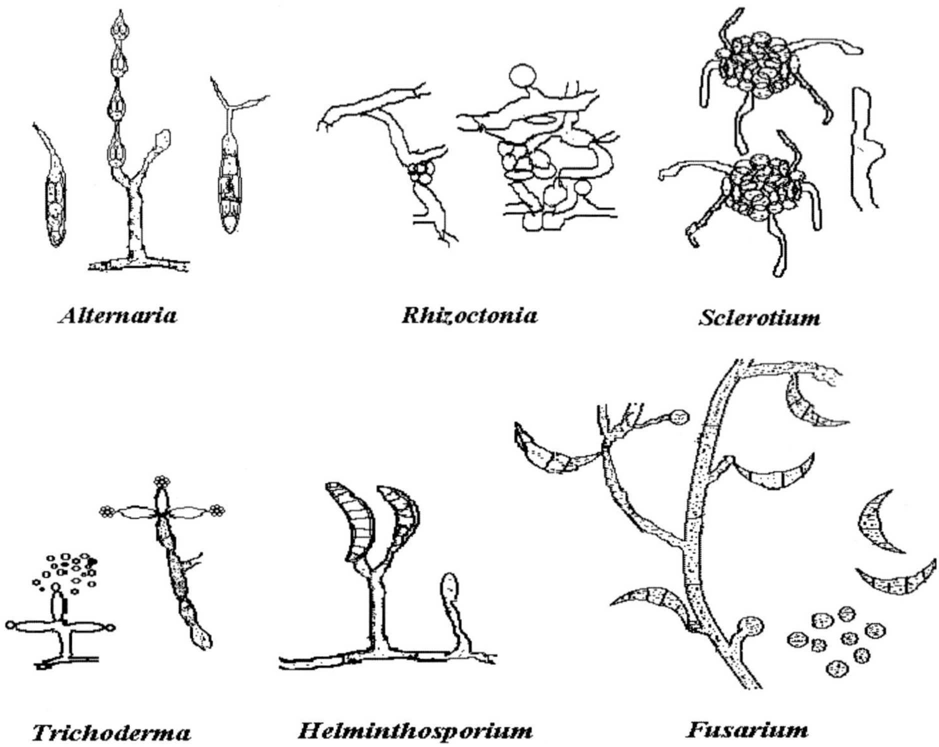

Fig. 2.9: Some important fungi imperfecti found in soil

Functional significance : One gram of soil generally contains mycelial filaments of 10-140 m length with an average number of about 7×10^6 hypae having a total volume of one mm^3 thereby play vital role in soil aggregation and conservation. Usually fungi are found in upper 0-30 cm soil surface. Fungi require water for nutrient uptake and are, therefore, restricted to moist environments. They occupy acidic environment between pH of 4.0 to 6.5 and grow at a temperature range between 5 and 40°C, although some can tolerate high and low temperatures as well.

Fungal distribution in soil is determined by the availability of organic carbon. Fungi require organic carbon compounds to satisfy their carbon and energy requirements. They obtain carbon by saprotrophy, symbiosis or parasitism. Soil fungi mostly grow as copiotrophs i.e. they grow when there is adequate supply of nutrients and a carbon source. Some are oligotrophic which thrive under low nutrient and carbon availability. Fungi like mushrooms have underground mycelium that absorbs nutrients and water and above-ground contain reproductive spores. Edible mushrooms are consumed as delicious foods. Fungi like ecto- and endo-mycorrhiza, through symbiotic association, play vital role in supplementing their host plant with essential nutrients in most soluble form besides providing protection against biotic and abiotic stresses. Furthermore, mycorrhizae help in conserving the soil nutrients and moisture and developing the soil. Symbiotic associations of fungi are also seen within algae in terms of lichens. Although many soil fungi such as *Rhizopus, Mucor, Actinomucor, Aspergillus,* etc. are responsible for food spoilage; many serve in autocontrol/biocontrol of plant pathogens and many are detrimental to plants for inciting major diseases like smuts, rusts, rots, mildews, blights, wilt, cankers, leaf spot, etc. Some soil fungi like *Darluca* are even carnivores, trapping and parasiting nematodes, etc. Many fungi like a strain of *Colletotrichum gloeosporioides* have biocontrol properties against weeds. However, main role of soil fungi is decomposition of organic matter, besides some being symbionts and others pathogenic. The ordinary soil saprophytic fungi efficiently convert substrate into microbial tissue and assimilate as much as 30-50% of the carbon present in organic matter in comparison to 5-20% by bacteria. The fungi release various degradative enzymes like cellulase, chitinase, protease, etc. into their environment. The fungal genera that are encountered commonly in soil include *Aspergillus, Botrytis, Cladosporium, Cephalosporium, Cunnighamella, Trichoderma, Monilia, Penicillium, Fusarium, Alternaria, Fomes, Gliocladium, Allomyces, Rhizoctonia, Agaricus, Pythium, Phytophthora, chaetomium, Trichothecium, Mucor, Rhizopus, Verticillium, Boletus, Lactarius, Amenita, Polyporus, Poria, Dictylostelium, Neurospora, Saprolegnia, Phycomyces, Zygorhynchus, Candida, Saccharomyces, Rhizophydium,* etc.

Many fungi of great economic importance have been isolated from soil, which include antibiotic penicillium producing *Penicillium chrysogenum* and immunosupressant cyclosporin producing *Cylidrocarpon lucidum* and *Tolycladium inflatum.*

2.8 ALGAE

Structure and Diversity : Algae are simplest of chlorophyllous eukaryotes that multiply by binary fission. Ubiquitous in nature, the algae are universally distributed in the surface layer of soil and found in damp terrestrial habitats wherever sunlight and moisture are available. Algae exhibit great diversity in form and size, ranging from a single microscopic cell to huge kelps of the ocean that are over 40 m in length. Algae may be motile or non-motile. In some species, daughter cells after division do not separate but segregate to form colonial, filamentous or membraneous thalli but never form true tissues. Cell wall in most of the algae, if present, contains cellulose, sometimes impregnated with silica, or composed of distinctive plates secreted from Golgi body. Only euglenoid algal cells do not posses cell wall but their cells are protected by a flexible proteinaceous pellicle beneath the plasma membrane (Fig. 2.10). The flagellate species have 1-2 flagella per cell inserted apically, laterally or posteriorly. The presence of eye spots near flagellar insertion point allows the cell to swim towards light. Movement may be either by lateral strokes or by spiral twists. The chloroplasts are variable in structure and shape. Chlorophyll pigments are membrane-bound within thylakoids. Like chlorophyll, accessory pigments such as carotenoids, phycobilin and xanthophylls also absorb light energy. Differences in pigmentation allow species of algae to occupy different parts of photic zones of their environment.

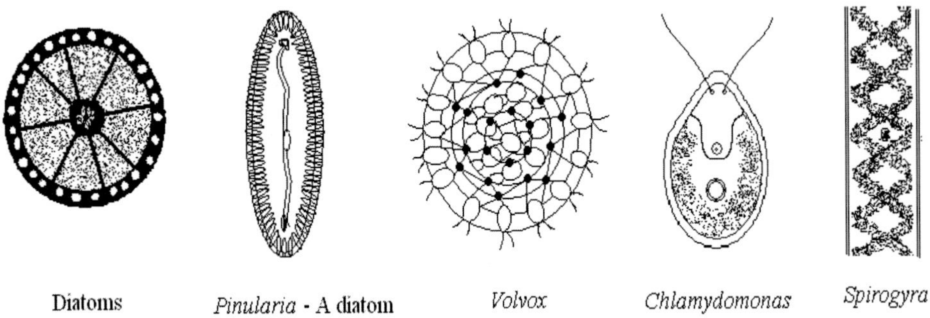

Fig. 2.10: Some representative algae found in soils

Asexual reproduction is associated with growth in unicells, with mitotic nuclear division occurring after a period of cell growth. In unicellular algae, the entire organism may function as gamete whereas in multicellular forms gametes produced are either uni- or multi-cellular gametangia with every gametangial cell producing a gamete. In asexual reproduction algae produce flagellate and/or non-motile spores. Sexual reproduction is associated with the formation of gametes followed by zygote formation. In organisms with haploid vegetative phase, meiosis immediately follows zygote formation. Diploid organisms do not undergo meiosis until gamete formation.

Algae are classified into ten major phylla on the basis of their cellular organization, the number of endoplasmic reticulum (ER) membranes that surround chloroplast and the number of thylakid stacks within them, pigmentation, storage products, cell wall constituents and life histories. Most soil algae, excluding blue green algae (now called cyanobacteria), belong to four major phylla whose characteristic features are given in Table 2.9. Green algae are more tolerant of acid conditions so dominate in temperate soils and blue greens are favoured by neutral to alkaline conditions while diatoms grow best in neutral soil.

Table 2.9: Main features of soil algal groups

Characters	Divisions				
	Chlorophyta Euglenophyta (green algae)	Heterokontophyta			Euglenophyta (euglenoids)
		Chrysophyceae (golden brown)	Bacillarophyceae (diatoms)	Xanthophyceae (yellow green)	
Cell structure	Unicellular, colonial, coccoid and palmelloids, some are multicellular (filamentous or thallose)	Unicellular, colonial or multicellular	Unicellular, colonial or coccoid	Unicellular, colonial or coccoid some multicellular filamentous form cysts	Unicellular, a few have stages during which cells are enclosed within a mucilage capsule
Flagella	1, 2-8, equal apical/subapical	1-2, +/- equal or unequal; apical/ subapical	1-2, +/- equal or unequal; apical/ subapical	1-2, +/- equal or unequal; apical/ subapical	1-3, +/- equal; apical
Cell wall	Cellulose, xylan, mannose, $CaCO_3$, protein (hydroxyl-proline glycosides)	Cellulose, silica, $CaCO_3$, chitin, or none of the above (only mucilagen-ous substances)	Cellulose, silica, $CaCO_3$	Cellulose, silica, $CaCO_3$	Cell wall absent proteinaceous or pellicle
Chloroplast-ER	0	2	2	2	1
Thylakoid stacks	3 -6	3	3	3	3 -6
Pigments	Chlorophyll a, b; β-carotene; zeaxanthin, siphonoxanthin neoxanthin, diatoxanthin, antheraxanthin, siphonein, lutein,	Chlorophyll a, c_1 & c_2; α, β, ε carotene; fucoxanthin, neoxanthin, diatoxanthin, violaxanthin, diadinoxanthin, α-cryptoxanthin neofucoxanthin zeaxanthin antheroxanthin	Chlorophyll a, c_1, c_2 & c_3; β, ε-carotene; fucoxanthin, neoxanthin, vaucheriaxanthin, canthaxanthin, echinenone, neofucoxanthin diadinoxanthin	Chlorophyll a, c_1 & c_2; β-carotene, diatoxanthin, diadinoxanthin, neoxanthin, heteroxanthin,	Chlorophyll a,b; β-carotene, diadinoxanthin, neoxanthin,

Contd...

Table 2.9. Contd...

Storage product	Sugar, starch fructosan	Chrysolamiranan (β-1,3 glucan) & oil	Chrysolamiranan (β-1,3 glucan)	Chrysola-miranan (β-1,3 glucan)	Paramylon (β-1,3 glucan), oil & sugar
Reproduction	All have vegetative phase, motile or non- motile; gametes by differentiation of vegetative cells; zygote follows meiosis	Vegetative phase diploid, meiosis and gamete formation preceed sexual reproduction	Vegetative phase diploid, meiosis and gamete formation precede sexual reproduction	Vegetative phase diploid, often lack flagella but have gliding motility on solid surfaces; sexual only in *Vaucheria*	Mitosis is followed by longitudinal cell division; sexual reproduction absent
Examples	*Chlamydomonas, Zygogonium, Chlorokybus, Trentopohlia, Prasiola, Chlorella, Chlorococcum Coccomyxa, Gleoemonas Ulothrix*	*Hydrurus, Rhizochrysis, Chrysocapsa, Ochromonas, Chrysosphaera, Synura, Thallochrysis*	*Nitzchia, Surirella, Achnanthes, Eunotia*	*Heterococcus, Vaucheria, Triomena, Botrydiopsis, Gleoechloris, Rhizochloris, Chloromeson*	*Euglena, Astasia, Paranema, Entosiphon*

Ecology and function : Many algae follow phagotrophic or saprophytic mode of nutrition. Some algae like *Cephaluros,* a green parasitic micro-alga, attack leaves of tea, coffee, pepper, etc. and cause red-rust disease in tropics. In algae, carbon and energy metabolism is via glycolysis and citric acid cycles. Almost all algae are aerobic and use mitochondrial respiration with oxygen as terminal electron acceptor. Nitrogen, phosphorus and silicon are frequently limiting nutrient. Algae show wide range of adaptability but do not survive severe desiccation. Besides wide pH and temperature (5-50°C) tolerance, most algae have high salt tolerance ranging from 0 to near 100%. Some algae are specialized to inhabit extremely acidic hot sulphur springs while others can complete their entire life cycle below 0°C. Some dormant stages of algae are capable of withstanding 100°C for several hours. The ability of algae to survive drying coupled with a rapid response to added moisture means that algae can make a significant contribution to microbial activity, even in the extremes of desert environment. Algae are important primary producers. Diatoms and coccoids account about 31% and blue green algae (cyanobacteria) about 10%of the annual net productivity.

Algal activities permit the retention of inorganic nutrients in such situations that might otherwise be leached away, thereby bring to the matrix the initial input of organic material necessary for the soil transformation to begin. Most algae are photoautotrophic; use atmospheric CO_2 and release O_2. They have a key ecological role as primary colonizers on bare surfaces. The release of carbonic acids, as a result of algal metabolism, accelerates mineral weathering and thereby makes substantial contribution to soil formation in its primary stages.

2.9 LICHENS

Some algae (like green algae) and cyanobacteria form symbiotic relationship or grow in close intimacy with many fungi, mostly ascomycetous ones, resulting in the formation of a single undifferentiated thallus called lichen. A lichen has a top layer of tightly interwoven fungal mycelium and a lower layer of algae or cyanobacterium and a bottom layer of fungal mycelium, attached to a substrate directly or anchored through hyphal strands. Some lichens are more complex. The algal partner (photobiont or phycobiont) in these symbioses captures energy and fungal partner (mycobiont) furnishes the mineral nutrients and helps its partner in the regulation of water and light requirement. This mechanism helps each partner to overcome harsh environments.

On the basis of habitat, lichens may be saxicolous (grow on rocks e.g. *Dermatocarpon)*, corticolous (grow on bark e.g. *Usnea*), lignocolous (grow on wood e.g. *Cyphelium*) or terricolous (grow in soil e.g. *Cladonia*). Lichens multiply by fragmentation, isdia and soredia. Isdia are superficial outgrowths comprising of both the partners. Sordia are microscopic propagules that develop in sori. Each soredium has one or a few algal cells and a fungal cell. Sexual reproduction and sexual spores are formed only by mycobiont. Lichens are considered primary colonizers on bare rocks and soil parent materials because they can withstand droughts and extreme temperatures. Many lichens grow at low temperatures and are found at higher altitudes or in polar environment. Lichens have a vital role in bio-weathering of rocks. Organic matter produced by the primary colonizers allows the growth of heterotrophic microorganisms and subsequent colonization by plants. However, lichens are sensitive to SO_2 and hence serve as indicators of pollution. Lichens produce large quantities of extracellular acids. Rock lichen *Parmelia conspersa* produces salacinic acid within a few hours when added to granite and mica. Dye orcein and litmus are extracted from lichen (*Rocella tinctoria*). Reindeer moss *(Cladonia rangifera*) and island moss (*Cetrasia islandica*) are staple diet of reindeer musk ox and other herbaceous arctic animals and cover many kilometers of land with ankle-deep growth. Some algae produce various toxins like domoic acid, saxitoxin, etc., which affect the animals consuming these.

2.10 SOIL FAUNA

Soil fauna or soil invertebrates are estimated to contribute 5-15% of the heterotrophic community metabolism over a wide range of ecosystem types. Conventionally, soil fauna are categorized into three size classes: microfauna (< 200 μm), which include nematodes and protozoa; mesofauna (<0.2 to 2 mm) such as enchytraeidae, collumbola (spring tail) and acari; and macrofauna (>2 to 20 mm) such as earthworms, ants, termites, spiders, centipedes, millipedes, snails, slugs, beetles etc. The variety and taxonomic complexity of soil fauna are very great. The soil fauna, in general, need a fairly well aerated soil for active growth. The small soil dwelling multi-cellular animals (metazoans) influence physical and chemical properties of soil as well as organic matter and the microbial communities present. The individual communities of this group may spend some or whole life cycle in the soil. Soil fauna may hardly contribute <5% of the total decomposer respiration but play a vital role in soil formation and in the breakdown of large sized plant remains into smaller fragments. Earthworm, nematodes, arthropods, centipedes, millipedes, spring tails, mites, molluscs (slugs and snails), rotifers, termites, ants as well as large burrowing animals (rodents, rabbits) help in comminuting the organic residues (surface litter), mix and convert it to excerta or/and transport fragmented products into deeper soil layers. The chewed plant tissues are thoroughly mixed with microorganisms in animal gut where conditions are ideal for microbial action. Mobile animals carry microorganisms with them and help the latter to disperse and find new food source to exploit. Generally 20-40% of annual litter fall is processed in this way through feeding, excretion and respiratory activities of these organisms so facilitate and accelerate decomposition of plant residues by microorganisms. It is estimated that microbivorous fauna consume approximately 50% of the annual production of microbial material. Besides, faunal burrowing influences the soil water balance and percolation rates. In areas of intensive faunal activities, plant nutrient and moisture availability may be increased and soil erosion reduced as a result of faunal interaction with the residues and soil organic matter. In nutshell the overall biological activity in soil is enhanced due to the coordinated actions of macro-fauna.

2.11 PROTOZOA

Soil protozoa are mobile unicellular microfauna having a diameter in the range of 5 to 100 μm (Fig. 2.11). They lack chlorophyll and true cell wall.

Protozoa are active in trophozoite stage but form cyst under adverse soil conditions like dryness. They may persist for years together in the form of cysts. Majority of protozoa reproduce sexually by cell fusion while others reproduce asexually by fission. The protozoa are diverse and essentially require water for their metabolic activities. More than 250 species have been isolated from soils. The principal free living soil protozoan groups are given in Table 2.10.

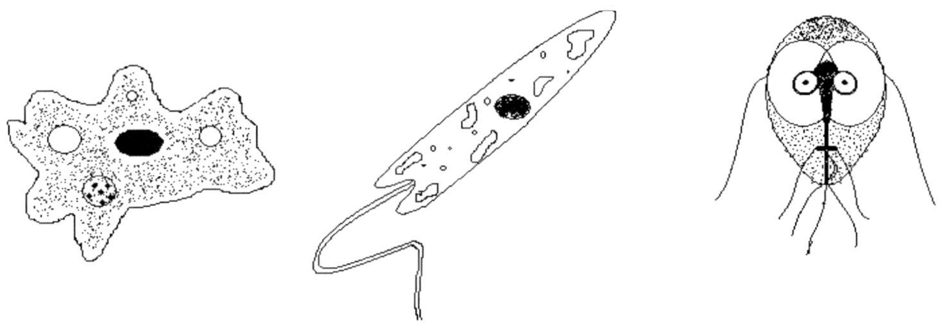

Amoeba ***Asteria*** ***Giardia***

Fig. 2.11: Some representative protozoa

In soil water flagellate and amoebae are more numerous. The protozoa in temperate arable soil enumerate 30,000 - 40,000 g^{-1} and are largely restricted to upper 20 cm soil profile. The organisms generally thrive best in moist well-drained soils. Free living protozoa in soil feed on dissolved organic substances and other organisms saprophytically. Many feed wholly by predation, engulfment and grazing of bacteria, yeast and other fungi etc. so are phagotrophic in nature. Bacteria form the staple diet of the protozoan population, though small algae, fungal spores, yeasts and flagellate and small amoebae are sometimes also ingested. A single protozoan may consume about 5000-cell min^{-1}. Protozoa often avoid feeding encapsulated bacteria, endospores, mycobacteria and streptomycetes. Protozoan feeding affects the structure and function of microbial communities. The rise in bacterial population after organic matter addition is always followed by protozoan rise. Selective feeding by protozoa keeps bacterial population physiologically young and thereby results in more efficient bacterial decomposition. It is now argued that protozoa accelerate nutrient cycling. Pathogenic protozoa apparently cause many diseases like phloem necrosis in coffee, hartroot in coconut, sudden wilt in oil palm, etc.

Table 2.10: Major soil protozoan groups

Groups	Characteristic features	Examples
Mastigophora	Small flagellate whose movement is through one or more whip-like flagella or by undulating membrane; reproduction by longitudinal binary fission	*Euglena, Giardia, Bodo, Cercomonas, Heteromita*
Rhizopoda or Sarcodina	Naked amoebae, have simple mass of protoplasm with flexible permeable cell membrane; exhibit movement through rhizopod or pseudopods; reproduction by simple binary fission	*Amoebae, Entamoeba, Naegleria, Hartmanella*
Ciliophora or Ciliata	Free swimming cells having short bristles or cilia (small hairs) on their surfaces; cell wall rigid; asexual reproduction by transverse binary fission and genetic exchange by conjugation	*Vorticella, Paramecium, Colpoda, Oxytricha, Leptopharynx*
Testatceous amoebae (Thecamoebae)	Have rigid chitinous or siliceous shell and pseudopodia; reproduction by binary fission	*Euglypha, Difflugia*

2.12 NEMATODES

Soil nematodes, known by many names such as eelworm or round worms or thread worms or hair worms, are most numerous of the soil metazoans. Of the 10,000 nematodal species known hardly 10% species are found in soil. Morphologically, they are tiny elongate, bilaterally symmetric with tripartite radial symmetry in the region of head and pharynx (Plate 2.2). External body wall is bounded by cuticle. The cuticle of most plant parasitic and soil nematodes is smooth and apparently non-segmented. It is cast off with each moult and a new layer is formed. Most soil nematodes, when adult, are 10-30 μm in diameter and 0.5-2.0 mm in length. The nematodes are obligate aerobes. They depend upon a thin film of water around soil particles for movement and completion of their life cycle. Nematodes usually produce 4-6 generations per year. Nematodes begin their life cycle as an egg and mature into an adult after around 4 months. Adults are sexually mature. Sexual reproduction is common but hermaphrodite (possessing both + and – sex organs) and parthenogenic (capable of reproducing without male) individuals are found in many species.

Nematodal distribution in soil generally follows that of organic matter. Most nematodes are found in upper 15 cm soil surface with largest nematode populations present in the vicinity of plant roots. Nematodes feed on plants, algae and many heterotrophic organisms so may be plant parasitic, microbivorous, carnivorous (feed on animal) or omnivorous (feed on diverse substrates). Most plant feeding nematodes (such as *Heterodera, Globodera, Meloidogyne, Ditylenchus, Pratylenchus, Xiphinema,* etc.) live in association with roots or other underground structures as ecto- or endo-parasites. In soil, majority of nematodes (70%) do not directly feed on plant roots but consume other soil organisms. Free living forms such as *Aphelenchus, Hexatylus, Dorylaimus,* etc. in soil are voracious feeders of microfauna and flora. For example, fungivorous nematode *Aphelenchus avenae* decrease the population of fungal pathogen *Rhizoctonia solani;* and bacteriovorous nematodes survive on bacteria like *Agrobacterium tumefaciens, Erwinia carotovora* etc.Bacterial intake by nematodes has been estimated to vary from 100 to 5,000 cells min^{-1} and total consumption of microflora by bacteriophagous and mycophagous nematodes is as high as 50% of the estimated annual production of microfloral biomass. Bacterial feeders are often abundant in decaying organic matter and in wounds or lesions caused by other unsects and pathogens.

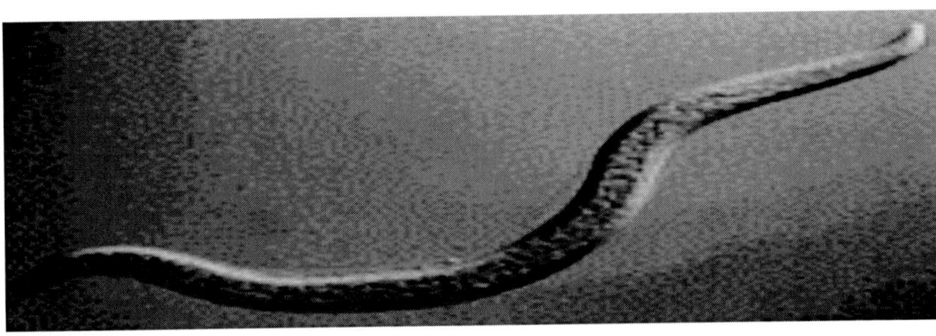

Plate 2.2: A nematode in a soil

Various nematodes are preyed by other nematodes and many species of fungi, which have well-developed nematode trapping mechanism. About fifty species of such fungi are quite widely distributed in field soils. Nematodes are active in water films in quite large soil pores (>50 μm) and generally abundant in moist coarse textured soil than fine textured one. The nematodes that encyst survive prolonged desiccation and other environmental adversities.

2.13 ARTHROPODS

A high proportion of soil animals is arthropods. They have an exoskeleton, jointed legs and usually well-developed nervous system. Arthropods are saprophagous animals which utilize dead debris; some are carnivores preying on other animals.

a) Micro-arthropods

The micro-arthropod population of soil is dominated by springtails (collembola) and mites (acarhnids). The micro-arthropods may be hemi-edaphic or eu-edaphic ones. The hemi-edaphic forms principally occur on soil surface or in litter layer, and are characterized by well-developed eyes, long appendages and chitinous body with marked pigmentation. The eu-edaphic forms are more specialized deep soil animals which lack pigment and eyes and have soft bodies with short appendages. The springtails are primitive insects, usually <1 mm in length. The springtails and most of the mites are entirely detritus or fungal feeders. Some mites are predatory preying other soil animal communities like nematodes, insect eggs, springtails, etc. Micro-arthropod populations attain their greater abundance (0.05 –0.2 million m^{-2}) and species diversity in the soil and litter of deciduous forests and old grasslands. Evidences show that micro-arthropod numbers are associated with particular species of plants in the ground cover. Microphytophagous microarthropods feed on fungi, bacteria and algae. Fungivores may engulf the prey and stimulate fungal growth by grazing senescent hyphae or may pierce host cell walls and drain the fluid, leaving a ghost hyphae in tact. Detritivores scavenge dead organic matter. Macrophytophagous micro-arthropod may be root sap feeders or root grazers. The role of soil micro-arthropods in organic matter decomposition is insignificant but, of course, play vital role in litter fragmentation.

b) Macro-arthropods

The macro-arthropods inhabiting soil are diverse and include terrestrial isopods (wood lice), chilopods (centipedes), diplopods (millipedes), and soil dwelling spiders and larvae and adults of a number of insects like beetle, wireworm, etc. passing some part of their life cycle in soil. Millipedes are initially saprophagous but later damage plant roots whereas most of the centipedes are carnivorous feeding on other soil fauna. Generally they help in processing and mixing of organic detritus in soil through their enzymatic activities and help in maintaining nutrient dynamics of forest ecosystem through grazing fungal saprophytes and dead organic matter.

Ants : Ants make nests in soil, some making visible mounds on soil surface. Their channeling and burrowing activities in soil improve soil aeration and drainage. They can survive even in the fringes of deserts. Some ants even cultivate fungi on which they feed and others collect seeds of grasses, etc. Many genera of ants are predacious feeders on mesofauna and some macrofauna. Ants dominate in temperate region. Harvester ants, considered as pests, denude the area surrounding their nests to a distance up to 3 m or more. They gather seeds for their consumption so prevent reseeding of natural grassland vegetation. They also transport materials from within soil and deposit on surface. Some mounds are a few feet high and more than 10 m in diameter.

Termites (white ants) : Termites are dominant component of soil fauna numbering about 16 x 10^6 ha^{-1} in drier regions of tropics and semi-tropics. About 2,000 species of termites are known, found in about 2/3rd of land area of the world and mostly prominent in grasslands (savannas) and forests of tropical and subtropical areas. Termites feed on all kinds of plant materials (wood, grass, roots etc.) both living and decomposing. The practice of copiophagy (eating faeces), nectrotrophy (eating dead) and cannibalism (eating living) leads to efficient recycling and utilization of nutrients.

A colony of termites forms a characteristic mound; often causes forest clearance or thinning. A single mound may contain 2.5 tonnes of earth, and termites build about 60 cm mound in one month. Termite mounds and their network of underground passage and above-ground covered-runways typically spread 20-30 m beyond the mound, forming essentially 'termite cities'. Depending upon the species involved and environmental conditions, termites may build mounds of 6 m or more in height and may extend them to even greater depth into soil in search of water or clay layer. Each mound provides home for one million or more termites. These mounds are usually abandoned after 10-20 years so can be used for crop production provided the queen termite is successfully destroyed. The central part of mound (hives) has a complex system of galleries leading to food source. Galleries radiate from the floor of nest to about 1 m depth and extend for about 3 m on either side of the mound. Galleries open at regular intervals into chambers of 90 cm wide. Termite mounds are often low in organic matter but have greater availability of P, K, Ca and moisture than surrounding undisturbed top soil because of soil movement from subsoil for mound building. Their mounding activities result, on an average, in the transfer of 1.0 t ha^{-1} subsoil to surface, in addition to the production of considerable amount of worm cast. While scavenging a large area around each mound, termites remove 2.0 to 4.0 t ha^{-1} of leaf and wood materials annually, a substantial portion of plant litter produced in many tropical ecosystems. These activities have significant impact on soil formation and current soil productivity and fertility. Their activities have both positive and negative effects on current land use. These accelerate decay process of dead trees and grasses and disrupt crop production. Several species such as *Macrotermes* sp. in Africa use plant residues to grow fungi in their mounds as a source of food. Bacterial metabolism in the guts of these insects accounts substantially to the global production of methane. Unlike earthworms, termites do not have a beneficial effect on soil productivity, because of their more efficient digestive system and localized actions (while denuding the remaining land area of the surface residues). A number of genera of termites feed on wood directly or utilize it as substrate for fungal garden cultivation so are mostly dominant in woodlands and forest soils.

2.14 OLIGOCHAETES

Oligochaetes include earthworms and enchytraeid worms. The earthworms, in particular, are efficient soil homogenizers in both temperate and tropical soils. They are intimately involved in soil organic matter decomposition, aeration, aggregation and water percolation besides influence nutrient and water movement in soil.

Earthworm : Earthworms are egg-laying hermaphrodites (Plate 2.3) that eat detritus and soil organic matter as well as the microorganisms present on these materials. They are most numerous in forest and pastures and much low in arable sites. Their population may vary from 1.0-8.0 million ha^{-1} weighing about 1200-2400 kg. The earthworms represent 80% of total soil animal biomass. About 1,800 species of earthworm are known world-wide. Earthworms are not uniformly distributed in a particular soil. For example, *Lumbricus rubellus*, litter feeder, is more frequently found beneath dung pats. Earthworm can migrate both horizontally and vertically in soil. They show a seasonal vertical migration. Soil moisture and temperature seem to be the major factors determining such activity.

Based on life style and burrowing habits earthworm may be epigeal or endogenous or anectic form. Epigeal or surface living forms such as *Lumbricus rubellus, Eisenia foetida,* etc. hardly burrow the soil and are confined to decaying organic matter of soil surface. Endogenous species such as *Allolobophora chlorotica* produce shallow branching and burrows in the organo-mineral layers of soil. Anectic forms are deep burrowing species (e.g. *Lumbricus terrestris, Allolobophora longa, A. caliginosa,* etc.) producing channels to a depth of one meter or more (sometimes to 2.5 m). Subterranean species are often paler in colour than surface feeders. The burrows of earthworm are 3–12 mm in dia. having smooth wall cemented with mucous secretions and ejected soil. Earthworms prefer well-aerated but moist habitat and thrive more on soil having fairly good amount of fresh or undecomposed organic matter. Mostly earthworms prefer neutral to slightly alkaline pH (5.5-8.5). Earthworms are capable to withstand starvation and water loss up to 70% of their body weight. Their populations are high in spring and autumn but low in winters and summers so migrate to deeper soil layers during these periods. They are extremely sensitive to soil disturbance. Earthworms may contribute only around 4% of the soil metabolism but provide a well-comminuted medium for microbial activities. They ingest a weight of soil equal to 2 to 30 times their own weight in a single day. The intestinal microflora of the worms assists in humification of feed material. Further, the mucus of worm acts as substrate for microflora. Earthworm activities greatly enhance soil fertility and productivity by altering both physical and chemical conditions in soils, especially in upper 15-30 cm soil. They increase nutrient availability through their enzymatic, ingestive and assimilative activities as well as cast excretions.

Plate 2.3: An earthworm

Earthworm wastes, called casts, serve as a substantial portion of stable aggregates in many soils. Besides, it helps in soil and water conservation, alleviate or reduce compaction problems and prevent erosion losses. Earthworms are estimated to produce 1.87–4.03 kg cast m^{-2} in English pasture. *L. terrestris* appears responsible for the fragmentation of most litter in woodland in temperate zone. It can remove 90% of annual leaf fall in an apple orchard during winter accounting for 120 g dry matter m^{-2}. They can consume 100-300 mg plant materials per gram of body weight per day. Earthworm may annually deposit 10 or more tonnes of casting on the surface so help to build up a 2.5 cm layer in 10 years. Earthworm cast is rich in ammonia and partially digested organic matter so provide a good environment. Intimate mixing of soil materials, creation of channels and production of casting leaves the soil more open and porous. Therefore, mixing and comminuting activities of earthworm contribute much to ecosystem productivity. The degree of selectivity, in terms of the type of plant litter processed by earthworms is variable.

Enchytraeids (Pot worms) : Enchytraeids are small worms about 0.1-5.0 cm long and 0.3-0.8 mm wide which feed on microorganisms, nematodes and plant litter. They tend to replace earthworm population in forest and acid soils of temperate latitudes. Pastures and coniferous forest soils contain 0.001 to 0.3 million enchytracids m^{-2} with a biomass of 30-100 kg ha^{-1}. The enchytraeids are smallest oligochaetes measuring <1 mm to several centimeters. Small enchytraeides may be, sometimes, confused with soil nematodes but they lack buccal stylets for piercing plants and a muscular oesophagus for sucking which are found in nematodes. Since they are sensitive to drought, therefore their population density is controlled by the water potential in soil. Humid habitats favour their high population densities due to relatively low summer mortality.

2.15 MOLLUSCS

Slugs and sails (gastropods) are major soil molluscs (Plate 2.4) and are considered of agronomic and ecological significance under damp soil conditions where they undergo intense herbivorous activity. Slugs are nocturnal animals normally active from about 2 hour after sunset and 2 hour before sunrise. Slugs shelter beneath soil surface. Their biomass in a garden soil may be 20- 40 g m^{-2}. Dusky slug *Arion subfuscus* feed mainly on fungi and animal faeces. Snails with their high cellulase enzyme activity cause considerable defoliation whereas slugs scavenge old plant materials and debris. Some molluscs secrete muco-proteins that help in forming water stable soil aggregates. Their population in soil varies from 10 to 25 m^{-2} with average population of 2 m^{-2}. Molluscs are sensitive to the changes in moisture and temperature.

Plate 2.4: Slug and snail

Exclusion of soil fauna from litter-enriched soil usually decreases the soil respiration rate. The soil fauna play role in accelerating the conditions favouring mineralization by microflora. Thus there is a tremendous degree of synergism between soil animals and microbes in this regard.

Selected References

Alexander, M. 1971. *Microbial Ecology.* John Wiley & sons, New York, USA.

Alexander, M. 1977. *Introduction to Soil Microbiology* (2nd edn.), John Wiley & sons, New York, USA.

Alexopoulos, C.J., Mims, C.W. and Blackwell, M. 1996. *Introductory Mycology* (4th edn.) Wiley & sons, New York, USA.

Andrew, J.H. 1984. Relevance of g and k theory to the ecology of plant pathogens. In: *Current Perspectives in Microbial Ecology.* (eds. M.J. Klug & C.A. Reddy), American Society of Microbiology, Washington, DC, USA.

Black, C.A. 1968. *Soil Plant Relationships.* John Wiley & sons, London, UK.

Bold, H.C. and Wynne, M.J. 1979. *Introduction to the Soil Algae.* Prentice- Hall, New Jersey, USA.

Brady, N.C. 1990. *The Nature and Properties of Soils* (10th edn.), MacMillan Inc., New York, USA.

Brady, N.C. and Weil, R.R. 1996. *The Nature and Properties of Soils* (11th edn.), Prentice-Hall, New York, USA.

Brocks, T.D. and Madigan, M.T. 1991. *Biology of Microorganisms* (6th edn.), Prentice-Hall, Englewood Cliffs, New Jersey, USA.

Buchanan, R.E. and Gibbson, N.E. (eds.). 1974. *Bergey's Manual of Determinative Bacteriology* (8th edn.), Williams and Wilkins, Baltimore, Maryland, USA.

Burns R.G. and Slater, J.H. (eds.).1982. *Experimental Microbial Ecology.* Blackwell Scientific Publication, Oxford, UK.

Campbell, R.E. 1977. *Microbial Ecology.* Blackwell Scientific Publications, Oxford, UK.

Clark, F.E. and Paul, E.A. 1970. Microflora of grassland. *Adv. Agron.,* 22: 375-435.

Coyne, M.S. 1999. *Soil Microbiology: An Exploratory Approach.* Delmar Publishers, New York, USA.

Edwards, C.A., Stinner, B.R., Stinner, D. and Rabatin, S. 1988. *Biological Interactions in Soil.* Elsevier Publication, Amsterdam, Holland.

Ellwood, D.C., Hedger, J.N., Latham, M.J. Lynch, J.M. and Slatert, J.H. (eds.). 1980. *Contemporary Microbial Ecology.* Academic Press, London, UK.

Fay, P. 1983. *The Blue Greens,* Edward Arnold, London, UK.

Fitter, A.H. (ed.).1985. *Ecological Interactions in the Soil- Plant, Microbes and Animals.* BES Special Publication, Balckwell, Oxford, UK.

Foth, H.D. 1990. *Fundamentals of Soil Sciences.* (8th edn.) John Wiley and sons, New York, USA.

Freckman, D.W. (ed.). 1982. *Nematodes in Soil Ecosystem.* University of Taxas Press, USA.

Garrett, S.D. 1981. *Soil Fungi and Soil Fertility.* Pergmon Press, London, UK

Glinski, J. and Lipec, J. 1990. *Soil Physical Conditions and Plant Roots.* CRC Press, Florida, USA.

Gray, T.R.G. and Williams, S.T. 1971. *Soil Microorganisms.* Oliver & Boyd, Edinburgh, UK.

Greenwood, D.J. 1967. *The Ecology of Soil Bacteria* (eds. T.R.S. Gray and D. Parkinson), Liverpool Univ. Press, Liverpool, U.K.

Griffin, D.M. 1972. *Ecology of Soil Fungi.* Chapman & Hall, London, UK.

Griffin, D.M. 1984. Water and microbial stress. pp. 91-136. In: *Advances in Microbial Ecology* (Ed. M. Alexander), Plenum Press, New York, USA.

Griffin, D.M. 1994. *Fungal Physiology* (2nd edn.), Wiley-Liss, New York, USA.

Hawker, L.E. and Linton, A.H. (eds.). 1979. *Microorganisms: Function, Form and Environment* (2nd edn.), Edward Arnold, London, UK.

Hawksworth, D.L. 1991. The fungal dimension of biodiversity, magnitude, significance and conservation. *Myco. Res.*, 95: 641-655.

Hendrix, P.F., Crossley, D.A. Jr., Blair, M.J. And Coelman, D.C. 1990. Soil biota as components of sustainable agricultural agro-ecosystems. pp. 637-654. In: *Sustainable Agricultural Systems.* (eds. C. Edwards and R. Lal). Soil Water Conservation Society of America, Akeny, Iowa, USA.

Hoek, C. van der, Mann, D.G. and Jahn, M.H. 1995. *Algae-An Introduction to Phycology.* Cambridge Univ. Press, Cambridge, UK.

Ingraham, R.E. 1994. Nematodes. pp. 459-490. In: *Methods of Soil Analysis, Part 2: Microbiological and Biochemical Properties* (eds. R. W. Weaver *et al.*) Madison, Wisconsin, USA.

Killham, K. 1994. *Soil Ecology.* Cambridge Univ. Press, Cambridge, UK.

Lavelle, P. and Pashanasi, B. 1989. Soil macrofauna and land management in Peruvian Amazone (Yuri-maguas, Peru). *Pedobiologia*, 33:289-91

Lynch, J.M.1983. *Soil Biotechnology: Microbiological Factors in Crop Productivity.* Blackwell Scientific Publications, Oxford, UK.

Martin, J.K.1977. Factors influencing the loss of organic carbon from wheat roots. *Soil Biol. Biochem.*, 9:1- 7.

Metting, F.B. (ed). 1993. *Soil Microbial Ecology*. Marcel Dekker, New York, USA.

Metting, F.B., Rayburn, W.R. and Reynaud, P.A. 1993. Algae and agriculture. pp. 335-370. In: *Algae and Human Affairs*. Marcel Dekker, New York, USA.

Moldenke, A.R. 1994. Arthropods. pp. 517-542. In: *Methods of Soil Analysis, Part 2: Microbiological and Biochemical Properties*. (eds. R. W. weaver), Soil Science Society of America, Madison, USA.

Newman, E.I. 1978. Root microorganisms: Their significance in the ecosystem. *Biol. Rev. Cambridge Philos. Soc.*, 53: 511-554.

Paul, E.A. and Clark, F.E. 1989. *Soil Microbiology and Biochemistry*. Academic Press, New York, USA.

Phillipson, J. 1968. *Ecological Energetics*. Edward Arnold Publication, England.

Postgate, J. 1994. *The Outer Reaches of Life*. Cambridge Univ. Press, Cambridge, UK.

Rather, A.Q. and Dar, G.H. 2000. Biodiversity of edaphic ecosystem ands its conservation. pp. 479-491. In : *Environment, Biodiversity and Conservation* (ed. M.A. Khan). APH Publ. New Delhi, India.

Richards, B.N. 1987. *The Microbiology of Terrestrial Ecosystems*. Longman, Harlow, England.

Ross, S. 1989. *Soil Processes: A Systematic Approach*. Routledge, Chapman & Hall, New York, USA.

Russel, E.W. 1973. *Soil Conditions and Plant Growth*. (10th edn.), Longmans Green & Co., London, UK.

Russel, S.R. 1977. *Plant Root System - Their Function and Interaction with the Soil*. McGraw-Hill, London, UK.

Sims, R.W. and Gerard, B.M. 1985. *Earthworms*. Linean Society, London, UK.

Stanier, R.Y., Adelberg, E.A. and Ingraham, J.L.1976. *General Microbiology* (4th edn.). Collier MacMillan, London, UK.

Stolp, H. 1988. *Microbial Ecology: Organisms, Habitats, Activities*. Cambridge University Press, Cambridge, UK.

Subba Rao, N.S. 1982. *Advances in Agricultural Microbiology*. Oxford & IBH Publishing Co., New Delhi, India.

Tate, R.L. III. 2000. *Soil Microbiology*. John Wiley, New York, USA.

Tinsley, J. and Darbyshire, J.J. (eds.). 1984. *Biological Processes and Soil Fertility*. Martinus Nijhoff/Dr. W. Junk, The Hague.

Tisdale, S.L., Nelsa, W.L., Beaton, J.D. and Hartin, J.L. 1993. *Soil Fertility and Fertilizers*. MacMillan Publ. Co., New York, USA.

Torsvik, V., Goksoyr, J. and Daae, F.L. 1990. High diversity in DNA of soil bacteria. *Appl. Environ. Microbiol.*, 56: 782-787.

Van Veen, J.A., Lilgeroth, E., Lekkerkerk, L.J.A. and Van deGeijin, S.C. 1991. Carbon fluxes in plant-soil systems at elevated atmospheric CO_2 levels. *Ecol. Appl.*, 1: 175-181.

Vancura, V. and Kune, F. (eds.) 1989. *Interrelationships between Microorganisms and Plants in Soil Developments in Soil Science* 18, Elsevier, Amsterdam, Holland.

Waksman, S.A.1957. *Soil Microbiology*. John Wiley & sons, New York, USA.

Wallwork, J.A. 1970. *Ecology of Soil animals*. McGraw-Hill, London, UK.

Wallwork, J.A. 1976. *The Distribution and Diversity of Soil Fauna*. Academic Press, London, UK.

Waterbury, J.B. 1989. Subsection II. Order Pleurocapsales Geitler 1925. emend. Waterbury and Stainer 1978. pp. 1746-1770. In: *Bergey's Manual of Systematic Bacteriology, Vol. 3. Section 19: Oxygenic Photosynthetic Bacteria* (eds. J.T. Staley, M.P. Bryant, N. Pfennig and J.H. Holt), Williams and Wilkins, Baltimore.

Waterbury, J.B. and Rippka, R. 1989. Subsection I. Order Chroococcales Wellstein 1924, emend. Rippka *et al.* 1979. pp. 1728-1746. In: *Bergey's Manual of Systematic Bacteriology, Vol. 3. Section 19: Oxygenic Photosynthetic Bacteria* (eds. J.T. Staley, M. Bryant, N. Pfennig and J.H. Holt), Williams and Wilkins, Baltimore.

Wood, T.G. Termites in soil environment. *Biol. Fert. Soil*, 6: 228-236.

Wood, M. 1989. *Soil Biology*. Blackie, Glasgow, UK.

Wrighton, J.W. and Newsham, J.C. 1919. *Agriculture*. Crosby Lockwood & sons, London, UK.

— ★★★ —

Chapter 3

Microbe — Soil Environment Interactions

Soil is a continuously developing entity with properties reflective of the site in which it has evolved and is still evolving.

(R. L. Tate)

Soil is the outer loose layer of earth consisting of layers of mineral material (regolith) whose properties differ from one another and these differences are largely due to the variation in biological activities. Soil micro-environment principally comprizes of inorganic particulates (sand, silt and clay) plus certain oxides and hydroxides, plant roots, organic matter, microbial biomass and aqueous and gaseous phases. The uneven distribution of these components provides a great variety of conditions from field to micropores. The interaction of physical and chemical factors contributes to the creation of varied soil habitat and hence determines the composition and activity of soil biota at a particular site and time. Over geological times, microorganisms have had great opportunity to become distributed worldwide. Such growth and development of microorganisms as well as species diversity is of considerable significance. The existing geographic distribution is reflective of long term adaptation of microorganisms to the various environmental conditions. The preferred environment of an organism is usually referred as 'habitat'.

3.1 DISTRIBUTION OF SOIL BIOTA

The versatility and diversity of soil biota are influenced by many factors including climate, vegetation and physico-chemical characteristics of soils. The specie-composition in an arid desert usually differs from that of a humid forest or arable soil. Water, wind and animals transport microbes over longer distances independently or alongwith dust or in aerosols. Microbes have been recovered from the snow flakes falling at high altitudes even after longer precipitation. Desert soils are teaming with microscopic life. Through ocean currents, migratory birds and dust particles, microbes traverse greater distances and become ubiquitous under similar set of environments prevailing at distant places. However, some geographic disparities in microbial distribution do occur. The fungus *Aspergillus* is more dominant in warm tropical and sub-tropical regions whereas *Penicillium* and *Furarium* dominate in temperate cold climates. Sporulating bacilli and actinomycetes are abundant in warm dry soils and extremely low in cold wet soils. Soil factors also influence their acclimatization in new environment. For example, *Fusarium* sp. causing wilt of banana is restricted to the soils having smectite clays. Cyanobacteria are more common in neutral to alkaline soil. Forested areas support a more diverse soil fauna than grasslands, although the total faunal weight per hectare and faunal activity is higher in grassland.

3.1.1 SOIL PROFILE AND MICROBIAL DISTRIBUTION

Microbial population and species diversity in a given soil is ever changing but by and large microbes remain in a dynamic equilibrium, provided no major fluctuations in soil conditions do occur. World-wide under similar set of environment almost similar soil biota do exist. The total soil biomass is generally related to the amount of organic matter present. The ratio of organic matter: detritus: microbial biomass: faunal biomass in soil is approximately 1000: 100: 10:1. Variation in microbial activity in surface soil is generally controlled by inputs received in the form of photosynthetically fixed C contained in or exuding from plant biomass. Distribution and amount of microbial activity in soil horizons may be related to the soil parent material, history of soil development, soil reaction and salinity. Generally soil microbial biomass and most individual groups of microorganisms decrease with increasing depth down the soil. However, there are certain exceptions with certain microorganisms and soils. For instance, in peat and forest soils acidity generated by surface litter decomposition increases the population of acid tolerant microorganisms. The physical and chemical characteristics of soil generally govern the distribution, growth and activity of microbes in a soil. *Penicillium nigricans*, for example, is usually restricted to upper 50 mm of the soil and is less tolerant of high CO_2 concentration than *Zygorrhinchus vuellemini*, a species which is more abundant below 100 mm. The nitrifying bacteria also show preference for growth below 60 mm soil depth because of greater generation of CO_2. The plate counts of microorganisms are maximum in upper 60 mm depth. Mostly phototrophs are restricted to upper layers of soil exposed to sun, provided other conditions are congenial. In lower depths either dormant propagules or some oligotrophs and anaerobes are present. Fluctuations in population density in lower zones are common features. In alluvial soils, populations fluctuate with textural changes. Organisms are more numerous in loamy or clay soils than in sandy soil. Microorganisms are numerous in soil profiles just above the water table than in higher zones. Similarly, microbial populations are more widely distributed in soil horizon where roots penetrate deeper than the soils with shallow root zone. Most fungi and algae show preference for upper zone. Bacteria are abundant in moist soils. Soil fauna, however, can migrate within profile selectively. The activities of some organisms have indirect influence on pedogenesis.

3.1.2 DISTRIBUTION WITHIN SOIL FABRIC

Soil fabric : The spatial arrangement of solid particles and associated pores or voids in micro-pedological terms is known as 'soil fabric'. The nature of soil constituents, the activity of soil organisms and various soil forming processes lead to the creation of a series of features with specific soil fabrics. The components of soil fabric include a series of organic and inorganic components, water and organic substances containing humus and organisms.

Soil aggregates : Soil aggregates are specific microsites in soil with porosity lower than the average value of soil. The soil aggregates formation is the result of random combination of organic and mineral components of soil. If such assemblage has mean particle diameter <50 μm, it is called a 'micro-aggregate'; and when mean particle diameter is >50 μm, it a 'macro-aggregate'. Micro-aggregates primarily comprize of clay and humified organic materials linked by polyvalent metals and are stabilized by persistent binding agents; whereas polysaccharides and biomass (root and microbes) result in less persistent association of micro-aggregates into macro-aggregate formation. The macro-aggregates are largely affected by soil management practices. The stability of soil aggregates depends on the organic matter content of individual soil and the nature of microbial products, which bind the particles together and are difficult to dislodge.

Significance of microbes in soil aggregate formation : Most microorganisms are believed to be placed at solid-liquid interface in soil. Usually bacteria remain adhered to or are adsorbed on organic and inorganic soil surfaces by extra-cellular mucilaginous polysaccharides (mucigels) or fine fibrallae. Most bacteria are larger than clay particles and both usually carry negative charge. The principal clays of soil i.e. kaolinite, illite, mica, montmorillonite and vermiculite carry charges of about 10, 25, 35, 100 and 150 m. eq. 100 g^{-1}, respectively. The bacterial cells interact with clays because the charges on cell and clay become polarized or alternatively these are bridged by ionic bonds. The stimulatory effects of montmorillonite on fungi and actinomycetes and the ability of clay to protect microorganisms against the inhibitory effects of osmotic potential are mainly due to the ability of clay minerals to modify the local physical and chemical conditions in the soil. Bacteria surrounded by clay crystals could form a micro-aggregate whereas fungi are probably responsible for binding larger soil particles. The network of roots and fungal hyphae hold larger macro-aggregates together. Fungal hyphae retain strength after their death thereby allowing a more permanent effect in soil aggregate. Mucigels are macromolecules that do not move around in soil, therefore probably fine clay particles migrate to mucigels and organisms in order to initiate aggregate formation.

Soil pore space : The arrangement of soil aggregates largely determines the amount of pore space available. Pore space varies from 35-50% in sandy surface soils to 40-60% in medium to fine textured soils. Glinski and Lipiec on the basis of diameter at the narrowest point classified soil pores into three categories (Table 3.1). The pore diameter in micro-aggregates ranges from 0.2 to 2.5 μm and in macro-aggregates >100 μm.

Table 3.1: Soil pore categories

Category	Diameter at narrowest point (μm)	Remarks
Micropores	<30	Primarily confined within structural peds or aggregates (intrapeds); air and water movement is slow.
Mesopores	30-100	Air and water movement medium.
Macropores	>100	Mostly occur between peds or aggregates and show free air and water movement.

Microbial diversity in pore space and soil aggregates : The size and distribution of pores in soil are amongst the most important factors which determine microbial life and diversity. Soil microorganisms reside in soil pores and, in general, occupy 0.2-0.4% of pore space in arable soils. The pore neck size screens the entry of micro-organisms into the soil aggregates. In soil micro-aggregates only bacteria invade and remain viable for a long period, even for decades together in stored and dried soils. In micro-aggregates mostly sugars of microbial origin (galactose, mannose and glucose) are present, whereas in macro-aggregate soil organic matter and high proportion of sugars of plant origin (xylose and arabinose) dominate. Clay-mucigel complexes are strong barriers against diffusion and usually anoxic. Therefore, these sites do not have intensive microbial colonization and metabolic activity. Nevertheless, the properties in soil vary over a few centimeters. A micro-aggregate may have an anaerobic center while its surface remains aerobic. Therefore, two distinctly different types of biological processes would be occurring even though the sites may be less than a millimeter apart.

Macro-aggregates permit aeration, water flow, diffusion of soluble products and microbial residence. These pores are colonized by microflora and microfauna and, sometimes, mesofauna as well. Mega-aggregates (>2 mm) show penetration of roots and are intensively colonized by associative microorganisms (Fig. 3.1). Macro-aggregates are metabolically most active sites than the soil as a whole. In macro-aggregates greater amounts of nutrients (C, N, P, S) and microbial and plant originated sugars are present, therefore to understand interactions between different soil biotic components the soil needs to be considered at microsite level. The microsite variation of soil is one of the greatest challenges to soil microbiologist and ecologists.

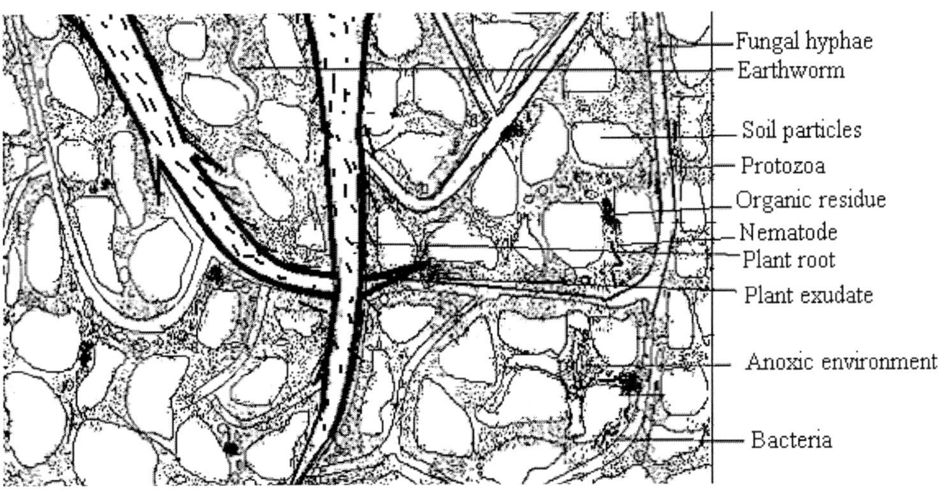

Fig. 3.1: Basic elements of soil constituents contributing towards aggregate formation

In soil microbes are always in constant competition for energy and nutrient supply. Their ability to secure growth supporting materials depends on physico-chemical conditions of soil like soil structure (aggregation), water activity, gaseous composition, pH, temperature, quantity and quality of organic matter present, cultural factors, microbial interactions etc. Osmotic pressure, surface tension, viscosity, adsorption phenomena and spatial constraints also influence microbial growth and activities. Each microbial species has optima for each physical and chemical factor, and its growth and activity usually declines on either side of the optima. Great diversity in microbial population do exist in productive soils, and the size of microbial biomass usually shows direct correlation with the amount of plant growth, primary productivity and soil organic matter level.

3.2 SOIL STRUCTURE AND MICROBIAL HABITAT

The relative dimensions of soil particles range from 2 mm or greater for soil aggregates to less than one µm for clay. The formation of clay-organic matter complexes and stabilization of sand, silt and clay particles into aggregates are dominant structural features of most soils. The proportion of each fragment in an aggregate determines soil texture (Fig. 3.2). The degree of soil aggregate formation is one of the most important factors controlling microbial activity and soil organic matter turnover by limiting such properties

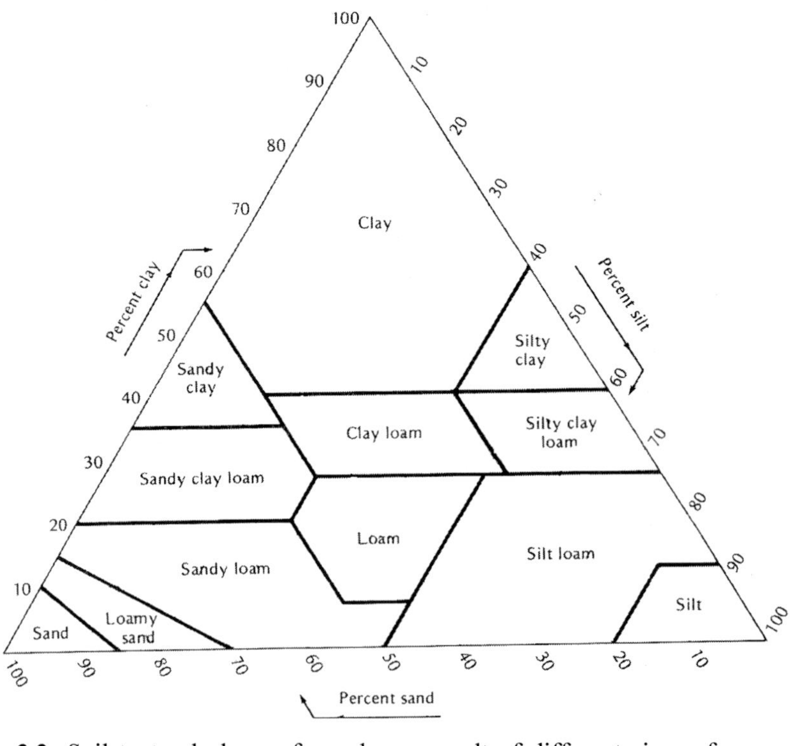

Fig. 3.2: Soil textural classes formed as a result of different sizes of aggregates

as water infiltration and availability, oxygen tension and nutrient movement etc. The aggregate dimension (size and arrangement of solids) usually determines neck size of pores, which in turn, influence accessibility of organisms to these pores. The average size or thickness of plant roots and other components of soil biota is given in Table 3.2. It reveals that size exclusion of components of soil biota operates in some soils. The pores as small as a few micrometers in diameter and filled with water are generally suitable for bacteria. Fungi invade somewhat larger pores. Microbial populations in small pores are often protected in contrast to larger pores especially during the periods of moisture stress, which allow localized sites for anaerobiosis influencing mineral nutrient transformation. Further, pore size limits the fauna to move about and so affects their grazing on microflora. Similarly, macropores provide refuge to microfauna and protect them from larger predatory organisms.

Table 3.2: The size of various soil constituents

Components	Size or thickness (μ m)
Inorganic constituents	
- Gravel	2,000-20,000
- Coarse sand	200-2000
- Fine sand	20-200
- Silt	2-20
- Clay	<2
Plants	
- Roots	>400
- Fine root	50-400
- Root hair	10-50
Soil microflora	
- Algae	1.0 μm - several m
- Bacteria	0.5-1.0
- Fungi	0.5-1.0
- Actinomycetes	0.5-1.5
- Viruses	0.05-0.2
Soil fauna	
- Protozoa	10-80
- Nematodes	50-200
- Mites	500-2000
- Earthworm	2-5000

Source: Glinski and Lipiec (1990); Killham (1994); Wood (1995)

Soil aggregates and their constituent clays and colloidal organic matter influence the interaction of enzymes with their substrates. Clays absorb enzymes and protect them from hydrolysis by other enzymes. Adsorption also makes catalytic sites less accessible. The physical protection of native organic matter through microbial exclusion as well as chemical protection (by processes such as clay-organic complexation) increases the turnover period of organic matter to hundreds and even thousands of years.

The organic matter, especially polysaccharides and gums, originating from live cells and native soil organic matter such as humic acids surround the soil particles and microorganisms. These materials serve as sources of adhesive forces that bind soil particles, which may also have a charged nature. The binding activity of polysaccharides is due to the length, linear structure and charge of these molecules, which allows them to bridge the space between soil particles. Microorganisms regulate porosity and disturbances caused by mechanical impedance of roots and aggregate stabilization at micro-level. For example, sloughed off root fraction may obstruct porosity, but its degradation by microorganisms reduces contact area of roots with soil and increases soil porosity. Thus microorganisms are probably major determinants of soil structure. It seems that microorganisms look after the wear and tear of soil and maintain its dynamic state. Reports indicate that water stability and cementation in clay soils is increased with added biomass. The gums and other excretory materials act as cement that exhibits some resistance to hydration. Fats and waxes also tend to give aggregates the ability to resist wetting and thawing. The cementing action of more resistant stable humus components, aided by similar action of certain inorganic compounds such as iron oxide, provides most of the long-term aggregate stability for smaller aggregates (<0.25 mm). The aggregating influence of different groups of soil microbes can be graded in the order fungi > streptomyces > gum producing bacteria > yeasts. Filamentous fungi such as *Rhizopus, Mucor, Chaetomium, Cladosporium, Fusarium, Aspergillus, Rhizoctonia,* mycorrhizae, etc. and some specialized rhizosphere bacteria such as *Azotobacter, Beijerinkia, Rhizobium, Xanthomonas, Bacillus,* etc. through their gum production activity have direct enmeshing effect thereby assist in larger stable aggregate (>25 mm) formation in uncultivated grasslands. Several species of algae that live near the soil surface are known to produce quite effective aggregate-stabilizing compounds. The application of small quantities of commercial preparation containing such algae may bring about a significant improvement in surface soil structure. The amount required may be very small because algae once established in soil can multiply. Earthworms through their burrowing activities influence water infiltration, root penetration and soil bulk density. The structural aspects of soil are, therefore, extremely important in governing the occurrence and activity of soil microorganisms.

3.3 SOIL WATER

Forms of soil water : Water in soil may occur in three forms - gravitational, capillary and hygroscopic form. The 'gravitational water' is drained through soil by gravitational forces after heavy rains or irrigation. When the downward movement of gravitational water (infiltration), immediately after the drainage, virtually ceases the soil water is said to have reached its 'field capacity'. The losses due to gravity at this time become negligible. At field capacity, water occurs in the megapores as thin films on the surfaces of soil particles and in wedges at the points of contact between soil particles. However, micropores still remain filled with water. The force holding water at field capacity varies from soil to soil and is approximately in the range of -0.1 to -0.3 bars (-10 to -30 kPa). The water content at field capacity ranges from 15 to 30% in sandy loam and 40 to 45% in clay soils. The water retained at field capacity by a soil is called its 'capillary water'. This water is held in soil by cohesive forces rather than gravitational forces and is available for plant growth. The 'hygroscopic water' is the water absorbed by dry soil from an atmosphere of high relative humidity. In simple words, hygroscopic water is the water remaining in soil after its air-drying. The hygroscopic water can be removed only by oven-drying the soil. It has to be kept in mind that soil water is never pure water but contains a wide range of solutes (hundreds of dissolved organic and inorganic substances) so is more accurately called 'soil solution'.

pF value : The supply of water to plant roots and microbes depends not only on the field capacity of soil but also on water tension or water suction created by the suction of water to the surfaces of soil particles. The water tension or water suction is expressed in terms of pF value. pF may be defined as the logarithm of soil water tension, expressed in centimeter height of a column of water. It is pertinent to mention that 1022 cm water column = 75 cm Hg column = approximately 1.0 atmosphere = 1 bar = 10^6 dynes cm^{-2} = 100 kPa = 0.1 MPa.

Wilting point : The living organisms and plants have to overcome the forces of adhesion, cohesion and osmosis to acquire water from soil. When the forces of adsorption and retention of water in soil become too strong that plants are unable to extract water to satisfy their needs, this point is called 'wilting point'. The force holding this water to soil is about 15 bars (-1500 kPa). The 'permanent wilting point' is the stage at which plants can no longer recover from wilt. This usually occurs at values >-1.5 MPa.

Soil water potential : The storage, retention and movement of water in soil and its uptake and translocation in plants and its loss to the atmosphere are all energy related phenomena. The mathematical description of water availability (tendency of water to flow) or water potential is 'ψ'. It is a measure of the potential energy of water at a point in a system relative to the potential energy of unbound pure water. In soil the attractive forces between solid and liquid phases, known as 'matric potential', enables the soil to hold water against processes such as gravity, evaporation, plant root uptake, etc. In soil ecosystem two components of water potential stress i.e. matric and osmotic potential, are more useful concepts to describe soil water availability to soil organisms and both have to be overcome for the uptake of water by plants. 'Osmotic potential' is due to the presence of solutes in soil solution so is less important than matric potential in most soils.

$$\psi_s = \psi_m + \psi_o + \psi_g$$

where, ψ_s = total soil water potential

ψ_m = matric potential ($2\ \pi r^2$)

ψ_o = osmotic potential ($-R.\ t\ .e.\ v.\ m.\ \phi\ .\ 10^{-9}$)

(R = gas constant [8.3×10^7]; t = temperature [°K];

e = density; v = ions mol^{-1}; m = molarity; and

ϕ = osmotic coefficient)

ψ_g = Gravitational potential

Water potential in soil fluctuates dramatically and microbes respond accordingly. Under dry conditions fungi and bacteria form resistant spores. The activity of some soil fauna, especially protozoa and nematodes, are adversely affected by soil drying. Soil protozoa move through water and graze soil microflora. So their locomotory activities are restricted under water stress conditions. The protozoa and nematodes, therefore, encyst and enter into dormant phase. The availability or flow of water is also influenced by soil moisture content. A good relationship exists between water content and soil biological activities upto near field capacity of soil but biological activities decrease as the soil becomes water-logged. The degree of displacement of air with water within soil pore together with the rate of microbial respiration produces anoxic or anaerobic microsites within the soil pore.

Microbial grouping based on soil moisture requirements : The soil microorganisms on the basis of their moisture requirement are classified into hygrophiles, mesophiles and xerophiles (Table 3.3). The growth of microorganisms is generally restricted to values greater than between -6,000 kPa and -800 kPa (-6.0 MPa to -0.8 MPa). At water potential below -14,500 kPa (-14.5 MPa) only some phycomycetous fungi such as *Aspergillus* and *Penicillium* species predominate.

Table 3.3: Soil microbial classification on the basis of moisture requirement

Classes	Soil moisture tension at which the activity of organisms disappear	Groups present
Hygrophiles	>7.1 MPa (pF 4.85)	Most bacteria, yeasts and protozoa
Mesophiles	7.1-30.0 Mpa (pF 4.85-5.48)	Majority of fungi, some bacteria
Xerophiles	30.0-48.0 MPa (pF 5.48-5.68)	Only some phycomycetous fungi e.g. *Aspergillus niger, A. glaucus, Penicillum glaucum,* etc.

Microbe - soil water relation : The microbial water relations are quite different from that of plants. Some soil microbes are able to tolerate water/ salt stress in greater magnitude than most plants. Microbial cells are able to counter balance the water potential stress of their soil environment partly through selective transport of soil salts. Clays offer some degree of protection to microorganisms against desiccation. The surface of these mineral particles retains water, which together with coordination water from exchangeable cations may assist microbes to survive dry conditions.

The typical range of tolerance to water stress for soil microorganisms is given in Table 3.4. Generally microbial activity in soil is optimal at -0.01MPa (field capacity) and activity decreases as the soil becomes either saturated (near zero water potential) or more droughty at large negative water potentials. Therefore, degree to which water space is filled up with water is of fundamental importance in determining soil biological activity.

Table 3.4: Typical ranges of tolerance to water stress for soil microorganisms

Organisms	Water potential (MPa)	
	Tolerance range	Upper range
Bacteria and protozoa	0 to -10	-10
Yeasts	0 to -20	-65
Fungi	0 to -60	-70
Actinomycetes	0 to -70	-80

Source: Harris (1981); Kilham (1994)

Fungi and actinomycetes are more tolerant of higher water potential (greater water stress) as compared to bacteria. Soil bacteria, protozoa and nematodes always tend to live in soil water whereas fungi grow across the air filled pore spaces. Bacteria appear to be readily motile in water films which are appreciably thicker than their diameter. However, bacteria are capable of showing growth even at lower moisture contents. Protozoa and nematodes are predatory so at a high moisture level their movements are greater. The water potential of below -0.5 bar is likely to stop their activities. For all motile microorganisms the rate of movement is dependent on soil moisture and so on soil structure. Their movements become restricted as the soil dries. Actinomycetes generally prefer dry conditions. The tolerance of actinomycetes to water stress is attributed to their ability to synthesize intra-cellular proline as part of their osmo-regulatory strategy, besides their capability of selective transport of salts. In tropical and subtropical areas of extreme water stress actinomycetes often totally dominate other soil microflora.

Like other factors, water availability affects specie diversity, survival, movement and the activity of microorganisms in soil. The nitrifiers, such as *Nitrosomonas*, are less tolerant to water stress than ammonifiers, such as *Clostridium* and *Penicillium*. Therefore, under dry conditions ammonia may accumulate as nitrifiers are less active. The soil water, besides having direct effect on microbial communities and their activities, indirectly influences the supply of nutrients to soil biota through effects on diffusion, mass flow and nutrient concentration. Soil fungi and bacteria take up their nutrients from water films in which they are growing. So thickness of water film influences their metabolic rate through its effect on food source. Under moisture stress conditions nutrient diffusion may be sufficiently slow so a limiting factor for microbial growth. The ample supply of soluble carbon root-exudates from root to rhizosphere microbes usually occurs through diffusion. This root exudate supplies the rhizosphere microbes with bulk of their growth substrates.

Alternate drying and rewetting of soil causes flush of CO_2, NH_4 and NO_3 with subsequent releases of less intensity. Similarly, freezing and thawing gives a flush of decomposition. Rewetting phenomenon causes desorption of organic matter from clay surfaces, rupture of microaggregates to expose protected organic materials and rapid decomposition of microbial cells killed during drying. Fluctuations in soil moisture and temperature have a more profound effect on soil microbial processes and cause grater oxidation of organic matter as a result of disruption of soil aggregate structure and death of more stringent microbial populations than constant extreme conditions.

3.4 SOIL AERATION

Gaseous diffusion in soil : In soil ecosystem there is always a continuous process of gas transport (diffusion) in the direction of decreasing concentration.

As per Fick's first law, the diffusion of gases is related to gas movement and soil characteristics like concentration of gases in soil and soil depth:

$$f_x = -D\frac{dc}{dz}$$

where

f_x = diffusion rate of gas through a unit surface area of the medium in which diffusion takes place in a unit time (g cm^{-2} sec^{-1})

D = diffusion coefficient (constant) is soil (cm^{-2} sec^{-1})

c = gas concentration in soil air (g cm^{-3})

z = soil depth (cm)

$\frac{dc}{dz}$ = gradient of concentration constituting the driving force of flow

In contrast to mass flow, the diffusion flow is not induced by the presence of external factors but only by the gradient of concentration which is a result of respiration process. The solubility of gases in water depends on the type of gas, temperature, salt concentration and partial pressure of gases in atmosphere. The CO_2, NH_3 and H_2S are most soluble and ionize in water. Oxygen (O_2) and nitrogen are markedly less soluble. The nitrogen may be limiting in N_2 fixing systems because of its low aqueous diffusion rate even though the atmosphere contains 80% N.

Oxygen diffusion in soil : Soil oxygen concentration is controlled by gaseous diffusion rates as well as by the respiration activity of soil biota. Oxygen diffusion rate (ODR) is at present the best indicator of soil aeration, although other methods of estimation include determination of CO_2 and O_2 contents and redox potential of soil. ODR largely determines the rate at which O_2 can be replenished when used by respiring biomass in soil or removed out of the soil pores by water. ODR in soils vary most frequently within a range of 0-200 μg cm^{-2} sec^{-1}. Exchange rates of air above 40 x 10^{-4} g m^{-2} min^{-1} seem to be fast enough for most of the plant roots and microbes. ODR decreases with increase in soil moisture content and compaction. Under high moisture conditions there is limited supply of air to microbes hence anaerobic or microaerophilic groups develop.

Soil air - microbe relation : Soil organisms show a wide range of adaptation to oxygen availability. Changing aeration in soil has profound influence on microorganisms and enzymes. Most aerobic microorganisms have a very efficient enzyme system, based on cytochrome oxidase as their terminal oxidase system, for absorbing free O_2. Greenwood suggested that due to the efficient terminal oxidases, possessed by most of the microorganisms, the soil could be considered aerobic as long as the soil solution can sustain an O_2 concentration of at least 5 μM, corresponding to 1/100th of O_2 concentration in air-saturated water. Many soil fungi have efficient cytochrome oxidase system. The poor tolerance of low O_2 conditions by fungi seems due to the adverse impact of high CO_2 levels. This may also be the reason why many fungi are restricted to superficial layers of soil. Some organisms are strict or obligate aerobes so must have easy access to oxygen in order to respire. Some are obligate anaerobes so O_2 is lethal to them. Yet a number of soil organisms appear to change their physiological activities to thrive in presence or absence of O_2.

Generally microbes require well-aerated soils i.e. the O_2 equivalent to that of air >8-20% and CO_2 equivalent to 0.1-1.0%. In such soils O_2 content seldom falls below 20% and CO_2 seldom increases greater than 1%. The typical range of concentration of major gaseous constituents (N_2, O_2 and CO_2) of soils is given in Table 3.5. CO_2 is commonly 10-100 times higher in soil air than in atmosphere but may be many thousand times higher in isolated pockets in wet soils.

Table 3.5: Diffusion constants of major gaseous components of the soil and their water solubility at 20°C

	O_2	CO_2	N_2
Concentration in atmosphere (%)	21.0	0.035	78.1
Concentration in soil air (%)	20-21	0.1-1.0	78.0
Diffusion constants in air ($cm^2\ sec^{-1}$)	0.205	0.161	0.205
Diffusion constants in water ($x10^{-4} cm^2 sec^{-1}$)	0.180	0.177	0.164
Solubility in water ($cm^3 L^{-1}$)	0.031	8.878	0.015

A soil having sufficient gas filled pores to allow gaseous exchange with atmosphere is considered good structured. Such soils have less chances of developing localized anaerobic pockets or microsites, which have varied effect on root growth with respect to its physiological processes, metabolism and ability to resist biotic and abiotic stresses. Reduced pore space means depletion of oxygen. When the concentration of oxygen falls to about 3×10^{-6} mol^{-1} (<1% v/v; the partial pressure of oxygen in free atmosphere) the microbial population appears to shift from being predominantly aerobic to anaerobic. That is, there is a shift from aerobic to anaerobic metabolism. The growth yield accordingly declines with the development of reduced environment, since energy yield per mole of fixed carbon oxidized anaerobically is far less than that produced by aerobic respiration. All soils have heterogeneous distribution of oxygen, therefore such small pockets do always exist. In many soils, water contains small but significant quantities of dissolved O_2. When all the soil pores are filled with water, soil microorganisms can extract most of the O_2 dissolved in water for their metabolic activities. This small amount of dissolved O_2 is soon exhausted and if excess water is not removed the aerobic microbial activities are jeopardized. Soil air is directly linked with soil moisture. Maximum biological activity generally occurs at soil water potentials approximating to 40% water holding capacity of soil. At soil moisture level greater than this, the rate of biological activity is generally limited by the rate of oxygen diffusion, which is about 10,000 times slower in water than air.

Soil respiration and gaseous exchange : The rate of soil respiration and rate of gas exchange in soil determine the composition of soil air. The respiratory demand of oxygen for roots and microbes in soil can exceed 20 g m^{-1} day^{-1}. Soil microbial respiration values range from 0.2-20 cm^3 O_2 kg^{-1} hr^{-1} [i.e. 0.05-5.0 mm^3 kg^{-1} sec^{-1} = 3-300 μ mol m^{-3} sec^{-1} = 0.1-10 mg m^{-3} sec^{-1}]. Bacteria and algae generally have much higher specific rates of O_2 uptake (1,200,000 and 40,000 cm^3 hr^{-1} kg dry body weight, respectively) than plant roots/seeds or fungi (27,100 and 10,000 cm^3 hr^{-1} kg dry body weight, respectively). Hence, they are always in competition for O_2 whenever O_2 supply slightly decreases. The specific oxygen uptake rate (qO_2) of lichen, insects and crustacea is comparatively low (900, 540 and 360 cm^3 hr^{-1} kg dry body weight, respectively). For example, qO_2 for *Azotobacter, Escherichia, Mucor* and *Penicillum* are 5000, 200, 150 and 20 μL mg^{-1} dry body weight h^{-1}, respectively, in comparison to 4.0 and 0.025 μL mg^{-1} dry body weight h^{-1} in case of sterile cereal roots and germinating barley, respectively.

Depletion of dissolved O_2 in soil water during respiration and chemical oxidation leads to the rapid development of anaerobic conditions. If water-logged conditions persist the anaerobic microorganisms replace aerobic organisms which respire via a sequence of components acting as alternative electron acceptor to O_2. The electron transport chain of cell is able to convert reducing equivalents such as reduced pyridine nucleotide or flavoproteins to energy as adenosine triphosphate (ATP). The terminal electron acceptor can range from O_2 to NO_2, NO_3, SO_4 or CO_2. The presence of gases like CH_4, N_2O, C_2H_4 and H_2 despite their low concentration is an important indicator of aeration status of soil. The soil respiration rates for different soils, measured on the basis of CO_2 evolution, are in the range of 1,800 to 47,000 kg of CO_2 ha^{-1} yr^{-1}. Measurements made on a soil at Rothamsted Experimental Station, UK have shown daily rate of O_2 consumption of 0.7-24.0 g m^{-2} and rate of

CO_2 production at 1.2-35.0 g m^{-2}. Soil contains enough O_2 to support respiration for not more than a few days. O_2 is unlikely to limit biological activity at low water matric potentials because most pores are filled with air in which O_2 diffusion rate is 3,00,000 times higher than that of water. The concentration of both O_2 and CO_2 are affected by microbial decomposition of organic residues. Soil aeration status determines the fate organic matter may face i.e. whether it will follow an aerobic or anaerobic pathway for decomposition.

3.5 SOIL REDOX POTENTIAL

Redox potential (E_h) : Redox potential (E_h) is a measure of the tendency of a substance to accept or donate electrons. It, therefore, describes the tendency of electrons to flow between compounds and is dependent on both oxygen content and pH. E_h is expressed in volts (V) or millivolts (mV). More highly a compound is in reduced form, greater will be the amount of energy released on its oxidation.

$$\text{Reduced state} \rightleftharpoons \text{Oxidized state} + e^-,$$

In terms of Nernst equation, redox potential

$$E_h = E^o + \frac{RT}{\ln F} \log E_h = E^o + \frac{RT}{\ln F} \log [Ox] / [Red]$$

since $\frac{RT}{F}$ is equal to 0.0592 at 25 °C, therefore

$$E_h = E^o + \frac{0.059}{n \,.\, \log [Ox]/[Red]}$$

where, R = gas constant;
F = Faraday's constant;
T = absolute temperature;
E^o = standard potential of the system (mV)
n = number of electrons in the system
[Ox] = electrons lost (oxidation state)
[Red] = elections gained (reduction state)

If E_h is high, the substance will accept electrons easily and is so called an oxidizing agent. Lower E_h (-ve) means greater reducing capacity of the compound. There is a broad E_h range in which each reduction step may occur because of possible variation in soil pH. A redox potential of 300 mV has been accepted as a boundary between reduced and oxygenated soil. The occurrence of a variety of microbial processes in soil is related to specific redox potential.

Redox intensity (P_e) concept : The concept of redox intensity (P_e) is often used in place of E_h. The redox intensity is written as P_e and mathematically P_e = -log [e^-]. A highly reducing substrate has high tendency to donate electrons so such substrates have low P_e. High P_e of a substrate means its more tendency to accept electrons (i.e. to get oxidized). The redox potential E_h is related to redox intensity P_e as:

$$P_e = \frac{E_h}{0.059}$$

Soil redox potential and microbial processes : The reduction-oxidation reactions explain both chemical and biological processes occurring in soil. All aerobic respirations require molecular O_2 to serve as electron acceptor as organic carbon is oxidized to release energy for life processes. O_2 can oxidize both organic and inorganic substances. Thus, O_2 rapidly accepts electrons and is a strong oxidizing agent. In soil, microbial activities of mutually exclusive nature do occur because of heterogeneity of soil system. There is enormous spatial variation in redox potential in soil particularly because of aggregate microsites of high biological activity. If organic matter is added to a soil, the oxygen present in soil gets depleted because of intense microbial activities and the potential drops. Under such a situation organisms switch over to alternate electron accepters provided they posses the enzyme systems capable to use more than one terminal electron acceptors. Denitrifying soil microbes (*Pseudomonas, Bacillus, Alcaligenes,* etc.) in the order of preference mostly use O_2, NO_3^-, and Mn^{4+} as electron acceptors so are facultative aerobes. The obligate anaerobes prevail only under sustained anaerobic soil conditions at E_h of about -0.2 V. Methanogenesis occurs only in organic soils under sustained water-logged conditions. The obligate aerobic bacteria and most soil fungi cannot survive for longer periods under anaerobiosis. The decline in soil fungi as a result of even short periods of anaerobiosis may explain why mycorrhizal development is low in poorly drained soils. Most soil animals are obligate aerobes and their number tend to be low in anaerobic soils. The changing aeration of soil ecosystem has profound influence on soil microorganisms and their enzyme activities.

Fungi, non-photosynthetic bacteria and soil animals obtain energy from the oxidation of reduced substrates which may be either organic (heterotropic metabolism) or inorganic (chemolithotrophic metabolism) in nature. These organisms during oxidation remove electrons from these substrates and the energy available from the electrons via respiratory or electron transport chain is used to produce ATP (adenosine triphosphate) from ADP (adenosine diphosphate). Phototrophs derive such energy directly from sunlight. Under aerobic soil conditions, oxygen is the terminal electron acceptor as organic carbon is oxidized to release energy for metabolic activities. But when all the oxygen, even dissolved in soil water, is consumed during respiration and chemical oxidation, the activities of aerobic microbes cease. At this stage, aerotolerant or anaerobic microbes come into fore which respire via a sequence of components acting as alternative electron acceptors to oxygen. Anaerobes use NO_3^-, Fe^{3+}, SO_4^{2-} and Mn^{2+}, depending upon the electron acceptor and the type of enzyme system available in their cells. As a consequence of microbial respiration, the redox potential (E_h) of a soil also declines. The range of E_h changes in soil from +800 to -400 mV (Table 3.6). Higher +ve E_h means greater oxidizing capacity and -ve E_h reflects the reduced state. For example, aerobic carbon oxidizers have E_h value >200 mV, denitrifiers 150 to 200 mV, methanogens -200 to -100 mV and sulphur reducers -200 to -10 mV. Similarly, NAD^+ is reduces to NADH at -320 mV, H^+ to H_2 at -420 mV and cytochrome B and C form oxidized to their reduced states at +70 and +250 mV, respectively. In moist non-flooded soil Eh is normally governed by O_2 concentration.

Table 3.6: Redox potential of some common electron acceptors in soil biological systems

Soil biota	**Redox couple (terminal e⁻ acceptor)**	**Redox potential Pe at pH 7 (mV)**	**Conditions**
Plant roots, aerobic microbes and animals	$O_2 + e^- \longrightarrow H_2O$	+820	Aerobic respiration
Bacillus/Clostridium	$NO_2 \longrightarrow NO_3^-$	+430	Nitrification
Pseudomonas	$NO_3^- + e^- \longrightarrow N_2$	+420	Denitrification
Bacillus	$Mn^{4+} + e^- \longrightarrow Mn^{2+}$	+410	Mn-reduction
Clostridium	Organic matter + $e^- \longrightarrow$ Organic acids	+400	Fermentation
Clostridium/Desulfovibrio	Fumerate $\longrightarrow$ Succinate	+033	Decomposition
Desulfovibrio	$SO_4^{2-} + e^- \longrightarrow SO_3^{2-}$	-060	SO_2-reduction
Pseudomonas	$Fe^{3+} + e^- \longrightarrow Fe^{2+}$	-180	Iron reduction
Achromobacter	$NO_3^- + e^- \longrightarrow NH_4^+$	-200	Dissimilatory nitrate reduction
Desulfovibrio	$SO_4^{2-} + e^- \longrightarrow H_2S$	-220	SO_4-reduction
Cellulomonas	$CO_2 + e^- \longrightarrow CH_3COOH$	-280	Acidogenesis
Methanobacterium	$CO_2 + e^- \longrightarrow CH_4$	-240 to -350	Methanogenesis

Sources: Paul and Clark (1989); Brock and Madigan (1991); Killham (1994); Tate (2000)

The changes in soil redox potential are related to high oxygen level. Therefore, redox potential can also be used as an indicator of aeration status in a soil. Well-aerated soils have E_h value of 400 to 700 mV or high; poorly aerated ones have 300 to -350 mV or low and water-logged or flooded soils have E_h <-400 mV. Plant and microbial respiration rapidly consume O_2 present in air, therefore O_2 content decreases with a decrease in E_h. At low E_h, dissolved O_2 in soil water is also consumed by microorganisms. Microorganisms capable of using elements other than O_2 such as N (in NO_3^-), Mn, Fe and S (in SO_4^{2-}) as their metabolic electron acceptors can function at very low E_h.

3.6 SOIL REACTION

The soil reaction is one of the outstanding chemical characteristic of soil solution that profoundly influences the soil properties.

Soil pH : In chemistry a reaction that binds a proton is a basic reaction whereas the one which releases the proton is an acidic reaction. So a soil reaction gives idea about the degree of acidity or alkalinity. It is expressed in terms of pH value. Soil pH is the measure of the concentration of H^+ ions in soil solution. The pH is logarithm of the reciprocal of hydrogen ion activity.

$$pH = \log \left(\frac{1}{AH}\right) = -\log (AH^+)$$

In strong acid soils besides H^+, Al^{++}/Fe^{++} also contribute to pH or acidity. In alkaline soils exchangeable basic cations, Ca^{++}, Na^+, Mg^{++} or K^+, tend to occupy the exchange sites by replacing H^+ and Al^{++} ions. Reactions of solids are largely limited to surface reaction i.e. greater the surface area, greater will be chemical reaction. Soil microorganisms and plants respond markedly to their chemical environments.

Microbial sensitivity to soil pH : A bacterial cell contains about 1000 enzymes whose activities are pH dependent. The enzymes influence the growth and development of a microbial population in a soil ecosystem. On the basis of pH sensitivity, microorganisms have been grouped as pH insensitive, acidophiles, neutrophiles and basidophiles (Table 3.7). The optimum pH of soils is near 7.0 that also is the optimum pH of microbial cytoplasm.

Table 3.7: Classification of microorganisms on the basis of pH sensitivity

S. No.	Microbial groups	Active at pH
i.	pH-insensitive	In wide range with optimum close to 7.0
ii.	Acidophiles	In acidic soils with optimum range 2.0 to 3.5
iii.	Neutrophiles	Require neutral soil reaction and are sensitive to acid or alkaline conditions
iv.	Basidophiles	Active at pH above 8.0

Influence of soil reaction on biological activities : Soil reaction influences biological activities through control of nutrient availability. High pH affects the soil biological diversity and versatility, including plants, by reducing the availability of Mn, P, Fe and B to rhizosphere system; whereas at very low pH Ca, Mg, K, P, N, S, Mo and B availability is curtailed and Al, Mn, Fe, Zn, Cu and Co dominate to toxic level.

Most relevant determinant of microbial activity in soil is the microsite variation in soil pH. The soil aggregate-attached microbes are protected by a thick layer of polysaccharide through which only protons can diffuse that too slowly. Besides, the pH optima of microbial enzymes are affected by soil adsorption phenomena. In soil matrix, adsorption of enzymes to soil humates shifts the pH optima of enzymes to higher values.

Bacteria and actinomycetes are less tolerant to acid conditions than fungi. Therefore, in acid soils dominant biological activity is generally the result of fungal community. Bacteria and actinomycetes preferably function at neutral to alkaline conditions. However, some acid tolerant bacteria (*Thiobacillus thiooxidans*) can thrive/grow even at pH 1.0 while most alkali tolerant streptomycetes can grow at pH 10.0. *Nitrosomonas* has optimum pH in alkaline range and is inactive below a pH of 5.5. Similarly, *Thiobacillus thioparus* grows only at near-neutral pH. The biological transformation of ammonium to nitrite is one of the most pH sensitive soil reactions.

In soil, there may be strain- and specie-diversity of microorganisms with respect to pH tolerance. The antibiotic producing actinomycetes are alkaline lovers but some species are routinely found in acid forest soils as well. This is because nitrogen rich plant residues are decomposed to produce localized ammonia sufficient to increase soil pH above 7.5. In wet soils, bacterial fermentation yields organic acids which cause decrease in soil pH. The interaction of pH and anaerobic conditions may contribute to the organic matter accumulation in swamps.

Acidophilic streptomyces are involved in decomposition process and the resultant ammonification can lead to the activity of neutrophiles. Fungi are particularly versatile and satisfactorily flourish over a wide pH range. Fungal activity tends to predominate in most acid soils. Soil animals generally have low pH requirement. Earthworms are highly sensitive to soil acidity.

Soil acidification : Plant root system can cause short term acidification in soil through cation removal in exchange of H^+. The uptake of NO_3^- by plant roots is balanced by the release of carbonates and hydroxyl ions in soil. The rhizosphere soil pH is partly controlled by the source of mineral nitrogen available to the plants. The rhizosphere pH decreases when NH_4^+ is predominant and increases when NO_3^- is predominant in soil. Use of acidic fertilizers, such as ammonium sulphate, ammonium phosphate or ammonium nitrate etc., and precipitation of acids through rains also causes soil acidification. Acid rains have pH of 4.0 to 5.0 i.e. lower than the normal rainfall pH of 5.0 - 6.0.

$$(NH_4)_2SO_4 + 4O_2 \longrightarrow 2HNO_3 + H_2SO_4 + 2H_2O$$

Microbial decomposition of organic matter tends to enhance soil acidity through production of organic acids. This occurs mostly under wet anaerobic conditions where fermentative microbes dominate especially in moorlands and forest soils producing acidic litter. Soluble organic acids such as fulvic acids are leached through soil profile causing acidification and weathering so help in podzol formation. Under anaerobic conditions, heterotrophs release CO_2 in soil, which reacts with water to form carbonic acid (unstable) and ultimately dissociates to yield hydrogen ions.

$$\underset{\text{Organic matter}}{R[C_2H_4ONS]} + 5O_2 + H_2O \longrightarrow \underset{\text{Carbonic acid}}{H_2CO_3} + \underset{\text{Strong organic acid}}{RCOOH} + \underset{\text{Strong inorganic acids}}{H_2SO_4 + HNO_3}$$

$$C_6H_{12}O_6 + 6O_2 \longrightarrow 6CO_2 + 6H_2O + \text{energy}$$

$$H_2O + 6O_3 \longrightarrow H_2CO_3 \rightarrow H^+ + HCO_3$$

Biological nitrogen fixation also creates acidity through hydrogen production during fixation process. Microbial oxidation of reduced sulphur in soil is potentially highly acidifying.

3.7 SOIL TEMPERATURE

Soil temperature depends on atmospheric temperature, input and losses of radiation, presence or absence of vegetation, soil water content and depth within soil. It has direct and profound influence on the physiological reactions occurring in biological components in soil ecosystem. Besides, it affects the physico-chemical characteristics of soil environment such as diffusion rates, redox potential, soil aeration, water activity, viscosity, surface tension, mineral weathering rates, etc.

Q_{10} concept : The activities of microbial cells are governed by the laws of thermodynamics. The rate of a chemical reaction is a direct function of temperature and microbial activity, expressed in terms of 'Q_{10}' which depicts that the reaction rate accelerates rapidly as the temperature rises. In most soils, microbial activities (Q_{10}) increase two fold for each 10°C rise between 0 to 35°C. Mathematically;

$$Q_{10} = \frac{kt}{kt-10}$$

The Q_{10} values for biological processes generally range from 1.5 to 3.0. Q_{10} is governed by Arrhenius equation:

$$\frac{A \log k}{\Delta T} = E/2.303\ RT^2$$

The equation can be written as:

$$k = Ae^{-E/RT}$$

where k = reaction velocity
A = frequency with which molecules collide
e = base of the natural logarithm
R = gas constant
T = temperature in Kelvin (°K)
E = activation energy of the reaction

The equation can be rewritten as: $\frac{k_2}{k_1} = \frac{\Delta E}{R(1/t_2 - 1/t_1)}$

A plot of reaction rate constant (k_1) v/s reciprocal of temperature in Kelvin (k^{-1}) at ambient temperatures gives a linear line with a slope. The slope represents the activation energy and must be overcome for any reaction to proceed.

Microbial sensitivity to soil temperature : Microorganisms vary greatly in their temperature maxima, minima and optima for growth (Fig. 3.3). The typical breadth of temperature supportive of microbial growth in soil is from approximately 0 to 70°C. Most soil organisms are mesophiles with temperature optima of 37°C in most organisms falling in this group. Only a few degrees above optimal temperature of an organism there is a drastic fall in activity as a result of thermal denaturation of proteins (especially enzymes) and changes in cell membrane fluidity which, in turn, is governed by the degree of unsaturation of fatty acids in the membrane. The specific cell structural proteins such as those of flagella, ribosomes etc. of thermophilic bacteria are more heat stable than those of mesophilic bacteria. The presence of unsaturated fatty acids in membrane lipids, selective membrane permeability and conformational changes in membrane proteins of psychrophiles also facilitate tolerance in physchropiles. Psychrophilic bacteria supposedly synthesize increased quantities of enzymes at low temperature to compensate for the reduction in the rate of metabolic processes at low temperature.

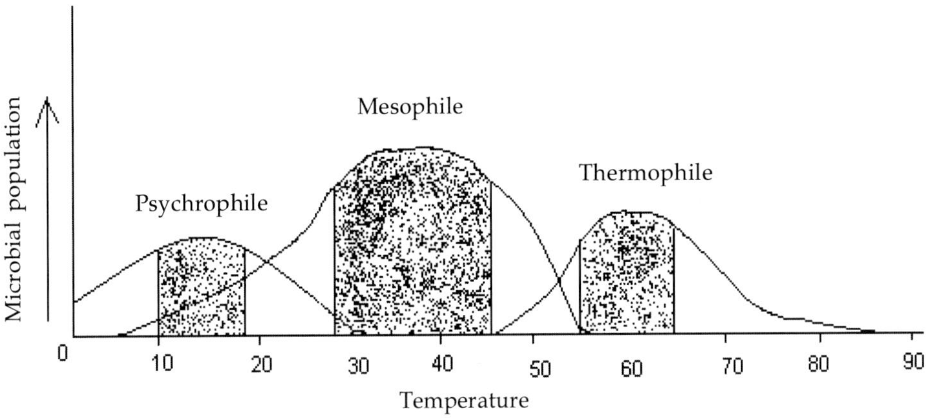

Fig. 3.3: Temperature tolerance range (°C) of soil microorganisms (darker areas depict optimum temperature in the respective ranges).

Influence of temperature on microbial diversity and function : Soil temperature influences the soil microbial species diversity and function. The temperature at which an organism has its maximum metabolic rate may be different from the temperature at which it is most abundant. This is because of the fact that rapid growth and survival often require slightly different conditions. Thermophiles can proliferate in tropical soils whereas physchropiles are well-adapted in arctic soils. Nevertheless, thermophilic bacteria and fungi are found in temperate soils, which appear more suited to phsychrophilic microorganisms. There are some exceptions as well. For instance, at polar ice caps some microbes can tolerate temperature as low as -50°C and some microbes thrive at 90°C in hot springs. The microbial populations vary quantitative and qualitatively under extreme conditions. In temperate soils the activity during winter when temperatures lies in the range of 0-5°C is lower than during summer. Various soil temperature-regime controlling factors like diurnal and seasonal variations, intensity and reflectance of solar radiation, slope and altitude, vegetative cover, moisture, depth, etc. also influence microbial biomass quality, quantity and diversity. Franz found that in mountain soils with an increase in altitude from 100 to 520 m above sea level (showing corresponding decrease in mean annual temperature from 25 to 0°C) about 100-fold drop in aerobic bacterial count and 10-fold reduction in O_2 uptake g^{-1} organic matter was encountered. Most soils are unable to maintain a uniform temperature in their upper layers because of the variations in atmospheric temperature. Diurnal fluctuations of upto 35°C can occur in well-aerated soil at surface. With increasing depth, submerged or wet soils are less susceptible to large diurnal changes than aerated or dry soil. Sub-soil temperature are less variable than surface soil temperature, generally warmer in late fall and winter and cooler in springs and summers so help to sustain biotic activities for longer periods in the rhizosphere of plantation crops.

Plant roots are particularly sensitive to changes in soil temperature probably due to the changes in water uptake, nutrient absorption, growth substance availability, root and shoot growth, partitioning of photo-assimilates within plant-soil system etc. Rhizosphere microflora is mostly dependent on carbon exudates from plant roots. Therefore because of its influence on plant root activities, soil temperature has profound influence on interaction processes occurring in soil ecosystem. However, such influence varies with the plant and microbial species involved, their growth stage, soil moisture, pH, soil fertility, etc. Although microbial activities like nitrification are negligible below 10°C still some of the nitrogen cycling processes, ammonification and denitrification do continue because of wide diversity of microbes involved in later processes. Soil fauna are generally very sensitive to overheating and usually migrate to deeper layers to avoid heat shock to their physiological processes, especially their respiratory enzymes.

3.8 SOLAR RADIATION

Solar radiation is the primary source of energy to heat soils and push biotic activities. Only about 10% of solar energy reaching the earth is absorbed by soil and used for its warming. This energy is of critical importance to soil processes as it provides the energy for photoautotrophic components of soil biota. About 5% of the net solar radiation is used for photosynthetic reaction in various components of soil biota viz., algae, photoautotrophic bacteria and plants. Photoautotrophic algae and bacteria are considered primary colonizers which help in soil formation, aggregate stabilization, nutrient cycling, etc. In fact, distribution and activities of microorganisms in soil surface are largely affected by solar radiation. Light provides basic energy to derive solar ecosystem. The input of chemical energy, as fixed carbon, is made available to soil as a result of death and decomposition of photoautotrophs and to some extent by chemoautotrophs. This enables soil heterotrophic community to proliferate. Soils rich in organic matter content are usually darker so capable of absorbing greater solar radiation.

3.9 SOIL NUTRIENTS

Soil microbes are dependent on environmental condition for their energy and nutrient flow. Generally soil biota derive their energy and nutrients directly from soil resources i.e. minerals, organic matter or living biomass. However, dominant source of food and energy for microorganisms is organic matter.

3.9.1 SOIL MINERALS

Weathering of soil minerals, especially secondary minerals such as silicates and oxides of Fe^{++}, Al^{+++}, Mn^{++} etc. release good amount of nutrients for use by soil biota. These nutrients may be retained/adsorbed by clay lattices or remain in soil solution. The soil mineral fraction serves as a nutrient sink. Clay particles are the smallest of the soil separates and mostly colloidal in nature. Clay minerals occur in soil as primary minerals or as weathered or synthesized secondary minerals, comprizing mainly of aluminium or silicates. A few clays are amorphous, but most are laminated into plates or lamellae (0.7-1.0 mm thick) with definite ionic arrangements. The variability in arrangement and chemical composition imparts different properties to clay including the charge development. Clays are chemically reactive and possess a net negative charge. Due to their extremely small particle size, the colloids possess more than 1000-fold greater surface area per unit weight than those of sand or silt (Table 3.8). The negatively charged clay minerals, depending upon soil conditions, attract and adsorb cations like Na^+, K^+, Ca^{++}, Mg^{++}, NH_4^+, etc. and through ion exchange phenomena release them into soil solution.

$$\underset{\text{(colloid)}}{\boxed{\text{Miscelle}}}\,Ca^{++} + \underset{\text{(soil solution)}}{2H^{+}} \rightleftharpoons \underset{\text{(colloid)}}{\boxed{\text{Miscelle}}\begin{matrix}H^{+}\\H^{+}\end{matrix}} + \underset{\text{(soil solution)}}{Ca^{++}}$$

Clays influence microbial activity by modifying the physical and chemical characteristics of the microbial habitat. The clay minerals serve as cation exchange sites for nutrients and play key role in controlling soil nutrient supply. Besides, these serve as sites for bacterial adhesion and adsorption of enzymes and organic molecules. Generally such adsorptions by clay decrease the microbial enzymatic activities in soil but simultaneously provide protection to the enzymes against predation and denaturating agents.

Table 3.8: Surface area and CEC of soil particles

Particles	Specific surface area ($m^2\,g^{-1}$)	CEC (C mol kp^{-1})
Sand	0.01- 0.1	0
Silt	1	0
Clay - Kaolinite	5-100	3-20
- Montmorillonite	700-800	100
- Vermiculite	300-500	100-150

Soil minerals and microorganisms interact with each other, mostly depending upon their size. Such interactions may involve either sorption of microbes on the surfaces of larger mineral particles or sorption of very small mineral particles on the surfaces of microbial cells or aggregate forming interaction between cells and mineral particles of similar size. Various soil microorganisms such as mycorrhizae, lichen, phosphate-solubilizing bacteria, etc. accelerate weathering of minerals through release of organic acids during normal organic matter decomposition processes; thereby enhance mineral solubilization at a faster rate than pure water. Further, inorganic acids (like sulphuric acid) produced by some chaemoautotrophs like *Thiobacillus* (S oxidizer) also help is mineral weathering and release of nutrients. The microbial activities like chelation of Fe, Mn, Cu, etc. enhance weathering of alumino-silicates through the removal of divalent cations and the solubilization of silica have been observed on newly exposed rock faces.

3.9.2 SOIL ORGANIC MATTER

Soil organic matter varies considerably in different soils from <0.5% in desert soils to 40% in peat soils and represents major source of microbial nutrition, especially for heterotrophs. The components of soil organic matter are presented in Table 3.9. The non-living components of soil organic matter makeup as much as 98% of the total soil organic matter. Heterotrophs decompose and depolymerize macromolecules into smaller units involving some specialized microbes, and depending upon the nature of substrate they secrete substrate-specific enzymes like cellulase, ligninase, chitinase, protease, etc. The softened substrate serves as nutrient base for majority of heterotrophs, which release organically bound nutrients through the process of mineralization. Various mineralization reactions like deamination, dephosphorylation, etc. are accomplished through enzymatic reactions by diverse microbes present in soil ecosystem. The nutrient mineralizing enzymes such as phosphatase, protease, sulphatases, etc. are abundant in the rhizosphere where resident microorganisms get ample root exudates and sloughed off root debris. Root exudations alone represent 5% of the total plant photo-assimilated carbon. Usually 95% or more of N and S occurs in organic form in most soil surfaces and essentially all the soil organic matter transformations that are necessary for stable and productive ecosystem development are catalyzed, at least in part, by soil microorganisms, mostly residing in upper 10 cm soil surface. The dynamic character of organic fraction contributes to a variety of essential soil processes including biogeochemical cycling, soil aggregate formation and mineral solubilization. In reality, the composition of organic matter is quite diverse and ever changing. Humus comprizes of well-defined classes of organic compounds, humic and non-humic substances, but because of their chemical complexity, heterogeneity and polydispersity their precise chemical structures are not fully defined. Soil physicochemical conditions influence the microbial habitat acting on soil organic matter. Neutral humus (organic) soils, for instance, are dominated by bacterial communities, earthworm, slugs etc. as compared to the acidic humus organic soils wherein fungal communities, collumbola, enchytraeids, mites, etc. dominate.

Table 3.9: The total soil organic carbon content in different components of soil organic matter

Components	% total soil organic carbon	Sub-components	Proportional amount of main component (%)
a) Living	2-4	i. Plant roots	5-10
		ii. Soil fauna or macroorganisms	10-30
		iii. Soil microorganisms	60-85
b) Non-living	96-98	i. Macro-organic matter (light fraction retained on 250 μm sieve	10-30
		ii. Humus (the fraction which pass through 250 μm seive)	70-90

The nature of soil organic matter is responsible for differential stimulations or variations in microbial population. In soil ecosystem it is, therefore, clay and organic matter which not only serve as sink for soil nutrients but also immobilize enzymes and organic substrates and modify the microbial communities in varied manners.

In nature, none of the above factors discussed work independently and separately but these all function in highly interactive manner. The interactions and influences on soil biological activities often occur at a microsite level. The high-tech computer modeling techniques may probably be able to describe all the possible interactions. The contribution of each of these factors can only be assessed in isolation which means field and laboratory investigations with more powerful and accurate techniques; and holistic approach need to be adopted to understand the complex soil ecosystem.

Selected References

Alexander, M. 1977. *Introduction to Soil Microbiology* (2nd edn.). John Wiley & sons, New York, USA.

Alexopoulos, C.J. and Mims, C.W. 1979. *Introductory Mycology*. Wiley & sons, New York, USA.

Atlas, R.M. and Bartha, R. 1993. *Microbial Ecology: Fundamentals and Applications* (3rd edn.). Benjamin/Cummings, Redwood City, USA.

Barak, R., Nur, I., Okon, Y. and Henis, Y. 1982. Aerotactic response of *Azospirillum brasilence. J. Bacteriol.,* 152: 643-649.

Black, C.A. 1968. *Soil Plant Relationships.* John Wiley & sons, London, UK.

Brady, N.C. and Weil, R.R. 1999. *The Nature and Properties of Soils* (12th edn.). Prentice-Hall, New Jersey, USA.

Brock, T.D. and Madigan, M.T. 1991. *Biology of Microorganisms* (6rd edn.). Prentice-Hall, New Jersey, USA.

Burns, R.G. 1986. Interaction of enzymes with soil, mineral and organic colloids. pp. 453-496. In: *Interaction of Soil Minerals with Natural Organics and Microorganisms.* (eds. P.M. Huang and M. Schnitzers), Soil Science Society of America, Madison, USA.

Burns, R.G. (ed.). 1978. *Soil Enzymes.* Academic Press, London, UK.

Burns, R.G. and Slater, J.H. (eds.).1982. *Experimental Microbial Ecology.* Blackwell Scientific Publication, Oxford, UK.

Campbell, R.E. 1977. *Microbial Ecology.* Blackwell Scientific Publications, Oxford, UK.

Curl, E.A. and Truelove, B. 1986. *The Rhizosphere.* Springer-Verlag, Berlin, Germany.

Dommergues, Y.R., Belser, L.W. and Schmidt, E.L. 1978. Limiting factors for microbial growth and activity in soil. pp. 49-104. In: *Advances in Microbial Ecology, Volume 2* (ed. M. Alexander). Plenum Press, New York, USA.

Ellwood, D.C., Hedger, J.N., Latham, M.J. Lynch, J.M. and Slater, J.H. (eds.). 1980. *Contemporary Microbial Ecology.* Academic Press, London, UK.

Foster, R.C. 1988. Microenvironment of soil microorganisms. *Biol. Fertil. Soil.* 6: 189-203.

Franz, G. 1976. The influence of altitude, precipitation and temperature in the investigated region on humus content and microbial activity in soil samples from Nepal. *Pedobiologia,* 16: 136.

Gilmour, C.M. and Allen, O.N. 1965. *Microbiology and Soil Fertility.* Oregon State Univ. Press, Corvallis, USA.

Glinski, J. and Lipiec, J. 1990. *Soil Physical Conditions and Plant Roots.* CRC Press, Florida, USA.

Greenwood, D.J. 1970. *Soil Sci.,* 21: 314.

Griffin, D.M. 1981. Water and microbial stress. pp. 91-136. In: *Advances in Microbial Ecology* (eds. M. Alexander *et al.*). Plenum, New York, USA.

Harris, R.F. 1981. Effect of water potential on microbial growth and activity. pp. 23-95. In: *Water Potential Relations in Soil Microbiology* (eds. J.F. Parr, W.R. Gardner and L.F. Elliott). Soil Society of America, Madison, Wisconsin, USA.

Harris, R.F., Chester, G. and Allen, O.N. 1966. Dynamics of soil aggregation. *Adv. Agron.*, 18: 107-169.

Hawker, L.E. and Linton, A.H. (eds.). 1979. *Microorganisms: Function, Form and Environment* (2nd edn.). Edward Arnold, London, UK.

Haung, P.M. and Schnitzer, M. (eds.) 1986. *Interaction of Soil Minerals with Natural Organics and Microorganisms.* Soil Science Society of America, Madison, Wisconsin, USA.

Killham, K. 1994. *Soil Ecology.* Cambridge Univ. Press, Cambridge, UK.

Lynch, J.M.1983. *Soil Biotechnology: Microbiological Factors in Crop Productivity.* Blackwell Scientific Publications, Oxford, UK.

Lynch, J.M. and Poole, N.J. 1979. *Microbial Ecology: A Conceptual Approach.* John Wiley & sons, New York, USA.

Marshall, K.C. 1976. *Interfaces in Microbial Ecology.* Harvard Univ. Press, Cambridge, Massachusetts, USA.

Metting , F. B. (ed.). 1993. *Soil Microbial Ecology.* Marcel Dekker Inc., New York, USA.

Papendick, R.I. and Campbell, G.S. 1981. Theory and measurement of water potential. pp. 1-22. In: *Water Potential Relations in Soil Microbiology.* (eds. J.F. Parr, W.R. Gardner and L.F. Elliott), Soil Science Society of America, Madison, Wisconsin, USA.

Parr, J.F., Gardner, W.R. and Elliott, L.E. (eds.). 1981. *Water Potential Relations in Soil Microbiology.* Soil Society of America, Madison, Wisconsin, USA.

Patrick, W.H. Jr. and Jugsujinda, A. 1992. Sequential reduction and oxidation of inorganic nitrogen, manganese and iron in flooded soils. *Soil Sci. Soc. Amer. J.*, 56: 1071-1073.

Paul, E.A. and Clark, F.E. 1989. *Soil Microbiology and Biochemistry.* Academic Press, New York, USA.

Richards, B.N. 1987. *The Microbiology of Terrestrial Ecosystems.* Longman Scientific & Technical, Harlow, England.

Russell, E.W. 1973. *Soil Conditions and Plant Growth* (10th edn.). Longmans Green & Co., London, UK.

Russell, S.R. 1977. *Plant Root System - Their Function and Interaction with the Soil.* McGraw-Hill, London, UK.

Stolp, H. 1988. *Microbial Ecology: Organisms, Habitats, Activities.* Cambridge University Press, Cambridge, UK.

Skopp, J., Jawson, M.D. and Doran, W. 1991. Steady state aerobic microbial activity as a function of soil water contents. *Soil Biol. Biochem.*, 17: 611-618.

Singer, M.J. and Munns, D.N. 1996. *Soil: An Introduction* (3rd edn.). Prentice-Hall, New Jersey, USA.

Stotsky, G. 1974. Activity, ecology and population dynamics of microorganisms in soil. pp. 57-135. In: *Microbial Ecology* (eds. A. Laskin and H. Lechevalier), CRC Press, Cleveland, Ohio, USA.

Subba Rao, N.S. 1982. *Advances in Agricultural Microbiology*. Oxford & IBH Publishing Co., New Delhi, India.

Swift, M.J., Heal, O.W. and Anderson, J.M. 1979. *Decomposition in Terrestrial Ecosystem: Studies in Ecology*, Volume 5. Blackwell, Oxford, UK.

Tate, R.L. III. 2000. *Soil Microbiology*. John Wiley & sons, New York, USA.

Tate, R.L. III. Microbial oxidation of organic matter of histosols. pp. 169-201. In: *Advances in Microbial Ecology*, Volume 4 (ed. M. Alexander). Plenum Press, New York, USA.

Tate, R.L. III.1978. *Soil Organic Matter, Biological and Ecological Effects*. John Wiley and sons, New York, USA.

Tan, K. H. 1993. *Principles of Soil Chemistry*. (2nd edn.). Marcel Dekker, New York, USA.

Waksman, S.A.1961. *Soil Microbiology*. John Wiley & sons, New York, USA.

Wood, M. 1995. *Environmental Soil Biology*. Chapman & Hall, London, UK.

Woomer, P.L. and Swift, W.J. 1994. *Biological Management of Tropical Soil Fertility*. Jhon Wiley and sons, New York, USA.

— *** —

Chapter 4

Microbial Exploration — Methods and Analysis

All difficulties are but easy when they are known

(W. Shakespeare)

The study of soil microorganisms involves various complicacies when compared with those of plants or animals aboveground. The complicated heterogeneous environment existing in soil and the varied growth behaviour of microbial populations in soil have necessitated the development of special techniques to study the microbes and their metabolic diversity, individually and holistically without disturbing their natural habitat. With the gradual improvement in microscopy since its discovery in 1667 and also with the greater understanding of the nature of soil microbes and their properties in relation to total ecosystem function, the concept of studying microbes in isolation has enormously changed. Microscopes have helped in magnifying these, otherwise invisible, minute creatures within the limits of naked eyes.

4.1 MICROSCOPY

Various types of microscopes are used in soil microbiology which vary from simple to electron/confocal microscopes. These microscopes are briefly discussed here and for detail study any text book on microbiology may be consulted.

i. **Compound microscopes:** Compound microscopes enlarge the object 1000-fold and enable the easy visualization of the objects of 0.2 μm size.

ii. **Dark field illuminations:** The dark field illuminations help to view the objects by providing a contrast (dark or bright field background). The phase microscopes have been designed to provide contrast to live specimen and detect cell inclusions under natural conditions.

iii. **Ultra violet (UV) microscopes:** The properties of certain fluorescent compounds have been exploited to view the organisms under ultra violet (UV) microscopes. The structural details of nucleic acid and proteins can be seen with the help of these microscopes. Acridine orange and auramine are used to stain bacteria directly whereas fluorescein isothiocyanate (FlTC) can be conjugated with antibodies for immunofluorscence-detection of particular organisms in any environment.

iv. **Electron microscopes:** Electron microscopes have helped in visualizing non-cultivable microorganisms as well as the finer structural inclusions in eukaryotes and prokaryotes. Under electron microscope the objects of 30°A can be viewed. In electron microscopy, 'transmission electron microscopes (TEM)' or 'scanning electron microscopes (SEM)' are often used. The specimens for such studies are first processed. For TEM, thin specimen (less than 80 nm size) is stained with heavy metal such as uranium, osmium or lead and then electrons scattered from the beam. This results in dark image on the fluorescent screen or in electron micrograph. Suitable fixatives are used for processing and staining the specimen. In SEM, the specimens are first dehydrated and then stuck on a support and vacuum coated with a metal like gold or platinum. The beams of electrons scan back and forth over the surface of specimen creating three dimensional images of specimen with magnifications of 15 to 100,000. SEM is especially useful to view fungal spores and internal cell structures.

v. **Confocal microscopy:** In confocal microscopy, the illumination emitted from a laser source is focused as a spot in one point of the specimen at a time in a sequential manner and complete image of entire specimen is obtained. This helps to visualize cellular organelles and other finer cell details as well as gives three dimensional pictures.

4.2 SOIL SAMPLING

It is very difficult to study soil environment without disturbing it. Therefore, while making the collection of a soil sample or viewing the soil for its physicochemical and biological properties *in situ* utmost care needs to be taken to avoid much disturbance. Depending upon the depth wherefrom the microorganisms are to be studied, samples are taken with the help of a soil auger for laboratory studies. In general, soil samples are collected upto a depth of 15 cm and placed in clean containers. Random sampling is essential to avoid heterogeneity in soil samples. Moreover to minimize heterogeneity in results from a field, composite samples are to be prepared by collecting soil samples between rows and within rows of a crop or across the field and mixing these sample uniformly to draw a composite sample. Comparison between different soils is complicated by difference in extraction procedures and in seasonal changes for a particular soil. If overall soil biological conditions are to be studied, special care is required to avoid sampling from hot spots. Hot spots are rich in organic matter, root molecules and growth favouring substances with heavy concentration of microorganisms. A minimum of 4-5 samples is required for each replicate study. The basic information of collection site, especially with respect to soil characteristics, is recorded. Such information may include location, topography, soil type, rainfall, soil temperature, vegetation, plant materials, specie growing, previous crops raised, soil moisture, air, pH, CEC, organic matter status, CO_2 and O_2 status, particle size, farm management practices adapted, etc. The samples need to be shifted to storage place (stored at 4°C), if the samples are not processed immediately for studies. Otherwise, samples are first thoroughly mixed and homogenous samples drawn as a representative sample of a particular replicate soil. The samples are usually ground, sieved (<0.2 μm) and plant roots, sand, etc. removed. Both axenic culture and soil sample are prerequisites to understand organism viz-a-viz resilience of the community. Interactions between soil organisms are exceedingly complex and difficult to study in isolation without creating simplistic or misleading situations.

4.3 CULTURE METHODS

Like chemical methods used for the determination of physicochemical properties of soil, microbiological methods used to analyze soil biological components do also vary. No single technique has been found able to count all types of microflora present in a soil sample because the physiological and nutritional requirements, essential for the growth of diverse microbial populations, vary. The use of various cultural methods is made to enumerate specific group or genus of microbes. Only relative number of bacteria, actinomycetes, algae, fungi, protozoa, etc. are determined that too show very little population than the actual one. For example, plate count of bacteria in general represents only 0.5-5% and at the most 50% of the population may be determined by direct microscopy. The soil serial dilution and plate count or Warcup method are widely used to have the idea about soil populations and for the isolation of a particular organism.

Serial dilution and most probable number method : Most probable number procedure is based on the determination of dilution of soil sample beyond which no propagule of the population to be quantified can be detected. Diluted soil is streaked or poured on Petri-plates containing molten nutrient agar media and incubated at a specific temperature and for a specific period. The relative preponderance of different soil microbial types and their relationships are then studied by direct microscopic examination. Population densities are determined by using prepared statistical probability tables which relate the presence of specific organisms to the number of positive samples for each dilution. In order to suppress the growth of other major groups interfering the growth on culture plates, specific antibiotics or chemicals are used. For example, to promote fungal growth, rose bengal or streptomycin or chlorotetracyclin or novobiocin or penicillin or chloramphenicol is incorporated into the culture medium to suppress bacterial growth. To suppress the growth of fungi, polyene antibiotics like nystatin is used. Similarly, specific nutrient medium or variations in physiological conditions such as temperature, pH, growth factors, etc. may be adopted to encourage the growth of specific types of microorganisms. Incubation is done as per the microbial growth requirements (1-2 days for fast growing bacteria, 2-7 days for slow growing bacteria and 7-14 days for fungi). For the isolation of thermophilic fungi, incubation need to be done at 55-65°C instead of traditional 27$\pm$1°C required for mesophilic fungi. Similarly, for the growth of acid tolerant microbes acidic conditions in culture media needs to be created.

Enrichment methods : Different media employed to support various groups of microorganisms include Kenknight and Monaier's medium or glycerol yeast agar or Jensen's medium for actinomycetes; potato dextrose or Sabouraud's agar or Czapek's Dox agar or Martin's medium for fungi; soil extract agar or nutrient agar or Thornton's medium for bacteria; malt extract agar or alkaline yeast extract peptone agar for yeasts; and Allen and Stainer medium or Pringshiem medium or Chu's medium for algae. Some microorganisms essentially require growth factors like vitamins, amino acids, etc. for growth and proliferation and do not generally grow on widely used media. Therefore, enrichment cultures are used to stimulate the growth of microbial population existing in very low densities in soil. Typically bacteria, actinomycetes, fungi and protozoa exist in soil at 10^{-8}-10^{-9}, 10^{-7}-10^{-9}, 10^{-5}-10^{-7} and 10^{-5}-10^{-6} cfu g^{-1} dry soil. Therefore, dilution plating is essential to restrict colonies on agar to 30-300 cfu per plate. Comparison of direct microscopic analysis with viable plate count shows that in most cases only 0.1-10% of observed bacteria grow in culture.

The use of selective media allows the growth of organisms with longer generation time. Soil extract agar, prepared by boiling 1.0 kg soil with water (1:1 w/v) followed by centrifugation or filtration of supernatant and solidification of the extract with 1.5% agar, yields the largest number of colonies. However, the presence of an organism does not necessarily prove its current participation in ecosystem-defining activities. Assessment of metabolic/enzyme activities and soil ecosystem properties controlling these activities is essential for better interpretations.

4.4 MICROSCOPIC EXAMINATIONS

Direct microscopic examinations are useful in determining the form and arrangement of soil microorganisms. Thin soil section technique, as used in soil macromorphological studies with normal light microscopy, has permitted observation of soil structure, root penetration and soil fauna but not soil microflora. Fluorescent stains like acridine orange, FITC, magnesium sulphonic acid, etc. are used to detect microorganisms on soil particles after focusing light on particle surface. Differential fluorescent stains (europhium chelate and a fluorescent brightener) have made it possible to demonstrate the presence of living and dead cells. Viable cells stain red and dead ones give green fluorescence. Similarly, the use of tetrazolium has been useful in locating metabolically active cells. A step-wise protocol for isolation and enumeration is given in Fig. 4.1.

Microorganisms, especially bacteria, live in micropores and within soil aggregates, so are difficult to enumerate or study. Soil crumbs need to be broken down by crushing. Russian scientists also use pedoscopes for studying the soil. These are thin, optically flat, capillary tubes which are buried inside the soil with or without added nutrient-coating inside the wall. These are left undisturbed for a specified period to allow steady state conditions to develop. Similarly, fine nylon meshes are buried in soil for a specified period and observed after staining.

***In situ* studies :** For *in situ* colonization of soil fungi and bacteria, Rossi-Cholodney buried slide method, perforated tube technique, soil plate method and fluorescent microscopy are some of the techniques which find limited use. In buried slide method the slides are buried in soil at various depths for some period and microorganisms adhere to the slide. The slides are then removed from the soil and stained to observe the qualitative nature of soil population. The addition of clear coating like cellophane makes it possible to observe the sequences of colonization of a complex substrate. The relative populations and cultural studies are made directly under microscope. However, to observe the behaviour beneath the soil surface, the method of enclosing the soil in boxes with glass slides which are kept darkened, except at the time of observation, has also been used. Soil biota are observed with the help of a horizontally mounted low power microscope and/or microphotographs taken at fixed intervals with a camera so as to monitor the changes in microflora in soil.

1g soil + 9 ml 0.015% agar solution
Shake well

↓

Transfer 1ml to a glass slide with help of a pipette

↓

Distribute sample evenly over 4 mm^2 area of the slide with a needle

↓ Dry the slide

Fix the slide in 0.1 N HCl for 1 minute

↓

Immerse the slide in water to rinse excessive acid

↓

Place slide on water bath of stain slide with rose bengal or erythrosin

↓

Examine under microscope

↓

Count the number of microorganisms per microscopic field

↓

Calculate the number of microorganisms g^{-1} soil

Fig. 4.1: Flow sheet for the isolation and enumeration of microorganisms from soil

In Jones and Mollison slide technique, a known dilution of soil particles is made after its dispersion in a wetting agent (sodium pyrophosphate or tween 80). The suspension is placed as such or mixed with agar on 1 cm^2 area of microscopic slide. The slide is stained after fixing and used for direct microscopic examination.

Use of immunofluorescence : To trace a specific microorganism in soil, the immunofluorescence technique is a powerful mean. Acridine orange is widely used in sandy soil, but clear differentiation is not achieved in clay soils. Acridine stains DNA depending upon the ratio of RNA to DNA within cytoplasm and the nature of cell wall. Living cells appear green and dead cells appear red under UV light. Fluorescein isothiocyanate (FITC), also used for bacterial identification, is absorbed to sulphydryl group of proteins. Other molecules used include chemiluminiscent moieties (such as acridinium ester that emit light when exposed to exciter substances), antigenic groups and affinity labels such as biotin, etc. Attaching a specific enzyme like phosphatase rather than FITC to the antibodies makes it more easy to use. For fungi, fluorescent water soluble aniline blue is used which is preferentially absorbed to β (1→3) glycan linkage in fungal cell. Fluorescein diacetate (FDA) stain is used to identify active hyphae. FDA is basically a non-fluorescent stain but on absorption splits into acetate and fluorescent "fluorescein" by an internal esterase. Fluorescein serves as a good indicator of hyphal tips and young hyphae. A flow sheet for the identification of microorganisms by using fluorescent technique is given in Fig. 4.2.

Specific microorganism injected into animal

↓

Antibody production

↓ FITC labelling

Conjugated antibodies

↓ Soil microorganisms incubated with antibodies and washed

Ab-Ag reaction with soil organisms

↓

Fluorescent cells identified under UV light

Fig 4.2: Flow sheet for the identification of microorganisms by using fluorescent technique.

Use of antibiotic markers : Antibiotic markers are sometimes used to recover and count inoculated organisms in soil. The inoculated soil is plated on a medium containing high concentration of antibiotic used for marking e.g. streptomycin. The marked organisms only grow on the plate.

Use of monoclonal antibody : In monoclonal antibody technology, an antibody-producing cell brings the ability to produce a particular type of antibody, while the cancer cell endows the hybrid immortality in culture. The resulting hybridoma cell, and all cells derived from it by cell division, produces large amount of a single antibody, a monoclonal antibody (MAb). Hybridoma cells can be grown in large fermentors or immobilized on solid surfaces. The antibodies are secreted into the growth medium, which allows easy harvesting of antibodies. MAbs are mainly used for diagnostic purpose to identify and quantify pathogenic organisms, metabolites, drugs, toxins, etc. For detection and quantification of antigen-antibody binding, the antibodies must be labelled with a marker molecule; usually an enzyme which catalyzes a specific reaction resulting in an easily identifiable product (enzyme linked immunosorbent assay or ELISA). The advantage of MAbs in diagnostic immunoassay is their relative speed, accuracy and specificity. These MAbs are undoubtedly superior to polyclonal antibodies. Advances in cell fusion and genetic modification have led to the development of a second generation of MAbs like bispecific MAb, chimeric Ab, antibodies with novel effector functions, single chain antibodies, etc.

4.5 ESTIMATION OF MICROBIAL DIVERSITY

Microbial diversity : The term 'microbial diversity' represents the versatility in microorganisms within and between species/microbial ecosystem. As already discussed, the estimation of total microbial number or species present in a soil ecosystem is a difficult task. However, as the soil is considered a living system so metabolic diversity of soil ecosystem can be assessed by assuming that a soil ecosystem is just-like a tissue wherein each cell works in a consortium and coordinated manner to perform a function.

Determination of metabolic pattern : To use a metabolic property in soil extract as an estimator for the changes in microbial consortium of a sample, the soil must have sufficient variability in physical, chemical and biological properties. Further, the biochemical processes occurring must be characteristic of that particular system but different from other systems. Common methods employed to determine biodiversity are based on redox system which indicates the existing capacity of microbe and enzyme communities to oxidize a large number of organic substrates. A microbial identification automated system based on substrate utilization is Biolog (Hayward, CA 94545) system. Biolog provides reproducible profile of microbial communities in a sample. The system utilizes 96 well microplates (specific for G +ve and G –ve bacteria) containing 95 substrates. Bacteria are first grown at 30°C for 24 hr on BUG agar (Biolog). The cells are then harvested and suspended in an inoculating fluid. Cell concentration in the suspension is standardized by using a spectrophotometer and cell suspension distributed individually into wells of microtitter plates containing a variety of organic substrates (>95) alongwith tetrazolium indicator. The substrate utilization is detected after 24 hr incubation by observing the changes in colour due to the oxidation of each substrate (purple colour resulting from a redox reaction of tetrazolium violet). The subsequent reduction of tetrazolium dye at various time intervals is quantified. Metabolic pattern for a particular soil ecosystem are analyzed and interpreted. The method gives metabolic fingerprint of unknown organisms. Biolog software with automated plate reader chooses the appropriate match and calculates similarity index.

Fatty acid profile estimation : The cytoplasmic membrane and/or outer membrane of soil microorganisms possess a wide range of fatty acids. For example, most bacteria contain 5-30 individual fatty acid and about 200 fatty acids have been identified in bacteria. This rich diversity in fatty acid profile has been exploited in the identification of soil organisms. Diversity of phospholipid fatty acid profile in soil samples also reveals the variations in biological diversity. Assuming that all phospholipid fatty acids in soil are equally accessible to the extractant and that phospholipid fatty acid profile of the extract is representative of total phospholipid fatty acid component of soil communities, the phospholipid fatty acid profile can show the changes in microbial community structure of whole soil sample. Soil microbial phospholipid fatty acids are extracted from soil with solvent mixture like chloroform, methanol, phosphate buffer, etc. The extracted fraction is concentrated and fatty acid separated on salicylic acid column. The polar lipid fraction (fatty acid methyl esters) is collected, saponified (alkaline hydrolysis) and methylated. The fatty acid methyl esters (FAME) are identified and quantified in GLC using known standards.

Fatty acid profile production in modern machines has gained popularity in determining the soil microbial diversity as it takes just 60-90 minutes to get the complete profile. The most widely used automated system for fatty acid analysis is the Microbial Identification System available through MIDI, Inc. (Newark, DE 19713) which quantifies 9-20 carbon chain lengths. MIDI identification machines use gas liquid chromatography of fatty acid methyl esters and are available from Hewlett Packard. A match is made of the most closely related entry in each database. A similarity index is calculated from MIDI libraries for bacteria. These machines can detect and compare the traces of fatty acid methyl esters in the computer. Similarity index is the numerical value that expresses how closely the data of a unknown matches with the mean fatty acid composition or substrate utilization pattern of a known strain. Similarity index is used to create the library entry. For instance, all *Xanthomonas* strains have several unique branched hydroxyl acids (11:0 iso 30 H and 3:0 iso 30 H) which is used as determinative test for *Xanthomonas*. The nomenclature for fatty acids commonly used with this procedure consists of a series of numbers and Greek letters. For instance, 16:ω7t in order refers to the total number of carbon atoms in the chain: the number of double bonds, the position of double bond from methylated end of the chain and the conformation. Different groups of microorganisms are represented in profile by quantification of specific fatty acids. 16:1ω5C is indicator for endomycorrhiza, 16:0:10 for nocardiform actinomycetes, 17:0:10 for actinomycetes and 8:2ω6C for fungi. The method shows overall changes in composition and groups of microbes.

4.6 BIOVOLUME AND BIOMASS ESTIMATIONS

Microbial biovolume : The microbial biovolume depicts the size (diameter and length) and number of organisms in a particular space. The calculation of bacterial number in soil, using fluorescent dyes such as fluorescein isothiocyanate, is done by using the following equation:

$$N_g = N_f.\frac{A_s}{A_m}.\frac{V_{sm}}{V}D\frac{W_w}{W_d}$$

where,

N_g = number of bacteria g^{-1} dry soil
N_f = number of bacteria $field^{-1}$
A_s = area of smear or filter (mm^2)
A_m = area of microscopic field (mm^2)
V_{sm} = volume of smear or filter (ml)
V = volume of sample (ml)
D = dilution made
W_w = wet weight of soil (g)
W_d = dry weight of soil (g)

Density of suspended cells is determined by centrifugation and sedimentation of the cells alongwith cesium chloride salt solutions of different densities. Since the cells with similar density cannot be sedimented separately, therefore, density of organisms is determined.

Microbial biomass : Microbial biomass is the living part of the soil organic matter, excluding plant roots and soil animals larger than the largest amoebae (about 5×10^3 μm^3), and represents the total weight of living microbial cells in the soil. The microbial biomass on the basis of volume and density is determined as under:

Microbial biomass = Number of cells x Density x Volume.

The average volume of spherical microorganisms in soil is estimated to be 0.71 $mm^3 g^{-1}$. Assuming a spherical bacterial cell has 1.0 μm diameter and 1.5 g cm^{-3} density; the weight of a bacterial cell works out to be 0.8 x 10^{-12} g. The ratio of biovolume to biomass has been assumed to be 0.22. Therefore, microbial biomass (M_b) will be:

$$M_b = Ng. V. D. S_c. 10^{-6}$$

where,

N_g = number of bacteria g^{-1} soil

V = average volume of bacterial cell ($\pi r^2 L$)

D = density (1.1 x 10^{-3} in liquid culture)

S_c = solid content (0.2 in liquid culture and 0.3 in soil)

$$\text{Microbial biomass carbon} = M_b \times \frac{\%C}{100}$$

where, %C is the carbon content in cell; about 45% by dry weight.

Fungal mycelia are assumed to be cylindrical (V = $\pi r^2 L$) with average diameter of 4 μm and density 1.5 g cm^{-3}. The soil fungi grown at moisture tension of -0.03 to -1.3 MPa have been found 50% heavier than those grown in laboratory shake cultures. Therefore, volume to biomass conversion of 0.33 is taken for fungi in calculations rather than 0.22.

$$\text{Fungal biomass} = \pi r^2 \cdot L. e. S_c. 10^{10}$$

where, r = radius of hyphae (1.3 μm)

L = length of hyphae (cm g^{-1} soil)

e = density (1.1 in liquid culture and 1.3 in soil)

S_c = solid content (0.2 in liquid culture; 0.25-0.35 in soil)

$$\text{Fungal biomass } C_f = \text{Fungal biomass} \times \%C$$

where C_f is the fungal carbon (μg C g^{-1} dry soil)

4.6.1 SOIL RESPIRATION MEASUREMENT

The microbial biomass can be estimated by measuring the soil respiration. Soil samples are placed in flasks and in each flask a test tube containing 1 N NaOH is placed so that the CO_2 released during microbial respiration is absorbed. The flasks are stoppered, sealed airtight and incubated at 27±1°C. To eliminate CO_2 evolution from abiotic sources, a control flask is incubated at 2°C. Flasks are opened at regular intervals so that CO_2 level in the flasks does not drop below 10% to influence microbial growth or create anaerobic conditions. During substrate decomposition carbohydrate, etc. are converted to $CO_{2,}$ which is absorbed by NaOH.

$$2NaOH + CO_2 \text{ -------------}\rightarrow Na_2CO_3 + H_2O$$

After the addition of $BaCl_{2,}$ which precipitates CO_2 as $BaCO_{3,}$ the excessive NaOH is back titrated with acid. The amount of CO_2 evolved is calculated using the relationship that one milli-equivalent of NaOH used absorbs 22 mg CO_2 or 6 mg C.

Differentiation of fungal and bacterial respiration in soil can be made by the amendment in soil sample with a population-specific antibiotic. For example, streptomycin and cycloheximide are generally used to inhibit bacterial and fungal populations, respectively. The CO_2 evolved in antibiotic supplemented samples is estimated. The utilization of glucose in soil respiration measurement provides a good estimate of the biomass of the organisms capable of oxidizing glucose as an energy source.

4.6.2 MICROBIAL BIOMASS ESTIMATION BY CHEMICAL METHODS

Measurement of mixed microbial population in soil ecosystem can be done by estimating the total microbial carbon content present in a soil. Soil is first subject to chloroform ($CHCl_3$) fumigation, which kills and destroys most of the microbial cells and releases the cytoplasmic contents into the soil. In chloroform-fumigation-extraction method, the soil is first extracted immediately after fumigation and the amount of organic C and N in the extract is determined. The amount of amino-N in the extract may also be measured by using ninhydrin reaction and accordingly the amount of biomass N in the soil may be estimated. The amount of cytoplasmic C, N, P or S released is determined from the soil extract by appropriate conventional method and related to the amount of living microbial biomass originally present in the soil. For biomass C and N estimation by fumigation-incubation method, the chloroform-fumigated soils are inoculated with a small amount of fresh unfumigated soil (having living populations) and subsequently incubated at 25$\pm$2°C for 10 days. The colonizing microbial population decomposes freshly killed microbial biomass and yields CO_2 and NH_4^+. The amount of C or N mineralized (CO_2 and NH_4^+ released) over a period of 10 days is used to estimate the amount of C in original microbial biomass. An unfumigated control is used to estimate the decomposition of non-living soil organic matter by colonizing population. The relationship between the amount of CO_2 or NH_2^+ released and biomass in a particular soil is determined separately with radioactive ^{14}C or ^{15}N labeled cells added to soil and then subject to $CHCl_3$-fumigation.

K_c and K_N: A conversion constant of carbon and nitrogen (k_C or k_N) for a range of microorganisms has been worked out. The k_c for fungi and bacteria is 0.41. The k_c represents the amount of microbial C evolved as CO_2. Therefore, microbial biomass carbon (B_c) or microbial biomass nitrogen (B_n) can be calculated as:

$$B_c = \frac{Fc}{kc}$$

$$B_n = \frac{Fn}{kc}$$

where B_c is biomass (mg C 100 g^{-1} dry soil); B_n is the biomass N; F_c or F_n is the CO_2 or NH_4^+ produced by fumigated soil in 10 days incubation minus the CO_2 or NH_4^+ produced by unfumigated soil and K_c represents the %biomass C mineralized to CO_2 (K_c of 0.41 is assumed for most of the soils).

Since the amount of N mineralized after $CHCl_3$-fumigation and incubation varies with the C:N ratio of the organism, therefore to take into account the remobilization of N after fumigation, the K_n can be calculated by equation:

$$K_n = 0.8 \left(\frac{C_f}{N_f}\right)^{-0.43}$$

where K_n is the %N mineralized after fumigation; C_f the flush of CO_2-C produced and N_f the flush of NH_4^+-N released.

Direct determination of C and N extracted after $CHCl_3$ treatment somehow overcomes the problem of remobilization of N on 10 days incubation. The amount of phosphorus held in soil microorganisms can be calculated from the difference between the amount of inorganic P extracted by 0.5 M sodium bicarbonate (pH, 8.5) from fresh fumigated and unfumigated soils. The chloroform-fumigation-extraction method is simple than direct microscopy and permits the use of ^{14}C or ^{15}N in chosen substrate (respirometry).

4.6.3 MEASUREMENT OF ATP

The metabolic reactions in cells involve participation of adenosine triphosphate (ATP). The ATP can donate PO_4-groups to other metabolic intermediates to convert them to activated form. In soil ATP does not persist in a free state. Therefore, it is the most ideal material for determining the life activity in a soil. If larger organisms are somehow excluded from a soil and it is presumed that all microorganisms contain same proportion of ATP in their cells; then the amount of ATP in soil can be related to the amount of biomass. ATP is extracted from a soil using paraquat and trichloro-acetic acid, and measured in a soil by using luciferin. The substrate luciferin (pure chemical or from fire-fly) can react with ATP and luciferase enzyme in presence of magnesium to give an enzyme-luciferin-AMP intermediate. This in presence of O_2 splits to liberate AMP, inorganic P and light. There is a very sharp emission of light during first 15 seconds of the reaction, followed by fairly steady state reaction for 1-3 minutes. The light generated is measured by a spectrophotometer or a scintillation counter and plotted against ATP content form a standard curve. A correlation factor is applied in extraction of ATP from soil. Data obtained can be used for biomass C estimation. In resting state, the ATP: cell nutrient ratio is as under:

ATP: C: N: P: S = 1: 250: 40: 9: 2.6

ATP measurement gives idea about both the microbial biomass and activity. The valve of ATP:C of 160 has often been used because during microbial growth C: ATP ratio varies from 40: 1 to 1000: 1. However, the principle does not hold true for the soils which may have freshly received large quantities of phosphorus-rich organic matter. The other methods of biomass estimation include measurement of muramic acid, phospholipids, chitin, nucleic acid etc. in soil.

4.7 SOIL ENZYMES AND THEIR MEASUREMENT

In soil a series of reactions are catalyzed by enzymes which are basically proteins acting as organic catalysts in biological systems. Enzymes are capable of acting independently and modify or activate chemical reactions, even though being in a very small concentration. Soil enzymes can be regarded as the sum total of all enzymes possessed by soil influencing the biochemical processes in it.

Types of soil enzymes : Microorganisms produce a variety of enzymes in soil. The types of enzymes produced vary with the environmental conditions. The biochemical versatility of soil organisms provides soil with the capacity to degrade all natural and most synthetic compounds which enter the soil deliberately or accidentally. Some bacteria or fungi produce substrate-specific enzymes like cellulase, phenolase, esterase, etc. only when a particular substrate is present. These enzymes are known as 'adaptive or induced enzymes'. Some enzymes such as dehydrogenase, urease, etc. are always present in cells and are produced without the influence of any substrate. These are called 'constitutive enzymes'. The enzymes primarily produced inside a cell are endoenzymes (intracellular). The endoenzymes are responsible for the release of energy for microbial activities and processes involved in the synthesis of various cell constituents. These are usually incapable of functioning or even surviving outside the intact living cell. Some enzymes are produced inside the cell but excreted through cell wall. These are called exoenzyme (extracellular enzymes). These function outside the cell and once released diffuse into the immediate vicinity of cell or remain bound to the exterior of cell. The extracellular enzymes are released into the surrounding soil matrix. These enzymes are capable of functioning independently and continue to do so even after the death of organism, which synthesized it. The extracellular enzymes catalyze depolmerization of high molecular weight compounds to simpler one. The accumulation of extracellular enzyme in soil depends upon the mechanism of stabilization of these molecules. Enzymes being proteins are themselves liable to degradation of proteases and may be adsorbed to clay mineral surface or retained between clay lamellae. Enzymes associated with organic matter are more resistant to degradation, perhaps due to the changes in their viscosity or H-bonding. The association of an enzyme with a solid surface such as clay particle may subsequently lead to the changes in substrate affinity due to unfolding or coagulation of protein chains or physical blocking of active sites. Alternatively, enzymes may become aligned in a manner which enhances their activities. More than fifty enzymes are active in soil, involved in various biochemical transformations. The important classes of enzymes active in soil include oxidoreductases, transferases and hydrolyases. The various reactions catalyzed by some of these enzymes is shown in Table 4.1.

Table 4.1: Some soil enzymes and the reactions these catalyze

Enzyme	Reactions catalyzed
Oxidoreductases	
i. Catalases	$2H_2O_2 \longrightarrow 2H_2O + O_2$
ii. Catechol oxidase (tyrosinase)	o-diphenol + $1/2\ O_2 \longrightarrow$ o-quinone + H_2O
iii. Dehydrogenase	$XH_2 + Y \longrightarrow X + YH_2$
iv. p-diphenol oxidase (laccase)	p-diphenol +$1/2\ O_2 \longrightarrow$ p-quinone + H_2O
v. Glucose oxidase	Glucose + $O_2 \longrightarrow$ gluconic acid + H_2O_2
vi. Peroxidase and polyphenol oxidase	A + $H_2O_2 \longrightarrow$ oxidized A + H_2O
vii. Urate oxidase (uricase)	Uric acid + $O_2 \longrightarrow$ allantoin CO_2
viii. Aldehyde oxidase	Indole 3-aldehyde + $O_2 \longrightarrow$ indole-3-carboxylate
Transferases	
i. Transaminase	$R_1R_2CH\text{-}N^+H_3 + R_3R_4CO \rightarrow R_3R_4CH\text{-}N^+H_3 + R_1R_2CO$
ii. Transglycosylase and levansucrase	$nC_{12}H_{22}O_{11} + ROH \rightarrow H(C_6H_{10}0_5)_n OR. + nC_6H_2O_6$
Hydrolyases	
i. Acetyl esterase	Acetic ester + $H_2O \longrightarrow$ alcohol + acetic acid
ii. α-and β-amylase	Hydrolysis of β (1$\rightarrow$4) glucosidic bonds
iii. Asparginase	Asparagine + $H_2O \longrightarrow$ aspartate + NH_3

Contd...

Table 4.1. Contd...

iv.	Cellulase	Hydrolysis of β (1→ 4) glucan bonds
v.	Deamidase	Carboxylic acid amide+H_2O → carboxylic acid +NH_3
vi.	α-and β-galactosidase	Galactoside + H_2O ⟶ ROH + glucose
vii.	α-and β-glucosidase	Galactoside + H_2O ⟶ ROH + glucose
ix.	Inulase	Hydrolysis of β (1→ 2) fructan bonds
x.	Lipase	Triglyceride + H_2O ⟶ glycerol + 3 fatty acids
xi.	Lichenase	Hydrolysis of β (1→ 3) cellutriose bonds
xii.	Metaphosphatase	Metaphosphate ⟶ orthophosphate
xiii.	Nucleotidase (nucleases)	Dephosphorylation of nucleotides
xiv.	Phytase	Inositol hexaphosphate + 6 H_2O → inositol + 6 phosphate
xv.	Protease	Proteins ⟶ peptides and amino acids
xvi.	Pyrophosphatase	Pyrophosphate + H_2O ⟶ 2 orthophosphate
xvii.	Sulphatases	Sulphate ester + H_2O ⟶ ROH + sulphate
xviii.	Urease	Urea ⟶ 2 NH_3 + CO_2
xix.	Phosphatase	Phosphate ester + H_2O ⟶ ROH + phosphate

Estimation of enzyme activity : Soil enzyme activities and microbial number have good correlation. However, most enzyme activities are independent of microbial growth because soil treatments with toluene, sodium azide, gamma rays, antibiotics, etc. usually destroy microorganisms as well as unbound extracellular enzymes without preventing enzyme activities. Further, soil enzyme assay is a complicated phenomenon and, in addition to cell lysis, soil factors especially sorption by soil colloids may affect the outcome. Soil enzyme activity is often measured under standard but artificial conditions. For example, a buffer suspension at pH 7.0 with either neutral or synthetic substrate may be provided in excess but not the activity which actually occurs in soil. The synthesis of requisite enzyme for previously unexpressed metabolic activities is induced primarily by the changes in the physical or chemical properties of soil. For instance, development of anoxic conditions in metabolizable C and nitrate rich environment stimulates denitrification process. Enzyme induction provides a mechanism for rapid short term response of microbial community to the changes in soil ecosystem. Within 30 to 60 minutes new enzyme activities are expressed on stimulation. The enzyme induction is, therefore, considered as rapid biochemical response for the maintenance of ecosystem equilibrium. Most soil enzymes are trapped in soil organic and inorganic colloids. Therefore, soils possess a good amount of extracellular enzymes, which are not often associated with the microbial biomass present at a particular time and site. The evaluation and interpretation of enzyme activities in a natural ecosystem makes it imperative to have a knowledge of enzyme adsorption to isolated soil colloids and the influence that adsorption will have on enzyme-substrate relationship.

Dehydrogenase activity has been measured by several workers and is considered a convenient laboratory technique for large scale comparative studies. About 2-5 g of fresh soil samples are treated with 0.5 ml of 3% 2,3,5-triphenyl tetrazolium chloride (TTC) dye and 2.5 ml of distilled water. After thorough shaking the mixture is incubated at 37°C for 24 hr. In absence of $O_{2,}$ TTC (a colourless dye) serves as the terminal acceptor for hydrogen evolved, and is reduced to triphenyl formazon (TPF) by microbial action. The TPF (extracted with methanol) of soil filtrate in reduced state is red and is measured at 485 nm with the help of a spectrophotometer. Similarly alkaline phosphatase activity can be assayed by using p-nitrophenyl phosphate dye. The p-nitrophnol released g^{-1} soil h^{-1} is measured.

4.8 MOLECULAR METHODS IN SOIL MICROBIOLOGY

The conventional methods of soil dilution, culturing, direct microscopy, chemical extractions and enzyme assays may not always provide the true picture of microbial diversity in soil. This is because some recalcitrant microorganisms do not grow even after prolonged incubation and the methods adopted only provide ideas about the specific groups or specific activities of the microbes. Recent molecular methods have helped to overcome the limitations and provided good understanding in evolutionary links, classification and quantification through *in situ* analysis.

4.8.1 DNA PROBE UTILIZATION

The methods for extraction, purification and characterization of soil DNA have been developed and are being improved day by day. DNA probes, which can identify specific genetic codes in cell, have successfully been used in organisms cultured from soil. DNA probes are short segments of DNA or RNA (often <500 base) that have a reported group (radioactive or fluorescent or enzyme labeled) attached and can bind with high specificity to complementary sequence of target DNA. Probes can be complementary to either DNA or RNA and their sizes vary from 20 to many thousand bases. Oligonucleotide probes (20-50 bases) can be chemically synthesized in any nucleotide sequence desired. The main advantages of DNA probes in diagnostics are their high specificity, sensitivity and speed.

The nucleic acid (DNA and RNA) from soils and microflora, after removal of interfering compounds, is extracted by chemical and physical methods. These treatments include lysozyme, sodium dodecyl sulphate (SDS), proteinase, guanidine thiocyanate, EDTA, acetone, heat, sonication, bead mill homogenization, microwave heating, mill-grinding, pronase, etc. DNA of more than 12 mg g^{-1} dry soil has been extracted by using SDS and bead-mill homogenization. This DNA can be used in molecular analyses.

4.8.2 DNA HYBRIDIZATION

DNA hybridization assays are aimed at detecting the presence of unique nucleic acid sequences in the genome of an organism or in DNA molecule of interest. These nucleic acid sequences represent the target DNA molecules which can be segments of DNA or entire molecule in a chromosome or a plasmid or even a mRNA or rRNA. In DNA hybridization, extracted DNA is first denatured with alkali or at higher temperature to separate the strands. The single strands of DNA are then attached to a solid support or immobilized on a nitrocellulose membrane in salt solution and then allowed to renature with complimentary or labelled DNA fragments. The percent hybridization or relatedness is measured spectrophotometrically. Labelling of DNA with radioactive ^{32}P, ^{35}S, etc. can also be done and labelled strands from unlabelled ones identified by selective adsorption to hydroxyapetite followed by column chromatographic separation. Single stranded DNA may also be degraded using specific nucleases and labelled DNA measured by a scintillation counter. Hybridization experiments have revealed that soils generally may contain $4x10^4$ species of microorganisms per gram of soil. DNA hybridization is useful in screening microbial isolates from soil for specific genotype strain or for specific gene sequence in DNA sample extracted directly from soil. Evaluation of unique DNA sequences of organism or strain such as nitrogen fixing one allows the estimation of their population densities in soil.

4.8.3 POLYMERAZE CHAIN REACTION

Polymeraze chain reaction (PCR) is used to amplify a specific DNA sequence, without the necessity of cloning them, with the help of a thermostable DNA polymeraze enzyme such as *Taq, BstE, Pfu, Tthi, UlTma,* etc. and specific primers (15-20 base oligonucleotide, complementary to target DNA). With the help of PCR any particular stretch of genetic material can be pinpointed and replicated several times simply by selecting a pair of primers that flank the desired stretch of DNA. Chain reaction involves three-step process of denaturation, annealing and extension carried out at 94, 37-60 and 72°C, respectively. The double stranded DNA is first denatured by heating to 94°C for a shorter period (15-30 seconds) and then primers added which anneal with single stranded DNA. The thermostable DNA polymerase enzyme initiates synthesis of complementary nucleotide chain at known primer sequence. This process is repeated or automated in a thermocycler and products of discrete length (amplicons) are formed exponentially to million-fold after 20 cycles and billion-fold after 30 cycles. The DNA fragment can be detected by staining with ethidium bromide an electrophoretic gel of the PCR products prepared by Southern blotter method or even by probes. The flow sheet diagram of PCR technique is presented in Fig. 4.3. PCR amplification enables the detection of even small number of cells (as few as 3 cells g^{-1} soil) or specific microbes in soil environment. However, humic acids are to be removed from soil samples as they inhibit PCR reaction.

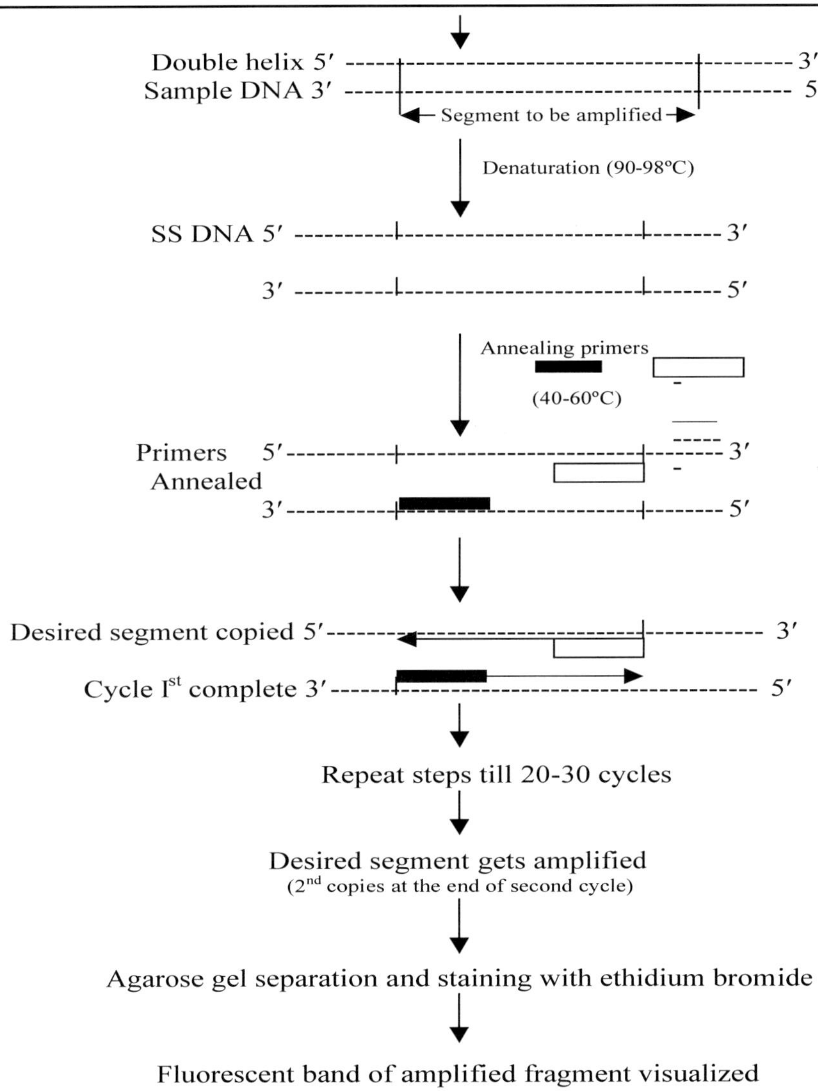

Fig. 4.3: Flow diagram of PCR technique

In another method DNA is cleaved with specific enzymes into various fragments. A desired nucleotide sequence (adapter) is ligated to these fragments and amplified through PCR reaction with primers complementary to adapters. Optimal band patterns are obtained by using specific restriction enzymes or primers. The related strains are identified by this way.

4.8.4 OTHER MOLECULAR TECHNIQUES

The large fragments (10 kb to 1.5 mb) of chromosomal DNA are analyzed by pulse-field gel electrophoresis (PFGE) by using restriction endonuclease enzymes and observing the digestion pattern of DNA fragments. The 16S ribosomal RNA (rRNA) having 1500 bases has been exploited to detect and identify new species in soil. 16S rRNA has certain nucleotide sequences specific for major groups of bacteria and others specific for species or genus. From 16S RNA complementary DNA is synthesized and amplified through PCR. The amplified DNA is cloned in *E. coli* and then sequenced and compared with available 16S rRNA in gene bank. The method has helped soil microbiologists to identify the bacteria difficult to culture and those very low in number in soil.

A variety of techniques have been developed for the characterization of DNA extracted from soil microbial community. Diversities can be estimated by assuming that the similarities in DNA sequences greater than 70% represent individual species. However, the presence of resting and inactive cells does not give clear picture about the active microbes in soil. The preparation of mRNA, only synthesized by active cells, from soil can provide idea about active microbes present. Specific enzyme activity expression in soil through reporter gene use (like *Lac Z* gene coding for β-galactosidases) may provide idea about the diversity of soil biological genome.

Selected References

Alef, K., Beck, T., Zelles, L. and Kleiner, D. 1988. A comparison of methods to estimate microbial biomass and N-mineralization in agricultural and grassland soils. *Soil Biol. Biochem.*, 20: 561-566.

Alef, K. and Kleiner, D. 1989. Rapid and sensitive determination of microbial activity in soil and in aggregate by dimethyl sulfoxide reduction. *Soil Biol. Fertil.*, 8: 349-355.

Anderson, J.P.E. and Domsch, K.H. 1978. Mineralization of bacteria and fungi in chloroform-fumigated soils. *Soil Biol. Biochem.*, 10: 207-213.

Anderson, J.P.E. and Domsch, K.H. 1975. Measurement of bacterial and fungal contributions to respiration of selected agricultural and forest soils. *Canadian J. Microbiol.*, 21: 311-322.

Bohlool, B.B. and Schmidt, E.L. 1980. The immunofluorescence approach in microbial ecology. *Adv. Microbiol. Ecol.*, 4: 203-241.

Brookes, R.E., Powlson, D.S. and Jenkinson, D.S. 1982. Measurement of microbial biomass phosphorus in soil. *Soil Biol. Biochem.*, 14: 319-329.

Burns, R.G. (ed.). 1978. *Soil Enzymes*. Academic Press, London, UK.

Cholodny, N. 1930. Uber eine neue methode zur Untersuchung de Bodenmikroflora. *Arch. Microbiol.*, 1: 620-652.

Findley, R.H., King, G.M. and Walting, L. 1989. Efficiency of phospholipids analysis in determining microbial biomass in sediments. *Appl. Environ. Microbiol.* 55: 2888-2893.

Halsted, R.L. and McKercher, R. B. 1975. Biochemistry and cycling of phosphorus. In: *Soil Biochemistry* (eds. E.A. Paul and A.D. McLaren). Marcel Dekker, New York, USA.

Jenkinson, D.S., Davidson, S.A. and Powlson, D.S. 1979. Adenosine triphosphate and microbial biomass in soil. *Soil Biol. Biochem.*, 11: 521-527.

Jenkinson, D.J. and Ladd, J.N. 1985. Microbial biomass in soil: Measurement and turnover. pp. 415-471. In: *Soil Biochemistry, Volume 5* (eds. E.A. Paul and J.N. Ladd). Marcel Dekker, New York, USA.

King, J.D. and White, D.C. 1977. Muramic acid as a measure of microbial biomass in estuarins and marine samples. *Appl. Environ. Microbiol.*, 33: 777-783.

Ladd, J.N. 1978. Origin and range of enzymes in soil. pp 51-96. In: *Soil Enzymes* (ed. R.G. Burns). Academic Press, London, UK.

Lynch, J.M.1983. *Soil Biotechnology: Microbiological Factors in Crop Productivity*. Blackwell Scientific Publication, Oxford, UK.

McLaren, A.D. and Skujiin, J. 1971. *Soil Biochemistry*. Marcel Dekker, New York, USA.

MIDI. 1995. *Skerlock Microbial Identification System Operating Manual, Version 5*. MIDI, Newark, DE., USA.

Paul, E.A. and Clark, F.E. 1989. *Soil Microbiology and Biochemistry*. Academic Press, New York, USA.

Paul, E.A. and Johnson, R.L. 1977. Microscopic counting and adenosine 5′ triphosphate measurement in determining microbial growth in soils. *Appl.Environ.Microbiol.*, 34: 263-269.

Paul, E.A. and Ladd, J.N. (eds.) 1981. *Soil Biochemistry, Volume 5*, Marcel Dekker, USA.

Russel, J.R. 1962. *Soil Conditions and Plant Growth*. Longmans Green & Co., London, UK.

Singh, A. and Gopinathan, K.P. 1998. Confocal microscopy: A powerful technique for biological research. *Current Sciences*, 74:841-851.

Smith, R.F. 1990. *Microscopy and Photomicrography*. CRC Press, Boca Raton, Florida, USA.

Sparling, G.P. and West, A.W. 1988. A direct estimation method to estimate soil microbial C calibration in situ using microbial respiration and ^{14}C labeled cells. *Soil Biol. Biochem.*, 20: 337-343.

Subba Rao, N.S. 1982. *Advances in Agricultural Microbiology*. Oxford & IBH Publishing Co., New Delhi, India.

Tabatabai, M.A. 1994. Soil enzymes. pp. 775-833. In: *Methods of Soil Analysis, Part 2: Microbiological and Biochemical Properties* (ed. R.W. Weaver). Soil Science Society of America, Madison, USA.

Thornton, J.I., Crim, D. and McLaren, A.D. 1975. Enzymatic characterization of soil evidence. *J. Forensic Sci.*, 20: 674-692.

Vance, E.D., Brookes, P.C. and Jenkinson, D.S. 1987. An extraction method for measuring soil microbial biomass C. *Soil Boil. Biochem.*, 19: 703-707.

Waksman, S.A. 1957. *Soil Microbiology*. John Wiley & sons, New York, USA.

Warcup, J.H. 1950. The soil-plate method for isolation of fungi from soil. *Nature*, 6: 117- 118.

Weaver, R.W. *et al.* 1994. *Methods of soil Analysis Part 2: Microbiological and Biochemical Properties*. Soil Science Society of America, Madison, USA.

West, A.W., Sparling, G.P., Speir, T.W. and Wood, J.M. 1988. Comparison of microbial C, N flush and ATP, and certain enzyme activities of different textures soils subject to gradual drying. *Aust. J. Soil Res.*, 26: 217-229.

Verstraete, W., van De werf, H., Kucnirowicz, K. and Illaiwi, M. 1983. Specific measurement of soil microbial ATP. *Soil Biol. Biochem.*, 15: 391-396.

— *** —

Chapter 5

Carbon Cycle and Soil Organic Matter Transformation

Organic matter constitutes the heart of soil.

(Wood)

Soil microorganisms in true sense are one of natural safety valves that regulate biogeochemical transformations in soil ecosystem. They act as natural agents for the disposal of organic matter. The various essential biological elements such as C, N, S and P, apart from H and O, form the primary building blocks of cells and their constituent elements are recycled through tiny microorganisms present in soil. More they are in quality and quantity, better recycling is expected by them. The organically bound materials are converted to simpler inorganic forms to complete biological cycle initiated during photosynthesis and other anabolic processes.

5.1 CARBON CYCLE AND GLOBAL CHANGES

Global carbon cycles are closely tied to biological productivity and soil organic matter turnover (Fig. 5.1). For steady state equilibrium, the production of CO_2 through decomposition and respiration should release an equal amount of carbon back to the atmosphere. However, the conservative estimates suggest that the rate of CO_2 release from soils is over two-fold than the rate of carbon input. The rapid degradation of organic matter due to the burning of fossil fuels, forest fires and decline in soil organic matter through forest felling, cultivation of virgin prairies and intensive cultivation have contributed to the increase in atmospheric CO_2 to the level sufficient to induce climatic changes. Annual CO_2 release from soil varies between 80 to 550 g C m^{-2}. This act of draining the soils of huge organic reserves has lead to an increase in atmospheric CO_2 estimated worldwide at 150-180 M t $annum^{-1}$. The present estimates of global C reservoirs indicate the presence of about 360 ppm CO_2 in atmosphere (0.036% by volume), which somewhat 150 years back was just 250 ppm. The annual increases of 0.5% in CO_2, 1.1% in CH_4 and 3.0% in chlorofluorocarbon are presently believed to have contributed to a rise in global temperature by about 0.5-1.0°C in past 100 years. Atmospheric carbon is dynamic but relatively small in global pool. The estimates indicate that 50% greenhouse effect or global warming is due to increase in atmospheric CO_2. The atmosphere at presents has 760 billion metric tonnes (= 760 x 10^9 Mg) of carbon. Almost 590 billion metric tonne carbon is locked up in dead organic matter. The soil humus and fossil fuel contribute to about 1,500 and 10,000 billion metric tonnes of global carbon, respectively. But much of the increase in atmospheric level comes from a net loss of organic matter from soil system.

The soil carbon pool is approximately three times the size of atmospheric pool. The carbon present in living plants naturally undergoes interchange with atmospheric carbon and the soil serves as a very dynamic part of this global carbon cycle. 'Carbon fixation', a key biological transformation process in soil, is a pivotal event in biosphere. Also, the other cycles in nature are dependent on the annual input of carbon into the soil from above- and below-ground sources. The annual carbon input from these sources is estimated to be 37.5 x 10^9 Mg. Autotrophs or primary producers on land and sea use solar energy, through chlorophyll, to fix CO_2 and water, and form more complex carbohydrates with the release of oxygen to the atmosphere.

$$nCO_2 + n\,H_2O \xrightarrow{\text{Reduced energy}} (CH_2O)_n + n\,O_2$$

5.2 PRIMARY PRODUCTION AND CARBON TURNOVER

Primary production is an essential link in soil carbon cycle as it is the main flux from the atmosphere to soil biota. By definition, 'primary production' is the rate at which energy is stored by photosynthetic and chemosynthetic activities of producer organisms in the form of organic substances. 'Net primary production' (NPP) is, therefore, the rate of organic carbon production by autotrophs in excess of respiration losses.

NPP = Total photosynthetic production - Biotic respiration

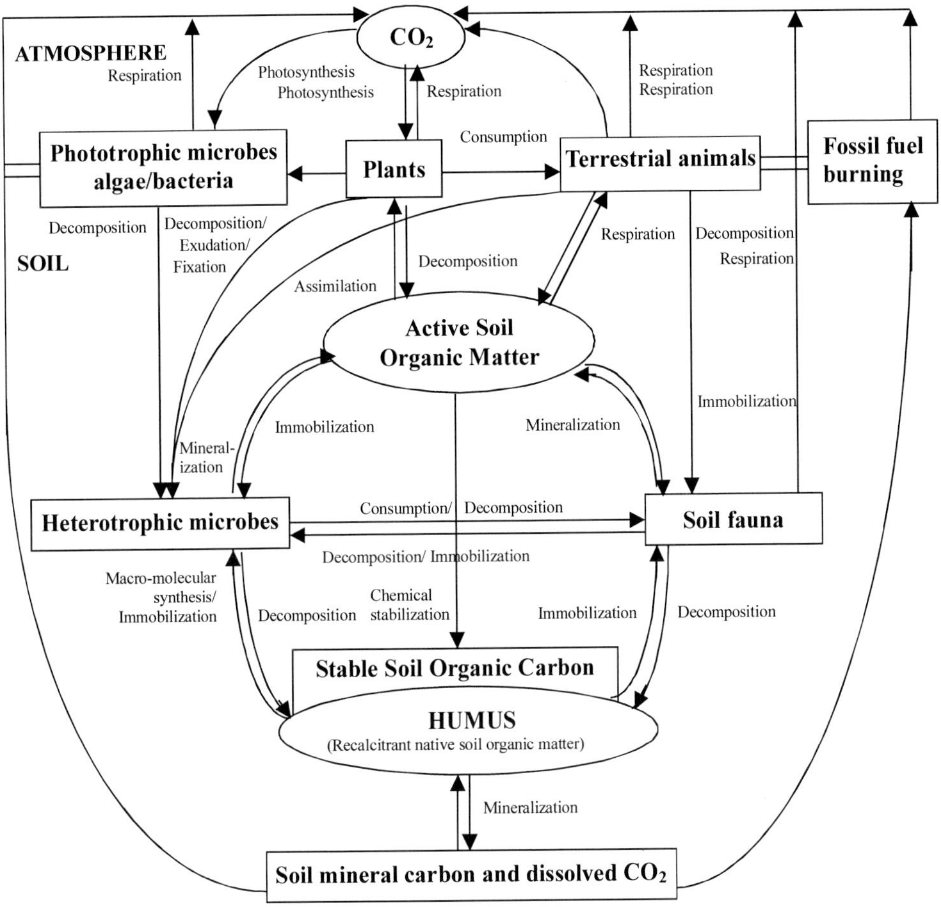

Fig. 5.1: The carbon cycle in nature

The annual net primary productivity on land in terms of carbon is 63 billion metric tonnes; more than double the annual net primary productivity in oceans (30 billion metric tonnes of carbon). Plants, photoautotrophic cyanobacteria, photo- and chemo-autotrophic bacteria and eukaryotic algae contribute to CO_2 fixation. As per the conservative estimates, of the 150 billion metric tonnes of carbohydrates generated annually through photosynthetic activities on the earth, about 80-90% is produced by algae alone. Thus they virtually contribute to the formation of all oxygen present in the atmosphere. However, on terrestrial ecosystem the major contributors are plants followed by algae. The net primary production of organic carbon by photosynthesis is about 55 x 10^9 tonnes, most of which eventually decomposes in soil. The conservative estimates indicate that primary productivity input of approximately 3540 kg C ha^{-1} yr^{-1} is made to the upper soil surface (5 cm) in arable soil. Of this, root decomposition, root exudates, straw residues and microbial biomass contribute 400, 240, 2800 and 100 kg C ha^{-1} yr^{-1}, respectively. This indicates that leaf litter contributes a lot to the above-ground organic matter production. Assuming that all this carbon is decomposed in soil, the global mean turnover time of soil organic matter works out to be approximately 27 years. The decomposition of primary production is assumed mostly to occur in top 0-50 cm soil layer. The decomposition of primary production above or below this layer leads to the longer turnover time. Further, the net primary production greatly varies from one ecosystem to another. The highest rate of annual primary production occurs in moist tropical forests (>8 t C ha^{-1} yr^{-1}or up to 3500 g m^{-2}), followed by cold temperate beach forests (Table 5.1). High turnover in tropical rain forests is because of high temperature, high humidity and action of termites which accelerate the process of decomposition. However, the losses also are quite high in tropical forests as depicted by less turn-over time of 9 years in comparison to 30–37 years turnover time in case of temperate forests, humid savannah or native prairie. The accelerated losses contribute to the rise in CO_2 level in atmosphere, the phenomenon well known as 'greenhouse effect'.

Table 5.1: Distribution of carbon and its turnover in soils of different ecosystems

Ecosystems	**NPP of C (t ha^{-1} yr^{-1})**	**Carbon entering soil (t ha^{-1} yr^{-1})**	**Organic C in soil (t ha^{-1})**
Cultivated	2.0-6.0	1.5	30
Continuous wheat (unamended)	2.6	1.2	2.6
Continuous wheat with NPK	5.1	1.9	30
Continuous hay	2.7-3.2	2.0-2.5	77
Native prairie	2.8	1.7	52
Humid savannah	5.0-8.0	1.5-2.0	56
Sub-humid savannah	1.4	0.5	17
Moist tropical forests	9.0-10.0	4.9	44
Cold temperate beach forests	7.1	2.4	72
Temperate grasslands	3.0-6.0	0.4-0.7	150-190
Tropical rain forests	11.0	11.0	80
Deciduous temperate boreal forests	12.0	12.0	135
Deserts	0.05	0.01	1

Source: Whittaker (1975); Whittaker and Likens (1975); Jenkinson and Ladd (1981)

Some part of fixed C is utilized by primary producers itself for energy release, some part consumed by animals and some released as CO_2 during respiration. The assimilated carbon enters soil directly either in the form of plant litter or dead microorganisms or sloughed-off roots or exudates or it may enter indirectly via body wastes or body tissues of primary and secondary consumers. The ultimate end product of aerobic decomposition of organic carbon in soil is CO_2.

$$(CH_2O)_n + n\,O_2 \longrightarrow n\,CO_2 + n\,H_2O$$

About 85, 75 and 67% of the carbon present in above-ground plant residues, manures and root residues, respectively, is lost in the form of CO_2. The remaining organic carbon is assimilated into soil as humus. Small grains and perennial plants approximately return 0.3-1.0 and 1.7-2.5 t C ha^{-1} each year, respectively, in the form of roots. Litter contributes a lot to the above-ground organic matter production. Once deposited in or on the soil, organic material is metabolized by soil microbial biomass, which eventually returns this carbon to the atmosphere. Varying types of microbial metabolites produced interact with lignin and form stable humus complexes, which remain in soil for centuries. However, in productive ecosystems the turnover of organic C in soil is rapid. The turnover of organic C in grassland soil lies between 30 to 40 years for 0-30 cm soil depth. Whereas, the turnover time of organic C is usually less in cultivated soils (16 to 22 years) than undisturbed soils like grasslands or forest soils (30 to 39 years).

5.3 FORMS OF ORGANIC CARBON

Composition of organic carbon entering the soil : The largest fraction of all the organic carbon entering the soil is that contributed by plant residues. Plants, in general, contain 15-60% cellulose, 10-30% hemicellulose, 5-30% lignin and 2-15% proteins (Table 5.2). In addition to these, water soluble substances such as simple sugars, amino sugars, organic acids, amino acids and ether soluble materials (lipids, wax, resins, pigments, etc.) constitute a 5-30% of dry weight. The content of proteins and water-soluble compounds decreases and more resistant fractions like cellulose, hemicellulose and lignin increase as the plant ages. The dead plants and animals are primary input for whole soil population. The bodies of soil population form secondary input of substrate in soil. Most abundant organic substrate in soil is cellulose. Rough estimates reveal that in an average soil the annual decomposition is approximately 3×10^3 t cellulosic carbon.

Table 5.2: Composition of plant tissues, soil organic matter and humus

Components (%)	Plant tissues	Active soil organic matter (Stable organic matter)	Humus
Cellulose	15-60	2-10	4
Hemicellulose	10-30	0-2	7
Lignin	5-30	35-50	45
Proteins and amino acid	2-15	28-35	35
Fats, pigments, waxm etc.	1-8	1-8	3
Phenolic glycosides & tannins, etc.	<1	1-2	6

Classification of organic carbon on solubility basis : On the basis of solubility, approximately half of the organic carbon in soil is in aromatic form. Although soil organic C exists in a great variety of forms (Table 5.3), it can be broadly classified into:

i) Insoluble organic C
ii) Soluble organic C
iii) Biomass C

Insoluble carbon comprizes of about 90% of the total soil organic carbon and includes main components of plant cell wall (cellulose and lignin), animal exoskeletons and fungal cell walls (chitin) and soil humus. The soluble carbon serves as immediate substrate for microbial decomposition in soil and comprizes largely of plant root exudates, microflora and faunal exudates, enzymatically degraded constituents, etc. The soluble organic carbon is easy assimilable substrate in soil and comprizes of <1% of soil organic C. The biomass in soil, excluding primary producers, comprizes about 1-2%, and mainly consists of soil microflora and micro- and meso-fauna. The soil microbial activity (decomposition) initiates the turnover of soil organic C. All organic matter passes through microbial pool before being redistributed in nature. Biomass carbon pool is, therefore, considered to be the driving force for nutrient cycles.

Table 5.3: Forms of soil organic carbon and their approximate relative amounts

Forms of carbon	Organic carbon (%)
Aromatic carbon	50
Carbohydrates	15
N-associated carbon	20
Fatty acids and alkene carbon	15

5.3.1 CARBOHYDRATES

Carbohydrates are polyhydroxy aldehydes, ketones or substances that yield one of these compounds on hydrolysis and have empirical formulae of $(CH_2O)_n$. These can be grouped into monosaccharides, oligosaccharides and polysaccharides depending upon the number of simple sugar units present i.e. 1, 2-6 or more than 6, respectively. Monosaccharides are simple sugars such as glucose, fructose, galactose, mannose, xylose, etc. These compounds serve as energy substrates for most of the soil microorganisms. Sucrose, lactose, cellobiose, maltose, etc. are the examples of disaccharides. The polysaccharides may be homopolysaccharides having a single kind of monosaccharide (such as cellulose, starch, glycan) or heteropolysaccharide having 2 or more kinds of monosacharride (such as hyaluronic acid). The properties of carbohydrates change significantly with increasing molecular complexity (length, linkage, branching, etc.). Carbohydrates account 10-20% of soil carbon. Originally carbohydrates enter the soil through debris but most of these are thought to be either the decomposition products or products of microbial synthesis *in situ*. These remain protected from microbial attack because of their interactions with other soil constituents such as clay, metal cations, etc. The carbohydrates present in soil include simple sugars, starch, cellulose, hemicellulose, lignin, pectic substances, etc. The carbohydrates especially polysaccharides act with lignin and amino acids and contribute to the formation of humic substances in soil. The soil carbohydrates influence the physical conditions, cation exchange reactions, anion retention, metal chelation and metabolic activities of soil. Further, these also encourage the aggregate formation with consequent formation of granular to crumb structure in soil. Oxidative reduction of soil polysaccharides results in 30-90% reduction in the stability of soil aggregates.

5.3.1.1 Starch

Starch serves as the major food reserve of plants and is actually admixture of two components i.e. amylose (20%) and amylopectin (80%). Both are long chain polymers having repeated units of glucose. In amylose the glucose molecules are linked together by $\alpha(1\rightarrow4)$ glucosidic bonds, whereas in amylopectin, besides $\alpha(1\rightarrow4)$ linkage, side chains of glucose are attached by $\alpha(1\rightarrow6)$ glucosidic bonds. Starch is more readily broken down than cellulose by soil microorganisms.

$$C_6H_{12}O_6 + 6O_2 \xrightarrow{\text{Anaerobic}} 6CO_2 + 6H_2O + \text{energy}$$

$$C_6H_{12}O_6 \xrightarrow{\text{Anaerobic}} 2CH_3CH_2OH + 2CO_2 + \text{energy}$$

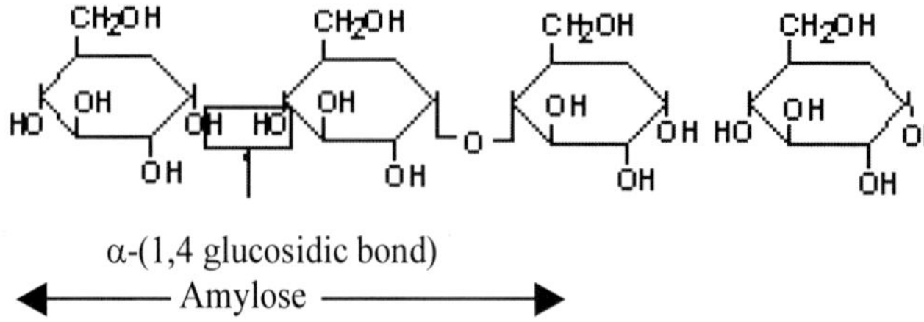

Monomer of starch

Many bacteria and fungi hydrolyze starch by their extracellular enzymes viz., α and β amylases. α-amylase reduces both amylose and amylopectin whereas β-amylase reduces amylose to maltose. Amylopectin is broken down to a mixture of maltose and dextrins. The enzyme α-glucosidase reduces maltose to glucose.

$$\text{Starch} \xrightarrow{\text{Amylase}} \text{Maltose} \xrightarrow{\alpha\text{-glucosidase}} \text{Glucose}$$

5.3.1.2 Cellulose

Cellulose is the most abundant organic matter in soil. It is often associated with hemicellulose and lignin. Cellulose is water insoluble semi-crystalline homopolysaccharide with its molecular weight varying from 2×10^5 to 2×10^6. It consists of glucose molecules with $\beta(1 \rightarrow 4)$ linkage. The decomposition of cellulose is catalyzed by an adaptive enzyme 'cellulase'. The enzyme acts in two distinct stages. First, there is loss of crystalline structure of cellulose, also known as 'conditioning', and then depolymerization. Thus cellulase splits long chain cellulose into short chain glucose units of cellotriose (3 glucose units) and cellobiose (2 glucose units). These, in turn, may yield single glucose units by hydrolytic activities of cellotriase or cellobiase enzymes, respectively. The cellulose decomposition, therefore, involves specialized microbial enzyme system but subsequent release of simpler units serve as substrate for all heterotrophic soil microbes, which possess glycosidase enzymes. The rate-limiting step in cellulosic carbon turnover in soil is the rate of depolymerization, not the breakdown of subunits. Diverse group of bacteria (e.g. *Bacillus, Cellulomonas, Chromobacterium, Clostridium, Corynebacterium, Cytophaga, Achromobacter, Pseudomonas, Vibrio, Cellulomonas, Nocardia, Microbispora, Micromonosporium, Micromonospora, Streptosporangium, Streptomyces, Sporocytophaga,* etc.) and fungi (e.g. *Aspergillus, Alternaria, Coprinus, Fusarium, Penicillium, Phanerochaete, Polyporus, Trichoderma, Chaetomium, Curvularia, Zygorrhynchus,* etc.) possess cellulolytic properties. Through their extracellular enzymes, cellulases, cellulose may be depoylmerized into endo $\beta(1\rightarrow4)$ gluconases, exo $\beta(1\rightarrow4)$ gluconases and cellobiose. Some anaerobic bacteria and thermophilic fungi do also have active cellulose degrading potential under certain environmental conditions. The physical comminutation or mechanical disruption of plant litter by soil animals provides greater surface area for microbial action, especially in the disruption of lignin and hemicellulose-protected cellulose. Similarly, the gut animals like earthworms possess cellulose, which help them in the decomposition of cellulose materials.

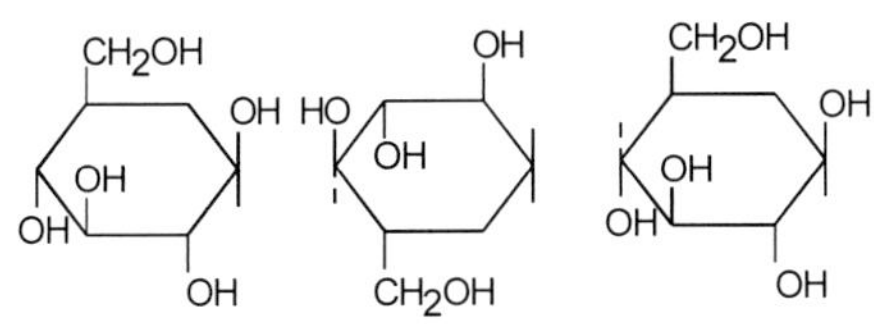

Monomers of cellulose

5.3.1.3 Hemicellulose

Hemicelluloses are complex water soluble polysaccharides (polymers of hexoses, pentoses and uronic acids). The commonly occurring monomers include xylose, mannose, glucose, galactose, arabinose, glucouronic acids and galacturonic acids. The hemicellulose pectin is a polymer of galacturonic acid with a molecular weight of about $4x10^5$ and is a key component of the middle lamella region of plant cell wall. Hemicelluloses are easily decomposed when in pure state. However, in nature they frequently exist complexed with other substances which render their breakdown more difficult. The decomposition of hemicellulose pectin involves several enzymes, collectively known as 'pectinases', which may be repressed by high sugar levels in fresh substrates. This repression is a key factor in determining the nature of microbial succession in plant tissue degradation. The fungi appear to initiate the decomposition activity but actinomycetes maintain the attack over a more prolonged period. Soil bacteria having pectinolytic activities include *Arthrobacter, Pseudomonas, Bacillus, Streptomyces,* etc. Pectinases are found in mutualistic symbionts, such as *Rhizobium* and mycorrhizas, as well as in pathogenic soil microbes.

5.3.2 PECTIC SUBSTANCES

Pectic polymers constitute a substantial proportion (about 35%) of primary wall in dicotyledonous plants but comparatively very less proportion (8-9%) occur in monocotyledonous plants. Pectic substances are polysaccharides which mostly consist of galactouranan molecules, interspersed with a much smaller number of rhaminose molecules, small chains of galacturonan and many other sugars. Several pectinases or pectolytic enzymes degrade various types of pectic substances (Fig. 5.2).

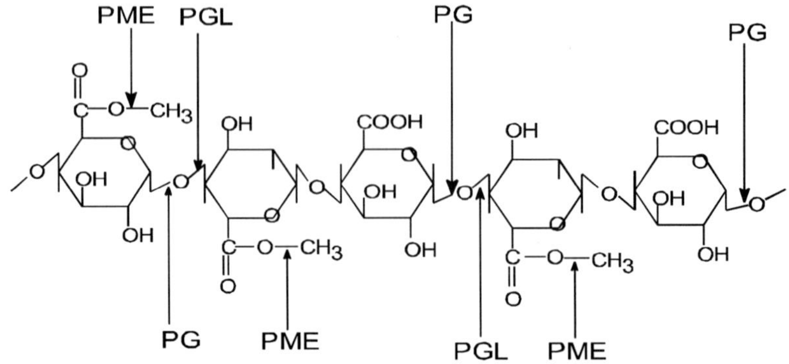

protopectinase: Protopectin → Pectin; pectin methyl esterase (PME): Pectin → Pectic acid; polygalcturonase: Pectic acid → Galacturonic acid

Fig. 5.2: Various pectinase enzymes cause cleavage of $\alpha(1\rightarrow4)$ polygalacturonide chain (site of action depicted by arrows) in pectin by endopolygalacturonase (PG), endopoly-galacturonide lyase (PGL) and demethoxylation by pectin methylesterase (PME)

Pectin methyl esterases remove small branches of the pectin chains, alter their solubility and soften the molecules by the action of chain splitting pectinases (polygalacturonases, pectin lyases or transeliminases, etc.). Three types of pectic substances, found in plant tissues, especially in cell wall are pectin, protopectin and pectic acid. Each of these requires a specific enzyme for its hydrolysis. Soil bacteria such as *Bacillus, Clostridium, Pseudomonas,* etc. and facultative pathogens such as *Erwinia, Streptomyces,* etc. secrete several pectinase enzymes.

5.3.3 LIGNIN

Lignin is the most resistant part of plant residues entering the soil carbon pool. Lignin is a three dimensional amorphous and highly branched molecules. It comprizes of complex non-uniform polymer of aromatic ring with basic building unit of phenyl-propane (C_6-C_3) type structure [coniferyl, coumaryl or sinapyl alcohol or guaiacyl propane monomers] (Fig. 5.3). The molecular weight of lignin varies from 10^4 to 10^6.

The lignin material does not show any specific order. Derived from an aromatic ring called phenyl propanoid (phenyl alanine or phenyl alanine and tyrosine), lignin is synthesized via shikimic acid pathway. Different plant residues contain lignin with different functional group characteristics. The functional groups may be -OH or -OCH_3 (methoxyl) groups. For instance, hardwood lignin consists of coniferyl and sinapyl building blocks in 1:1 ratio. Moss and grass have lignin of mainly coumaryl type while soft wood has

80% coniferyl precursors. Lignin is formed as an encrusting material on the cellulose and hemicellulose matrix. The random disorderly cross-linked structure of lignin makes it more resistant to microbial attack. Lignin thereby protects cellulose and hemicellulose from the enzymatic attack. The decomposition of lignin involves a three-phased process:

i. Esterification of exposed methoxyl groups
ii. Depolymerization
iii. Splitting of phenolic ring after side chain removal.

Grass | Hard wood (dicotyl angiosperm) | Soft wood (gymnosperm)

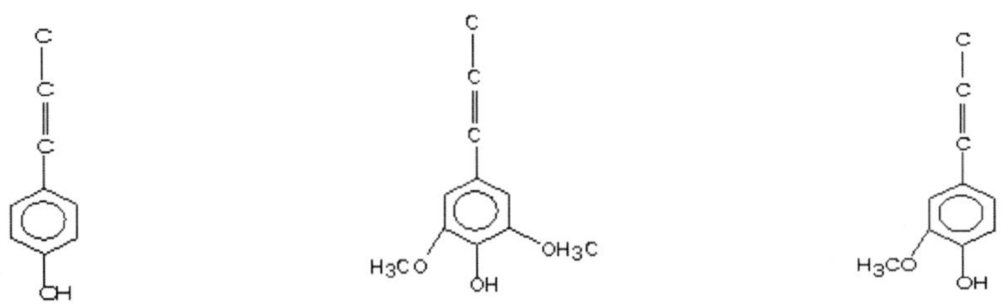

4-hydroxyphenyl propane (coumaryl) | 3,5 methoxy 4-hydroxyl phenyl propane (sinapyl) | 4-hydroxy 3 methoxyl phenyl propane (coniferyl)

Fig 5.3: Chemical structure of basic phenyl propanoid precursors of lignin from grass, softwood and hardwood dicotyledonous angiosperms.

Only a few soil microbes have the ability to degrade the entire lignin structure. These mostly belong to fungal group Basidiomycotina and are commonly known as 'white rot fungi'. Ligninolytic white rot fungi include *Coriolus versicolour (Polyporus), Pleurotus ostreatus* and *Phanaerochaete chrysosporium*. No organism has so far been isolated with the ability to use lignin as sole energy source. The white rot fungi perhaps degrade lignin only in presence of some readily degradable substrates (as the primary energy source) and adequate oxygen. Presence of high nitrogen inhibits lignin degradation. The lignin degradation by white rot fungi is strictly a secondary metabolism. The lignolytic fungi possess wide range of special enzymes viz. esterases, phenolases, peroxidases, laccase, tyrosinase, etc. The enzymatic breakdown of lignin may result in the formation simpler aromatic compounds such as syringaldehydes, vanillin, p-hydroxy-benzaldehyde, ferulic acid, caffeic acid, etc. The final cleavage may involve other fungi and bacteria to yield low molecular weight organic acids, CO_2, methane etc. Brown rot fungi such as *Gloeophyllum, Poria* etc. degrade polysaccharides associated with lignin and cleave methoxyl and R-side chains leaving phenol behind. This phenol on oxidation turns brown. Similarly, soft rot fungi such as *Chaetomium, Preussia,* etc. degrade hardwood lignin under wet conditions.

The turnover of aromatics in nature is a major factor for determining soil organic matter dynamics. The portion of lignin entering soil does not undergo complete degradation but reacts with microbial products in the formation of soil humus or humic matter.

5.3.4 PROTEINS AND OTHER NITROGENOUS COMPOUNDS

Of the total soil nitrogen, at least one-third remains structurally unidentified. Amino acids and amino sugars comprize 20% of soil carbon and 30-45% of soil nitrogen (Table 5.4). Amino acids, the fundamental structural units of protein, contribute much towards soil nitrogen mineralization than amino sugars. Chitin is major contributor of amino sugars in soils.

Table 5.4: Distribution and forms of soil nitrogen

Components	Percentage		
Amino sugars	5-15		
Amino acids	30-45	- Acidic acids (Asp. & Glu)	- 15-25%;
		- Basic acids (Asp, His, Lys & Orn.)	- 20-25%
		- Neutral acids	- 50-55%
		- Sulphur containing amino acids	- 2- 3%
Unidentified acids			
- Soluble organic N	10-20		
- Unhydrolyzed acid	20-35		

5.3.4.1 Proteins

Proteins are the most abundant nitrogen containing constituent of plant and animal residues. These are readily attacked by proteolytic enzymes (such as proteinases and peptidases) which hydrolyze the peptide linkages and ultimately release aliphatic, aromatic and heterocyclic amino acids. Aliphatic amino acids have straight or linear carbon side chains. Aromatic amino acids have ring shaped C side chain whereas heterocyclic amino acids have at least one ring containing both C and N. The amino acids are utilized by a variety of soil microbes which produce ammonia, carbon dioxide, various organic acids and alcohols.

$$\text{Glutamine} + NAD^+ + H_2O \longrightarrow \alpha\text{-ketoglutarate} + NH_4^+ + NADH + H^+$$

The mineralized nitrogen compounds are either incorporated into aggregates/adsorbed by soil clays/organic matter or remain in free solution. Basic amino acids with more than one positively charged amino groups readily react with other organic constituents (like reducing sugars, quinones, etc.) than acid or neutral amino acids. Most amino acids in soil are thought to exist as peptide chains. These peptide chains remain stabilized against enzyme action by linkage to the fractions of soil organic matter or by reactions with clay or by the reactions between amino acid containing organic matter and polyvalent metal cations such as Fe, Al, etc. Under anaerobic conditions various amino acids and mercaptans are produced thereby give an offensive odour. Fibrous proteins like keratin are resistant to microbial actions although most actinomycetes and some fungi like *Penicillium* eventually are able degrade them. The nitrogen incorporated in organic constituents of soil persists for longer periods in soil ecosystem.

5.3.4.2 Chitin

Chitin is the second most abundant organic compound found in nature. It has long chain polysaccharides with amino sugar as basic unit and empirical formula of $(C_6H_9O_4.NHCOCH_3)_n$. It is present in fungal and insect cell walls as well as in several plants and animals. Chitin comprizes of repeated subunits of N acetyl glucosamine joined through $\beta(1\rightarrow4)$ linkage (Fig. 5.4). The acetyl glucosamine structure of chitin contains nitrogen and its degradation is limited by environmental factors as well as by the materials associated with it. The cell wall of bacteria have two amino sugar derivatives viz., N acetyl glucosamine and N acetyl muramic acid having peptide cross linkages, often composed of L-alanine, D-glutamic acid, mesodiaminopimelic acid and D-alanine. Microbially produced organic nitrogen is stabilized in soil than individual amino acids and proteins from plant residues. This is because of relatively slow turnover of microbial biomass and its stabilization within soil aggregates and association with resistant organics. Soil microbes such as *Streptomyces, Nocardia, Micromonospora, Aspergillus, Penicillium, Trichoderma* and *Thaminidium* possess chitinase enzyme which reduce chitin to diacetyl chitobiose. The enzyme chitobiase further degrades diacetyl chitobiose to acetyl glucosamine, ammonia, acetic acid and CO_2.

5.3.4.3 Nucleic Acids

All living or dead biotic materials reaching the soil possess nucleic acids which are either deoxyribonucleic acids (DNA) or ribonucleic acids (RNA). Nucleic acids make up 5-15% of dry weight of all cells and are involved in the transmission of genetic information and translation of proteins. Nucleic acids

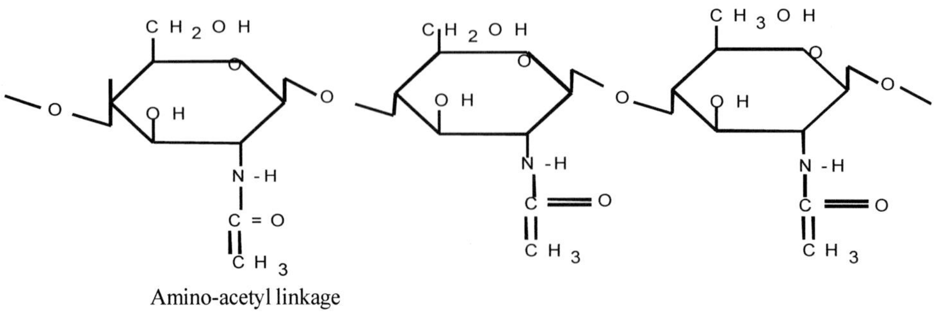

$$\text{Chitin} \xrightarrow{\text{chitinase}} \text{N-acetyl glucosamine} \xrightarrow{\text{kinase}} \text{Glucosamine glucose-PO}_4 \xrightarrow{\text{deaminase}} \text{Gulcose-6-PO}_4 + \text{HN}_4^+$$

Fig. 5.4: Structure of chitin and the reaction involved in its degradation

have polymer chains of nucleotides as their building block, which are linked to each other via 3′→5′ phosphodiester bonds. The nucleotides are phosphorylated derivatives of nucleosides which, in turn, have nitrogen containing bases, pentose sugar (ribose or deoxyribose). The bases are purines (adenine and guanine) or pyrimidine (cytosine and thymine or uracil). In most cells, nucleic acids are conjugated with proteins, so form nucleoproteins. Nucleic acids are degraded by a number of organisms possessing nuclease and other enzymes, thereby split nucleic acid into sugar, phosphate and nucleoside units.

5.3.5 LIPIDS AND OTHER RELATED CONSTITUENTS

Lipids are heterogeneous compounds of fatty acids, waxes, resins and oils. Lipids have limited solubility in water and exhibit hydrophobic characters but are soluble in organic solvents. Many plant and animal lipids are associated with proteins and carbohydrates such as glycolipids, lecithins, etc. Oils and fats of plant and animal origin which reach the soil are hydrolyzed to glycerol ($C_3H_8O_3$) and fatty acids. Glycerol is readily oxidized to CO_2 and water whereas fatty acids produce certain resistant and toxic substances. The transformation of fatty acids under anaerobic conditions is a process believed to contribute to the formation of petroleum in nature. Other complex carbon compounds such as gum, inulin, etc. are selectively attacked by soil fungi and bacteria through their exo-enzymes and degraded into smaller units and then mineralized.

5.4 INCORPORATION OF ORGANIC MATTER IN SOIL

The organic matter which enters the soil primarily comprizes of plant materials, animal residues and dead microorganisms or their products. Most of the net primary production is decomposed in upper 50 cm layer of soil. The approximate annual input of primary productivity to surface (0-5 cm) of an arable soil is about 3550 kg C ha^{-1} with approximately 250 kg C ha^{-1} contributed by microbial biomass. Litter contributes the bulk of the aboveground organic matter production (more than 90%). When carbon from organic residue or microbial source is subject to decomposition in soil, it either is lost in the form of CO_2 or incorporated into cell constituents or it enters into relatively stable and physically/ chemically protected carbon pool. Physical protection depends upon the structural characteristics of soil. The physically protected carbon is accessible to microbial decomposition only when soil is physically or mechanically disrupted. The chemically protected carbon pool is often referred as 'humus'.

5.4.1 DECOMPOSITION PROCESS

The decomposition of organic matter involves complex interactions of microbes and soil animals. The initial breakdown and comminutation of large objects is carried out by soil animals (termites, earthworm, etc.). This is followed by oxidation of organic carbon by saprophytes for their metabolic activities. Decay proceeds as long as oxygen, water, temperature, nutrients, etc. are adequate for decomposers. For example, in desert soils absence of water greatly reduces the oxidation rate of organic matter in soil. During decomposition three processes proceed simultaneously:

- Mineralization
- Increase in microbial biomass and immobilization
- Liberation and/or accumulation of end products

When fresh organic matter is added to soil, the number of microorganisms developing on these substrates is small owing to the resistance offered by the fresh residue. Moreover, at the time of organic matter addition the principal active metabolizing soil microorganisms are small proportion of autocthonous microbes. But as the substrate is softened or conditioned by specific enzymatic action, there is appearance of easily decomposable compounds such as sugars, starch, cellulose, etc., which stimulate increased metabolic activity among the microbes. This causes an abrupt rise in microbial population, especially those of heterotrophs (zymogens) that usually remain dormant in the absence of readily available substrates. Microbial growth and nutrient uptake leads to the immobilization, the process just opposite to mineralization. The microbial biomass at this point may account for as much as one-sixth of the organic matter in soil. During decomposition enormous heat is generated. The auto-heating in soil plays vital role is organic matter decomposition. Decomposition is optimum at 37°C but declines on either side of optimal level because of inactivation of microbial population. However, decomposition increases at 60°C because of chemical oxidation.

Priming effect : The peak of microbial activity during organic matter decomposition often stimulates the breakdown of even the original resistant soil organic matter. This phenomenon is known as 'priming effect'. Newly added materials may sometimes retard the decomposition of organic matter present in soil. Therefore, the effect of fresh addition (priming action) may be positive or negative. Such effects may be either because of direct activation of dormant microorganisms or indirect alteration in microbial environment, the pH change, O_2 deficiency, nutritional limitation, etc.

Succession of microorganisms during decomposition : Classically, organic matter decomposition leads to the development of microorganisms in succession i.e. primary saprophytic colonization is by bacteria and sugar fungi (moulds) followed by cellulolytic and lignolytic fungi which depict comparatively slow growth rate. Also, non-successional microbial populations build up which are interdependent and form communities to derive mutual benefits from each other. The secondary colonizers depend on the metabolized or lytic products of primary species to sustain growth. The communities seem to be stable but there exist changes in population size of component organisms and residues. Decomposition products of secondary colonizers provide further substrate for primary colonizers so there is variation in population. The changes observed in C:N ratio, immobilization and mineralization during organic matter decomposition are depicted in Fig. 5.5.

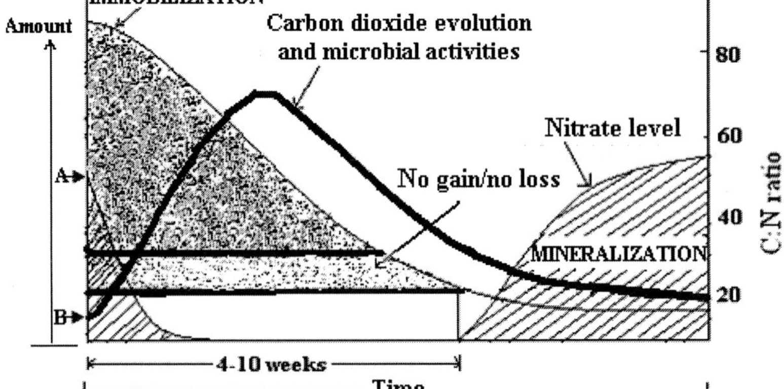

Fig. 5.5: Changes during process of organic matter decomposition; (A = Original soil nitrate level; B = Original soil carbon dioxide level).

Half life of fresh organic matter decomposition : The approximate half-life of fresh organic matter in temperate regions is about three to four months but only 3-4 weeks in humid tropics. Studies on labelled cellulose decomposition in soil revealed that more than half of the cellulosic carbon is evolved within first 30 days of decomposition. At the same time, labelled carbon derived from cellulose appeared as amino acid carbon presumably through activities of microbial population. The second stage of organic matter decomposition is slow and has half-life of 5.2 and 1.6 years in temperate and humid tropical regions, respectively. Lignin is slowly degraded because the nature and amount of lignin affects the decomposition process. In initial stages of decomposition heterotrophic bacteria and fungi dominate and cause breakdown of organic matter. Other bacteria, actinomycetes and fungi capable of attacking intermediary products become active at later stage, utilize acids and CO_2 etc. and synthesize various carbon compounds. These oxidize S, NH_4^+, etc. to obtain energy for their build-up processes. The final stage of decomposition is the gradual loss of more resistant plant materials, like lignin, and dominance of more resistant humic materials as major carbon fraction in which actinomycetes and fungi play important role. However, in general carbon becomes increasingly resistant to decomposition with time.

5.4.2 PRODUCTS OF DECOMPOSITION

The decomposition process results in the production of labile and stable fractions of organic matter.

i. **Labile soil organic matter :** It is unstable and vulnerable to changes so is readily degraded to microbial products. It comprizes of 10-20% of soil organic matter and serves as an important sink of nutrients to plants. The labile organic matter tends to disappear fast under favourable soil conditions.

ii. **Stable soil organic matter :** It consists of resistant compounds in decomposing residues and microbial products and the compounds formed as a result of either interaction of organic compounds with each other or with the mineral components of soil, especially clay. The stable organic matter is equivalent to clay in behaviour as the nutrients contained in it are recycled very slowly. Radioactive studies have revealed that certain stable organic matter fractions are commonly more than a few hundred to thousands of years old. The addition of relatively small amounts of easily degradable organic matter such as manure, organic wastes, plant residues, sewage sludge, etc. may contribute more to the available nutrient supply than slow release of nutrients from much larger amounts of stable organic matter. Stable organic matter has significant role in soil structure formation and water absorption than nutrient release.

Major organic matter fractions in soil : On the basis of turnover time organic components can be categorized into three major fractions:

- Decomposing plant residues and associated biomass with turnover period of a few years;
- Microbial metabolites and cell wall constituents that become stabilized in soil with half life of 5-25 years; and
- The resistant fraction of 250-2500 years or more in age.

Despite the stability of resistant fraction of soil organic matter, 50-80% of freshly added organic matter is lost in most temperate soils during the first year of its application. Thus plant or animal residues must be added to soil continuously if favourable effects of organic matter on soil properties are to be maintained. However, increase in the soil organic matter content through organic matter additions is a difficult task because the rate of decomposition of soil organic matter is proportional to the addition rate i.e. the more we add, more rapidly it disappears.

Soon after the first year of decomposition, all residual C is either conserved in microbial (dormant) pool or humified. Farm yard manure appears to be more resistant to decomposition than straw because its initial rapid phase of decomposition is almost over by the time it is applied to soil. It is reported that aromatic plant constituents (polyphenols, lignin, etc.) are not essential for the formation of resistant form of organic matter in soil. The actual amount of stabilized carbon left when a given compound decomposes in soil depends on the initial pathway of decomposition in the early stage of decomposition. Biomass data on decomposition of litter reveals complex interactions of microbes and soil animals in soil. Most soil animals are the consumers of primary decomposers. Studies on estimation of biomass in deciduous woodland soil at Meathop, Lancashire (brown earth with a mull humus) revealed annual litter production of 764 kg ha^{-1} (on dry weight basis) and total soil animal and microbial biomass of 497.5 kg ha^{-1}. The individual groups in terms of kg ha^{-1} were bacteria 7.3, actinomycetes 12.0, fungi 454, protozoa 1.0, earthworm 12.0, enchytraeidae 4.0, molluscs 5.0, acari (mites) 1.0, collumbola 2.0, diptera 3.0 and other arthropods 6.0. Thus microbial biomass was more than 90% in soil.

5.5 FACTORS AFFECTING THE SOIL ORGANIC MATTER DECOMPOSITION

The factors that govern microbial activities also influence organic matter decomposition in soil. These include temperature, moisture, pH, oxygen availability and C:N ratio of substrate, lignin content, soil type, clay content, management practices etc. The amount of biomass and metabolite formation during the first phase of decomposition is greatly influenced by soil texture. Clay soils retain more organic matter than sandy soil. Moreover, clay protects the organic matter through adsorption of readily available substrates so makes them less accessible to microbes as well as increases the longivity of soil organisms. Tillage, weathering and drying enhances CO_2 evolution by breaking up of soil aggregates thereby release encapsulated organic matter.

Soils from warmer climate contain less organic matter than those from cooler areas. Obvious reason for this variation is fast decomposition process in tropical area (4 times than temperate areas). Temperature is an important factor governing the accumulation of mats in hill pastures. Similarly, decomposition is optimum in soils near field capacity. A pH of 6-8 is most appropriate for microbial action. Of course, adequate aeration boosts aerobic growth. Microbial decomposition proceeds most rapidly under plenty of O_2. Under low O_2 or in its absence, anaerobic or facultative microorganisms become dominant and decomposition proceeds at much more slow rate, that is why large organic matter accumulates in partially decomposed form such as organic acids, alcohols, methane gas, etc. Some of the products of anaerobic decomposition, such as methane gas, are major environmental concern and contribute towards greenhouse effect.

C:N ratio - A vital factor in organic matter decomposition : Organic matter content in plants vary widely with respect to their C:N ratio. Wheat straw, for instance, has <1% nitrogen and legumes have 2-5% nitrogen. The C:N ratio in soil and in the organic residues applied to soil, apart from other factors, is of paramount importance in determining the rate of decomposition because intense competition for available N between microorganisms occurs when a residue of wide C:N ratio is added to a soil. The soil microorganisms have certain dietary requirement for their growth and maintenance. An optimum C:N ratio in the range of 20 to 25 (1.4-1.7% N) seems to be ideal for maximum decomposition. At this ratio no immediate release of mineral nitrogen occurs from residues over and above the amount required for microbial synthesis. In other words, a favorable soil environment is created to bring about equilibrium between immobilization and mineralization processes. This critical balance may be upset if C:N ratio of organic matter added is less than

25 because mineralization will dominate immobilization leading to the accumulation of NH_4^+ and NO_3^-. When the C:N ratio of various microbial tissues, plant residues and soil organic matter was compared, it was observed that bacteria usually have C:N ratio of 5-6:1 and fungi 10:1, plants 15:1 to 100:1 and soil 15:1 to 20:1. So microbial action narrows down the C:N ratio of plant residues as a result of CO_2 release during initial decomposition. However, decomposition may be delayed or lessened if enough nitrogen is not available to microbes in soil. Incorporation of plant residues having a wide C:N ratio will cause immobilization, unless and until adequate nitrogen is supplemented with it. Suppose oat straw (C:N ratio 80:1) is added @ 5.0 t ha^{-1} to a soil, it means the addition of 2,000 kg C (assuming 40% C in the straw) and 25 kg N. If a soil has enough nitrogen and it is assumed that microbes use 75% of the carbon input for energy and rest (25% C) for growth and synthesis purpose, then the resulting soil microbes will contain 500 kg C (i.e. 25% of plant C x 2000 kg). The resultant microbial population consuming the whole straw will require 62.5 kg N (the average microbial C:N ratio is 8:1). This means 62.5 - 25.0 = 37.5 kg of N is to be immobilized by microbes from soil.

If a soil does not contains sufficient available N, the decomposition process will be limited to the formation of soil microbial population containing 25 kg of N (i.e. the total available N in plant residues) only or consuming only 200 kg of plant C (8:1 C:N ratio). In other words, only 200 kg C (used for population build-up) + 600 kg C (used for energy) = 800 kg C will be utilized. Thus the rest plant residues i.e. 1200 kg C shall have to wait for next cycle of decomposition i.e. when microbes die and their bodies decompose. Thus under such a situation there will be immobilization for a longer period. The wide C: N ratio also holds some C that would otherwise be mineralized rapidly. The fixation of atmospheric nitrogen by some bacteria provides an alternative strategy for utilizing the substrates with a wide C:N ratio.

Now instead of oat straw suppose legume residue containing C:N ratio of 16:1 is added. The situation will be entirely different. The soil microbial population build-up will be rapid and maximum as there will be no nitrogen limitation for microbial growth. This will favour greater mineralization and release sufficient inorganic N to soil. The one reliable measure of mineralization in soil is to find out the total available ammonium, nitrate and nitrite forms of N at a given time since the processes of an ammonification, nitrification and denitrification occur in cultivated soil simultaneously. The lignin content of plant residues may also serve as an index of vulnerability of organic residues to microbial attack.

5.6 HUMUS AND HUMIC SUBSTANCES - THE STABLE SOIL ORGANIC FRACTION

The soil organic matter is composed of non-humified (decomposing residues, excretions and by-products of decomposers and decomposing biomass) and humified (resistant soil humates) substances. The non-humified substances have definite/recognizable structural, molecular and other characteristics. These materials are subject to microbial degradation/ decomposition reactions. Sometimes, these may persist through adsorption by clays only for a brief time or remain as such under anaerobic or unfavourable conditions, therefore, are protected from decomposition. The humified fraction is known as 'humus', or sometime 'humic substances', which are formed by secondary synthesis reactions. Humic substances may vary from 33-75% of the total soil organic matter in a grassland soil.

5.6.1 HUMUS: DEFINITION, COMPOSITION AND TYPES

Humus : The humus is an insoluble highly heterogeneous polymeric material left after extensive and exhaustive decomposition of plant and animal remains. It can be defined as the amorphous, dark coloured fairly stable lignoprotein complex or amino acid-lignin complex formed by susceptible and microbially-resistant compounds. Humus represents total organic compounds in soil excluding undecayed plant and animal tissues and the soil biomass. Humus is relatively high in molecular weight and contains thousands of variable and complex organic compounds cross-linked together.

Composition of humus : The humus, on an average, contains 45% lignin, 35% amino acids, 11% carbohydrates (7% hemicellulose, 4% cellulose), 6% phenolic glycosides or tannins, 3% fats, wax and resins, and 3 to 6% other compounds. Organic matter content in soils may be as low as 0.5% in mineral soils and as high as 80% in organic soils. With the exception of podsol and palaeosol soils, the organic matter concentration usually decreases with soil depth. The annual decomposition and renewal of humus is 2-5 t ha^{-1} yr^{-1} in most soils. Barring environmental changes, the amount of humus decomposed each year remains approximately same as the amount of new humus formed during the year. Temperate regions have an average of 3% annual turnover of soil humus.

Humus formation : Humus is formed by two general types of biochemical reactions viz. decomposition and synthesis (Fig. 5.6). The nature of humus, therefore, depends on:

- The composition of organic matter added;
- The microflora acting on them; and
- The condition under which humus is formed.

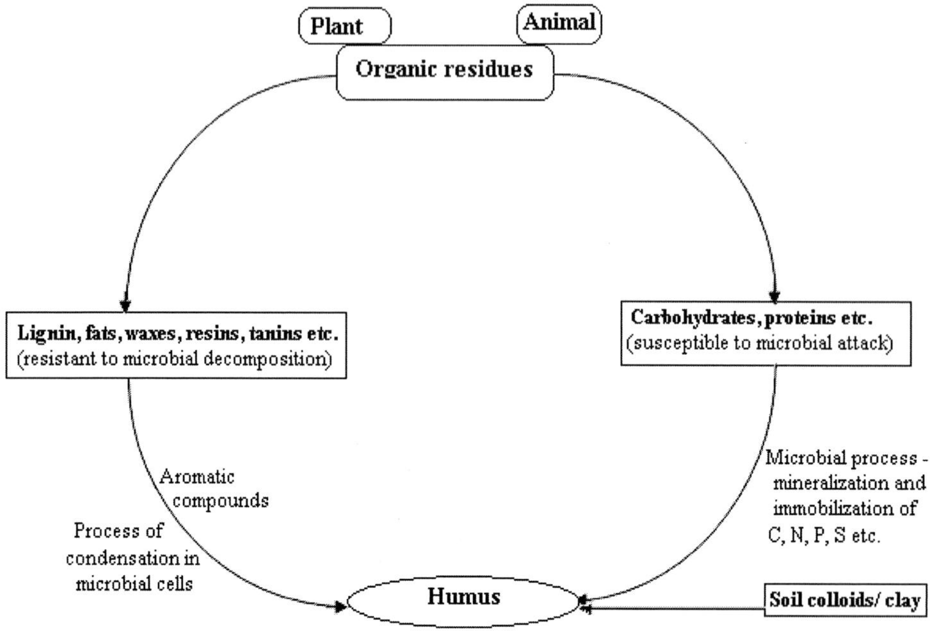

Fig. 5.6: Formation of humus

Types of humus : The stability of humus in soil is mainly due to its interactions with the mineral constituents and entrapment within clay aggregates. Humus may be either mull or mor type. 'Mull humus' have humified organic matter that is well mixed with mineral soil during decomposition process. These are formed under moist but well-aerated and neutral or mildly acid conditions. Such soils have rapid rate of humification so there is little or no organic matter accumulation on the surface. However, 'mor humus' are formed under waterlogged/low aeration conditions with slow organic matter breakdown leading to the formation of peat (an organic) soil. In mor humus there is no mixing of organic and inorganic soil. Highly acidic conditions generated inhibit the incorporation of organic matter into inorganic soil fraction so soil surfaces show large buildup/accumulation of organic matter essentially in inorganic soil.

5.6.2 PROPERTIES OF HUMUS

1. Humus is highly colloidal in nature with colloidal size ranging from 1.0 to 0.001 μm and forms colloidal complex with the clay particles.
2. Humic substances are hydrophilic and consist of globular particles, which in aqueous solution contain hydration water.

3. The specific surface area and absorptive capacities of humus are greater than those of silicate clay minerals. The specific surface of well developed humus may be 800 to 900 m^2 g^{-1} and its exchange capacity may range from 150 to 300 cmoles kg^{-1} with total acidity of 300-1400 cmoles kg^{-1}. About 50-70% of cation exchange in most soils is because of the presence of humus.
4. Humus carries negative charge derived from ionization of acidic functional groups so it exhibits base-exchange properties thereby conserves and releases the nutrient elements of the plant.
5. Humus is largely insoluble in alkali, partially soluble in acid and dissolves in dilute alkali
6. It generally has a C:N ratio of 10-12:1 with carbon about 55-60% and nitrogen 4-6%.
7. Humus improves physical conditions of soil aggregates, aeration and water holding capacity and also serves as a store-house for essential plant nutrients. The water holding capacity of humus on weight basis is 4-5 times greater than that of silicate clays.
8. Humus buffers the soil against rapid changes in acidity, alkalinity or salinization; besides it has capacity to absorb pesticides and other organic chemicals. Through its chelation property, it forms stable complexes with heavy metals such as Cu^{2+}, Mn^{2+}, Zn^{2+} and other polyvalent cations.
9. Humus is quite resistant to microbial attack and contains various active functional groups exposed to surrounding environment. The typical dark colour of many soils is caused by organic matter. Thus humus plays a vital role in soil ecosystem development.

5.6.3 STRUCTURAL CHEMISTRY OF HUMUS

Humic substances are considered to be the partially degraded products of lignin. Several mechanisms have been proposed to explain the formation of humus in soil; some of them are described as under:

1. **Lignin theory :** Selman Waksman gave this classical lignin theory proposing that humates are modified or partially degraded lignins that remain after microbial death. The residues undergo demethylation, oxidation and condensation with nitrogen compounds such as proteins to form humus (Fig. 5.7). The modified lignins are characterized by loss of methoxyl (OCH_3) groups and the presence of O-hydroxy phenols and oxidation of aliphatic side chains to form COOH groups. The lignins undergo more modification and form humic acids followed by fulvic acids.

2. **Polyphenol mechanism :** In this pathway phenols (catechol), phenolic aldehydes and phenolic acids, derived from partial degradation of lignin, are postulated to be oxidized to quinines. The quinine, in turn, react either within themselves to polymerize or with other degradation products such as amino compounds, amino acids, etc. Polyphenols may also be synthesized from non-lignin sources such as cellulose, etc. by the microbes present in soil and then enzymatically oxidized to quinones. The lignin-derived phenolic products together with those synthesized by microorganisms polymerize alone or in the presence of amino compounds (amino acids, etc.) to form amino-quinone intermediates (Fig.5.7). These condense to form brown coloured polymers (marginal humates) having variable molecular weights and contents of acidic functional groups.

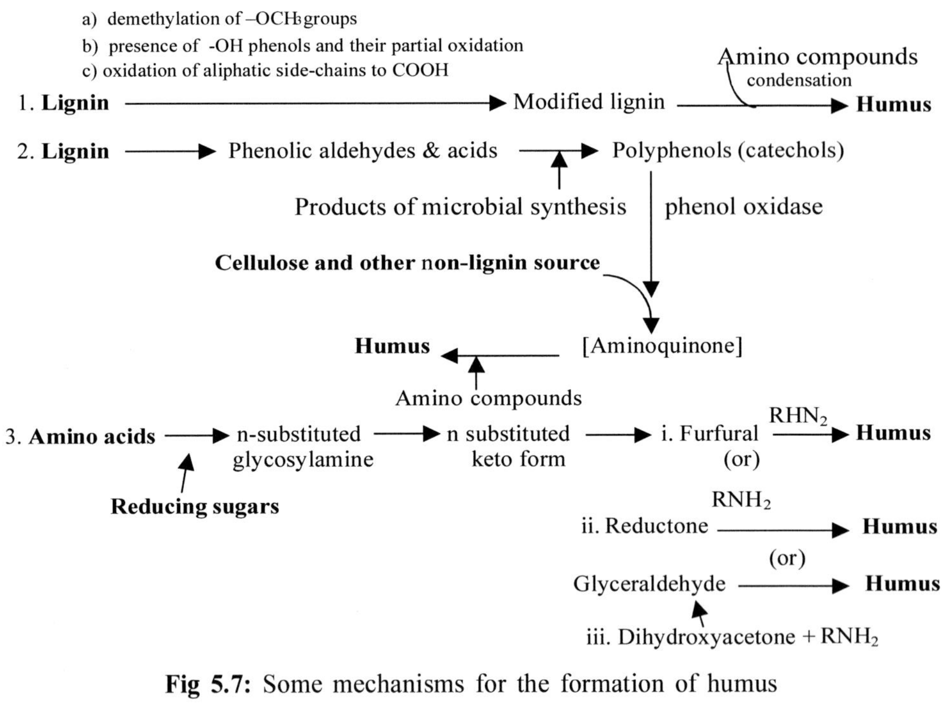

Fig 5.7: Some mechanisms for the formation of humus

3. **Browning reaction :** It postulates that the reducing sugars react with amino acids or related substances, formed as by-products of microbial metabolism, and form glycosylamines, etc. The subsequent rearrangements and fragmentation result in the formation of three intermediates, a three carbon aldehydes and ketones, reductone and furfurals (Fig. 5.7). These intermediates may react with amino compounds and

undergo non-enzymztic polymerization to form dark brown coloured nitrogenous polymers.

5.6.4 FRACTIONATION OF HUMUS

Humus needs to be separated from mineral colloidal matrix of clays and sesquioxides and dispersed in a liquid before being studied. Several methods using acids such as HCl, HF, H_3BO_3, etc., bases and alkali like NaOH, Na_2CO_3, $Na_4P_2O_7$, Na_2-EDTA, $Na_2B_4O_7$, etc. and organic solvents such as oxalic acid, formic acid, benzene, chloroform, hydroxymethylamine, phenols, etc. are available for the extraction and isolation of humic substances from the soil. However, classical fractionation technique involves dispersion in either 0.1-0.5 M NaOH or $Na_4P_2O_7$. $Na_4P_2O_7$ is used frequently for the extraction of humic fractions from soils high in sesquioxides. The fractioned portions during several treatments are indicated in Fig. 5.8. The fractions not dispersed either by the peptizing action of Na^+ or by the chelating action of pyrophosphates or by the H-bond breaking activity of very alkaline pH is known as 'humin'. The dispersible material precipitated at acidic pH values is 'humic acids' and the material unprecipitated is 'fulvic acids'. No definite chemical structure is assigned to these fractions.

In addition to extraction and precipitation procedures, gel permeation, chromotography, ultrafiltration membrane, adsorption on hydrophobic resins/charcoal and Al_2O_3 and centrifugation methods are also employed for soil organic matter fractionation.

5.6.5 CHEMICAL CHARACTERIZATION OF HUMUS

Humic substances comprize about 60-80% of soil organic matter and are classified on the basis of their solubility in acid or base as humic acid, fulvic acid and humin. The rest organic matter is non-humic substances which include polysaccharides, polyuronides and the acids synthesized by soil microbes. Humus is resistant to oxidation and their interactions with clay minerals provide means of stabilizing soil nitrogen and soil organic matter. The basic properties of humic substances are given in Table 5. 6.

i. **Humic acids :** Humic acids are the fractions extracted by dilute alkali but precipitated or insoluble at pH 2. These are higher molecular weight compounds comprizing of aromatic ring and nitrogen in cyclic forms and in peptide chains. These chemically reactive molecules contain both free -OH groups and quinones ($O{=}C_6H_4{=}O$) linkages. This type of structure results in an unusually high density of reactive functional groups like carboxylic acid (RCOOH), phenolic (ArOH), alcoholic (-OH), carbonyl ($R_2C=O$), hydroquinones (Ar=O) and some ketonic (RCOR). Individual humic acid molecules vary in their structure and density of functional groups, but the basic structure is presumed to remain approximately same. The actual structure of humic acids has not been determined as yet. These compounds, however, are formed by poly-condensation of similar but unidentical humic acid units. These contain a few carbohydrate residues. Humic acids are usually rich in carbon but low in total acidity or exchange capacity. Electron spin resonance (ESR) analysis has revealed that humic acids have an aromatic core, containing physically or chemically bonded proteins, polysaccharides, simple phenols, semiquionones and metals. The linkage to core renders considerable stability to amino acids, proteins and polysaccharides towards biochemical attacks.

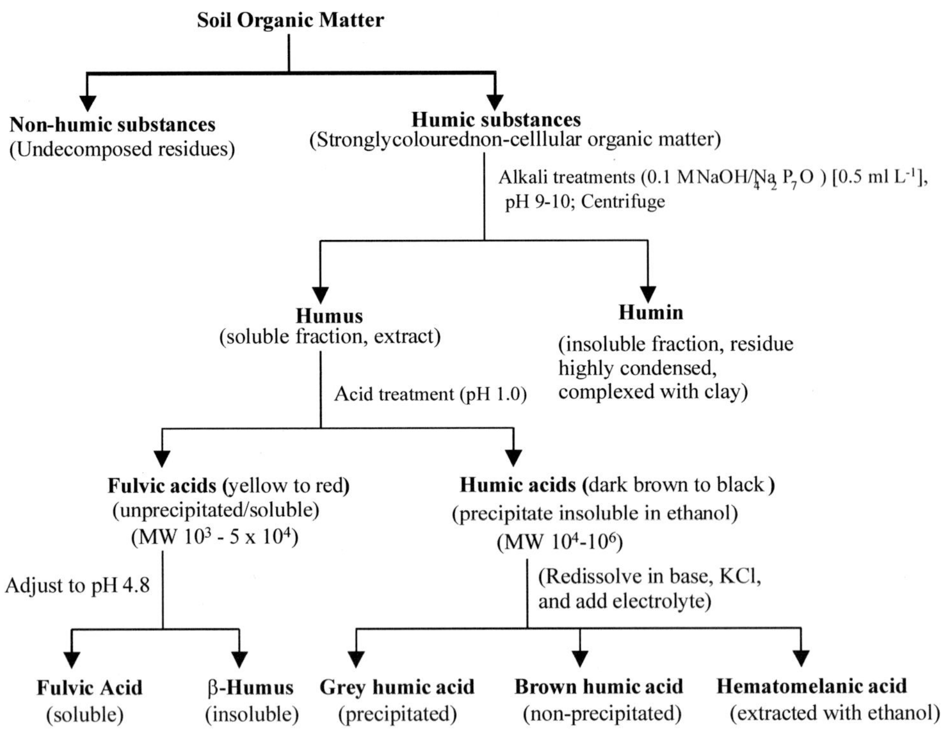

Fig 5.8: Flow sheet for the fractionation of humic compounds into different humic fractions.

The humates (water insoluble salts of humic acids) are adsorbed by the clay minerals through polyvalent cations such as Ca^{2+} and Fe^{3+}, and these also form associations with Fe and Al hydrous oxides either through co-ordination ligands or anion exchange with positive sites.

ii. **Fulvic acids :** This is the fraction extracted by alkali but soluble at pH 2. These acids comprize of a series of highly oxidized aromatic rings having large number of side chains. The basic building blocks are benzene carboxylic acids and phenolic acids, primarily held together by hydrogen bonds or van der Waal forces or ionic bonds. The fulvic acids because of their flexible nature and relatively open structures are capable of fixing organic as well as inorganic compounds that fit these openings, provided these are complementary. Fulvic acids have higher oxygen and low H and N content as compared to humic acids. The composition of various elements is presented in Table 5.6. The functional groups largely are carboxylic (RCOOH) or phenolic-OH groups. The fulvic acids have higher acidity which is attributed to higher carboxylic content. The more soluble fulvic acids are usually responsible for brownish-yellow colour of many natural waters. Fulvic acids present in lower soil horizons are formed by leaching and usually have little N whereas fulvic acids of mollisols contain high N concentration as well as fatty acids and cell constituents of microorganisms.

Table 5.6: Basic properties of humic substances

Components	Fulvic acid	Humic acid	Humin
Av. molecular formula (without S)	$C_{12}H_{12}O_9N$	$C_{10}H_{12}O_5N$	nd*
Molecular weight	10^3-$5x10^4$	10^4-10^6	$\geq 10^6$
Carbon (%)	42-51	46-60	≥60
Oxygen (%)	42-54	33-40	≤30
Nitrogen (%)	0.7-3.5	2.0-5.5	≥3
Hydrogen (%)	3.3-4.6	4.3-7.7	2.5-4.9
Sulphur (%)	0.1-3.6	0.1-2.5	0.3-0.7
Ash content (%)	2.5	1.0	1.5
O:C ratio	0.7	0.5	nd*
Soluble in	alkali and acid	only in alkali	insoluble acid and alkali
Exchange capacity (m eq 100 g^{-1})	100- 120	50-60	≤50
Mean resistance time (years)	10-50	1220	180
Response to microbial attack	Susceptible	Intermediate	Resistant

*nd = not determined

Source: Schnitzer (1976); Steelink (1982); Mathur and Farham (1985); McBride (1994); Sparks (1995)

iii. **Humin :** It is non-dispersible in NaOH and most soils release fulvic acids, humic acids and humin in almost equal proportion. Humin is composed of fulvic acid, humic acids and non-soluble plant and microbial constituents like undecomposed cellulose, lignin, cell walls, etc. These are high molecular weight compounds with low exchange capacity. Being similar to humic acid, humin is slightly less aromatic and contains higher polysaccharide content. Alcoholic groups are higher in humin as compared to other fractions.

Behaviour of humic substances in soil : The chemical behaviour of humic matter, in general, is determined by carboxylic and phenolic groups whose dissociation yields about 80-90% of the negative charge of humus. The carboxylic group initiates dissociation of protons at pH around 3.0 so making humic molecules electro-negatively charged. Dissociation increases with the increase in pH and initiates the dissociation of other hydroxyl groups also. At pH around 9.0, phenolic group dissociates so molecules attains high negative charge. Because of this variable pH-dependent charge, various reactions or exchange phenomena, metal complexation, etc. are initiated. At lower pH, humic molecules attract cations leading to cation exchange reactions. Humic substances sorb organic compounds through temporary association. Humic substances affect biological processes to a greater extent by interacting with soil minerals in soil aggregate formation and by their role in removal of biodegradable substances from the pool of easily metabolized substances.

5.6.6 ROLE OF HUMUS IN SOIL FORMATION AND PRODUCTIVITY

Soil organic matter or humus plays vital role in soil structure, fertility, profile development and aggregate stability. These effects are often interrelated e.g. fertility affects the type of plants which grow and, in turn, type of plants influence development of soil profile.

Ped formation : Humus acts as an adhesive, binding the particles of inorganic fractions of soil together into relatively robust pellets called 'peds'. The formation of ped is generally accompanied by the appearance of voids which serve as channels for effective drainage and air movement. Moreover, humus enables the soil peds to resist crushing and breakdown from tillage and raindrop impact. The accumulation of organic matter leads to horizon differentiation in most soils.

Humus and soil productivity : The humus influences root development, water percolation and nutrient availability. Small quantities of both fulvic and humic acids in soil are known to enhance seed germination, stimulate root initiation and elongation and increase plant growth. Major elements and many minor nutrients are retained in soil by humus, which serves as a storehouse of nutrients. It acts as a centre of activity around which chemical reactions and nutrient exchanges do occur. These protect essential nutrients from leaching through attraction of ions to their surface humus temporarily and release them slowly for plant use. Humus is a significant source of S and P and has C:N:P:S ratio of 100:10:1:1. The soil organic matter supplies almost whole nitrogen, 50-60% phosphate, about 80% S and other nutrients absorbed by plants from unfertilized temperate region soil. The harmful effects of K and P are offset by complex formation with organic ligands. Evidences suggest that humic and fulvic acids increase the release of K fixed in the intermicellar spaces of clay. The availability of micronutrient cations is strongly affected by soil organic matter. Humic acids effectively bind and slowly release the micronutrients like Fe, Zn, Mn, etc. Soil organic matter gives dark colour to the soil, therefore facilitates soil warming. Clay formation and chelation play an important role in improving soil fertility. Various low molecular weight components of organic matter form stable complexes or chelate with Fe^{3+}, Cu^{2+}, Zn^{2+} and other polyvalent cations which shield the cations from hydrolysis and precipitation reactions. Besides decreasing aluminum toxicity in soil through humic acids, soil organic matter bind toxic metals, such as Cd^{++}, Mn^{++}, Ni^{++}, Al^{+++}, pesticides, etc. and reduce their availability.

Because of large surface area humus absorbs huge quantities of water, about 20 times its weight. Humus in medium-textured soils may be responsible for one-third of the moisture retaining ability while representing only 5% of its mass, thus helps in preventing drying and shrinkage of soil. The buffering action of humus helps in maintaining uniform soil reaction as well as the availability of trace elements to plants.

Role in soil formation : The quality of soil formation in soil depends upon five soil forming factors viz. time, biotic factors, vegetation pattern, seasonal climate, topography and parent material. This 'soil factor theory' proposed by Jenny has recently been modified to include anthropogenic influences.

a) **Time :** In early stages of soil development input of organic matter is greater than loss. This leads to the rapid accumulation of soil organic matter. As the time progresses such accumulations slowly tend to decline and reach an equilibrium state. The equilibrium may be attained in 110 years in case of fine textured parent material and >1500 years in case of sandy materials. The stability of humus is due to its interactions with polyvalent cations and clay, and low amount of one or more essential nutrient which limit the quality of stable humus that can be synthesized by soil organisms. The amount and depth of soil organic matter is the greatest single component controlling the productivity of both cultivated and uncultivated soils. A soil is considered to be organic if either its organic matter content exceeds 30% or organic matter influences the soil properties about five times greater than the equal amount of clay. High organic matter in soil is the result of slow decomposition rate rather than high rate of organic matter addition. Organic soils are very common in tundra region where the rate of plant growth is very slow. During soil tillage and other farming operations organic matter is oxidized faster than its replenishment. Subsidence (dropping of surface level) at the rate of 2.0-2.5 cm per year is fairly common in organic soils. An organic soil 3 m thick may be nearly exhausted or gone after 100 years of farming.

b) **Vegetative pattern :** The quantity of soil organic matter in a soil depends on the balance between primary productivity and the rate of decomposition. Grassland soils are higher in soil organic matter than forest soils because of greater amounts of plants produced, inhibition in nitrification that preserves N and C, higher humus synthesis in rhizosphere and high base content which promotes ammonium fixation. A succession of plant species is often observed in natural ecosystems like grasses and sedges followed by taller plants and complex structures. Changes in vegetation pattern alter organic matter accumulation within a soil. In grasslands most organic matter accumulation is confined to 0 and A_1 horizons, whereas in forest soils leaf fall/litter organic matter accumulation occurs up to $2\text{-}6^2$ layer and root penetration is deeper. Podzol soils are formed where forest areas are under acid leaching conditions.

c) **Climate :** During early phase of soil formation, the organic matter content in soil increases. After a few centuries, soil attains a dynamic equilibrium i.e. the content of organic matter addition compensates the losses. Even though the organic matter remains constant for years together, yet there is significant turnover of organic matter and recycling of nutrients.

The highest accumulations of 700 Mg C ha^{-1} to a depth of 1 m occurs in highly productive swamps and marshes where decomposition is inhibited because of anaerobic conditions. Grasslands with wet-dry seasons, and especially if stabilized by Ca^{++}, have higher organic carbon level than the equivalent boreal and temperate forests, although their carbon inputs is almost half of the 12 Mg C ha^{-1} yr^{-1} in forests. Tropical forests, with annual average production of 19 Mg C ha^{-1}, have less organic carbon than temperate forests with a productivity of 12 Mg C ha^{-1}. This organic matter is distributed deeper throughout the profile with C:N ratio of 10:1, similar to the cultivated soils. However, in temperate forests C:N ratio is mostly in the range of 15 to 20. The organic matter contents of grassland soils of drier climates average less than the grassland under sub-humid temperate conditions. This is because of less plant growth and less organic matter turnover. Freezing weather during winters of cooler climates tend to preserve organic matter by preventing microbial activities. Forest soils do not show much drastic increase in organic matter with increased precipitation as do grasslands. In grasslands, there is increased leaching which decreases the fertility of soil thereby affects plant growth. Moisture and temperature interaction is known to control the accumulation of soil organic matter. On an average, tropics are 15°C warmer than temperate regions. This causes an increase in the rate of chemical and biological reactions where moisture is not a limiting factor. In humid forests, for example, the rate of primary production and decomposition is several times more rapid than in temperate region. Under moist conditions, the rate of carbon accumulation through photosynthesis is larger than the opposite process of decomposition. Highest carbon accumulation occurs in wet rain forests of temperate zones. The accumulation of nitrogen in drier areas shows lower C:N ratio. It seems that nitrogen losses are higher and N_2 fixation low in wet rain temperate areas.

d) **Topography :** Topography modifies microclimate and influences vegetation with a profound effect on organic matter. Further, it influences soil and water movement, evaporation and transpiration. Steep soils have more runoff making less water availability to plants. Also, erosion of soil removes organic matter to the foothills of slope. Soils facing northern slopes are wetter and lower in temperature as compared to those facing south which are hotter and drier, hence are higher in soil organic matter. The average temperature of a soil that faces towards warmer early afternoon sun may be several degrees above that of a soil on opposite slope. So on these soils less accumulation of organic matter occur.

e) **Parent material :** Parent material of soil influences the kind and amount of vegetation. A good relationship exists between organic matter and clay in soil. The presence of silt and clay generally increases soil organic matter for a particular set of environmental interactions and the presence of minerals such as those in certain volcanic soils also result in greatest retention of soil organic matter.

f) **Cultivation :** Cultivation of soil affects the soil organic matter content. Organic matter accumulation generally increases if a cultivated field is allowed to revert to its original native vegetation. Organic matter decline is due to decrease in plant biomass input and disruption of soil physical structure in upper surface horizon. But cultivation of a virgin soil begins to slow down organic matter accumulation and after several decades and centuries it reaches a constant or stable status under a particular environmental condition. Cropping practices decrease soil organic matter. In soils under maize cultivation about 25% N was lost during first 20 years of cultivation, 7% during next 20 years and only 3% in third 20 years. So soil organic matter content attains a new equilibrium after cultivation. The rate of humus oxidation in undisturbed forest is considerably lower than the tilled field. Wetting and drying of soil also causes respiration, which reduces soil organic matter content.

The sloughed off root materials such as root hair and mucilages constitute major carbon source in continuously cropped soils. Perennial natural grassland under field conditions retained 52% of assimilated carbon aboveground and 36% in shoot bases of roots. The association of mycorrhiza with plants contributes much to aboveground litter to enhance soil fertility. The mucigels secreted by bacteria in the rhizosphere reportedly bridge the gap between roots and nearby soil particles. Most of the plant nutrients are probably passed to plant through a layer of mucigel. Thus, the presence of microbes influences soil organic matter content and its accumulation as well as plant growth. The beneficial role of humic acids has now been established. It improves seed germination, root growth and nutrient uptake. Further, it is reported to cause decrease in phosphate fixation particularly in calcareous soils.

5.7 DYNAMICS OF ORGANIC MATTER DECOMPOSITION

The basic concept to describe decomposition (decay) rate of plant residues, microbial biomass and soil humus is to interpret the turnover or cycling of nutrients like C, N, P, S, etc. in soil. The decomposition of plant materials in soil can be studied directly by estimating the losses in weight (i.e. the disappearance of a substrate) or by monitoring the radioactive carbon dioxide ($^{14}CO_2$) produced from ^{14}C-labelled materials. The decay curve is prepared with respect to time which gives idea about the fate of a nutrient in a particular time period under specified conditions. Microbially-mediated decay processes, involved in soil organic matter decomposition and nutrient cycling, can be best understood by using mathematical analysis of tracer and non-tracer data, measuring the reaction rate of decomposition of different fractions and use of computer models to predict their fate. Many environmental and soil physico-chemical factors govern the kinetics of decay. The rate of organic matter decay in soil is influenced by the composition and amount of substrate, the inoculum potential of decomposers and the interaction between decomposers and soil physico-chemical conditions (especially temperature, water potential, acidity etc.). Therefore, the analysis of decomposition concerns basically the substrate disappearance which is very much related to the microbial growth dynamics while interpreting decomposition kinetics.

5.8 MICROBIAL GROWTH RATES IN SOIL

Microbial growth rate : 'Microbial growth rate' may be defined as change in the amount of microbial biomass or number per unit time. The microbial growth rate follows an exponential curve provided there are no constraints on microbial growth in terms of carbon substrate or nutrients and growth conditions. All populations have the capacity to increase exponentially, however microbial populations do not continue exponential growth after some time because growth conditions become less favourable due to the limitation of nutrient resources or some toxic product formation. So organisms follow S-shaped or sigmoid growth curve (Fig. 5.9).

Growth rate in soil : Assuming the soil to be a closed system, the generation time of fast growing soil bacteria such as *Pseudomonas* or *Bacillus* under most favourable growth conditions is 20-30 seconds. The soil fungi double after every 1-4 hours and protozoa and algae after 12 to 24 hours or more. Each organism has its own 'maximum specific growth rate'. The maximum growth rate of various soil microorganisms is shown in Table 5.7.

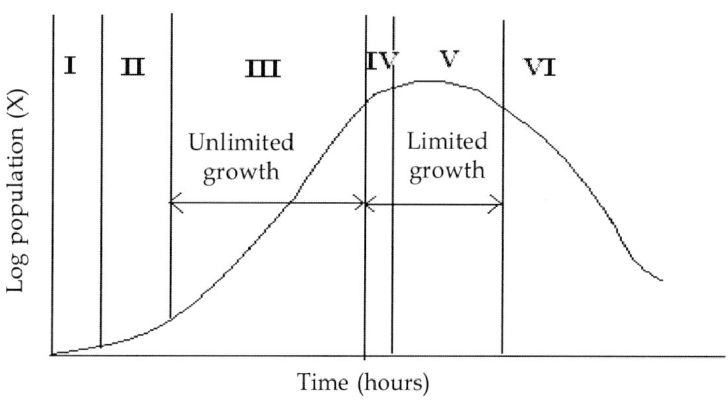

Fig 5.9: Population dynamics of soil microbes in a closed system (I. lag phase, II. acceleration phase, III. exponential phase IV. retardation phase V. stationary phase VI. death phase).

Table 5.7: Maximum recorded growth rates for certain soil bacteria and yeast, measured at/or near their temperature optima grown in complex media

Organisms	Temperature (°C)	Generation time (h)
Bacillus stearothermophilus	60	0.14
Escherichia coli	40	0.35
Bacillus subtilis	40	0.43
Pseudomonas putida	30	0.75
Nitrobacter agilis	27	20.00
Nitrosomonas sp.	27	30.00
Saccharomyces cerevisiae	30	0.50
Schizosaccharomyces pombe	30	0.22
Neurospora crassa	27	0.58
Gibberella fujikuroi	27	0.21
Achyla bisexualis	-	0.81

Assuming x individuals were initially present during exponential growth then growth in various generations will follow the following pattern:

Number of generations →	0	1	2	3	4............n
Number of microbial cells →	x	$2x$	$4x$	$8x$	$16x$..........ax
This can also be written as →	2^0x	2^1x	2^2x	2^3x	2^4x..........2^nx

The rate of change in the biomass or population size (x) with time (t) or growth rate ($\frac{dx}{dt}$) will be :

$$\frac{dx}{dt} = \mu\, x_o$$

where μ is a constant term depicting intrinsic rate of increase termed as "specific growth rate"; and x_o the initial population size under optimal condition.

$$dx = \mu_{max} . x_o \quad (\mu_{max} = \text{Maximum specific growth rate})$$

A soil bacterium *Bacillus subtilis* weighing 10^{-12} g, with doubling rate of 30 minutes, if continues its maximum specific growth for only 72 hours, it will produce progenies of 2.2 x 10^{31} g or roughly 4,000 times the mass of earth. As per first law of growth:

$$\frac{dx}{dt} = \mu x$$

If there are x organisms at time t, then $\frac{dx}{dt} = \mu x$ after integration and using natural logarithms will be:

$$\ln x_t = \ln x_0 + \mu t$$

$$\text{or} \quad x_t = x_o e^{\mu t}$$

When the population doubles in size;

$\ln \frac{x_2}{x_0} = 2$, and $t = t_d$ (t_d = doubling time)

Then, $\ln 2 = \mu\, t_d$

If the initial inoculum concentration is $x_{0,}$ then growth yield Y can be calculated by knowing the amount of substrate consumed, Δs, the second law of growth will be:

$$Y = \frac{x_1 - x_0}{\Delta s} = \frac{\text{Microbial growth (g)}}{\text{Substrate utilized (g)}}$$

$$Y = \frac{dx}{ds}$$

$$\text{or} \quad ds = \frac{dx}{Y}$$

The metabolic quotients, Q, will be:

$$\frac{ds}{dt} = Qx \quad \text{or} \quad ds = Qx.dt$$

However, as per the first and second laws of growth:

$$ds = \mu_x.dt$$

$$\text{Therefore,} \quad Q = \mu_x . \frac{dt}{Y}$$

$$\text{we know that} \quad -\frac{ds}{dt} = \frac{\mu_x}{Y} \quad \text{because,} \quad dx = dt.\mu_x$$

$$\text{or} \quad -\frac{ds}{dt}.x = \frac{\mu}{Y} \quad \text{but} \quad Qx = -\frac{ds}{dt}$$

$$\text{therefore,} \quad Q = -\frac{ds}{dt}.x \quad \text{or} \quad Q = \frac{\mu}{Y}$$

This type of situation is observed in closed system i.e. when there is no fresh input of substrate or no dilution of culture is done. But this does not happen in soil because growth of microbes is always under tight constraints such as substrate limitation, environmental conditions, etc.

Soil population dynamic studies : Soil is an open system. This emphasizes more realistic study of population dynamics under 'open' conditions of continuous culture using chemostat. Under this condition, the population will achieve a steady state (i.e. the population is kept constant with the production of cells balanced by their washout under a growth limiting substrate concentration with a fixed dilution rate).

Therefore, change in organism concentration = Growth - Outflow

For a unit volume of culture of the interval *dt* will be equal to:

$dx = \mu_{x}.dt - Dx.dt \qquad D = \text{dilution rate } (h^{-1})$

$= dt\,(\mu_{x} - Dx) \qquad D = \dfrac{\text{Flow rate (F)}}{\text{Volume (V)}}$

Volume displaced in time $t = \dfrac{V}{F}$

$$\frac{dx}{dt} = \mu x - Dx = x\,(\mu - D)$$

In a steady state, $\dfrac{dx}{dt} = 0$ or $\mu = D$

The equation does not consider the energy used in the maintenance of cells which is important when μ is small. Therefore,

Rate of substrate utilization = Rate of substrate used for biomass production + rate of substrate used for maintenance

$$\frac{ds}{dt} = \frac{\mu x}{Yg} + mx$$

where, Y_g = actual growth yield

However, in soil environment washouts occur to a limited extent only. Most of the cells are retained by the soil. The population balance in soil will be between cell growth and death rather than through dilution or washout in chemostat. On death, cells become a source of substrate and an addition to the substrate available. Further, for maintenance some energy needs to be utilized. In the soil system, however, the environmental stresses are great. Thus diversion of substrate for the maintenance of cell integrity becomes considerable, and even dominates the partitioning of the substrate. It is, therefore, difficult to predict '*m*' in soil where growth is very slow. Under these circumstances, the equation of growth yield is modified to take into account the additional demand of substrate required for maintenance:

$$\text{Growth yield} = \frac{\text{Cell growth}}{\text{Substrate used for growth}} + \text{Substrate used for maintenance}$$

$$\text{or } Y_g = \frac{dx}{d(Sg + dSm)}$$

In soils more complexity arises in a situation where some organisms may not be dependent on exogenous energy as they may have accumulated their own reserves. Also, the other organisms form resting structures such as spores, cysts, etc. which require extremely small maintenance. Therefore, growth theory, as described, does not allow for the release of substrate during cell death and its utilization by new generations.

Decomposition of organic matter by soil microbes *per se* is determined by weight loss in the substrate contained in nylon mesh bags. Preferably, bags should contain some soil to serve as inoculum. Different sizes of mesh are used to separate contribution of soil animals and microorganisms in decomposition. The rate of decomposition (reaction rate) can be expressed as a function of the concentration of one or more substrates being degraded. This is termed as the order of reaction and is the value of exponential used to describe the reaction.

5.9 REACTION KINETICS IN ORGANIC MATTER TRANSFORMATION

Reaction kinetics : Kinetics deals with the study of the mechanisms of rates of reaction by which reactions occur and reactant concentration (e.g. substrates, inhibitors, activators and enzymes). The speed with which reactants are converted to products is 'rate of reaction' and may be expressed in terms of rate of disappearance of any of the reactants or rate of appearance of any of the products. The basic purpose of kinetics is to describe time-dependent processes of reactants. The subject of enzyme kinetics in organic matter decomposition is very large, crucial and complex.

Zero order reactions : When the change in substrate concentration is independent of the rate of transformation of a substrate, it is a 'zero order reaction'. This happens when the rate of reaction is determined by a factor other than the substrate concentration e.g. enzyme concentration. The rate of change of a substrate *S* to product *P*, also known as decomposition rate, will be:

$$\frac{dS}{dt} = -k\,[S]^0$$

where S = concentration of added substrate at any time
t = time from initiation of reaction (in hours or days)
k = decomposition rate constant (hr^{-1}or day^{-1})

Integration of this equation gives:

$$S_t = So - kt$$

$$\text{or}\quad \ln\frac{S_t}{S_0} = -\,kt$$

where S_0 = concentration of substrate '*S*' at zero time.
S_t = concentration of substrate '*S*' remaining at time *t*.

The values of '*k*' can be determined by plotting $\ln\frac{S_t}{S_0}$ against time *t* (Fig. 5.10).

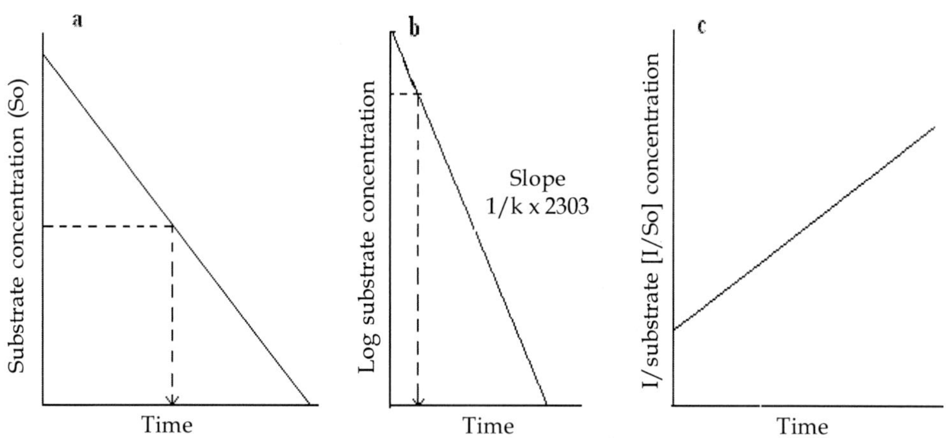

Fig 5.10: Zero (a), first (b) and second (c) order reactions (graphic representations)

The reaction kinetics can be described by considering the time required to utilize 50% (half life time of the substrate) or 100% (mean turnover time) of the initial substrate. Therefore, when 50% substrate is utilized $S_t = \frac{S_0}{2}$ and $t_{0.5}$ will be:

$$t_{0.5} = \frac{S_0}{2k}$$

The disappearance time also called 'residence time (t_r)' is the time required to decompose an amount of substrate equal to the initial amount t_r will be equal to:

$$t_r = \frac{S_0}{k}$$

At high substrate concentrations where concentration of substrate is not a limiting factor, enzymatic reactions are of zero order; for example nitrification of high ammonium content or denitrification of high nitrates content.

First order reactions : It is generally accepted that organic matter decomposition follows the first order reaction kinetics. As per the first order reaction, the rate of decomposition of substrate (S) is substrate dependent or proportional to the substrate concentration. For example, the rate of denitrification {soil N $\rightarrow N_2 + N_2O$} depends on nitrate concentration in soil solution. Any change in pH, incubation temperature, ionic strength of incubation solution, adherence in concentration of minor elements or cofactor required etc. may have a significant impact on reaction rate. Therefore, the rate of change of substrate to P will be governed by:

$$S_t = \frac{dS}{dt} = -kS$$

Upon integration and using log, the equation becomes:

$$S_t = So\ e^{-kt}$$

where S_t = concentration of substrate remaining at time t

k = decomposition rate constant (hr^{-1} or day^{-1}or $year^{-1}$)

Dividing the above equation by *So* and taking natural logarithm of both the sides, we get

$$\ln\frac{S_t}{S_0} = -kt$$

The values of k can be determined by plotting $\ln\frac{S_t}{S_0}$ verses time t. A plot of $\ln\frac{S_t}{S_0}$ v/s t gives a linear slope (-k). By plotting the equation in natural logs, t is a straight line with slope $-\frac{k}{2.303}$. The rate constant k is independent of the concentration of substrate or product since the slope does not change with time. For calculating the time required for half of the substrate to be consumed or transformed, we use the equation:

$$\ln \frac{S_0/2}{S_0} = -kt_{0.5}$$

where $$t_{0.5} = \frac{0.693}{k}$$

The mean turnover time for first order reactions is equal to $-\frac{1}{k}$. The $\ln \frac{S_t}{S_0}$ v/s t plot gives a straight curve if the reaction is of first order. Radioactive decay, rate of population growth (if there is no change in birth or death rate) and growth rate of bacterial culture until nutrients become limiting are some of the example of first order processes.

As per the first order reaction kinetics, in most of the studies decomposition of a specific compound is determined by using ^{14}C and measuring $^{14}CO_2$ evolution or ^{14}C remaining in the soil. However, neither method gives real decomposition rate. When microorganisms utilize C, the carbon compounds enter various pools viz. CO_2, biosynthesis and to some extent low molecular weight metabolites. Let X be the amount of CO_2-C evolved during decomposition of a compound and Y the efficiency of C-use for biosynthesis expressed as %total carbon uptake under aerobic conditions. Also, assuming that non-cell metabolite production is negligible then the actual amount decomposed will be:

$$X\left(\frac{1+Y}{100-Y}\right)$$

However, problems in assessing actual decomposition are microbial production and recycling of labelled materials which interfere in the determination of real decomposition rate as in case of sugars. Organic matter contains a wide variety of compounds so decomposition rates usually show gradual decrease with time. For example, in straw decomposable soluble fractions with high k values (0.03 day^{-1}) decompose first and later show decline in decomposition when resistant and recalcitrant fractions such as lignin and phenols are left (k = 0.003 day^{-1}). Rate of carbon and biomass decay follow first order kinetics.

Complex reaction kinetics in soil : Reaction rates in soil are more complex than those in pure systems. Relative contribution of each reactant is to be determined. For example, rate of loss of nitrogen during denitrification is also dependent on microbial activity and energy source. Therefore,

$$\frac{dS}{dt} = -k[(S_1)^n + (S_2)^m]$$

where S_1 and S_2 are the concentration of the two substances. The reaction is of n + m order or second degree order.

When biochemical reactions or microbial growth are studied with time, two situations are commonly encountered. If enzyme concentration or biological population in effect remains constant, Michaelis-Menten kinetics (often termed as 'Monods equation' in microbial growth studies) becomes operative. In a biochemical reaction it is assumed that enzyme (E) and substrate (S) combine to form an enzyme-substrate complex (ES) that dissociate to yield either original substrate and enzyme or a new product (P) and enzyme (E).

$$S + E \underset{k_2}{\overset{k_1}{\rightleftharpoons}} ES \underset{k_4}{\overset{k_3}{\rightleftharpoons}} E + P$$

The rate constant $k_{1,}$ $k_{2,}$ k_3 and k_4 describe the rates associated with each step of catalytic process. From the observations of the properties of many enzymes involved in decomposition it was assumed that the initial velocity V_o at low substrate concentration is directly proportional to [S]. However, at higher substrate concentration the velocity tends to attain a maximum value i.e. the rate becomes independent of [S] (Fig 5.11). This maximum velocity is called 'V_{max}' (μ mol min^{-1}). The Michaelis-Menten equation describes the rate of disappearance of substrate as under:

$$\frac{\Delta S}{\Delta t} = \frac{k_3[E][S]}{k_m + [S]} + [S]$$

$$V = \frac{V_{max}.[S]}{k_m + [S]}$$

where V is the rate of reaction (concentration time^{-1}) and V_{max} is the maximum transformation rate. Brackets indicate concentrations and k_m is Michaelis affinity constant (concentration time^{-1}) or half saturation constant. The equation can be described by the curve of Fig. 11A. It assumes that all non-specified rate factors are at steady state during the period of measurement. However, when substrate concentration is very small in comparison to $k_{m,}$ then denominator of equation is reduced to k_m at constant enzyme. Therefore, equation with respect to the change in substrate concentration will be:

$$\frac{\Delta S}{\Delta t} = \frac{k_3[E][S]}{k_m + [S]} = kS$$

$$\text{or } V = kS \quad where \ k = \frac{V_{max}}{k_m}$$

This equation describes the steep portion of curve near the origin in Fig 5.11.

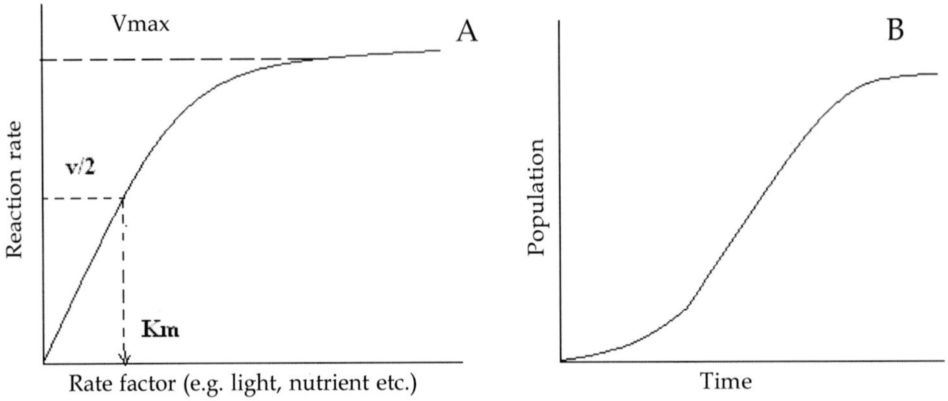

Fig 5.11: Changes in reaction rate (A) with a rate factor according to Michealis-Menten kinetics (B) with time and population response

When substrate S is too larger in comparison to km, then the equation becomes: $\frac{\Delta S}{\Delta t} = k.$

Lineweaver-Burk double reciprocal plot (Fig. 5.12) shows the relationship between substrate concentration (s) and initial reaction velocity (V_o).

$$\frac{\Delta S}{\Delta t} = k_3 [E]$$

$$or \quad V = Vmax$$

$V = V_{max}$ describes a zero order reaction. At low substrate concentration, S is less than k_m and $V = kS$ which shows a first order reaction when $k = \frac{V_{max}}{k_m}$. That is the rate is independent of (or zero order with respect to) S. The rate then depends only upon the steady state concentration of enzyme or in broader sense on microbial population.

The second situation is more common in experimental work i.e. the biological population increases during the study period in response to some growth factor. This requires time and causes a characteristic lag before the biological reaction begins. During lag period, the growth of organism and the corresponding reaction rate are slow but increase exponentially with time and later slows to zero when a new set of limitations come in the way. This results in a sigmoid curve which can be described as:

$$\frac{\Delta x}{\Delta t} = \frac{\Delta S}{\Delta t} = \frac{d_{bd}\, x_0 (x_m - x_0)}{x_m}$$

where x_o is the initial population of organisms; x_m the maximum population of organisms and d_{bd} is difference between the coefficient of birth and death of the organism. This type of equation applies to the situation like CO_2 evolution in soil from freshly added organic matter. The first order curve is best solved when plotted as a straight curve. Similarly, hyperbolic curves are best analyzed when transformed into linear ones. The Lineweaver-Burk double reciprocal plot transforms the hyperbolic equation in such a way that it becomes linear on ordinary graph paper. Computer use has simplified the analyses of hyperbolic curves to obtain V_{max} or $k_{m.}$

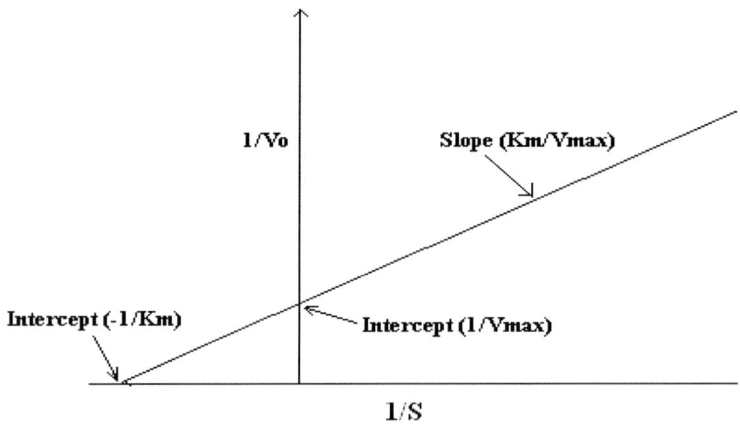

Fig. 5.12: Lineweaver-Burk double reciprocal plot showing relationship between substrate concentration [S] and initial reaction velocity [V_0]

5.10 MODELLING IN ORGANIC MATTER TRANSFORMATIONS

Simulating the turnover time of organic matter : The soil organic matter contains the fractions of widely different turnover time viz., fresh undecomposed plant and animal residues, soil microbial biomass and humified materials of varying ages. Therefore, it is necessary to determine the turnover time of organic input to understand organic matter transformation in soil. The turnover of organic matter is expressed as:

$$\text{Turn over time (yr)} = \frac{\text{Total soil organic matter}}{\text{Annual input of organic matter}}$$

The simulation of turnover of organic matter in soils under different agricultural management practices have effectively been presented through models. Multi-compartmental or multi-sequential models are, therefore, necessary if the turnover processes are to be represented in a realistic way. The percent decomposition of organic residues added is relatively independent of the rates of addition but dependent on its form and composition. The proportion of organic matter decomposed is nearly always independent of the quantity added, if the carbon addition does not exceed 1.5% of the soil dry weight and if decomposition is allowed to continue at least for 3-6 months. However, higher additions cause alteration in the soil characteristics and slow down the decomposition process. The organic matter decomposition depends on its composition especially C, N, carbohydrate, lignin content, etc. Herman *et al.* have used the following formulae to estimate such dependence:

$$CO_2 \text{ evolution} = \frac{1}{\text{lignin - C loss}} = \sqrt{\frac{\%\text{ carbohydrate}}{\text{C : N ratio in straw}}} \quad \text{x \% lignin}$$

Different soil types with varying concentration of clay and organic matter resulted in the carbon mineralization rates ranging from 30 to 81% for uniformly labelled plant materials added at the rate varying from 77 to 84% mg C 100 g^{-1} soil. The turnover times of uniformly labelled immature rice straw ranged from 0.8 to 3.4 depending on loading rate and soil type. It has been observed that the proportion of different plant materials retained in various soils under different agro-climatic conditions is almost similar. It often approaches 20% of the carbon present in the original substrate. This is due to the similarity in microbial production and stabilization of soil organic matter even though the initial decomposition rates may vary widely.

Mathematical models developed for soil organic matter : Many models have focussed on early stage decomposition over period of days, weeks or months. There are fewer models which show simulation over longer periods of years and centuries. SCUAF, Rothamsted carbon model and CENTURY are some examples of such models. These models vary in their complexity but nearly all models partition organic matter into different fractions, and assume a turnover time for each fraction. Jenkinson and Rayner on the basis of 30 years long term experimental data, obtained from Rothamsted, England, developed simulation models for soil organic matter turnover. The data used included:

a) Incubation experiments over 1 to 10 years for decomposition data using ^{14}C labelled plant materials;

b) Radiocarbon dating of soil organic matter which gave idea about the apparent age of old fraction (100-3000 years);

c) Effect of thermonuclear radiocarbon on radiocarbon age (bomb effect), produced through explosions and distributed throughout the atmosphere during the period 1950-1970. This provided information about the materials that had been in soil for 5-20 years;

d) Chloroform-fumigation-incubation data on microbial biomass;

e) Information about plant input levels; and

f) Long term soil organic matter levels used in the experimental plots of 10-100 years.

They identified five soil carbon fractions with different decomposition rates:

a) Decomposable plant materials with calculated half-life $k = 4\ yr^{-1}$
b) Resistant plant materials $k = 0.03\ yr^{-1}$
c) Microbial biomass $k = 0.41\ yr^{-1}$
d) Physically stabilized organic matter $k = 0.014\ yr^{-1}$
e) Chemically stabilized organic matter $k = 0.0003\ yr^{-1}$

Decomposable plant materials had half life of 0.17 yr^{-1} whereas the chemically stabilized soil organic matter had half life of over 3000 years. The decomposition rate of each of these components was assumed to be proportional to its actual content (first order). Further, during decomposition each of the five components presumably decomposed to CO_2; microbial biomass and physically and chemically protected (stabilized soil organic matter) in similar ratio. The model was useful to describe the turnover of soil organic matter in cropped soil. The input from Jenkinson-Rayner model showed a microbial turnover time of 2-5 years with 34 kg N ha^{-1} yr^{-1} flowing through microbial biomass annually in these unfertilized low yielding wheat plots (Table 5.8).

Mean turnover or residence time : The mean turnover or residence time for the reaction is equal to $\left[\frac{1}{k}\right]$. Simple sugars have larger decomposition rate constants (k) of <1 day under laboratory conditions but when same materials are added to soils, the decomposition rates greatly increase. Decomposition curves of complex substrates like straw usually yield multi-slope decomposition curve. This is owing to the presence of varied components in straw having different rate constants (Table 5.9). In straw glucose disappears fast followed by cellulose and hemicellulose during the first year of decomposition while lignin is spatially inaccessible and also, probably, chemically recalcitrant to microorganisms during early decomposition period. So these substrates can be graphically analyzed by curve-splitting techniques which involve subtraction of data obtained for decomposition of most stable components like lignin from overall decomposition curve. Plotting these new points give curve for cellulose and so on. It can be repeated to get curve for sugars and other components therein. Presently the work has been eased by use of computers.

Table 5.8: Carbon and nitrogen turnover in three different ecosystems

Components	Rothmsted (continuous wheat without fertilizers)	Canada (wheat fallow)	Brazil (sugar cane-soil-plant system)
Soil weight (Mg ha^{-1})	2200	2700	2400
Organic carbon (Mg ha^{-1})	26 (1.2)*	65 (1.6)*	26 (1.3)*
Soil carbon turnover (yr)	22	40	2.0
Microbial carbon (kg ha^{-1})	570	1600	460
Microbial nitrogen (kg ha^{-1})	95	300	84
Microbial turnover (yr)	2 - 5	6 - 8	0.24
N flux through microbial biomass (kg $ha^{-1}yr^{-1}$)	34	53	350
N removed by crops (kg $ha^{-1}yr^{-1}$)	24	40	220

Source: Jenkinson and Ladd (1977); Paul and Voroney (1984)

The high input of carbon from sugar cane (1.3 Mg ha^{-1} yr^{-1}) resulted in a calculated turnover time of 0.24 for microbial biomass. This was 28 times faster than the Canadian mollisol under wheat-fallow crop rotation. This also demonstrated the situation in which microbial biomass acts more as a catalyst and as a short term reservoir than a major sink for nutrients. The above models prepared describe the long term effects of crop production on soil organic matter synthesis. However, a more mechanistic model is required to simulate the degradation of complex materials.

---

* Carbon input (Mg ha^{-1} yr^{-1})

Table 5.9: Decomposition rate (k) and turnover time for major cell constituents, fungal components, organic matter and microbial products

Components	Decomposition rate (d^{-1})	Turnover time ($1/k$ in days)
Glucose	1.0000	1
Hemicellulose	0.0070	143
Cellulose	0.0075	133
Amino sugars	0.0045	224
Lignin	0.0005	2000
Fungal cytoplasm	0.0040	250
Fungal cell wall	0.0020	500
Easily decomposable organic matter (unprotected)	0.2000	5
Decomposable organic matter (protected)	0.0080	125
Ligneous organic matter (unprotected)	0.0008	1250
Ligneous organic matter (protected)	0.000008	125000
Decomposable microbial products	0.0800	12.5
Active soil organic matter (protected)	0.0003	3333

Source: Van Veen and Paul (1981); Paul and Clark (1989)

Models in decomposition of complex substrates : A model may provide a better understanding of the quantitative aspects of decomposition of a complex substrate (such as crop residues) when one considers the production of microbial materials from the original substrate as well. A complex substrate like straw was considered to have three fractions.

- Easily decomposable sugars and amino acids;
- Slowly decomposable cellulose and hemicellulose; and
- Resistant lignin.

The knowledge about 'k' (decomposition rate constant) and microbial growth efficiency gives idea about the flow of each fraction. Van Veen and Paul in a study observed organic carbon added to a grassland soil through decomposable organic matter, ligneous materials, plant roots (as residue) and soil organisms was 47.4, 24.8, 8.4 and 1.1 t ha^{-1}, respectively, with decay rate of 0.008-8.0, 0.000008-0.0008, 8.8 and 3.0% per day. The above respective materials contributed 58.0, 30.0, 9.2 and 1.3% of the total organic carbon present in upper 40 cm soil.

The decay rate constant of microbial biomass in this model is function of all processes that require substrate other than for growth. These include maintenance of cell integrity and nutrient uptake etc. In soil, environmental stresses are greater and different species/genus occur simultaneously with varying level of their activities so chemostat concept is not applicable. Measurements of maintenance of energy of a soil population that is essentially in a steady state are much lower than the values for chemostat. Further, natural populations feed on each other (cryptic growth). The concepts of maintenance rates and cryptic growth are, therefore, taken into consideration by incorporating them into death rate constants. These microbes on death become part of easily decomposable part (i.e. their cytoplasmic contents) as well as more resistant pool (i.e. their cell wall).

Growth efficiency of 60% has often been found for a soil population utilizing a simple substrate like glucose. More complex substrates have lower efficiencies of even 20%. The decomposition of glucose follows first order reaction kinetics. The degradation rate for lignin was 25% than that of the associated cellulose indicating that even closely associated materials decompose at different rate. The turnover of aromatics is a major factor in determining soil organic matter dynamics. Aromatics are co-metabolized by fungi but require the presence of other easy/slow decomposable substrates.

The residue decomposition rate has been found insensitive to soil type but stabilization of soil organic matter and microbial population are soil type-dependent. Clay type, such as allophane, and soil silt and clay content control the stabilization rates. Soil organic matter components show great variation in decomposition rates. In native grassland soil about 9% soil organic matter consisted of roots and plant residues and 4% protected and decomposable labile organic matter with decomposition rate of 8% day^{-1} under optimal conditions. Soil organisms were 1.3% with decay rate of 3% day^{-1}. Other fractions comprizing about 84% showed much slow decay rate. Bulk of soil organic matter (54%) consisted of decomposable organic matter that was protected by attachment to clay and other mineral components and had decay rate of 0.08% day^{-1}. Ligneous fractions had much lower decay rates. Thus decomposition rate of most readily decomposable material was about 1 million times than that of most stable fraction.

Mathematical modelling has proved valuable tool in predicting long term changes in soil organic matter owing to different management practices. Such data are extremely important in management decisions about soil and are applicable in estimating the interactions of soil organic matter and global carbon cycle, CO_2 content in atmosphere and climatic changes due to global warming etc. Since quality of organic matter retained in a particular soil is the net result of balance between organic inputs and mineralization as affected by physicochemical properties of site, therefore modelling data may help in maintaining soil ecosystem as a healthy renewable resource.

Selected References

Aiken, G.A., McKnight, D.L., Wershaw, R.L. and MacCarthy, P. 1985. *Humic Substances in Soil, Sediment and Water: Geochemistry, Isolation and Characterization.* John Wiley & son, New York, USA.

Alexander, M. 1977. *Introduction to Soil Microbiology* (2nd edn.), John Wiley & sons, New York, USA.

Alloway, B.J. and Ayres, D.S. 1997. *Chemical Principles of Environmental Pollution.* Chapman Hall, London, UK.

Allison, F.E. 1973. *Soil Organic Matter and its Role in Crop Production.* Elsevier, New York, USA.

Anderson, J.P.E. and Domsch, K.H. 1980. Quantities of plant nutrients in the microbial biomass of selected soils. *Soil Sci.,* 130: 211-216.

Atlas, R.M. and Bartha, R. 1993. *Microbial Ecology: Fundamentals and Applications* (3rd edn.), Benjamin/Cummings, Redwood City, USA.

Black, C.A. 1968. *Soil Plant Relationships.* John Wiley & sons, London, UK.

Bolin, B. Degen, E.T., Kempe, S. and Ketner, P. (eds.). 1979. *The Global Carbon Cycle.* John Wiley & sons, New York, USA.

Chen, Y. and Avnimelech, Y. 1986. *The Role of Organic Matter in Modern Agriculture.* Martinus Nijhoff, The Hague, Netherland

Coleman, D.C., Oades, J.M. and Gorollehara. 1989. *Dynamics of Soil Organic Matter in Tropical Ecosystems.* NIFTAL, College of Tropical Agriculture and Human resources, University of Hawaii, Honolulu, Hawaii, USA.

Cranwell, P.A. and Haworth, R.D. 1975. The chemical nature of humic acids. pp. 13-18. In: *Humic Substances: Their Structure and Function in Biosphere* (eds. D. Povoledo and H.L. Golterman), Centr Agric. Publ., DOC, Wageningen, the Netherlands.

Fenchel, T. and Blackburn, T.H. 1979. *Bacteria and Mineral Cycling.* Academic Press, London, UK.

Flaig, W., Beutelspacher, H. and Reitz, E. 1975. Chemical composition and physical properties of humic substances. pp.1-211. In: *Soil Components Vol. 1. Organic Components* (ed. J.E. Gieseking), Springer Verlag, New York, USA.

Herman, W.A., McGill, W.B. and Dormaar, J.F. 1977. Effects of initial chemical composition on decomposition of roots of three grass species. *Can. J. Soil Sci.,* 57: 205-215.

Jackson, A.R.W. and Jackson, J.M. 1996. *Environmental Science: Natural Environment and Human Impact,* Addison-Wesley Longman Ltd., Edinburg, Harlow, Essex, UK.

Jenkinson, D. S. 1981. The fate of plant and animal residues in soil. pp. 505-561. In: *The Chemistry of Soil Processes.* (eds D. J. Greenland and M. H. B. Hayes). John Wiley and sons, New York, USA.

Jenkinson, D.S. and Ladd, J.N. 1981. Microbial biomass in soil measurement and turnover. pp. 415-471. In: *Soil Biochemistry Vol. 5.* (eds. E.A. Paul and J.N. Ladd), Marcel Dekker, New York, USA.

Jenkinson, D.J. and Rayner, J.H. 1977. The turnover of organic matter in soil of the Rothamsted classical experiments. *Soil Sci..,* 123: 298-305.

Jenny, H. 1941. *Factors of Soil Formation.* McGraw Hill, New York, USA.

Jones, D.G. 1987. *Plant Pathology: Principles and Practices.* Open Univ. Press, Milton, Keynes, UK.

Kay, B.D. 1998. Soil structure and organic carbon: A review. pp. 169-197. In: *Soil Processes and the Carbon Cycle* (eds. R. Lal, J.M. Kimble, R.F. Follett and B.A. Stewart), CRC Press, Boca Raton, Florida, USA.

Killham, K. 1994. *Soil Ecology.*Cambridge Univ. Press, Cambridge, UK.

Lespinat, P.A., Heiter, J.M., Thomann, C., Chone, T. and Dimon, A. 1976. Utilisation de maismur uniformement marque au ^{14}C pour l'etude de la matiere organique de trios sols (andosols, sol burn, soil ferrasutique). *Science du Sol, Bulletin de l'A. F. E. S.,* 1: 53-66.

Lisk, D.J. 1991. Environmental effects of landfills. *Sci. Total Environ.,* 100: 415-468.

Lynch, J.M. 1983. *Soil Biotechnology: Microbiological Factors in Crop Productivity.* Blackwell Scientific Publications, Oxford, UK.

Lynch, J.M. and Panting, L.M. 1980. Cultivation and the soil biomass. *Soil Biol. Biochem.,* 12: 29-33.

Manhahan, S.E. 1991. *Environmental Chemistry.* Lewis Publishers, Chelsea, UK.

Martin, J.P. and Haider, K. 1971. Microbial activity in relation to soil humus formation. *Soil Sci.,* 111: 54-63.

Mathur, S.P. and Farnham, R.S. 1985. Geochemistry of humic substances in natural and cultivated peatlands. pp. 53-85. In: *Humic Substances in Soil, Sediment and Water: Geochemistry, Isolation and Characterization.* (eds. G. R. Aiken, M. Mcknight, R. L. Wershaw and P. MacCarthy), John Wiley & sons, New York, USA.

McBride, M.B. 1994. *Environmental Chemistry of Soils.* Oxford Univ. Press Inc., New York, USA.

Metting Jr., F.B. (ed.). 1993. *Soil Microbial Ecology.* Marcel Dekker, New York, USA.

Patton, A.R. 1968. *Biochemical Energetic and Kinetics.* W.B. Saunders Co., Philadelphia/ London, UK.

Paul, E.A. and Clark, F.E. 1989. *Soil Microbiology and Biochemistry.* Academic Press, New York, USA.

Paul, E.A. and Van Veen, J.A. 1979. The use of tracers to determine the dynamic nature of organic matter pp. 75-130. In: *Modelling Nitrogen from Farm Wastes* (ed. J.K.R. Gasser). Applied Science Publishers, Essex, London, UK.

Paul, E.A. and Voroney, R.P. 1984. Field interpretation of microbial biomass activity measurements pp. 509-514. In: *Current Perspectives in Microbial Ecology* (eds. M.J. Klug and C.A. Reddy). Am. Soc. Microbiol., Washington, DC, USA.

Robinson, H.D., Barber, C. and Morris, P.J. 1982. *Water Pollut. Cont.*, 54: 465-478.

Russel, J.R. 1962. *Soil Conditions and Plant Growth*. Longmans Green & Co., London, UK.

Russel, S.R. 1977. *Plant Root System - Their Function and Interaction with the Soil.* McGraw-Hill, London, UK.

Schlesinger, W.H. 1997. *Biogeochemistry: An Analysis of Global Change* (2nd edn.), Academic Press, San Diego, USA.

Schlesinger, W.H. 1984. Terrestrial vegetation in the global carbon cycle – Measurements by remote sensing. pp. 111-127. In: *The Role of Terrestrial Vegetation on the Global Carbon Cycle: Measurement and Remote Sensing* (ed. G.M. Woodwell), John Wiley & sons, Chichester, UK.

Schnitzer, M. 1976. The chemistry of humic substances pp. 89-107. In: *Environmental Biogeochemistry Vol. 1. Carbon, Nitrogen, Phosphorus, Sulphur and Selenium Cycles,* (ed. J.O. Nriagu). Proc. 2nd Int. Symp. Environ. Biogeochem., Hamilton, Ontario, Canada, 8-12 April, 1975, Ann Arbor Science, Ann Arbor, Mich., Canada.

Schnitzer, M. and Khan, S.U. (eds.). 1978. *Soil Organic Matter*. Elsevier, New York, USA.

Schulten, H.R. and Schnitzer, M. 1993. A state of art structure concept for humic substances. *Naturwissonschaften,* 80: 29-30.

Sparks, D.L. 1995. *Environmental Soil Chemistry*. Academic Press, San Diago, California, USA.

Steelink, C. 1982. Elemental characteristics of humic substances. pp. 457- 476. In: *Humic Substances in Soils, Sediments and Water: Geochemistry, Isolation and Characterization.* (eds. G. R. Aiken, M. Mcknight, R. L. Wershaw and P. MacCarthy). John Wiley & sons, New York, USA.

Stevenson, F.J. 1982. *Humus Chemistry: Genesis, Composition and Reactions.* John Wiley & sons, New York, USA.

Stevenson, F.J. 1986. *Cycles of Soil Carbon, Nitrogen, Phosphorus, Sulphur, Micronutrients.* John Wiley & sons, New York, USA.

Subba Rao, N.S. 1982. *Advances in Agricultural Microbiology*. Oxford & IBH Publishing Co., New Delhi, India.

Swift, M.J., Heal, O.W. and Anderson, J.M. 1979. *Decomposition in Terrestrial Ecosystem. Studies in Ecology Vol. 5,* Blackwell Scientific Publication, Oxford, UK.

Tan, K.H. 1993. *Principles of Soil Chemistry* (2nd edn.). Marcel Dekker, New York, USA.

Tate, R.L. 1987. *Soil Organic Matter: Biological and Ecological Effectors.* John Wiley & sons, New York, USA.

Van Veen, J.A. and Paul, E.A. 1981. Organic carbon dynamics in grassland soils 1. Background information and computer simulation. *Candian J. Soil. Sci.,* 61:185-201.

Vaughan, D. and Malcolm, R.E. 1985. *Soil Organic Matter and Biological Activity. Developments in Plant and Soil Sciences Vol. 16,* Nijhoff & Junk, Dordrecht.

Whittaker, R.H. 1975. *Communities and Ecosystem* (2nd edn.). Macmillan, New York, USA.

Wolf, D.C., Legg, J.O. and Bouton, T.W. 1994. Isotopic methods for the study of soil organic matter dynamics. pp. 865-906. In: *Methods of Soil Analysis, Part 2: Microbiological and Biochemical Properties* (eds. R.W. Weaver *et al.*), Soil Science Society of America, Madison, USA.

Woomer, P.L. and Swift, M.J. (eds.). 1994. *The Biological Management of Tropical Soil Fertility.* John Wiley & sons, Chichester, Sussex, UK.

Zibilske, L.M. 1994. Carbon mineralization. pp. 835-863. In: *Methods of Soil Analysis, Part 2: Microbiological and Biochemical Properties* (eds. R.W. Weaver *et al.*), Soil Science Society of America, Madison, USA.

— *** —

Chapter 6

Transformation of Nitrogen in Soil

Someday we are destined to become a part of the future as we enter into something or someone in the never-ending and intriguing continuity of pattern set in motion by nature's incredible forces.

(Stanley E. Wedberg)

Nitrogen ranks fourth element in cell composition after C, H and O. It forms main components of amino acids, the building blocks of peptides and proteins. Besides, it is present in the genetic material of cell as well as in the chlorophyll and cell wall in the form of chitin and mucopeptides. Plant growth in soil is often limited by less supply of nitrogen, therefore a major nutrient limiting factor in crop production. About 26 x 10^6 Mg N is globally applied to soil each year to enhance crop productivity.

Nitrogen sources in nature : About 98% of earth's nitrogen (1.63 x 10^{11} Mg) exists in lithosphere (contained in igneous rocks) and hardly 1.2% of the total nitrogen exists in the atmosphere (3.86 x 10^9 Mg). The hydrosphere and biosphere share 2.30 x 10^7 and 2.50 x 10^5 Mg nitrogen, respectively. The atmospheric air contains 79% N by volume and 75% N by weight. This accounts for approximately 85, 000 t N present over each hectare of land. However, this N is inert so cannot be used unless chemically combined with H and

O. The largest biologically potential available reserviour of N is atmosphere. Annually 400 x 10^6 t N is transferred between various nitrogen pools from atmosphere to soil and vice versa (Fig. 6.1).

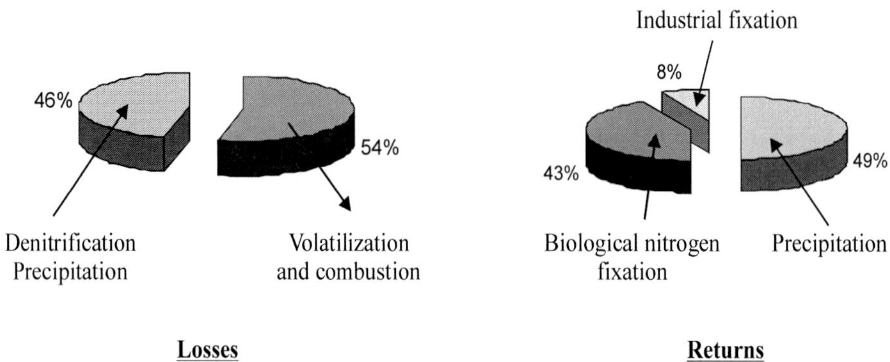

Fig 6.1 : Graphic presentation of annual losses and returns of nitrogen to soil

The soils, whether used for forest or agriculture, possess 10-20 times nitrogen than the standing vegetation including roots. Plants take up only 30-60% of the available mineral nitrogen in soil at the utilization efficiency of 40%. The estimated photosynthetic fixation of carbon on global basis is 70,000 Mg (1 Mg = 1 metric tonne) which means plant N uptake of 14,000 Mg (C :N ratio 50 :1). Therefore, mineral N available in soil works out to be 3,500 Mg. A soil having 3% stabilized organic matter containing about 5% N (on dry weight basis) will have 0.15% N i.e. 3,000 kg N HFS^{-1} (per hectare furrow slice - 15 cm layer). But the quantity of nitrogen liberated from soil organic matter is less than 5% of the total. It amounts less than 150 kg N annually. Most of this N is usually lost either through denitrification (in the form of NO, N and N_2) or leached as NO_3 or volatilized as ammonia (NH_3) gas or gets fixed in clay lattice (as NH_4^+). Therefore, soils always remain deficient in nitrogen so far as the plant requirement is concerned. At a time only about 1% or less of the total N in soil ($\leq$15 kg N ha^{-1}) is available to the plants and microorganisms as nitrate or ammonium.

6.1 NITROGEN CYCLE IN NATURE

About 90% of soil N is present in unavailable form, which exist as a component of rocks, sediments etc. Like that of soil, the amount of nitrogen in biosphere is very small. This nitrogen is mobile and keeps on recycling. The annual rate of removal of N from atmosphere by fixation and its addition to soils is approximately balanced by an equal amount of N returned to the atmosphere by denitrification, volatilization etc. Thus, overall quasi-equilibrium exists between the amount of N in the atmosphere and the amount of N in soils on global basis. In fact, nitrogen has many biological states. This means that biologically-mediated redox reactions are central to the nitrogen cycle. Alteration in chemical speciation in these reactions often results in the movement of nitrogen from one reserviour to another in different physical locations.

Nitrogen is recycled to the soil through electric discharges, combustion, industrial and biological processes. The dry and wet deposition of nitrogen in the form of nitrogen oxides, ammonia, nitrates, etc., dissolved in rain waters, including lightening discharges to soils is estimated to be 200 x 10^6 t yr^{-1}. The industrial fixation, human-induced interference in nitrogen cycle, is based on Haber-Bosch process in which H_2 and N_2 gases are forced to react under huge pressure (500 atmospheric pressure) and temperature (1,200°C) in presence of finely divided Fe_2O_3 catalyst to break the triple covalent bond between N≡N atoms, and form ammonia. The ammonia so produced is directly used as fertilizer in soil. This fixation contributes as much as 30 x 10^6 t yr^{-1}.

At normal temperature and pressure only tiny microbes are capable of fixing inert nitrogen, present in atmosphere, through their specialized enzymes. The inputs of nitrogen to soil from biological fixation vary greatly from one ecosystem to another e.g. tropical legumes may fix as much as 400 kg N ha^{-1} whereas symbiotic nitrogen fixation in temperate may yield 5-30 kg N ha^{-1} yr^{-1}. About 12.8 kg N ha^{-1} is naturally added to soil annually, of which 72% comes from biological fixation and 28% from other sources. The nitrogen passes through several processes during recycling (Fig. 6.2). It may be taken up by the plants either directly from symbiotic microorganisms harbouring in their roots, stems or leaves or indirectly through soil. The plants absorb ammonium as well as nitrate from soil through their roots for growth and development. Plants on their death directly or through secondary consumers i.e. animals contribute to soil nitrogen and become part of soil organic nitrogen on microbial decomposition and proteolysis. Besides, microbes independently contribute to soil organic N. As much as 85-90% of fixed N incorporated into new plant biomass is derived from the mineralization of soil humus and indigenous biomass. It is through microbial nitrogen pool that organic nitrogen is released in inorganic form (NH_4^+) after mineralization or it may become a part of stabilized soil organic matter or humus. The ammonium may be fixed by clay lattice on exchange sites or become organically bound. Much of the mineralized ammonium is converted to nitrate by nitrifying bacteria through nitrification process. The nitrate may be either taken up by plants or because of being most soluble in water and highly mobile may leach through percolating waters to underground waters. It may also be converted to nitrogen gas and nitrogen oxides (N_2O and NO) that escape into the atmosphere thereby complete the nitrogen cycle.

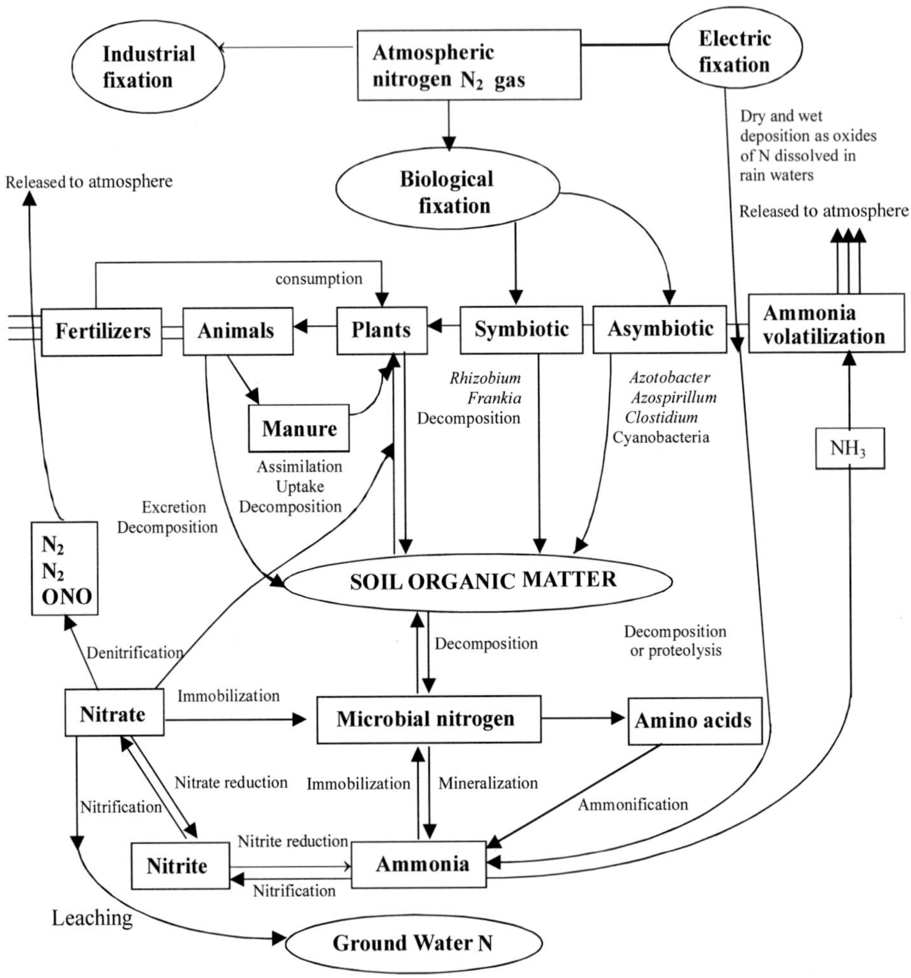

Fig. 6.2 : The nitrogen cycle in nature

The plant assimilated nitrogen (proteins, etc.) is consumed by animals and ultimately returned to soil in the form of organic residues or manure which then follows same cycle as those of plant residues. Volatilization of ammonia and combustion also contribute in the return of nitrogen from soil to atmosphere.

6.2 ORGANIC NITROGEN IN SOIL

The total nitrogen content in soils varies from 0.2% in sub-surface arable soil to 2.5% in peat. In top soil the nitrogen content is 0.03-0.4%. Most of this nitrogen (95%) occurs in organic form so is less available to the plants because of its association with certain silicate clays or resistant humic acids. Typically only about 2-3% N in soil organic matter is released as inorganic N. Different forms of nitrogen found in soil and their approximate relative amounts are given in Table 6.1.

Table 6.1 : Contribution of different forms of nitrogen to soil nitrogen

Forms of nitrogen	Relative amount (% of total N)
Proteins, peptides, etc.	20-40
Labile N i.e. NH_4^++ exchangeable N	25-30
Hydrolyzable unknown N forms	15-20
Acid insoluble N associated with aromatics	10-15
Amino sugars N	5-10
Nucleic acid (purines and pyrimidine derivatives)	1

Organic N is rendered available in soil by microbial decomposition. In soils the processes of mineralization and immobilization do occur simultaneously, one trying to supersede the other depending upon the prevailing soil and environmental conditions and the nature and amount of substrate availability. The net effect is an increase or decrease in available nitrogen.

6.2.1 NITROGEN MINERALIZATION

The conversion of organically bound nitrogen such as proteins, enzymes, nucleic acids, amino sugars and cell wall constituents like chitin, peptidoglycan etc. to simple mineral (inorganic) nitrogen forms, NH_4 or NO_3, is called 'mineralization' and reverse process is 'immobilization'. The mineral nitrogen added to soil through various fixation processes is subject to uptake by plant roots and microorganisms and assimilated into organic form.

$$\text{Organic N} \underset{\text{Immobilization}}{\overset{\text{Mineralization}}{\rightleftharpoons}} \text{Inorganic N}$$

$$R\text{-}NH_2 \underset{-2H_2O}{\overset{+2H_2O}{\rightleftharpoons}} OH^- + R\text{-}OH + NH_4 \underset{-O_2}{\overset{+O_2}{\rightleftharpoons}} 4H^+ + \text{energy} + NO_2^- \underset{-1/2[O_2]}{\overset{+1/2\,[O_2]}{\rightleftharpoons}} \text{energy} + NO_3$$

A wide variety of heterotrophic organisms (bacteria, actinomycetes and fungi) are involved in the process of mineralization and immobilization of substrates in soil. Both processes may operate concurrently in a soil, hence are sometimes referred to as mineralization-immobilization turnover. The mineralization process involves two reactions : a) amminization and b) ammonification.

Aminization : Aminization means the enzymatic conversion of proteins to amines and amino acids through hydrolytic reactions so the process leads to proteolysis.

$$\text{Proteins} \xrightarrow[H_2O]{\text{Proteinases}} \text{Polypeptides} \xrightarrow[H_2O]{\text{Peptidases}} \text{Amino acids}$$

The proteinases depolymerize large macromolecules (proteins) into smaller units i.e. polypeptides, each having more than two amino acid units. Further degradation is accomplished by peptidases converting di-, tri- or oligo-peptides into amines, amino acids and urea. Heterotrophic organisms are involved in this process. Powerful proteinases are secreted by fungi and by some bacteria, particularly certain species of *Bacillus, Clostridium* and *Pseudomonas*. The production of proteinases by microorganisms depends on several factors. When proteins reach soil, these are first attacked by fungi and reduced to polypeptides. Further breakdown is accomplished by bacteria as well as fungi. Spore forming bacteria, *Pseudomonas*, actinomycetes and fungi readily attack amino acid and amino sugars in soil. In neutral and alkaline soils mostly bacteria dominate and in acid conditions fungi predominate. Approximately 10^5-10^7 micro-organisms are active mineralizers in a single gram of soil which typically release 1-20 mg NH_4^+-N day^{-1} kg^{-1} soil under temperate conditions.

$$R\text{-}NH_2 + H_2O \longrightarrow NH_3 + R\text{-}OH + \text{energy}$$

$$NH_3 \xrightarrow{H_2O} NH_4^+ + OH^-$$

The proteolytic breakdown may involve oxidative or reductive deamination or decarboxylation depending upon the aerobic or anaerobic conditions as well as the type of microorganisms acting on the substrate.

$$\underset{\text{Asparatic acid}}{COOH\text{-}CH_2\text{-}\overset{NH_2}{CH}\text{-}COOH} + O_2 \xrightarrow{\text{Aerobic}} COOH\,CH_2COOH + NH_3 + CO_2 + \text{energy}$$

$$\underset{\text{Asparatic acid}}{COOH\text{-}CH_2\overset{NH_2}{\overset{/}{C}}HCOOH} \xrightarrow{\text{Anaerobic}} COOH\text{-}CH{=}CH\text{-}COOH + NH_3 + \text{energy}$$

$$\underset{\text{Valine}}{(CH_3)_2CH_2\overset{NH_2}{\overset{/}{C}}HCOOH} \xrightarrow{\text{Decarboxylation}} (CH_3)_2CH_2CH_2NH_2 + CO_2 + \text{energy} \xrightarrow{+H_2O} (CH_3)_2\,CH_2\,CH_2\,OH + NH_3$$

Ammonification : The conversion of amines and amino acids produced during aminization into ammonia, known as ammonification, is accomplished by diverse types of aerobic and anaerobic fungi, bacteria and actinomycetes. For example, deamination of alanine can occur as under :

$$\underset{\text{Alanine}}{CH_3\overset{NH_2}{C}HCOOH} + 1/2O_2 \xrightarrow{\text{Alanine deaminase}} \underset{\text{Pyruvic acid}}{CH_3COCOOH} + NH_3$$

The breakdown of amino acids releases ammonia, carbon dioxide and a variety of low organic acids like aldehydes, ketones, alcohols, etc. Some soil bacteria like *Bacillus, Micrococcus, Sarcina, Aerobacter,* etc. are capable of converting urea to ammonia using enzyme urease.

$$(H_2N)_2CO + 2H_2O \xrightarrow{\text{Urease}} (NH_4)_2CO_3 \longrightarrow 2NH_3 + CO_2 + H_2O$$

The amount of available mineral nitrogen ($NH_4^+ + NO_3$) found in soil largely depends on the difference between the rates of immobilization and mineralization. In many well-drained mineral soils about 1.5-3.5% organic N is mineralized annually. The mineralization depends largely on the temperature, moisture and aeration status of the soil. For soils with 0.05 to 0.10% N, this amounts to 25-50 kg N ha^{-1}-20 cm layer.

6.2.2 FATE OF AMMONIA IN SOIL

The ammonia, produced as a result of enzyme-catalyzed reactions, is a polar molecule and readily combines with a proton to form ammonium (NH_4^+). It, however, does not remain as such in soil but is rapidly transferred to other nitrogen pools in soil-plant system (Fig. 6.3).

Mineral nitrogen is removed from soil through assimilation by plants and microbes or may be denitrified or leached from root zone. The change in mineral nitrogen (ΔN_i) is represented as :

$$\Delta N_i = N_m - (N_a + N_p + N_v + N_i + N_d)$$

where N_i = inorganic nitrogen in soil; N_m = nitrogen mineralized; N_a = nitrogen assimilated by microbes; N_p = nitrogen assimilated by plants; N_v = nitrogen volatilized; and N_d = nitrogen denitrified.

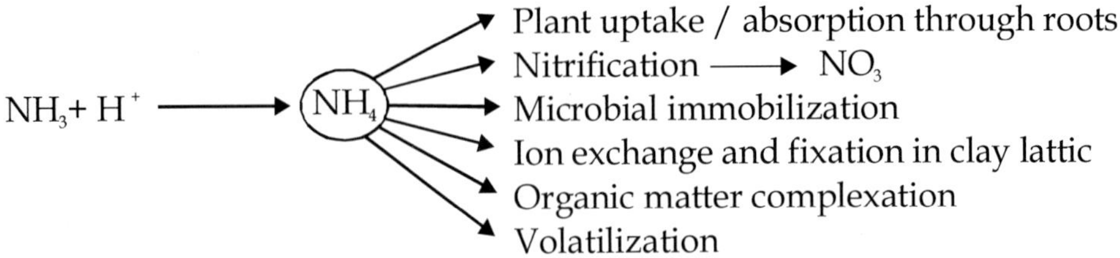

Fig. 6.3 : Fate of ammonia in soil

6.2.3 AMMONIUM FIXATION BY CLAY MINERALS

Ammonical nitrogen appears in soil solution and can be taken up by plants. It is often a preferred nitrogen source in solution. However, in most soils owing to the positive charge of ammonium ion, the ammonical N is adsorbed in clay lattice and humus where these are held in exchangeable form; so partially protected from leaching. Therefore, available amount for plant uptake temporarily gets decreased. Having approximately the similar ionic radii and energy as that of K^+ ion, it readily competes with K^+ for fixation sites and gets fixed in clay lattice of certain expanding type of clay minerals. Certain 2 :1 type clay minerals especially vermiculite and illite fix NH_4^+ by replacement of their interlayer cations in expanded lattice of clay minerals. The fixed NH_4^+ can be replaced by cations that expand the lattice (Ca^{++}, Mg^{++}, Na^+ and H^+) but not those that contract it (K^+). Fixation of freshly applied NH_4^+ can occur in clay, silt or sand-size particles, if they contain substantial amount of vermiculite. Coarse clay (0.2-2.0 μm) and fine silt (2-5 μm) are most important fractions involved in the fixation of NH_4^+. Clay fixation of fertilizer-NH_4^+ provides some degree of protection against rapid nitrification and subsequent leaching. The presence of large amounts of K^+ often restricts NH_4^+ fixation. Therefore, adding potassium fertilizers prior to NH_4^+ is a practical way of reducing NH_4^+ fixation in such problematic soils. The upper limit of fixation in field soils is about 1-2 m. eq 100 g^{-1} soil. About 10% nitrogen may be fixed and its distribution in soil profile parallels that of the clay. The fixed NH_4^+ is slowly released back to the soil solution. Some ammonium may react with soil organic matter to form quinone-NH_3 complexes so help in the stabilization of soil organic matter. Whether an amino acid is used for deriving energy or for synthesis of building blocks depends on a complex series of feed back controls. Usually carbohydrate carbon is used as an energy source. However, under extreme stress conditions amino acid carbon may also be used as energy source. Only a few bacteria are known to utilize amines both as carbon and nitrogen source for their growth.

The non-protein sources, including cell wall constituents of bacteria and fungi, involve specific enzymes like chitinase, kinase etc. to decompose and deaminate them to ammonium. The hydrolysis of nucleic acids (DNA and RNA), though not readily, is carried out by aerobic bacteria and some fungi through their exo- and endo-nuclease enzymes (Fig. 6.4). Nucleic acids are cyclic nitrogen compounds connected to phosphate groups by ester linkages. The breakdown of purines and pyrimidines has not been well understood so far. Anaerobic bacterium *Clostridium* sp. seems to attack purine and yield methane.

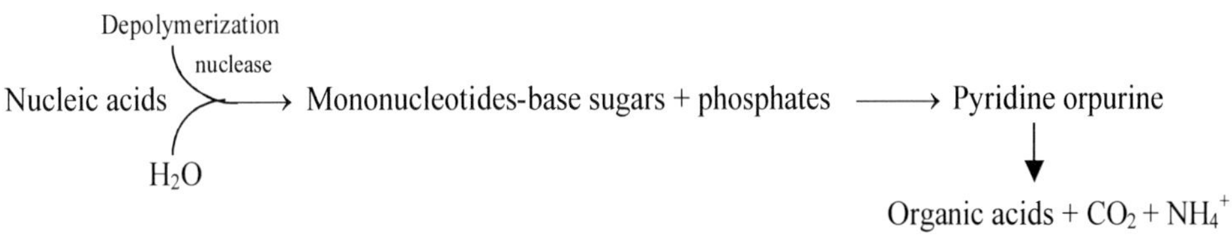

Fig. 6.4 : Reactions involving the breakdown of nucleic acids

Only about 50% the nitrogen present in soil organic matter has so far been characterized. The nitrogen frequently attached to phenolic compounds is much more resistant to mineralization process. Another fate of NH_4^+ may be its immobilization by microorganisms or conversion to NO_2 and NO_3 by nitrifier microorganisms and either taken up by plants or leached to ground waters alongwith percolating soil water. The nitrate may be denitrified if anaerobic conditions prevail. Presence of larger amounts of NH_4^+ leads to ammonia volatilization so loss of soil N is high, especially when temperatures are high.

There appears to be a consensus that N mineralization occurs only if nitrogen content of the substrate is >2%, otherwise immobilization prevails. Further, the amount of nitrogen mineralized tends to be related to the type and amount of soil organic matter and environmental conditions. Soil organic matter contains approximately 5% nitrogen, and more than 90% of soil nitrogen is in organic form. Annually about 1-3% of organic nitrogen is mineralized to inorganic nitrogen. Therefore, the amount of nitrogen mineralized during a growing season can be estimated as under :

$$SOM \times S_w \times N_p \times N_m$$

where, SOM = Soil organic matter content in soil (%)

S_w = Weight of hectare furrow slice soil (kg ha^{-1})

N_p = Nitrogen content present in soil organic matter (%)

N_m = Amount of nitrogen mineralized (%)

6.2.4 FACTORS AFFECTING NITROGEN MINERALIZATION

The optimum temperature, better aeration, greater amount of proteins in organic matter, neutral soil condition and optimum water content favour the process of mineralization. However, the presence of more carbohydrates and less amounts of nitrogen (i.e. wide C :N ratio) leads to immobilization. The accumulation or immobilization of NH_4^+-N in soil depends on the microbial requirement of nitrogen for growth, the C :N ratio of substrate and the competition from plant roots. Cultivation invariably decreases soil nitrogen content.

The optimum temperature range for mineralization is 40 to 60°C. The rate of metabolism of microorganisms increases about 2-fold for each 10°C rise in temperature in the range of 20 to 60°C. In general, temperatures favourable for root growth also favour mineralization. Wetting and drying appears to stimulate mineralization and causes rapid mineralization or a flush in nitrogen availability after each re-moistening of a dry soil. Optimal soil water conditions for aerobic transformations of nitrogen is 50-70% water holding capacity or around -0.3 MPa. Freezing followed by thawing also stimulates mineralization.

There is a good positive correlation between clay content and soil organic matter suggesting that clays inactivate the proteolytic enzymes. Decomposition is fast in warm well-aerated moist soils, such as in sandy soil in sunshine, but slow in clay soil in cold springs. Acidic conditions usually reduce the rate of organic matter decomposition, release mineral N and cause accumulation of partially decomposed plant residues on the surface of acidic soils. Liming increases the rate of mineralization. Anaerobic conditions also decrease microbial activity and cause accumulation of peat in flooded areas with reduced supply of mineral N. Ploughing of a well-fertilized or legume-rich grassland or cleaning forest in temperate accelerate mineralization.

6.2.5 NITROGEN IMMOBILIZATION

Immobilization, also known as 'assimilation', is biochemical conversion of inorganic nitrogen (NH_4^++ NO_3^-) to organic form such as amino acids, proteins, nucleic acids and minor nitrogenous components of cell. Organic nitrogen is a major by-product of microbial growth. Immobilization in soil is carried out by rapidly growing microbial populations and plant roots. The N immobilization or assimilation is catalyzed by all groups of soil microorganisms. Mineralization and immobilization are interdependent processes that strongly control the availability of nitrogen in soil. Nitrification can occur between mineralization and immobilization and the prevailing conditions control the available form of nitrogen i.e. whether NH_4^+ or NO_3. If decomposing organic matter contains low nitrogen relative to carbon, the microorganisms will immobilize NH_4^+ or NO_3 in soil.

Pathways of nitrogen immobilization : Microorganisms follow two major pathways for the assimilation of NH_4^+-ions. In first pathway NH_4^+ is directly incorporated into glutamate by enzyme glutamate dehydrogenase (GDH), which operates at relatively high NH_4^+ concentration. The other pathway operates at low NH_4^+ concentration and involves the use of glutamine synthetase (GS)-glutamine synthase {glutamine α-oxoglutarate amino-transferase (GOGAT)} enzyme system. Glutamate is a key substance in the formation of other amino acids.

GDH pathway makes the use of coenzymes NAD or NADP during the release or uptake of NH_4^+. This allows the organisms to control immobilization-mineralization processes. The enzyme GDH is involved in the formation of glutamate from 2 α-ketoglutarate molecules and ammonia. The reaction is reversible and has a low affinity for ammonia. So GDH is relatively inefficient in picking up the ammonia molecule. The GDH pathway incorporates NH_4^+ only at relatively high NH_4^+ level of >1 mM in solution or >0.5 mg NH_4^+ kg^{-1} soil.

Bacteria are frequently obliged to exist in conditions of shortage of ammonia and other precursors like nitrite or nitrate. Therefore, at low NH_4^+ level <1 mM in solution or <0.5 mg NH_4^+ kg^{-1} soil, most frequently observed in soils, microorganisms follow glutamine synthetase-glutamine synthase (GS-GOGAT) pathway which acts in two steps. The enzyme glutamine synthetase first binds ammonia to glutamate to form glutamine. GS is very efficient in picking up ammonia that requires more energy than the earlier pathway so one ATP is lost during each reaction. In second step, glutamine formed reacts with α-ketoglutarate in presence of another enzyme glutamine synthase. Two glutamates are formed from one glutamine, α-ketoglutarate and NH_4^+ (Fig. 6.5). α-ketoglutarate plays major role in the flow of NH_4^+ into amino acids. It is subject to cumulative feedback control by about eight different inhibition reactions. Once nitrogen is incorporated into cell, further transamination reactions, involving transaminases, transfers NH_4^+ from glutamate to pyruvate to make amino acid alanine and so on.

Nitrate Assimilation : Various microorganisms assimilate inorganic nitrogen as nitrate (NO_3^-). These organisms first use the assimilatory nitrate reductase enzyme for the reduction of nitrate (NO_3^-) to nitrite (NO_2^-) and then assimilatory nitrite reductase enzyme to reduce nitrite (NO_2^-) to ammonium (NH_4^+). These reductases are molybdo-proteins and are synthesized only when soil N is limiting or when nitrate is present. Once nitrate is reduced to ammonium (NH_4^+), it probably is assimilated via GS/GOGAT pathway of nitrogen assimilation rather than by GDH pathway.

Pathway I: GDH pathway

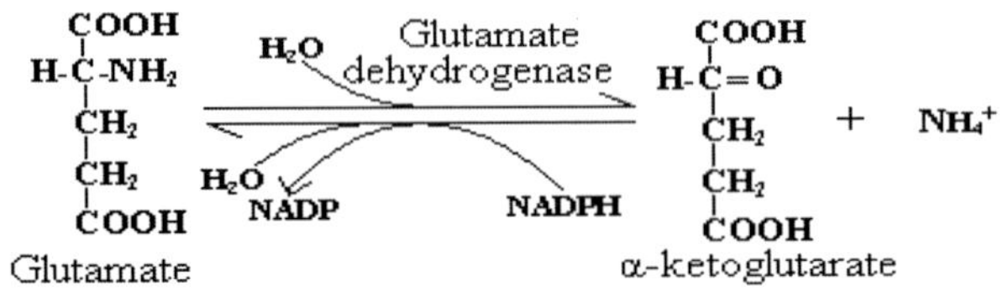

Pathway II: GS-GOGAT pathway

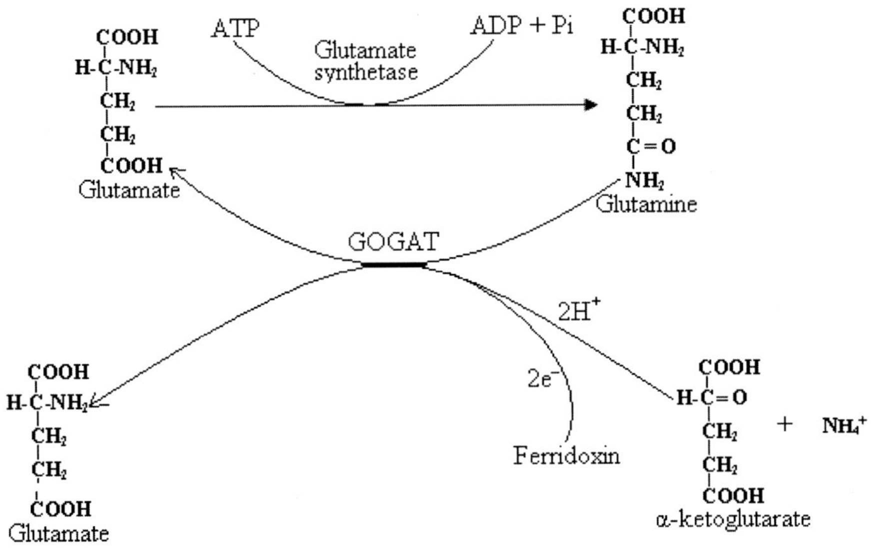

Fig. 6.5 : Pathways for glutamate synthesis; GDH = glutamate dehydrogenase; GOGAT = glutamine α-oxoglutarate amino-transferase (glutamine synthase); GS = glutamine synthetase

Microbial growth and C :N ratio : Most microbial cells contain approximately 45% C and 6-9% N. Fungal cells have wide C :N ratio of 10 :1 to 15 :1 whereas bacterial cells have a C :N ratio of 4 to 5 :1. Therefore, microbial growth, on an average, needs a C :N ratio of 8 :1 (Table 6.2). Due to wide C:N ratio in fungi it seems that they may immobilize less nitrogen per unit substrate assimilated than bacteria. Knowledge of growth efficiency of microorganisms on simple substrate like glucose, ammonium sulphate, etc. allows the calculation of nitrogen requirement for growth. For example, 100 g glucose (contains 40 g C) when used by microbes at a growth yield efficiency of 60% means 24 g C in microbial biomass (= 5.4 g N). Fungi containing nitrogen content of 3% would require about 1.6 g N while bacteria with N content of 11% shall require about 5.9 g N. In soils mixed populations do occur so intermediate amount of N will be required.

C :N ratio and nitrogen mineralization : The soil organic matter generally has C :N ratio of 10 :1 to 12 :1. The addition of organic residues containing C :N ratio >25 :1 generally will be needed for microbial growth. The continuous loss of carbon decreases C :N ratio and when it reaches <25 :1, the accumulation of mineralized products will start. The N content of the residues being added to soil can be used to predict the time when N will be mineralized or immobilized. The quantity of inorganic N immobilized by microbes can be estimated. For example, if 2 t residue having 45% C and 0.75% N is incorporated in a hectare soil, it means that 900 kg C and 15 kg N (i.e. 2,000 x 0.75 kg N) has been added. If the microorganisms use 35% C present in the residue through their activities and the remaining 65% is respired as CO_2, then microbes shall have used 900 x 35% = 315 kg C present in the residues. The increased microbial biomass will have taken :

$$N = \left[\frac{315\,\text{kg C}}{\text{C : N ratio of 8 : 1 (for soil microbes)}}\right] \quad \text{i.e.} \quad \frac{315}{8}\ \text{kg N or 39 kg N.}$$

But N present in plant residues being only 15 kg N, therefore, the amount to be added to maintain microbial growth will be 39-15 = 24 kg i.e. the amount required to balance immobilization.

Table 6.2 : C :N ratio in selected soil organisms, soil and plant materials

Components	C : N ratio
Soil bacteria	5 :1
Soil fungi	15 :1
Soil microorganisms	8 :1
Soil organic matter	10 :1
Rice straw	80 :1
Bean plant	15 :1
Corn/sorghum stover	60 :1
Clover (fresh)	12 :1
Saw dust	400 :1

Under aerobic conditions, substrates having 1.5-1.7% N will have less immobilization of nitrogen. Under anaerobic conditions in submerged soils, the nitrogen requirement for decomposition of crop residues may be only about 0.5%. To balance this large immobilization of nitrogen in straw usually 1 kg N should be added per kg of straw for rapid decomposition. Tree residues usually have high lignin content so degradation is slow and is not substantially incorporated into microbial biomass. Therefore, net mineralization may be observed even at a C:N ratio of 50 :1. Plant nitrogen uptake may account for 100 kg ha^{-1} yr^{-1} in high yielding crops. Regardless of vegetation type, the annual uptake of nitrogen by growing plants represent only a minute proportion of total nitrogen reserves in soil which occur largely in a variety of organic forms. A C :N ratio of 20 :1 for arable soils is generally considered as the approximate threshold between net mineralization and net immobilization. The C :N ratio of undisturbed top soil in equilibrium with its environment is 10 :1 to12 :1, but narrow in subsoil because of low amount of carbon. Disturbance of soil through tillage causes immediate rise in mineralization.

Continued cultivation without new inputs of crop residues ultimately leads to a decline in humus content of soil which is about 50% decline in soil organic matter in 40 to 70 years. A soil containing 5% organic matter (with 5%N) when put under cultivation supposedly loses 4% organic matter annually, then the quantity of nitrogen mineralized during first year mineralization will be :

$$\left[\frac{5}{100}\right] \text{x } 2 \text{ x } 10^6 \text{ x } \frac{4}{100} \text{x } \frac{5}{100} \text{ i.e. 200 kg N or 4,000 kg of organic matter loss yr}^{-1}.$$

Suppose after 50 years of cultivation, organic matter showed decline of 2.5% N. Also, assuming that 2% organic matter is oxidized per year, then the quantity of N mineralized will be 2.5% x 2 x 10^6 x 2% organic matter loss yr^{-1} x 5% N. This will be equal to 50 kg N HFS^{-1} yr^{-1}. A soil mineralizing 2% soil organic matter (humus) of its 4% organic matter content would release 93 kg N ha^{-1} as NH_4^+. This is only $\frac{1}{3}$-$\frac{1}{4}$ of the N needed by cereals.

6.3 KINETICS OF NITROGEN MINERALIZATION

The net nitrogen mineralization in native soil is generally considered of the first order reaction kinetics :

$$d_N/d_t = -k_N$$

where N is the concentration of mineralizable organic N; k is the rate constant which is negative because substrate constantly disappears, and t is the time.

Upon integration of this equation we get :

$$N_t = N_o e^{-kt}$$

where N_t is the amount of organic N remaining at time t and N_o is the amount of organic N at initial time.

Since the N at time t is equivalent to initial N content - N mineralized (N_m); therefore, the equation becomes :

$$N_t = N_o - N_m$$

$$\text{or } N_o - N_m = N_o e^{-kt}$$

$$\text{or } N_m = N_o(1-e^{-kt})$$

These calculations provide an estimate of the original organic N and its mineralization rates under a particular set of agro-climatic conditions. However, under the situations where N mineralization is independent of organic N concentration, zero order reaction kinetics applies :

$$\text{i.e. } d_N/d_t = k_N$$

Such situations can occur where organic N rich substances are added to soil. In these situations, the capacities of mineralizing population are saturated. Increased mineralization cannot occur without an increase in enzyme or population densities of the organisms catalyzing the process.

The most commonly occurring alternative mathematical model used to describe N mineralization in soil samples is the double exponential model. Because some organic N compounds are easily decomposable by microbial biomass, and others are inherently biochemically resistant or physically occluded (in reality each class or specific organic N has a characteristic decay rate). The above model can be expanded to include both slow and fast degrading N reservoirs.

$$N_t = N_o S(1-e^{-ht}) + N_o(1-S)\ (1-e^{-kt})$$

where N_t is N mineralized in time t; N_o is the potentially mineralizable N; S and 1-S are labile and less accessible organic N fraction of the total organic pool, respectively, and h and k are decomposition rate constants for labile and less accessible organic N fraction, respectively.

6.4 NITRIFICATION

Nitrification is the process of conversion of ammonia to nitrite and nitrate through biochemical oxidation. These oxidative reactions are catalyzed by two mutually exclusive groups of organisms i.e. ammonium oxidizers and nitrate oxidizers. Chemoautotrophic nitrifiers are obligate aerobes. These specialized bacteria rapidly oxidize ammonium in aerobic soil, with pH 6.0 or higher, in two sequential steps (Fig. 6.6) :

$$\underset{(-3)}{NH_4^+} + O_2 + H^+ + 2e^- \xrightarrow[\searrow H_2O]{\textit{Nitrosomonas}} \underset{(-4)}{[NH_2OH]} \xrightarrow[\searrow 5H^+]{\textit{Nitrosomonas}} \underset{(+3)}{NO_2} + 4e^- \xrightarrow[H_2O \nearrow]{\textit{Nitrobacter}} \underset{(+5)}{NO_3}$$

Fig. 6.6 : Two step ammonia oxidation in aerobic soils

The first step results in the conversion of ammonia to nitrite by specific autotrophic bacteria *Nitrosomonas* and in second step another group of autotrophs, *Nitrobacter,* immediately act to convert NO_2 to NO_3. The biochemical reactions involved in nitrification are given in Fig. 6.7.

$$\underset{\text{(Ammonia)}}{2NH_4^- + O_2} \xrightarrow[a]{-2H} \underset{\text{(Hydroxylamine)}}{2NH_2OH} \xrightarrow[b]{-4H} \underset{\text{(Hyponitrite)}}{[HON{=}NOH]} \xrightarrow{+O_2} \underset{\text{(Nitrite)}}{2HON{=}O} \xrightarrow{+2H_2O} \underset{\text{(Intermediate product)}}{2[HO\underset{\substack{| \\ OH}}{N}.OH]} \xrightarrow{-4H} \underset{\text{(Nitrate)}}{HNO_3}$$

Fig. 6.7 : Reactions involved in nitrification process : a = ammonia monooxygenase, b = hydroxylamine oxido-reductase

The oxidation of NH_4^+ to NO_3 is an endothermic reaction. The oxidation catalyzed by enzyme ammonium monooxygenase involves hydroxylamine formation. The oxidation of nitrite to nitrate by *Nitrobacter* appears to be a cytochrome-linked single step hydrolytic reaction. A small proportion of electrons, transported across the membrane by this reaction, is transported back (reverse electron transport) to generate reductants to allow conversion of ammonia (NH_4^+) to hydroxylamine (NH_2OH). In case of *Nitrosomonas* the oxidation state of N changes from –3 to +3 and in case of *Nitrobacter* from +3 to +5. The energy released is sufficient to enable bacteria to assimilate their C requirement from CO_2. The energy yield to chemoautotrophs are approximately 66 kcal (= 8.8 ATP molecules) per molecule of NH_4^+ oxidized by *Nitrosomonas* and 18.2 kcal (= 2.5 ATP molecules) per molecule of NO_2 oxidized by *Nitrobacter*. The energy yields in chemoautotrophic organisms are low in comparison to heterotrophic metabolism. This partly explains the slow growth characteristic of nitrifiers. Generation time for nitrifiers is 20 to 40 hours. During oxidation of ammonia to nitrate, O_2 is utilized for electron transport. Also, nitrification is associated with the production of hydrogen ions ($2H^+$ moles per molecule of NH_4^+) resulting in the acidification of soils, especially in weakly buffered soils. Nitrification may significantly decrease the soil pH when large amounts of ammonium fertilizers are applied. So liming in such cases is essential to restore soil productivity because nitrifiers and earthworms are acid-sensitive. Various chemo-autotrophic and heterotrophic microorganisms involved in nitrification are given in Table 6.3.

Table 6.3 : Various nitrifier microorganisms isolated from soils

Organisms	Processes in which involved	Source
Chemoautorophic		
Nitrosomonas europaea	Oxidation of ammonia to nitrite	Soil, water, sewage
Nitrospira briensis	"	Soil
Nitrosococcus mobilis	"	Marine
Nitrosococcus tenuis	"	Marine, soil
Nitrosococcus nitrosus	"	Soil
Nitrosococcus oceanus	"	Marine
Nitrosococcus nitroxydans	"	Soil
Nitrosovibrio sp.	"	Soil
Nitrosolobus multiformis	"	Soil
Nitrobacter winogradskii	Oxidation of nitrite to nitrate	Soil
Nitrobacter agilis	"	Soil, water
Nitrospira gracilis	"	Marine
Heterotrophic		
Arthrobacter	Oxidation of ammonia to nitrate	Soil, water, sewage
Aspergillus flavus	"	Soil, sewage
Absidia sp.	"	Soil, water

The hetertrophic nitrification is carried out by both bacteria and fungi. These include *Arthrobacter, Aerobacter, Mycobacterium, Streptomyces, Aspergillus* and *Penicillium.* In acidic soils where autotrophic populations are low, heterotrophic nitrification accounts for half of the nitrification. Autotrophic nitrification is closely linked to cell growth and is proportional to nitrifier biomass whereas heterotrophic nitrification is independent of cell yield. The rate of nitrification for heterotrophs is 10 times less than those of autotrophs. The nitrifier population in soil is very low about one-tenth to one-hundredth times less than ammonium oxidizers which may be as high as 10×10^6 g^{-1} soil.

6.4.1 NITRATE LEACHING & ITS ENVIRONMENTAL CONSEQUENCES

Soils rarely contain nitrite because it is immediately oxidized to nitrate thereby toxic effects of nitrite to plants are greatly reduced. Nitrate anion is readily produced and is very mobile. It is, therefore, subject to leaching losses.

The annual leaching rate of nitrate-nitrogen in UK soils has been found to be about 100 kg ha^{-1}. Nitrate leaching may enhance eutrophication i.e. excessive aquatic plant growth and algal bloom in waters thus may reduce water quality. Further, its accumulation in ground waters beyond 45 ppm is considered hazardous to health when these waters are consumed for drinking purpose. The World Health Organization has set a limit of 11.3 mg NO_3-N L^{-1} (i.e. 50 mg nitrate L^{-1}) for good quality drinking water. The levels greater than this limit cause methaemoglobinaemia or 'blue baby' syndrome in animals, especially in infant babies in first 6 months of life. The syndrome is caused by defective carriage of O_2. For haemoglobin, iron must be present in ferrous [Fe(II)] form to carry O_2 in blood streams. Nitrate in blood oxidizes Fe(II) to ferric [Fe(III)] form. When O_2 transport fails the symptoms of cyanosis, bluing of lips, appears. Older children and adults are protected from this kind of oxidative stress by an enzyme 'methaemoglobin reductase' which develops gradually in the body after birth. Nitrate reduces the oxygen supply to blood streams due to nitrite complexation with haemoglobin. The nitrite can also react with secondary amines in soil to produce nitrosoamines which are carcinogenic in nature.

$$\underset{\text{Secondary amine}}{\underset{\underset{R}{/}}{\text{R-NH}}} \quad + \quad \underset{\text{Nitrite}}{NO_2^-} \longrightarrow \underset{\text{Nitrosoamine}}{\underset{\underset{R}{/}}{\text{R - N = NO}}} + OH^-$$

6.4.2 FACTORS AFFECTING NITRIFICATION

In comparison to ammonification or N mineralization, which is quite insensitive to soil environment, nitrification is much more sensitive and dependent on soil environment. Ammonium production is ubiquitous whereas its nitrification largely depends on soil reaction and oxygen supply.

a) **Soil reaction :** The nitrifying microorganisms are generally sensitive to hydrogen (H^+). Optimum nitrification occurs at pH 6.6 to 8.5. The activity of nitrifiers is reduced below or above the optimal range and becomes negligible below pH 4.5. In alkaline soils (at higher pH >8.5) ammonium inhibits the transformation of NO_2 to NO_3 because ammonia may raise to toxic level for *Nitrobacter* growth and without active *Nitrobacter* nitrite may accumulate. The nitrifying bacteria need an adequate supply of Ca^{2+} and H_2PO_4 and a proper balance of micro-nutrients as well. The influence of soil pH and available Ca^{2+} on the activities of nitrifying microorganisms depicts the significance of liming in farming. Evidences suggest that some nitrifier organisms may tolerate acidic pH in soil through surface attachment, slime production or their location close to the sites of mineralization or localized cluster formation of hundreds of cells associated with soil aggregates. In acid soil, certain fungi and heterotrophic bacteria have a role in nitrification, especially in acid forest soils. The nitrifying fungi can nitrify organic nitrogen directly to nitrate without formation of ammonium.

b) **Soil aeration :** Nitrifying organisms are obligate aerobes and require oxygen to form NO_2^- or NO_3^- ions, hence depletion of oxygen in soil diminish its activity. Diffusion of O_2 into soil or aeration is controlled by factors like soil moisture and structure etc. Nitrification can occur at O_2 concentration as low as 0.3 mg ml^{-1}. Nitrification is optimum when soil water content is 50-60% of water holding capacity. Nitrification is mostly confined to the outer part of an aggregate because oxygen diffuses very slowly through water. In comparison, denitrification is vigorous in the interior part of soil aggregates.

c) **Soil moisture :** Nitrifier organisms are sensitive to soil moisture. Nitrification proceeds readily at -0.1 to 1.0 MPa moisture tension, and at above -0.6 MPa dehydration effects become critical. Similarly, water-logging limits the diffusion of O_2 and suppresses nitrification. Nitrification continues to decline gradually from wilting point (15 bars) to air dryness. Autotrophic bacterial nitrifiers are generally much more sensitive to water stress whereas fungal nitrifiers are less sensitive. Soil beneath coniferous forests are often subject to extreme water stress, therefore, fungal nitrifiers dominate such situations. Generally, mineralization reactions producing NH_4^+ are less sensitive to both water stress and low temperature. Ammonium, therefore, accumulates in water-stressed or cold soils.

d) **Soil temperature :** Nitrification is a temperature sensitive process. Low temperature often limits nitrification in temperate soils. Despite this, nitrification proceeds in forest soils at even 0°C. It seems that there is more predominance of low temperature-adaptive nitrifying soil fungi. In general, the process of nitrite oxidation is slow when soils are cool and virtually ceases at temperature below 5°C and above 40°C. The optimum range lies between 25 and 35°C. Temperature fluctuations in soil determine the extent of nitrification during winter and summer months. In temperate areas nitrification is greatest in spring and fall.

e) **Soil organic matter :** Nitrifiers being autotrophs do not require organic matter either as a carbon source or energy source. Previously organic matter was considered toxic to nitrifiers but it is now accepted that organic matter as such is not inhibitory to nitrifiers. Since organic matter decomposition may require inorganic nitrogen and oxygen, the supplies of NH_4^+ and oxygen to nitrifiers may temporarily get limited. Malhi and

McGill found a concentration of 200 μg g^{-1} NH_4^+-N supports rapid nitrification, but subsequent increase to 300 μg g^{-1} inhibits the process. High C :N ratio of residues which prevents the release of ammonium may prevent nitrification. Shift to mineralization, of course, increases the nitrifier population, provided temperature, moisture and O_2 supply are adequate. The presence of high level of ammonia can inhibit nitrification process. The more a complex substrate is added to soil, more specialized will be the microbial mineralization.

f) **Population of nitrifier organisms :** Soils differ in their ability to nitrify NH_4^+ even under similar conditions of temperature, moisture and level of added NH_4^+. It has been attributed to the variation in population of nitrifying organisms.

g) **Cultural practices :** Nitrification may be stimulated by the presence of adequate levels of essential elements. Nitrifiers are quite sensitive to most of the pesticides, thus pesticide application generally reduces nitrification. Soil aeration and good soil drainage promote nitrification. Nitrification is generally somewhat slower in minimum tilled than good tilled soil. Possibly certain cropping systems have inhibitory effect on nitrification For example ploughed grasslands usually result in ample nitrate production.

6.4.3 MEASUREMENT OF NITRIFICATION IN SOIL

Soil nitrification is measured in laboratory by adding NH_4^+ to soil. The rate of oxidation of NO_2 and NO_3 is estimated. The soil slurry is made to help the diffusion of substrate to the sites of nitrifying microorganisms in these assays. The soil perfusion columns have been used where soil is continuously oxygenated by a fresh perfusion (percolation) of solution that contains ammonium substrate. The repeated perfusion permits direct study of transformation kinetics and the effects of environmental changes. Soil columns receiving water and substrate through multi-channel pumps and collected to a computer controlled sampling devices are also used. These perfusion systems have considerable advantages and can be used to partition different forms of nitrification into the soil.

6.4.4 INHIBITION OF NITRIFICATION

Nitrates do not accumulate in soils due to its high solubility, fast mobility and non-adsorption to soil colloids. In many coniferous forests available N is usually in NH_4^+ form with little or no NO_3-N. This has been attributed to allelopathic inhibition exerted by these plants on nitrifiers. Climax ecosystem such as undisturbed grasslands or mature forest plants produce organic flavonoids etc., which are toxic to nitrifiers. This mechanism is a sort of autocontrol (natural control) over nitrogen losses. The more widely accepted alternative argument is that the competition for NH_4^+ by plant roots, mycorrhizal fungi and nitrogen-immobilizing microorganisms keeps inorganic N contents low. Various inhibitors have been developed to induce slow release of nitrogen from fertilizers. These chemicals are substituted pyridines, pyrimidines, acetanilides, anilines, isothiocyanates, etc. These include thiourea, Nitrapyrin [brand name N-Serve (2-chloro-6 trichloromethyl pyridine)], Tetridiazol sold commercially as Dwell or Terrazole (5, ethoxy-3-(trichloromethyl)-1,2,4 thiadiazole), AM (2 amino 4 chloro 6 methyl pyridine, sulfathiozole (nitropyrin), Didin or DCD (dicyandiamide), ATC (4 amino 1,2,4-6, triazole HCl), urea pyrolyzate, urea formaldehyde, etc. Nitrapyrin is effective when applied @ 0.5 kg ha^{-1} or less alongwith anhydrous ammonia or mixed with urea. Nitripyrin @ 1.0 ppm inhibits the growth of *Nitrosomonas europea* and *N. agilis*. Oxidation of ammonium is suppressed at this concentration. These inhibitors restrict the activity of *Nitrosomonas* rather than *Nitrobacter*, thus allowing the accumulation of ammonium rather than phytotoxic nitrite is soil. DCD breaks down in soil to form NH_4^+, NO_3^-, water and CO_2. The seeds of neem tree (*Azadirachta indica*) contain certain allelopathic compounds (lipid associated), which also act as nitrification inhibitors and thereby increase the efficiency of urea fertilizers.

6.5 DENITRIFICATION

Denitrification is the process of biological reduction of nitrate and nitrite to dinitrogen or nitrous oxide. The process, operative in absence of O_2, is mediated by a wide range of facultative aerobes and anaerobic bacteria. The facultative anaerobic bacterial species use the oxygen present in air as hydrogen acceptor (aerobic) and utilize nitrite or nitrate in place of oxygen (anaerobically) as an alternative electron acceptor. Denitrifying bacteria are primarily heterotrophs. The most common ones are *Pseudomonas, Micrococcus, Thiobacillus* and *Alcaligenes*. Denitrification is often referred to as 'dissimilatory nitrate reduction' because it is associated with the dissimilation (oxidation) of organic matter. Often NO_3^- is converted to NO_2^-, which then enters into the assimilatory pathway of nitrogen cycle. However, it differs from assimilatory nitrate reduction, carried out by green plants, bacteria, fungi, etc. wherein nitrate is immobilized and reduced to ammonia for the biosynthesis of cell constituents like amino acids, proteins, nucleic acids, etc. Secondly, although nitrite and hydroxylamine (NH_2OH) are intermediate products, the assimilatory nitrate reduction is not inhibited by oxygen but the process is inhibited/repressed by the presence of ammonium or reduced nitrogenous organic metabolites. Thirdly, in assimilatory nitrate reduction the enzyme reductase is linked to NADPH, and molybdenum is an essential co-factor. Fourthly, ammonium is terminal product. Denitrification also differs from dissimilatory reduction wherein process occurs in highly reduced condition to produce ammonium or nitrite as terminal end product. So the process of dissimilatory reduction does not lead to the loss of nitrogen from the system.

Denitrification leads to the terminal end product formation of nitrous oxide (N_2O) or N_2 gas and the process is favoured by anaerobic conditions. There is considerable decomposition or putrefaction of organic matter in soil during denitrification. Some organisms have the ability to obtain O_2 from NO_2^- or NO_3^-, so in some cases denitrification may be a part of assimilatory process of organism. Some microorganisms may denitrify NO_3 completely in absence of O_2 and other do so partially. Anaerobic bacteria like *Clostridium* reduce NO_3 to NH_4^+ directly.

$$HNO_3 + 4H_2 \longrightarrow NH_3\uparrow + 3H_2O$$

$$HCOOH + HNO_3 \longrightarrow CO_2\uparrow + HNO_2 + H_2O$$

$$2HNO_3 \longrightarrow 2HNO_2 + O_2$$

$$HNO_2 + H_2^+ \longrightarrow NH_3 + O_2$$

$$5C_6H_{12}O_6 + 24HNO_3 \longrightarrow 24H_2CO_3 + 6CO_2 + 18H_2O + 12N_2$$

Denitrification occurs when oxygen diffusion rates in soil are insufficient to meet the demands from microbial respiration. This situation arises when soil is near or at water saturation level and oxygen diffusion is slowest; and also when a recent addition of available carbon substrate to soil has stimulated heterotrophic microbial activity. Denitrification also occurs in microsites in aerated soils as long as there are water saturated aggregates larger enough to restrict the diffusion of oxygen to the zone of denitrification potential. Greenwood observed that the concentration of dissolved O_2 in solution surrounding bacteria must fall to about 4×10^{-6} M before reduction begins. However, soils which apparently appear well-aerated may reduce NO_3 to nitrous oxide and nitrogen, if organic matter is high; because O_2 demand in such soil is very high so interior portions are likely to become anaerobic. Aggregates having radii >9 mm are likely to have denitrification sites even though such soils may appear to be well-aerated. In most soils, N_2 is the dominant gaseous product of denitrification, although N_2O evolution becomes increasingly important in more acid soils. Denitrification process causes a net loss of biological N from the soils and water, and globally around 2×10^8 t N is lost each year through denitrification in the form of N_2.

6.5.1 BIOCHEMISTRY OF DENITRIFICATION

The nitrogen loss through denitrification, in the form of nitrogen (N_2), nitrous oxide (N_2O), nitrite (NO_2) or nitric oxide (NO), is a base forming reaction because of the production of hydroxyl ions (OH^-) equivalent to the amount of nitrate that has been denitrified. Denitrification results in the formation of nitrate (NO_3), nitrite (NO_2) nitrous oxide (N_2O), nitric oxide (NO) and nitrogen gas. It is not necessary that all intermediary products would be converted to N_2; some may escape to atmosphere during production. A variety of electron donors are used in electron transport mechanism involved in denitrification. The quinone component of electron transport system is perhaps the only one influenced by oxygen. In absence of oxygen, as electron acceptor, electron transport system follows b-type cytochrome path to have oxidized nitrogen as terminal acceptor. Specific nitrogen oxide reductase enzymes have been isolated for each step of reaction. Four reductase enzymes have been found involved in the process (Table 6.4) at each acceptor level.

Table 6.4 : The reactions catalyzed by the enzymes involved in denitrification

S. No.	Enzymes	Reaction catalyzed
1.	Dissimilatory nitrate reductase	$NO_3 \longrightarrow NO_2$
2.	Dissimilatory nitrite reductase	$NO_2 \longrightarrow$ NO or N_2O
3.	Nitric oxide reductase	$NO \longrightarrow N_2O$
4.	Dissimilatory nitrous oxide reductase	$N_2O \longrightarrow N_2$

1) **Dissimilatory nitrate reductase :** This is a membrane bound enzyme containing molybdenum and iron, besides labile sulphur groups and is often detected in insoluble fraction of cytoplasm. Low pH or O_2 inhibits its activity. It involves phosphorylation during reduction process.

2) **Dissimilatory nitrite reductase :** The enzyme acts at second stage of denitrification and is often found in soluble fraction of cytoplasm. It is perhaps a membrane bound enzyme. Low pH and presence of O_2 inhibits its activity. Wide *km* values of enzyme reflect the dependence of nitrate and nitrite reductases on anion transport within the cell or inhibition by end products.

3) **Nitric oxide reductase :** It is a membrane bound enzyme which, unlike the first two, is inhibited by the presence of nitrate. Nitrite reduction to nitrous oxides is supposed to involve nitric oxide (NO) formation after the reduction of nitrosyl to a free nitroxyl (HNO) which spontaneously dimerize to N_2O. It is also argued that N_2O (nitrous oxide) is produced

by nucleophilic intermediate rather than via free nitroxyl. The scheme for N-N bond formation by nucleophilic attack of nitrite on an enzyme bound nitrosyl is as under :

$$E\text{-}Fe^{II} + NO_2^- \rightleftharpoons E\text{-}Fe^{II}.NO_2^- \rightleftharpoons E\text{-}Fe^{II}.NO^+ \xrightleftharpoons{+H_2O} E\text{-}Fe^{II}(N_2O_3) \xrightarrow{+NO_2^-} N_2O + E\text{-}Fe^{II}$$

[E = enzyme; Fe^{II} = iron group]

4) **Dissimilatory nitrous oxide reductase :** It is a unique enzyme in denitrification, which is inhibited by acetylene (C_2H_2) and sulphide (HS^-). It is found in insoluble fraction of cytoplasm, involves phosphorylation and is membrane bound. Oxygen greatly represses its activity.

Generally the proportion of gaseous nitrogen products composed of nitrous oxide increases with increasing soil acidity, reduced soil temperature and augmented soil nitrate level. However, in soils N_2O and N are produced in varying ratios depending upon the substrate, the environmental conditions, the microorganisms involved and the time elapsed since the onset of denitrifying activity. The reactions involved in denitrification are shown in Fig. 6.8.

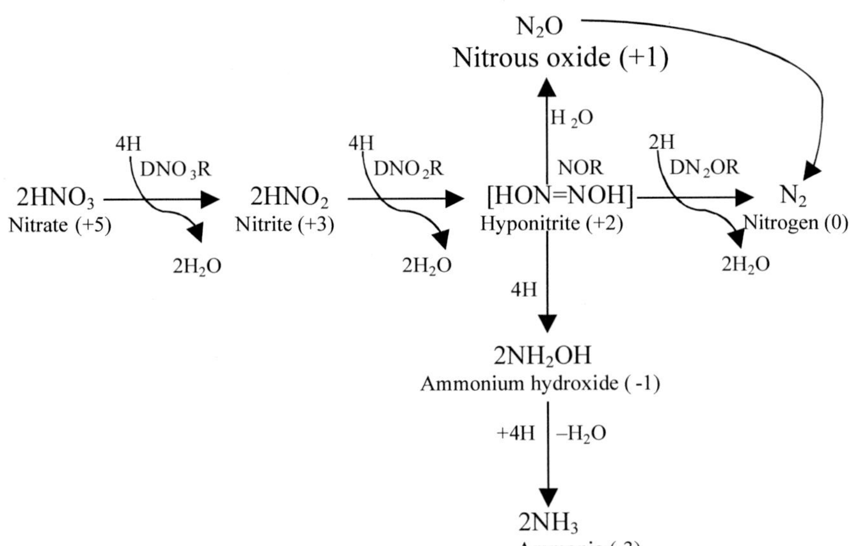

Fig 6.8 : Reduction of Nitrate to nitrogen. Each step is triggered by specific reductase enzyme DNO_3R = dissimilatory nitrate reductase; DNO_2R = dissimilatory nitrite reductase; NOR = nitric oxide reductase and DN_2OR = Dissimilatory nitrous oxide reductase (bracketed figures show the oxidation state of compound)

6.5.2 CHEMODENITRIFICATION

Nitrite (NO_2^-) in soil is also converted to N_2 by non-enzymatic pathways under well-aerobic conditions. Losses of NO_2 nitrogen because of its instability and dismutation to other oxides of nitrogen and to N_2 are commonly referred to as 'chemodenitrification', although production of nitrite in soil is a biotic process. Chemodenitrification is rarely a major route of fixed nitrogen loss from soils.

Aerobic dismutation of nitrite occurs through several pathways which are as under :

i) $R\text{-}NH_2 + HNO_2 \longrightarrow ROH + H_2O + N_2\uparrow$

ii) Also known as "Van Slyke reaction", here a-amino groups yield nitrogen gas.

$HNO_2 + NH_3 \longrightarrow 2H_2O + N_2\uparrow$

Nitrite also reacts with ammonia, urea, purine, pyrimidine and methyl-amine to yield N_2.

iii) Sometimes chemical decomposition of nitrous acid may occur spontaneously to give NO.

$3HNO_2 \longrightarrow HNO_3 + H_2O + 2NO$

iv) NO may escape to air or react with oxygen and water to form NO_3^-

$2NO + O_2 \longrightarrow 2NO_2 \longrightarrow N_2O_4 + H_2O \longrightarrow HNO_2 + HNO_3$

v) It is argued that $NH_4.\ NO_2$ instability leads to chemodenitrification of NO_2 to N_2 gas

$NO_2^- + NH_4^+ \longrightarrow NH_4NO_2 \longrightarrow N_2 + 2H_2O$

vi) N and N_2O gases may be lost from the soil arising from lignin and phenols as under :

$HNO_2 + \text{lignin} \longrightarrow N_2\uparrow + 2\,N_2O + CH_3ONO$

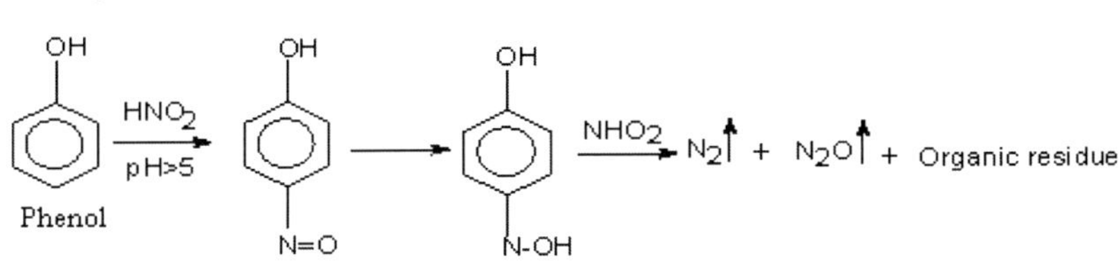

vii) Sometimes decomposition of nitrous acids with transition metal cations also leads to N and N_2O loss :

$$3HNO_2 \longrightarrow 2NO + HNO_3 + H_2O$$
$$Mn^{2+} + HNO_2 + H^+ \longrightarrow Mn^{3+} + NO + H_2O$$
$$Fe^{2+} + H\,NO_2 + H^+ \longrightarrow Fe^{3+} + NO + H_2O$$

Chemodenitrification is seldom a major route of fixed nitrogen loss from soils.

6.5.3 MICROORGANISMS INVOLVED IN DENITRIFICATION

Most denitrifiers in soil are heterotrophic bacteria requiring preformed organic carbon as their energy source. Some chemo- and photo-autotrophic bacteria also cause denitrification. All these microorganisms reduce nitrate or nitrite in absence of oxygen. The population of denitrifiers in common soils is $>10^6$ g^{-1} soil (merely 0.1-5% of the cultured soil population). Genera capable of denitrification causing reduction to N_2 or N_2O are given in Table 6.5. Most denitrifiers belong to genera *Pseudomonas, Bacillus, Alcaligenes* and *Paracoccus.* Most of the genera are aerobic but capable of growing anaerobically in presence of nitrogen oxides only.

Evidences of aerobic denitrification suggest that denitrification is not restricted to anaerobic microsites in aerobic soils only. The ecological significance of aerobic denitrification is not known. There are large populations of denitrifiers in arable soils and are most numerous in the vicinity of plant roots. The rhizosphere rarely has reducing conditions but even then has a large number of denitrifiers, higher than in the bulk soil. This means that whenever anaerobic conditions in soil prevail, higher denitrification will occur provided enough nitrate is present in soil solution. Further, denitrification occurs in soil aggregates in rhizosphere if aggregate pores are water filled and diffusion of O_2 is restricted. Carbonaceous exudates from actively functioning roots are believed to support the growth of denitrifying rhizobacteria. The potential for denitrification is immense in most field soils provided the conditions are favourable to induce organisms to shift or switch over from aerobic respiration to denitrifying type of metabolism. Some species of *Azospirillum* and *Rhizobium* have both denitrifying and nitrogen fixing capabilities. However, the two processes do not occur simultaneously as the presence of nitrate represses nitrogenase activity. Non-sulphur photosynthetic bacteria of genus *Rhodopseudomonas* have both denitrifying and N_2 fixing capabilities. *Thiobacillus denitrificans* is sulphur oxidizing anaerobic bacterium capable of causing denirification. The reaction induced by this organism is as under :

i) $5S + 6KNO_3 + 2H_2O \longrightarrow 3N_2 + K_2SO_4 + 4KHSO_4^-$

ii) $5K_2S_2O_3 + 8KNO_3 + H_2O \longrightarrow 9K_2SO_4 + H_2SO_4 + 4N_2$

T. denitrificans oxidizes sulphur compounds chaemoautotrophically and reduce nitrate to molecular N_2. Source of energy is either S or thiosulphate.

Table 6.5 : Bacteria involved in denitrification

Genus	Characteristics	Energy source
Pseudomonas	Mostly heterotrophic; some chemoautotrophic	Organic matter
Alcaligenes	Heterotrophic; some chemoautotrophic	Organic matter
Bacillus	Heterotrophic; some thermophilic	Organic matter
Agrobacterium	Phytopathogenic	Organic matter
Micrococcus	Facultative anaerobe, heterotroph	Organic matter
Achromobacter	Facultative anaerobe, heterotroph	Organic matter
Flavobacterium	Heterotrophic	Organic matter
Halobacterium	Heterotrophic, halophilic	Organic matter
Halobacterium	Heterotrophic	Organic matter
Hypomicrobium	Grows on single C substrate	–
Paracoccus	Chemoautotrophic/heterotrophic	Hydrogen
Inlobacillus	Chemoautotrophic	Reduced sulphur
Rhodopseudomonas	Phototrophic, dizotrophic	Reduced sulphur
Thiobacillus	Chemoautotrophic, S oxidizer	Reduced sulphur
Rhizobium	N_2 fixing symbiont	Root exudates/ organic matter
Azospirillum	N_2 fixing free living	Root exudates/ organic matter

6.5.4 MEASUREMENT OF SOIL DENITRIFICATION

Many approaches have been used to measure soil denitrification. One approach is to estimate the difference between the nitrogen occurring in soil system and the sum of that previously occurring therein plus any amendment. The balance nitrogen left is considered to be denitrified. Another approach is measuring the nitrogen balance using ^{15}N. This technique provides indirect but valuable information, especially when leaching losses can be accounted for, as in lysimeters or monitored watersheds. The difficulty arises in the spectrometry of mixture of N_2, N_2O, NO and NO_2. For example, $^{15}N_2$ and NO show mass peak 30 in mass spectrometry. Acetylene inhibition of nitrous oxide reduction is a quick and reliable procedure for the estimation of *in situ* level of denitrification enzymes in soil slurries, soil cores and in field. Because of good detectors and packing materials, gas chromatographs have proved valuable asset to measure the denitrified fractions. Such measurements involve the use of seal soil cores and addition of acetylene gas to these cores. Acetylene can also be injected through probes inserted in field soil to the required depth. Acetylene rapidly diffuses through soil to the sites of denitrification where it blocks the terminal nitrous oxide reductase enzyme, ensuring that N_2O is the sole product of denitrification.

$$NO_3 \xrightarrow{\text{Nitrate reductase}} NO_2 \xrightarrow{\text{Nitrite reductase}} NO \xrightarrow{\text{Nitric oxide reductase}} N_2O \xrightarrow[C_2H_2\text{ block}]{\text{Nitrous oxide reductase}} \!\!\!\!\!\!\!/\!/\;\; N_2$$

The C_2H_2 blockage is reversible at C_2H_2 concentration below 0.02 atmosphere. The gas samples must, therefore, be taken from a soil column enclosed in a container or under a canopy. One of the disadvantages of "C_2H_2 blockage technique" is that acetylene also blocks primary oxidase of autotrophic nitrification; therefore the technique may under-estimate the true denitrification. However, in acid forest soils where denitrification is of fungal origin acetylene blockage gives good estimates of denitirfication. The nitrous oxide produced during assimilatory reduction of nitrate to ammonium by both fungi and bacteria may, however, causes over-estimation of denitrification measured.

6.5.5 FACTORS INFLUENCING DENITRIFICATION IN SOIL

The magnitude and rate of denitrification are strongly affected by several soil and environmental factors. The important determinants of the rate of denitrification in soil are :

- Level and form of inorganic nitrogen,
- Nature and amount of metabolizable carbon as source of energy to denitrifiers,
- Soil O_2 status,
- Soil water,
- Soil reaction,
- Soil temperature,
- Soil physical characteristics, and
- Presence of plants.

a) **Level and forms of inorganic nitrogen :** The presence of NO_3^- and NO_2 are pre-requisite for denitrification. Higher concentration of nitrates in soil increases the rate of denitrification and exerts a strong influence on the ratio of N_2O to N_2 in gases released from soil by denitrification. High nitrate concentration inflicts inhibition on nitrous oxide reductase so has greater depressive action on the conversion or reduction of N_2O to N_2 than nitrite. At low nitrate concentration the reaction kinetics appears to be of first order (substrate dependent) i.e. at low levels of nitrate the rate of NO_3^- diffusion to the sites of denitrification will be determinant. At NO_3^- level >20 μg N ml^{-1} denitrificaiton follows zero order kinetics i.e. it is independent of the amount of nitrate present. At higher NO_3^- level, N_2 is predominant and at low levels N_2O is often predominant. The reaction rate of nitrate concentration may be determined by the carbon availability for metabolism rather than NO_3^- level.

b) **Nature and amount of metabolizable carbon (organic matter) :** The amount of readily decomposable soil organic matter strongly influences denitrification in soil. Most of the denitrification is carried out by heterotrophic bacteria which show strong dependence on carbon availability. There is a general correlation between the total soil organic matter content and denitrification. The available carbon substrates supply electrons and organic matter decomposition produces CO_2 and reduces O_2 supply, thus increases demand for NO_3^- as electron acceptor during microbial growth. Drying and air storage of soils greatly enhances their ability to denitrify nitrate under anaerobic conditions because these treatments increase the amount of readily available organic matter to be used by denitrifiers. Freshly added crop residues stimulate denitrification process. However, if nitrogen is limiting gaseous nitrogen losses may be reduced due to immobilization. The amount of water soluble carbon measurable in soil accounts for about 71% of its denitrificaion potential, and the amount of carbon mineralizable during 7 days of incubation has been found more than adequate to meet this requirement.

c) **Soil O_2 status :** Aeration influences denitrification in two ways :

   i) Formation of nitrite and nitrate is dependent on good O_2 supply and denitrification occurs when O_2 supply is too low to support microbial requirements. Denitrification process can occur in apparently well-aerated soil at anaerobic microsites where biological O_2 demand exceeds the supply or at localized microsites of low oxygen in the centre of soil aggregates. Large losses of nitrogen during denitrificaion are possible with simultaneous occurrence of low rate of O_2 diffusion into soil having greater respiratory demand.

ii) Oxygen inhibits denitrifying enzyme synthesis as well as the electron flow to denitrifying enzyme. O_2 represses the all nitrogen oxide reductases but nitrate reductase is slightly affected. Therefore, decreased partial pressure of oxygen increases denitrification losses. The losses do not become appreciable until O_2 level is drastically reduced to a concentration of <10% or <0.2 mg L^{-1} of O_2 dissolved in the solution.

d) **Soil water :** Water and soil structure control O_2 diffusion to the sites of microbial activity. Water logging enhances denitrificaion by impeding diffusion of O_2 to these sites. The interrelationships between water and O_2 in soils are complicated by the existence of soil aggregates, soil pores and channels of diverse size and shape. Wet soils are often cold so have less biological oxygen demand. Denitrificaiton increases with soil saturation. Rapid conversion of NO_3^- to N_2O and N_2 is induced when air saturates a warm but biologically active soil. Potential denitrificaiton losses up to 16 kg N ha^{-1} have been measured on the first day following soil saturation. Generally in fields, provided there is no abnormally high O_2 consumption, denitrification becomes insignificant at moisture level below 60% water holding capacity of soil which is approximately equivalent to -0.01 MPa moisture stress. Above this value, denitrification generally correlates with soil moisture. The duration of snow cover on fields and the time when snow melting takes place are the two factors that seem to influence denitrification associated with spring thawing.

e) **Soil reaction :** Most denitrifiers show growth optima at near neutrality (pH 6.0 to 8.0). Denitrification is negligible in soils having pH below 5.0 and absent below pH 4.0. The degree of soil acidity also influences N_2O :N_2 ratio in the evolved gases. Acidity regulates the sequence and relative abundance of gases NO, N_2O and N_2 during denitrification. At pH less than 6 .0, N_2O predominates and represents more than 50% of the nitrogen gases released. The occurrence of N_2O under acid conditions is believed to be due to its resistance to further reduction to N_2. Formation of NO is confined to low pH soils usually <5.5. NO_2 may be the first detectable gas in neutral or slightly acid soil reaction but is reduced microbiologically so N_2 tends to dominate at pH >6.0.

f) **Soil temperature :** Denitrification is directly influenced by soil temperature and rate of denitrificaion increases rapidly from 2 to 35°C with optimum temperature 25-35°C. Temperature indirectly affects both O_2 solubility and O_2 diffusion in water so affects denitrification. The minimum temperature for denitrification is 5°C and maximum 75°C. At soil temperatures near the upper limit, thermophilic microorganisms play major role in gaseous losses. Also, at temperature above 50°C, chemical decomposition of NO_2^- may become significant. Greater denitrification losses during spring thawing may mostly be due to the accelerated rates of denitrification when soils are quickly warmed from 2-5 to 12°C or higher.

g) **Soil physical characteristics :** The physical characteristics of the soil like texture and structure influence denitrification. In fine textured soils high rates of denitrification have often been noticed, probably due to higher surface area availability for microbial attachment or due to other natural differences in capacity of soils to moisture retention and supply of carbon and nitrogen etc. Usually there is increase in the rate of denitrification with soil texture in the order of sandy loam < loam < clay loam.

h) **Presence of plants :** Under field conditions, denitrification rates are generally increased by plants because of their release of readily available carbon in the form of root exudates and sloughed off roots. Therefore, rhizosphere region is more conducive for denitrification than non-rhizosphere region. Denitrification in most of the fertilized-soil cropping system is believed to be controlled by the supply of organic carbon. The studies on the effects of dead roots and living roots on denitrification have revealed a loss of 9% NO_3-N in former case and 15-37% NO_3-N in living roots. The rhizosphere rarely experience reduced conditions but serves as natural zone of proliferation for heterotrophs, especially facultative anaerobes, hence increased population of denitrifiers as compared to non-rhizosphere soil is observed in this zone. Therefore, whenever anaerobic conditions prevail, high rate of denitrification will occur in rhizosphere even when the system is well-aerated, aggregate pores water filled and oxygen diffusion restricted. Further, plants are presumed to enhance denitrification by consuming O_2 through their root activity and stimulate high microbial populations in the rhizosphere. Simultaneously plants also restrict denitrification process by their uptake of NO_3^- and NH_4^+. The reduction in soil water content by plant absorption improves O_2 supply and directly increases O_2 level in the rhizosphere of certain plants that transport oxygen (e.g. rice). Therefore, instances of decreased denitrification potential in cropped soils, as compared to fallow soils, are attributed to NO_3^- and water uptake by plants.

6.5.6 SIGNIFICANCE OF DENITRIFICATION

Conservative estimates reveal that 5-15% of fertilizer nitrogen is lost to the atmosphere through denitrification from loam and clay loam soils. Denitrification as high as 1 kg N ha^{-1} day^{-1} can occur due to the microbial respiration in soils having high water and nitrate content under warm temperature. A loss of 30-60 kg N ha^{-1} yr^{-1} has been observed in agricultural systems. Denitrification is probably the process responsible for the return of N to the atmosphere, thus off-setting the gains from biological nitrogen fixation. Rapid extensive flushes of nitrogen losses are associated with heavy rains, irrigation and snow-melts and also, there are continuous small losses of nitrogen over extended periods in anaerobic microsites. Increased use of fertilizer N substantially enhances the emission of N_2O from soils. Nitrogen oxide is an ozone depleting gas so leads to the partial destruction of the stratospheric ozone layer protecting the biosphere from biologically harmful ultraviolet radiations from sun. Both NO and NO_2 produced as gaseous products of denitrification either react in soil or fail to persist in lower atmosphere. However, N_2O is relatively more stable in troposphere and can diffuse into the stratosphere. In stratosphere it photo-dissociates to form NO (Fig. 6.9). The NO, in turn, catalyzes the destruction of O_3, thereby reduces ozone shield against UV radiations (The N_2O content in atmosphere is presently about 0.3 ppm). The adverse effect of denitrification also occurs through phytotoxic nitrite accumulation in soil. This is because of a lag in the establishment of nitrate reduction compared with that of nitrite reduction.

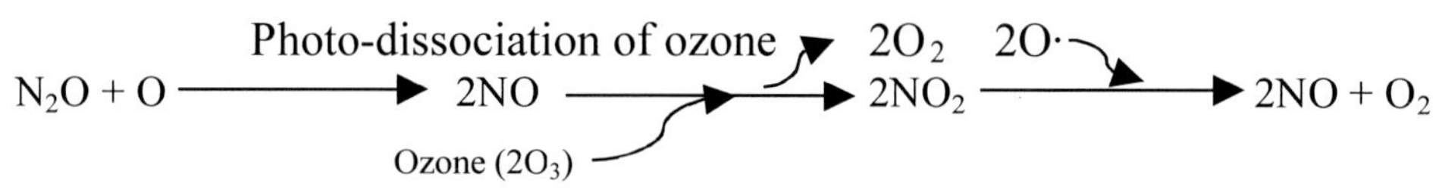

Fig. 6.9 : Reaction sequence leading to ozone destruction by nitrous oxide

The consequences of denitrification are not always adverse. The process helps in the removal of excessive nitrate from irrigation water and waste waters. For direct treatment of waste waters, it may be necessary to inoculate denitrifying organisms and provide sufficient waste metabolizable C (e.g. methanol). Had there been no denitrification in anaerobic soils, it might emit products such as hydrogen sulphide and methane, which can cause greater harm to agriculture and environment than what the products of denitrification do. Therefore, denitrification seems to act as 'safety value' by releasing nitrogen from soils that are receiving heavy pollutant loads of nitrogen and/or acid rains.

To balance the losses and gains of nitrogen, it is essential to follow proper management. The soil nitrogen mineralized is as susceptible to denitrification process as is added fertilizer N. Therefore, effective measures such as use of nitrification inhibitors and N_2-fixers, optimum time and doze of fertilizer, controlled mineralization and immobilization etc., need to be followed as per the soil and environmental conditions and crop requirements. In flooded rice fields and natural wetlands, soils have aerobic and anaerobic zones to allowing both nitrification as well as denitrification to proceed simultaneously. Therefore, N loss can be very high. About 60-70% of fertilizer may be lost because of volatilization as oxides of N or N_2. To avoid these losses, fertilizer in paddy fields be applied in deeper reduced zones so that N remains in ammonium form and is not lost by denitrification.

6.6 PLANT UPTAKE OF INORGANIC NITROGEN

Plants absorb nitrogen either in NH_4^+ or NO_3^- ionic form. Nitrogen in reduced form (NH_4^+) can be incorporated readily into amino acids so is a preferred source of nitrogen, because energy is saved. Oxidized form [NO_3] needs reduction which is an energy consuming process and uses 2NADH molecules per NO_3^- molecule reduced. Microorganisms also show similar preferential uptake of NH_4^+ over NO_3^-. Most of the nitrogen uptake in plants occurs through mass flow (80%) and diffusion (20%). In soils, NH_4^+ ions are held on exchange sites/complexes so its movement through solution is restricted. However, nitrate ion can move through water to roots. Absorption of NH_4^+ depresses Cl^- and $SO_4^=$ uptake. Preference of plants for either NH_4^+ or NO_3^- is determined by the age and type of plant, the prevailing environmental conditions and other soil factors. When NH_4^+ fertilizers are applied, the rhizosphere pH gets decreased. Also, exudation of H^+ ions by roots to maintain electro-neutrality or charge balance inside the plant also decreases. Plants tolerate NO_3^- if in excess and accumulate it to comparatively higher levels in their tissues. NO_3^- fertilizers promote Mg^{++} and Ca^{++} uptake but decrease PO_4^{2-} uptake. Excessive NH_4^+ causes excessive toxic reactions in plant and retard growth.

6.7 NITROGEN LOSSES THROUGH NON-BIOLOGICAL WAYS

a) **Leaching losses :** Nitrate nitrogen is easily leached from upper soil profile to deeper layers and ground water. An annual rate of nitrogen leaching from an unfertilized soil in UK has been estimated to be 0.2 g m^{-2} under grass, 3.0 g m^{-2} under white clover and 13.7 g m^{-2} under bare fallow. Leaching rates are higher in fertilized soils. The major source of nitrate leaching from arable soil is mineralization of organic nitrogen rather than unused chemical N. The half life of this organic N has been estimated to be 41 years so is a persistent long term problem. This abiotic process is driven by both convection and diffusion. About 20-50% of the fertilizer N applied to arable lands is lost through leaching. In rice soils more than 50% N losses occur through leaching. Amongst the various factors that affect leaching are the rate of nitrification, the amount of rains, soil permeability, soil water holding capacity and crop growing on the soil etc. Leaching periodically removes most of the NO_3-N from the profiles of permeable soils in humid regions. Soils of drier regions do not suffer leaching losses if water does not percolate beyond the depth of solum. Very sandy soils are particularly susceptible to NO_3-N leaching losses.

b) **Erosion losses :** Nitrate is lost from soil by wind and water. Each ton of soil lost from field carries 0.5-1.5 kg nitrogen with it. During water erosion, NO_3^- removal does not parallel with that of bulk soil removal. Nitrate is easily removed through run off, therefore removed from top soil to down layers. Water loss can be avoided by improving soil infiltration and decreasing water run-off. Nitrate may also be transported off the site during wind erosion. Loss is limited to, and proportional to, particulate movement. Drought usually precedes wind erosion and during drought capillary movement of water to soil surface may enrich the soil surface with NO_3.

c) **Ammonia volatilization :** The gaseous loss of NH_3 from soil occurs in significant amounts under certain circumstances, especially in sandy and calcarious or alkaline soils. Dry conditions and high temperature, alkaline soil pH and greater air movement above the soil favour volatilization. Volatilization of NH_3 ultimately depends on the quantity of NH_3 and NH_4^+ in the soil solution which is highly pH dependent.

$$NH_4^+ \longrightarrow NH_3 + H^+ \text{ (pka 9.3).}$$

Appreciable quantities of NH_3 appear in soil solution when pH is >7.5. At pH 8.0 or 9.3 NH_3 is 10 and 50%, respectively, of the total NH_3+NH_4^+-nitrogen in soil solution. Thus NH_3 loss is favoured by high soil pH. However, in acid or near neutral soils, the equilibrium between NH_3 and NH_4^+ is strongly towards NH_4^+. This is the reason for higher NH_3 losses in arable soils as compared to forest soils. Gaseous ammonium loss from N fertilizers applied to flooded rice fields and fish ponds can be appreciable even in slightly acidic soils. The applied fertilizers stimulate algal growth in paddy waters. Algae during photosynthesis extract carbon dioxide from water and reduce the amount of carbonic acid formed. This increases pH of paddy water markedly especially during daylight hours to the levels above 9.0. At these pH values, ammonia is released from fertilizer into atmosphere. The NH_4^+ containing fertilizers decrease soil pH when hydrolyzed so there is low or no NH_3 volatilization.

$$\underset{\text{Urea}}{CO(NH_2)_2} + H^+ + 2H_2O \longrightarrow 2NH_4^+ + HCO_3$$

In alkaline and calcareous soils NH_3 volatilizaiton is greater. The clay and organic matter contents of soil also influence the ammonia volatilization as these serve as exchange sites for NH_4^+. Volatilization increases with increase in soil temperature upto about 45°C. Maximum NH_3 losses occur when soil surface is at or near the field capacity moisture content and when slow drying exist for several days. Water evaporation encourages NH_3 loss. Presence of surface crop residues accelerates NH_3 volatilization by maintaining wet and humid conditions at soil surface. Besides, crop residues have high urease activity. Ammonia volatilization in soil is seldom a problem but significant losses of ammonia are more likely to occur from a manure pile or from manure left on the surface of a soil or when heavy ammonia or urea fertilizer in soil is applied. The presence of vegetative cover significantly reduces NH_3 loss by plant absorption of NH_4^+ through roots as well as folia. Many trees and grasses are reported to serve as important sinks for volatilized ammonia. Field crops exposed to air containing normal atmospheric concentration of NH_3 may meet their N requirement by direct absorption of NH_3.

Excessive NH_3 volatilization can cause direct damage to plants, microbes and soil animals because cell membranes are highly permeable to NH_3 and local pH rise in the cell is too rapid for normal compensation mechanisms to operate. Though microbes recover from such shocks but ammonia damage to plants may be irreversible. The loss is serious to the extent of 25-75% when organic materials are applied to the soil surface. Upto 25% of the ammonia applied in soil as fertilizers or formed microbiologically may be lost as ammonia. However, such losses are insignificant below soil pH of 7.0.

6.8 BIOLOGICAL NITROGEN FIXATION

The conversion of inert nitrogen (N_2) molecule into combined nitrogen is brought about by either reduction of N_2 to ammonia (NH_3) or oxidation to nitrate (NO_3^-). This conversion, also referred to as 'fixation', is a highly energy consuming process. Bulk of diazo nitrogen is converted to reduced form by biological means involving enzymatic activities. This capability is confined to a diverse nutritional group of microorganisms belonging to prokaryotes. This diverse group contains enzyme nitrogenase which is responsible for the fixation of N_2.

6.8.1 NITROGENASE ENZYME : PROTECTION MECHANISMS AGAINST MOLECULAR O_2

The nitrogenase enzyme is inhibited by O_2 so organisms have developed special mechanisms for protecting the enzyme (Table 6.6). For example, in *Azotobacter chroococum* molecular oxygen under unfavourable conditions represses nitrogenase synthesis. Other mechanisms include protection of nitrogenase complex with membranes within cell, more slime production and clump/aggregate formation by cells, etc. These mechanisms reduce O_2 diffusion rates to enzyme. In presence of O_2, nitrogenase enzyme undergoes conformational changes leading to a protected but non-fixing enzyme. These changes are reversible when intracellular O_2 is removed. *Azotobacter* cell respires quickly at the cell surface (so reduce O_2 tension to water) to keep the interior of cell O_2-free. Some aerobic N_2 fixing bacteria like *Beijerinckia* frequently form large quantities of extracellular polysaccharides. The slime layer impedes diffusion of O_2 to bacteria and serves a barrier against excessive oxygenation.

Table 6.6 : Means for exclusion of molecular oxygen from environment of nitrogenase

Mechanisms	Organisms involved
1. Respiratory protection	*Azotobacter vinelandii*, cyanobacteria
2. Conformational protection	*A. vinelandii*, *A. chroococcum*
3. Oxygen repression of nitrogenase synthesis	*Klebsiella pneumoniae; A. chroococcum*, some strains of *Rhizobium*
4. Gum/capsule production	*Azotobacter, Derxia gummosa*
5. Heterocyst formation	Cyanobacteria
6. Leghaemoglobin production	*Rhizobium, Bradyrhizobium*
7. Migration to suitable environment	*Azospirillum*
8. Clumping of cells	Cyanobacteria
9. Partitioning of O_2 generating and N_2 fixing processes	Cyanobacteria

6.8.2 ORGANISMS INVOLVED IN NITROGEN FIXATION

Diazotrophic microorganisms include organotrophic and phototrophic bacteria including cyanobacteria and some actinomycetes. Among the several thousands of species of bacteria classified/named so far, diazotrophy is confined to 100-200 prokaryotic species. At present, organisms from more than 33 separate genera are recognized as diazotrophs. These organisms based on their associations involved and the nature of organism can be classified into various groups as presented in Table 6.7. Diazotrophic bacteria are ubiquitous in soils. On an average, the relative contribution of symbiotic associative and free living N_2 fixing systems is in the order of 70% symbiotic and 30% non-symbiotic.

Table 6.7 : Microbial genera and associations involved in diazo fixation

Trophic level	State of living	Organisms
Organotrophs (heterotrophs)		
Free living	i. Aerobes	*Azotobacter, Azotococcus, Azomonas, Achromobacter, Arthrobacter, Beijerinckia, Derxia, Mycobacterium*
	ii. Facultative aerobes (aerobic when not fixing nitrogen)	*Bacillus, Klebsiella, Aerobacter, Escherchia, Enterobacter, Citrobacter, Pseudomonas, Propionibacterium*
	iii. Microaerophilic (normal aerobes when not fixing N)	*Aquaspirillum, Azospirillum, Xanthobacter, Thiobacillus, Methylosinus, Bradyrhizobium, Herbaspirillum, Burkholderia*
	iv. Anaerobes	*Clostridium, Desulfovibrio, Desulfomaculatum, Methanosarcina Rhodomicrobium, Methanococcus, Methanobacterium*
Rhizosphere		*Azospirillum, Azotobacter, Bacillus, Erwinia, Pseudomonas, Klebsiella, Enterobacter*
Phyllosphere		*Klebsiella, Beijerinkia, Bacillus, Enterobacter*
Symbiotic with legumes		*Rhizobium, Bradyrhizobium*
Symbiotic with non-legumes		*Rhizobium, Frankia*
Phototrophs		
Free living		
	i. Cyanobacteria	
	a) Aerobic	*Aulosia, Nostoc, Anabaena, Scytonema, Cylindrospermum, Tolypothrix, Trichodesmium, Gloeotheca, Plectonema, Calothrix*
	b) Microaerophilic	*Lyngbya, Oscillatoria, Spirulina*
	ii. Purple non sulphurbacteria (facultative)	*Rhodopseudomonas, Rhodospirillum, Erwinia Rhodomicrobium*
	iii. Purple and green sulphur bacteria (anaerobic)	*Chromatium, Chlorobium, Ectothiospira, Thiocapsa, Thiopedia, Thiospirillum, Pelodictyon*
Associative	i. Lichens	*Nostoc, Calothrix, Stigonema*
	ii. Water fern (*Azolla*)	*Anabaena*
	iii. Liverworts	*Nostoc*
	iv. Mosses	*Halosiphon*
	v. Gymnosperms (cycas)	*Nostoc*
	vi. Endocynoses (oocystis)	*Nostoc*

6.8.2.1 Free Living Organotrophs

The free living bacteria having the ability to fix molecular nitrogen may be obligate aerobes, facultative aerobes and anaerobic microorganisms. The free living bacteria capable of fixing nitrogen independently in soil are distributed over at least 26 genera, 11 families and 3 orders of bacteria.

a) **Free living aerobes :** The free living aerobes utilize organic substrates as a source of energy. Bacteria of family Azotobacteriaceae constitute the majority of heterotropic free living nitrogen fixers and have three genera viz. *Azotobacter, Beijerinckia,* and *Derxia. Azotobacter* is mostly found in neutral and alkaline soils, whereas *Beijerinckia, Azomonas* and *Derxia* are comparatively more tolerant to acidic conditions. Various species of *Azotobacter* recognized are *A. chroococcum, A. vinelandii, A. agilis, A. insignis, A. beijerinckia, A. macrocytogenes* and *A. paspali.* There are three species of *Beijerinckia* (*B. indica, B. mobile* and *B. fluminis*) and *Derxia* has one specie, *D. gummosa. Azotobacter, Beijerinkia* and *Derxia* are mostly dominant in tropical areas. They are most sensitive to O_2 stress. These organisms have high level of respiration. In aerobes, respiration itself performs a protective function for nitrogenase against O_2 damage. *A. chroococcum* and *A. vinelandii* are adopted to aerobic N_2 fixation but show exceptionally high respiration rate 10-50 times than some aerobes. The fast respiration process scavenges O_2 from reaching N_2 fixing sites. In *A. chroococcum* and *A. vinelandii* nitrogenase protein exhibits conformational protection in which enzyme is inaccessible to O_2. In protected state, nitrogen fixation is totally disallowed but the activity is restored when O_2 tension is reduced to a permissible level. The organisms showing respiratory protection involve enormous consumption of carbon (1 g sugar to fix 10-15 mg N), and if the organism is starved of O_2, upto 45 mg N g^{-1} sugar can be fixed. *Derxia gummosa* has a normal rate of respiration but makes a thick gummy or polysaccharide capsule that surrounds the cell. This capsule slows down the diffusion of O_2 from air to cells. *Derxia gummosa* behaves like a microaerophile when fixing nitrogen in air and shows dimorphic growth in air. A few organisms form massive glutinous (polysaccharidous) colonies, which fix N_2 vigorously and remainder do not. At the centre of these massive colonies, O_2 tension is low enough for the initiation of N_2 fixation.

The population of nitrogen fixing microorganisms vary from a few to some hundreds g^{-1} soil. Application of nitrogen fertilizer drastically reduces the *Azotobacter* population in soil whereas phosphate additions enhance the ability of *Azotobacter* to fix more N_2. Application of

Azotobacter is made through seed or soil and has proved effective in enhancing crop yields in well-manured soil with high organic matter content. Besides N_2 fixing ability, *Azotobacter* species are known to synthesize several growth promoting substances such as vitamin B, indole acetic acid, gibberellins etc. Also, the organism possesses fungistatic properties on certain pathogenic species such as *Alternaria* and *Fusarium*. *Azotobacter* inoculations are reported to enhance plant growth and yield by 10-50% and supplement 10-15 kg N ha^{-1} yr^{-1} to plant under temperate conditions. The organism has been isolated from the rhizosphere of monocotyledonous and a few dicotyledonous plants.

b) **Free living facultative aerobes :** Some facultative aerobic organisms have the ability to fix N_2 under reduced oxygen tension. These N_2 fixing bacteria of genera *Klebsiella, Enterobacter* and *Bacillus* and *Rhodospirillum rubrum* recognize the presence of free O_2 and switch off nitrogenase synthesis. Aggregation or dense population and collective respiration lower O_2 tension in their own vicinity to zero.

c) **Free living microaerophiles :** *Azospirillum,* a free living microaerophile, is of considerable importance as it forms association with roots of grasses and to some extent in root tissues. *Azospirillum* participates in all steps of the nitrogen cycle, except nitrification. The nitrogenase enzyme need not to be protected in this case as energy (in the form of ATP) is produced via oxidative pathways. Besides, *Azospirillum* shows aerotactic movement and migrate to suitable microenvironment for nitrogen fixation. *Azospirillum,* beside N_2 fixation, enhances root hair number and root hormone excretion. The inoculation of *Azospirillum* causes 10-30% yield increase and compensates 10-25 kg N ha^{-1} demand of the plant. Rhizobia especially *Bradyrhizobium* can grow at low O_2 tension and fix nitrogen even in the absence of symbiont.

d) **Free living anaerobes :** Anaerobic nitrogen fixing heterotrophs like *Clostridium* and sulphur reducing bacteria (e.g. *Desulfovibrio* and *Desulfomaculatum*) use organic compounds as electron donors. High substrate availability and anaerobic conditions, like water logging, result in extensive N_2 fixation by these microorganisms. *Clostridium pasteurianum* or *Desulfovibrio gigas* (SO_4 reducers) are obligate anaerobes and avoid oxygen to moderate extent to protect their nitrogenase from air. Generally free living diazotrophs fix a few kg N ha^{-1}. *Desulfovibrio* is considered to be the principal contributor of biologically fixed nitrogen in marine sediments. Diazotoprhic anaerobes living in the gastroenteric cavities of soil animals or decaying organic matter may contribute to increase in nitrogenase activity in organic residues added to soils.

Presently at least three strains of thermophilic diazotrophs are well-established. They are *Mastigocladius* (a cyanobacteria) growing in hot springs at 60°C, *Methanococcus thermolithotrophicus,* an archaeon which fixes N_2 at 64°C, and *Streptomyces thermoautotrophicus* which grows at 65°C.

e) **Phyllosphere organotrophs :** The surface of the leaf wherefrom fungi or bacteria are able to utilize leaf exudates is called 'phyllosphere' and the association involved is called 'phyllocoenoses'. The free living diazotrophs are found abundantly on phyllosphere. These include *Klebsiella,* some cyanobacteria, some *Bacillus* sp., *Beijerinckia* sp. etc. In temperate region *Enterobacter* or *Bacillus* are found on aerial plant parts whereas a large number of *Azotobacter* or *Beijerinckia* are found in tropical areas. In many plants of wet tropics like sugarcane phyllosphere is populated with *Beijerinckia,* which probably upgrade the local soil after tropical rainstorms. Certain tropical plants such as *Ardesia, Psychotria* etc. have tiny nodules in their leaves, which are colonized by N_2 fixing bacteria such as *Klebsiella* so subsequently benefit the plants. It seems that perhaps the benefits arise not because of nitrogen fixation but owing to the formation of hormone like substances in the leaf nodules. Estimates of annual fixation rates of diazotrophic bacteria in the phyllosphere of tropical rainforests trees are in the range of 10-20 kg N ha^{-1}.

f) **Rhizosphere organotrophs :** The six bacterial genera which are living in association with roots are *Azospirillum, Azotobacter, Bacillus, Klebsiella, Enterobacter* and *Pseudomonas*. *Azospirillum* and *Azotobacter* are dominant in tropics whereas *Klebsiella* and *Enterobacter* in temperate region because of their high and lower temperature optima, respectively. These organisms have better access to root exudates as energy source for nitrogenase activity.

6.8.2.2 Symbiotic Organotrphs

a) Rhizobium symbiosis : In various legume and non-legumes diazotrophs form nodule in roots and stem. Approximately 700 genera and more than 1300 species of Leguminoseae are known and a majority of which fix atmospheric nitrogen. However, only about 50 species have been exploited for agriculture purpose. Of the three sub-families in Leguminoseae family, the Papilionoideae has 80-90% species, which form nodules, Mimosoideae 25% and a few in Caeselpinoideae. All nodulated legumes have nodules on root, except in tropical weed *Sesbania* where it occurs on stem. Even non-leguminous plants like *Parasponia* are known to be colonized by rhizobia. Both legume plants and species of *Rhizobium* show a degree of specificity.

Rhizobia are G -ve, non-spore forming, aerobic, rods, mostly motile, metabolize several carbohydrates and produce acids but no gas. At present three genera of rhizobia viz. *Rhizobium, Bradyrhizobium* and *Azorhizobium* are recognized on the basis of their rate of multiplication, acid production etc. *Rhizobium* is a fast growing bacterium with a mean generation time of 2-4 hours on yeast extract manitol agar (YEMA) medium. It is an acid producing organism and excretes large quantities of extracellular gum. *Bradyrhizobium* is a slow growing bacterium with a mean generation time of 6-8 hours, non-acid producing but produce less gums. *Azorhizobium* nodulates *Sesbania.* Rhizobia are capable of fixing atmospheric nitrogen even in the absence of nodulation. These organisms infect root hairs and cortical cells and ultimately induce the formation of root nodule. The process of colonization of host plants by rhizobia involves a sequence of mutual interactions involving genes of both partners.

Parasponia regosa, P. parviflora and *P. andersonii,* all natives of Southeast Asia, occur as non-legume weed in tea plantation and possess nodules containing rhizobia. The nodules are devoid of haemoglobin, and rhizobia are contained in highly branched infection threads. Thus these rarely form any bacteroids. This reveals that bacteroid formation and leghaemoglobin are not essential for nitrogen fixation.

Stem nodules of *Sesbania* develop in response to a relatively simple exchange of chemical signal between plant and bacteria after rhizobia gain entry into plant through stem. The only species *Azorhizobium calinodari* forms stem and root nodules on *Sesbania rostrata.*

b) Actinorhizal symbiosis : About 200 plant species, mostly tree and herbs, distributed in 25 diverse genera in 10 families are now known to harbour diazotrophs other than rhizobia in their root nodules. They include alder (*Alnus*), Australian pine (*Casuarina*), California lilac (*Ceanothus*), *Eleagnus,* bayberry/bog myrtle (*Myricia*), seabackthorn (*Hippophae*), bitter brush (*Purshia*), raspberry (*Rubus ellipticus*), *Cerocarpus* and *Dryas.* The organism involved *Frankia,* isolated first in pure culture in 1978, is G+ve to G variable, aerobic to microaerophilic filamentous spore forming actinomycetous bacterium (thallobacterium) which forms swelling (vesicles) on the tips of hyphae and forms actinorhizae with its symbiotic partner. The major actinomycete-nodulated non-legume angiosperms are given in Table 6.8.

The root colonization by *Frankia* induces root hair deformation. The hypha penetrates these deformed root hairs within a crypt formed at the junction of several root hairs. Modifications in root hair tip lead to the formation of an infection thread. The hypha penetrates the root at the fold and grows in the channel between host's cells. *Frankia* proliferates within host cortical cells, which become hypertrophied with large nuclei and many mitochondria. Phenolic compounds, inhibitory to *Frankia,* may accumulate in nearby uninfected cells, depicting a partial defense response by the host plant. Nodules remain viable for 3-4 years in field and new lobes are formed each year. Nitrogenase activity is maximum in summer, when nodules complete their growth. Spores formed in senescent nodules form major inoculum for soil. In absence of host plant the nodulation capacity of a soil decreases with time. Certain plants may get infected and develop nodule containing filament with swelling called 'vesicles'. The vesicles appear to be the sites of nitrogenase activity. *Casuarina* does not form vesicles and nitrogen fixation is carried out by hyphae. Haemoglobin-like protein has been detected in actinorhizal nodules, which, unlike legume nodules, are modified lateral roots. In *Casuarina-Rhizobium* system woody nodules may reach a diameter of 10 cm or more. Unlike *Rhizobium,* these organisms can fix nitrogen in absence of host plant. The nitrogenase in root nodules of *Frankia* is protected from free oxygen mainly by cell wall around the microsymbiont (vesicle). Variation in thickness and presence of multiple lipid layers impart protection to cell wall against ambient oxygen pressure.

Table 6.8 : Major actinomycete-nodulating non-legume angiosperms

Genus	Family	Nodulated species	Geographic distribution
Alnus	Betulaceae	94	Cool region of northern hemisphere
Ceanothus/Discaria	Rhamnaceae	88	North America
Myrica/Camptonia	Myricacaea	74	Tropical, sub-tropical and temperate areas
Casurina	Casiaromaceae	46	Tropical and sub-tropical areas
Elaeagnus/Hypophae/Shepherdia	Elaeagnaceae	35	Asia, Europe and America
Coriaria	Coriariaceae	86	Mediterrarian to Japan, Newzealand and from Chile to Mexico
Cercocarpus/Purshia/Dryas	Rosaceae	9	Temperate and sub-tropical areas

Source : Torry (1978); Tate (2000)

The distribution of *Frankia* is widespread in nitrogen poor habitats of temperate region, although these occur in warm climates at higher altitudes in tropical and sub-tropical zone as well. Actinorhizal plants appear to be more acid tolerant. Alder tree *Alnus glutinosa* can grow at pH 4.5-5.5 and bog myrtle (*Myrica gale*) in acid soils having pH 3.5-5.0. *Casuarina* plays an important role in agriculture associated with shifting cultivation and in aforestation of eroded lands. *Alnus* alone reportedly adds 100 kg N ha^{-1} yr^{-1} to soil just from leaf fall. The plant is used to upgrade poor soil. *Frankia* can fix N_2 either within or outside nodules. In either case diazotroph forms often have vesicles which assist in O_2 exclusion by acting as diffusion barrier.

c) Cyanobacteria symbiosis : In contrast to rhizobia and *Frankia* which form association with only most advanced plants, the dicotyledonous angiosperms, cyanobacteria favour association with more primitive members of plants. Lichens depict most primitive association between cyanobacteria and fungus. About 8% lichens do have phycosymbionts, often *Nostoc* or *Calothrix.* Light is necessary for lichens, though they may tolerate extreme conditions. Antarctic species, for example, fix N_2 at 0°C and survive extended freezing. They also survive drying but do not fix N_2 until dampened again. They colonize most unpromising environments like rocks, stones, tree trunks, walls, roofs of building etc. Another such association involves liverwort including species *Blasia pusilla* which have cyanobacteria of genus *Nostoc* associated with the underside of their thallus.

Azolla is a tiny aquatic fern which grows often to a high density in tropical and subtropical waters. It is only 2-3 mm across and harbours *Anabaena* in the cavity at the base of its leaf-like frond. The microbe is particularly rich in heterocysts, which presumably protect its nitrogenase from oxygenation. Fern can grow well even in absence of *Anabaena* provided nitrogen is limiting. Azolla can be used as a green manure. Under suitable conditions, the *Azolla-Anabaena* combination can double in mass after every 3 to 5 days and fix atmospheric nitrogen at a rate exceeding that of legume-*Rhizobium* symbiotic relationship. The *Azolla-Anabaena* association contributes 50-80 kg N ha^{-1} yr^{-1} in paddy soils.

Family cycads among gymnosperms form root-like structures on roots beneath soil surface. *Macrozamia* in its roots develops club-shaped excrescences (coralloid roots), which include a deep green zone. These roots can be colonized by diazotrophic cyanobacteria such as *Nostoc*. Similarly, *Gunnera,* a subtropical plant, is invaded by a species of *Nostoc*. Nodules are formed at the bases of leaves and the organism is usually rich in heterocysts in its symbiotic state.

6.8.2.3 Free Living Phototrophs

The free living cyanobacteria, besides carbon, fix atmospheric nitrogen. Cyanobacteria, cosmopolitan in nature, are distributed in tropics and subtropics and play vital role in wet land agriculture so occur abundantly in aquatic areas and high moisture soil surfaces. Cyanobacteria add a good amount of organic matter to soil through rich blooms. They also synthesize/ secrete several vitamins and growth substances (vitamin B_{12}, auxins, ascorbic acid, etc.). Rice fields usually show abundant cyanobacterial growth and some species have proved neurotoxic once consumed by animals. The cyanobacteria evolve oxygen and have photosynthetic potential. The cyanobacteria have three major morphological groups (Table 6.9).

Table 6.9 : Major morphological groups of cyanobacteria

Cyanobacteria groups	Examples
i) Unicellular	*Gloeotheca (Gloeocapsa), Synchoccus, Gloeobacter, Synchocystis, Chamaesiphon,* etc.
ii) Filamentous non-heterocystous	*Spirulina, Oscillatoria, Lyngbya, Phormidium, Plectonema,* etc.
iii) Filamentous heterocystous	*Anabaena, Nostoc, Nodularia Scytonema, Calothrix, Cylindrospermum,* etc.

Mostly unicellular cyanobacteria protect their enzyme system in the same fashion as those of *Azotobacter*. Heterocysts in filamentous ones are specialized thick-walled cells produced in the filaments after every 10-15 cells. These protect the nitrogenase enzyme from O_2 damage by physical means such as membranes. The heterocysts are formed only in nitrogen deficient environments. Heterocysts lack photosystem II, responsible for the evolution of O_2. The transition or metamorphosis of a vegetative cell to heterocyst in cyanobacteria is a gradual process leading to the conversion of a CO_2 fixing and O_2 evolving cell into an anaerobic cell, conducive for active N_2 fixation. In non-heterocystous forms, compartmentalization of photosynthesis and N_2 fixation has been suggested as a mechanism to overcome O_2 inhibition of nitrogenase enzyme. Some suggest the presence of antioxidants, O_2-linked H_2 uptake or presence of certain internal membrane system to serve as intracellular O_2 protective mechanisms. Compartmentalization helps in the restriction of nitrogenase enzyme to special locations like heterocyst in filamentous cyanobacteria. These do not appear when bacteria are grown with adequate fixed N (ammonia). The O_2 evolving component is absent in heterocysts but not in vegetative cells. *Plectomena* (filamentous bacteria) are sensitive to O_2

and fix nitrogen only if the O_2 tension and illumination are low. *Gloeotheca*, a non-filamentous and non-heterocystous cyanobacterium, shows a complex intracellular membrane structure for nitrogen fixation. In this case diazotrophic form often has vesicles which supposedly assist in the exclusion of O_2 by acting as diffusion barrier.

The cyanobacteria are inoculated to paddy fields to supplement N_2 requirements of plant. They reportedly fix 35-200 kg N ha^{-1}, provide a good amount of organic matter and growth promoting substances to plants and help the soil in aggregate development. The contributions made largely depend on edapho-climatic conditions in the area.

The phototrophic purple non-sulphur bacteria (e.g. *Rhodospirillum rubrum, Rhodobacter capsulants*) and purple sulphur bacteria (e.g. *Chromatium, Thiopedia*) and green sulphur bacteria (e.g. *Chlorobium*) are aquatic organisms living at sediment-water interface under highly reduced conditions wherein during photosynthesis no O_2 is evolved. Instead H_2S/S or organic compound are used as electron acceptors. It is possible to grow these organisms heterotrophically (e.g. on sugars, alcohols, etc.) and aerobically, but only in dark. The organisms fix nitrogen only in microaerobic conditions.

6.8.2.4 Associative Symbiosis

The root zone of plants has long been regarded as an active site for microbial activities including nitrogen fixation. During 1970's a group of Brazilian scientists lead by Dr. Johanna Dobereiner highlighted the association and abundant colonization of the rhizosphere of sand grass genus *Paspalum notatum* with *Azotobacter paspali. Paspalum* grass is widespread and persistent in tropical and sub-tropical areas. *A. paspali* grows around the roots of the plant, though the dominant N_2 fixing microbes in the surrounding soil are usually different. In due course *Azotobacter* forms a sheath, partly mucilaginous, over the roots and ceases to grow though they do not cease N_2 fixing activity. Later works rediscovered the previously described diazotrophic spirilli, now named as *Azospirillum*. Two species of *Azospirillum* are associated with a diverse host plants such as forage grasses (*Pennisetum purpureum, Brachiaria mutica, Brugulosa* sp., *Digiteria decumbens, Panicum maximum, Melinis minutiflora, Hyparrhenix rufa, Spartina* sp. etc.), weeds (*Heracleum sphondylium, Anthriscus sylvestris, Mercurialis perennis, Rumex acetosa, Convolvulus arvensis, Viola caniana* and *Stachys sylvestris*, etc.) and field crops (sugarcane, rice, maize, sorghum, wheat, millet, rye, tomato, pepper, soybean, etc.). Dobereiner called the association as "associative symbiosis".

The six species in this genus, identified so far, are *A. lipoferum*, *A. brasilense*, *A. amazonense*, *A. halopraeferans*, *A. viakinse* and *A. seropedica*. The nitrogen fixing bacteria having associative symbiosis are presented in Table 6.10. They are microaerophilic in nature, Gram -ve and contain poly β-hydroxy butyrate granules. The bacteria are widespread in both tropical and temperate soils. The attachment of *Azospirillum* to the roots is essential for the establishment of efficient association with host. The polar flagellum and extracellular polysaccharides play key role in its attachment to the plant roots. The diazotrophs grow in rhizosphere immediately adjacent to the roots and invade intercellular space of root cortex. The nitrogen fixation varies from 1 to 20 kg ha^{-1} yr^{-1}. Organisms are also known to produce growth hormones such as IAA, kinetin and gibberellins so affect root development. Several species of genera *Herbaspirillum*, *Azoarcus* and *Burkholderia*, and *Acetobacter diazotropicus* are micro-aerophilic diazotrophs, most of these actually enter the host plant tissues and multiply in the sap.

Table 6.10 : Nitrogen fixing bacteria having associative symbiosis

Organisms	Association with
Azospirillum lipoferum	Grasses and cereals (maize)
A. brasielense	Grasses and cereals
A. amazonense	Grasses, coconut
A. seropedica	Grasses
A. Halopraeferans	Kallar grass *(Leptochloa fusca)*
A. viakinse	Grasses
Azotobacter nitrocaptans	Sugarcane
Azotobacter paspali	*Paspalum* grass
Bacillus azotofixans	Grasses, wheat, sugarcane
Acetobacter diazotrophicus	Sugarcane
Campylobacter nitrofigilis	*Spartina alterniflora* (salt marsh grass)
Pseudomonas diazotrophicus	Rice, *Deschampsia caespitosa*
Herbaspirillum seropedicae	Maize, rice, sorghum, sugarcane

6.8.3 NODULE FORMATION AND HOST SPECIFICITY IN RHIZOBIA

The successful *Rhizobium*-legume symbiosis depends on the entry of micro-symbiont (*Rhizobium*), present in the rhizosphere, into the root system of macro-symbiont (legume). This involves a series of events. The infection in most nodulating species occurs on developing root hairs at the junction of lateral root through structurally altered cell walls of the root cortex or at the base of the stem. In stem nodulating species such as *Sesbania rostrata* infection occurs preferentially at the sites where lateral root primordia may pierce the epidermis. In some species temporary flooding is required for nodulation to occur. Compared to root nodules, stem nodule formation and its activity is less depressed by high levels of soil mineral nitrogen.

Infection and nodule formation are host specific phenomena and involve:

a) Recognition and attachment;
b) Penetration and migration; and
c) Bacteroid formation and nodule development

a) **Recognition and attachment :** The bacteria present in the rhizosphere need proper recognition of the host. The roots of host plant secrete distinct attractant organic metabolite like phenolics called flavonoid, which serves as a signal for the recognition of specific rhizobia for symbiosis. A concentration of 'luteolin' flavonoid as low as 10^{-9} M is sufficient to signal for bacterial chemotaxis whereas a concentration of 10^{-6} M luteolin even stimulates the expression of nodulation (*nod*) gene. Bacteria sense the flavonoid and migrate towards roots where flavonoid initiates a genetic programme in bacteria. This sets the process of symbiosis in motion. The transfer of information between plant and *Rhizobium* is a bidirectional process. Induction of *nod* genes is also required for the production of lectin (glycerol-proteins) by host which is recognized by the capsular polysaccharides/glycoprotein receptors on specific rhizobial cell membrane and is meant for the attachment of bacteria on root hair. The attachment of rhizobia to root hair surface may be one of the critical recognition stages prior to the successful infection (Fig. 6.10).

b) **Penetration and migration :** Only a small and variable proportion of root hair becomes infected. *Rhizobium* first converts tryptophane, present in root exudates, into IAA which brings about the production of non-specific curling signal substance, a complex sugar derivative called lipo-chito-oligosaccharide. The signalling substance diffuses diffuse to nearby root hair and causes them to curl inwards. Pectin methyl esterase and IAA loosen the host cell wall material at the site of infection. Host cell wall is dissolved by bacterial enzyme cytase and plant pectic enzyme. Polygalactouronase production induces invagination of root hair to form tubular structure called 'infection thread'. This allows bacteria to penetrate into the root. Root hair curling, cortical cell division induced by bacterial signal (probably cytokinins) and branching of root hair are the first visible responses by host legume. This is followed by the development of the nodule meristem and invasion by rhizobia, mediated by host plant-derived infection thread. A marked degree of curling by root hair (360° or more) appears to be the characteristic of a compatible interaction.

c) **Bacteroid formation and nodule development :** The plant cell nucleus controls the inward growth of infection thread. The base of root hair is reached usually 48 hours after infection. The infection thread penetrates through cortex and migration of *Rhizobium* continues until it reaches pre-formed tetraploid cell i.e. the site of developing nodule in cortex. Mitotic cell division of host vascular becomes differentiated to form nodule in cortex. Infection thread bursts and liberates bacteria into host cytoplasm. Each bacterium liberated gets surrounded by a membrane, pinched off from the infection thread. Inside the dividing nodule cells, rhizobia are released into the plant cell cytoplasm. However, bacteria remain covered by a plant derived coat called the peribacteroid membrane within which they cease multiplication process and loose their shape. Bacteria are several times larger than original bacteria, loose flagella and become devoid of cell wall. These are now called bacteroids. Infected root cell may contain up to 20,000 bacteroids. The space between bacteroid and membrane gets filled in with leghaemoglobin. The amount of leghaemoglobin and bacteroid production has direct relationship with the amount of nitrogen fixed by legumes.

Leghaemoglobin : Leghaemoglobin has a red coloured pigment which gives a characteristic pink colour to the nodules. Leghaemoglobin has a central iron atom in the cyclic tetrapyrrole porphyrin ring, identical to cytochrome. Globin part of this monomeric O_2 binding protein is synthesized by the plant whereas the heme group is synthesized by *Rhizobium*. Leghaemoglobin has high affinity for O_2 and delivers O_2 to *Rhizobium* at a very partial pressure to nitrogenase. In addition, nodules seem to posses a barrier which slows down the diffusion of O_2 into them. So nodule itself is a highly specialized compartment in which N_2 fixation and O_2 consumption are rendered physiologically compatible. During nodule development and maintenance, a number of plant and bacterial genes are expressed sequentially in a well coordinated manner. Nodulins are nodules specific proteins encoded by bacterial genes and produced during the certain stages of nodule development.

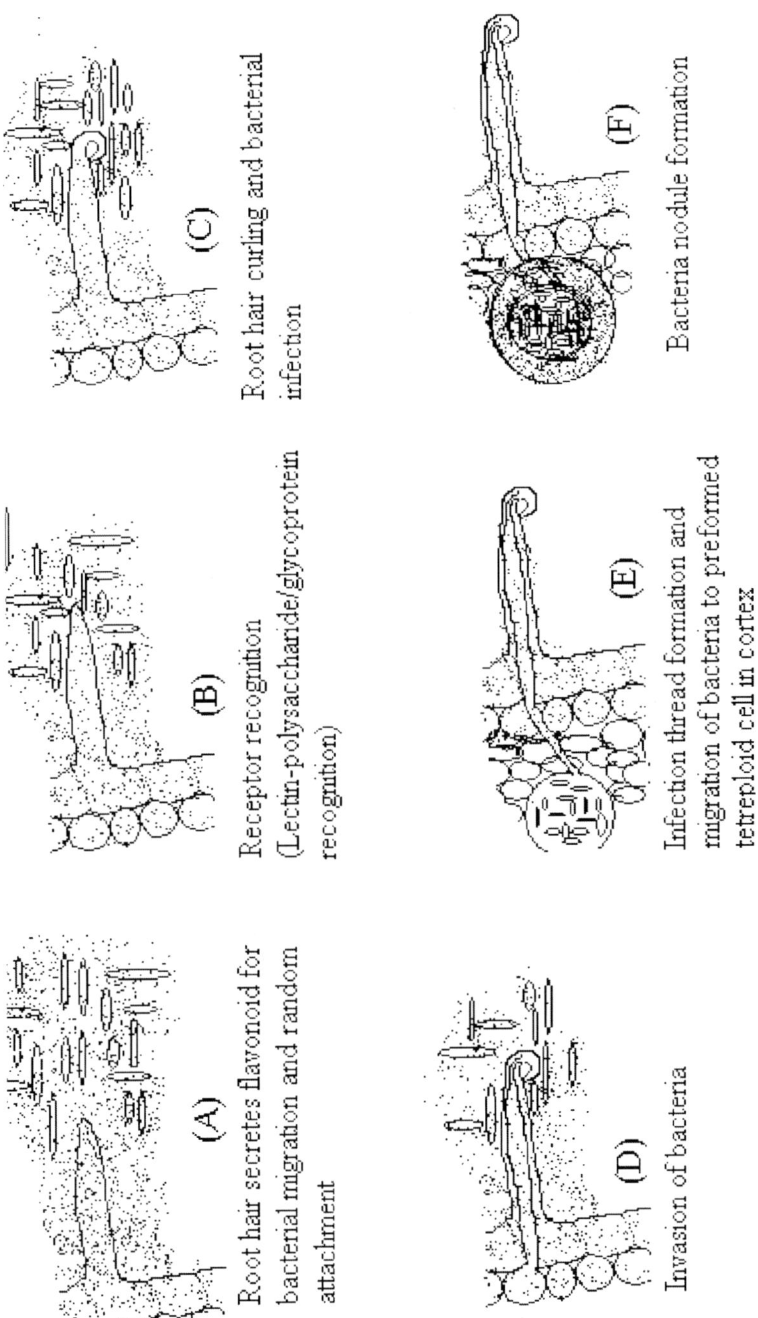

Fig 6.10 : Nodule formation in rhizobia

Transformation is accompanied with the synthesis of haemoglobin, nitrogenase and other enzymes involved in N_2 fixation. The concentration of leghaemoglobin is closely but not linearly linked with N_2 fixing capacity of root nodules. The nodules become rich in nitrogenase and initiate nitrogen fixation. At the same time plant forms nodule tissues around these bacteroids and surround them with leghaemoglobin so facilitates N_2 fixation by keeping O_2 concentration too low to damage nitrogenase. Peribacteroid membrane allows movement of certain fatty acids, and some nutrients inside for bacteroid consumption. Moreover, it allows release of ammonia to plant once fixed by bacteria.

Root nodule and nitrogen fixation : The presence of nodules on a plant does not necessarily mean nitrogen fixation by active *Rhizobium*. The amount of N_2 fixed by the bacteria is not only regulated to match the requirements of plant precisely but also differs with bacterial strains, host plant and environmental conditions. Fixation of N_2 continues till seed formation and at this stage nodule becomes senescent but rhizobia remain viable. Short season annual plants fix 50-200 kg N ha^{-1} yr^{-1}. Effective strains form healthy looking numerous small nodules (<2 mm dia.) with distinct pink colour which are scattered on the entire root system (Plate 6.1). Interior of cut nodule is distinctly red. In some cases nodules may be large (>8 mm dia.). Effective cultures specific to a particular crop are used for pelleting the seeds.

Plate 6.1 : A) Nodules in root system of common bean; B) a view of nodules and root system in soil

The part of the fixed N either remains in soil as root residues and nodules or returns to the soil with litter fall etc. The fixed nitrogen, in annual plants becomes available for next crop after harvest. In mixed stands of legumes and non-legumes direct transfer of fixed nitrogen during growing season may also occur e.g. *Phaseolus vulgaris* genotypes derive 5.6-21.1% N from fixed nitrogen in monoculture and 18.2-56.6% when intercropped with maize. In temperate climates, mixed stand of *Trifolium* sp. with *Lolium* sp. may annually fix 232-308 kg N ha^{-1}.

Stem nodules : About 86% of the species in leguminoseae are nodulating. Three legume genera *Aeschynomene, Sesbania* and *Neptunia* are known to bear nodules on stem, of which first two are used as green manure. *Aeschynomene* has about 250 species ranging from herbs to medium sized trees whereas *Sesbania* has about 170 species ranging from annual to perennial herbs and fast growing trees. In both legume and non-legume systems symbiosis is characterized by a more or less distinct host preference or even host specificity.

Cross inoculation groups in Rhizobium : The effective N_2 fixation can be achieved by inoculation of appropriate *Rhizobium* species. The cross inoculation groups are recognized on the basis of ability of rhizobia to form nodules with legume plants. Cross inoculation group are a collection of leguminous species that develop nodules when exposed to *Rhizobium* obtained from the nodules of any member of that group but do not nodulate the members of other group (Table 6.11). More than 20 cross inoculation groups are established. The various cross inoculant groups may overlap. Even there are large differences between strains of bacteria and host genotypes in relation to infection, nodulation and affectivity in N_2 fixation. This necessitates strain selection and genetic manipulation for increasing potential N_2 fixers in soil.

Table 6.11 : Cross inoculation groups of rhizobia

Organism involved	Cross-inoculant groups	Host plants
Fast growing type		
i) *Rhizobium leguminosarum* bv. *phaseoli*	Phaseolus group	Phaseolus bean
ii) *Rhizobium leguminosarum* bv. *trifolii*	Clover group	Clover (*Trifolium*)
iii) *Rhizobium leguminosarum* bv. *viceae*	Pea group	Pea (*Pisum*), lentil (*Lens*), vetch bean, (*Vicia*), sweet pea (*Lathyrus*)
iv) *Rhizobium meliloti*	Alfalfa group	Sweet clover (*Melilotus*), alfalfa or lucerne (*Medicago*), fenugreek (*Trigonella*)
Slow growing types		
i) *Bradyrhizobium japonicum*	Soybean group	Soybean (*Glycine*)
ii) *Bradyrhizobium lupini*	Lupin group	Lupins (*Lupinus*), leucaena, trifoils (*Lotus*), chickpea (*Cicer*)
iii) *Bradyrhizobium arachis*	Groundnut group	Groundnut (*Arachis*)
iv) *Rhizobium* sp. (cowpea)	Cowpea group	Cowpea (*Vigna*), acacia
Others		
i) *Azorhizobium* sp.	Stem-nodulating	*Sesbania rostrata*
ii) *Sinorhizobium* (intermediate in growth)	Soybean	Soybean (*Glycine*)

6.8.4 MEASUREMENT OF NITROGEN FIXATION

The development of methods permitting the use of radioactive and stable isotopes of nitrogen, ^{13}N or ^{15}N, has helped to identify and quantify nitrogen fixation by diazotrophs. The amount of isotope fixed is measured by mass spectrometry. Assay for ^{15}N in soil nitrogen pool are sensitive and accurate in natural soils. The use of alternate substrate like acetylene in nitrogenase has offered an important tool for estimating nitrogenase activity and nitrogen fixation. The product of acetylene i.e. ethylene can be rapidly and easily detected with great sensitivity by gas chromatographic technique. The acetylene assay is extensively used to measure nitrogen fixing ability of organisms. There are some problems associated with the use of this technique to measure rates of nitrogen fixation in the field which include endogenous ethylene production by soil microorganisms, inhibition of H_2 evolution from nodules and non-linear production of ethylene with time. Such measurements provide only short term estimates of nitrogenase activity. Preferably the nitrogen fixation may be estimated by tracer techniques using $^{15}N_2$.

6.8.5 STRUCTURE OF NITROGENASE

Nitrogenase enzyme system : The N_2 fixation is catalyzed by the key enzyme nitrogenase. Nitrogenase has been isolated from about 30 diazotrophic microbes and these nitrogenases differ slightly from species to species. Purified nitrogenases have been made from about 8 microbes including legume nodules bacteria. X-ray crystallography have confirmed that the three dimensional structure of molybdo-protein nitrogenase, purified from *Azotobacter vinelandii,* has 2 pairs of distinct subunits with FeMoCo group buried in a cleft in one of them, relatively inaccessible to water molecules. FeMoCo is a complex inorganic cluster containing 7 Fe and Mo atoms linked by S atoms, itself bound into a protein by way of a cysteine and a histidine group. The other Fe-S groups within the protein are present as pairs of clusters composed of 4 Fe and 4 S atoms each. The nitrogenase enzyme complex consists of two soluble non-heme-iron proteins viz., dinitrogen reductase and nitrogenase.

i) Dinitrogenase reductase (iron protein) : Also called 'azo-ferridoxin' or 'component II', it provides electrons with high reducing power. The dinitrogen reductase is a 64 kDa (52-73 kDa) dimer of identical subunits that contains one Fe-S cluster and two ATP binding sites. The dinitrogen reductase has 4 Fe and 4 labile S in a cubical structure.

ii) Nitrogenase (Mo-Fe protein) : Also called 'Mo-ferrodoxin' or 'component I', it uses electrons to reduce nitrogen to ammonia. Nitrogenase is a larger protein of 220 kDa (210-245 kDa; 210 kDa in *Clostridium pasteuranium* and 245 kDa in *Azotobacter vinelandii*). It has 4 subunits; 2 α and 2 β subunits and contains iron-molybdenum complex. The nitrogenase has 28 to32 Fe, 28 labile S and 2 Mo atoms per Mo-ferridoxin component I.

Both the above components of nitrogenase must be present for nitrogenase complex to function. These two proteins catalyze the reduction of N_2 to NH_3 and half life of these proteins in air varies from 0.5 to 10 minutes. Nitrogenase is also capable of reducing acetylene (C_2H_2) to ethylene (C_2H_4), H^+ to H_2, cyanide $[C{\equiv}N]^-$ to methane and NH_3, cyanogen (C_2N_2) to CH_4, hydrogen azide (N_3) to N_2, NH_3 andN_2H_4 (hydrazine); nitrous oxide to N_2 or NH_3 and cyclopropene to cyclopropane or propane. In order to function, all nitrogenases require ATP. The transfer of electrons from reductase to nitrogenase protein is coupled to the hydrolysis of ATP by reductase. Electron flows occur through normal metabolic route to ferridoxin or flavodoxin and then to the iron atoms in Mo-Fe protein. At least one electron is reduced for each e^- transfer.

Alternate nitrogenase : Paul Bishop and his associates in 1980 obtained evidence that *A. vinelandii* when starved of Mo formed a different kind of nitrogenase. Later studies revealed that when normal genes for nitrogenase in *A. vinelandii* and *A. chroococcum* were detected it still fixed nitrogen. The 'classic' form of this nitrogenase found in some diazotrophic bacteria contains vanadium instead of molybdenum. Yet another nitrogenase enzyme seemed to use an iron atom in place of Mo or V in diazotrophs, *Rhodospirillum rubrum* and *Rhodobacter* (*Rhodopseudomonas capsulatus*). Alternative nitrogenase is generally less efficient than Mo-nitrogenase and is formed only in absence of available Mo. The alternative nitrogenase is functionally and structurally almost similar. They consume ATP, generate H_2 during operation and reduce analogues of dinitrogen such as acetylene or cyanides, etc. However, they produce ethane in addition to ethylene during acetylene reduction. The alternative nitrogenases have been detected in *Clostridium*, cyanobacteria, *Azotobacter* and photosynthetic diazotrophs. Vanadium nitrogenase is repressed by Mo suggesting that it provides physiological back up nitrogen fixing system for use in case of shortage of Mo.

6.8.6 BIOCHEMISTRY OF NITROGEN FIXATION

The mechanism of N_2 fixation appears quite similar in most of the N_2 fixing prokaryotes. Nitrogenase reaction requires energy in the form of ATP and reducing equivalents (electrons). Nitrogenase enzyme when complexed with Mg-ATP reduces Fe-S protein (reductase) and releases ADP and Pi.

Protonation of molybdenum is the first step required for the binding of N_2. Nitrogen is bound to the active sites of Mo-Fe protein and reduced through a series of enzymes to dinitrogen hydride intermediate and eventually to ammonia which is assimilated to glutamate and glutamine. The Fe-S protein passes electrons to N_2 or other reducible substrates. Activity of Mo-Fe-S enzyme (nitrogenase) results in co-reduction of H^+ to H_2, although the proportion varies. At low electron flux H_2 is predominant product. At high electron flux and high ATP content, N_2 is effective competitor and results in the formation of high amount of NH_3. ADP is, however, inhibitor and on accumulation blocks some steps in the reaction. Mg^{2+} is essential for the functioning of nitrogenase enzyme. The reduction of N_2 to NH_3 is only a six electron-process and involves the reaction shown in Fig. 6.11.

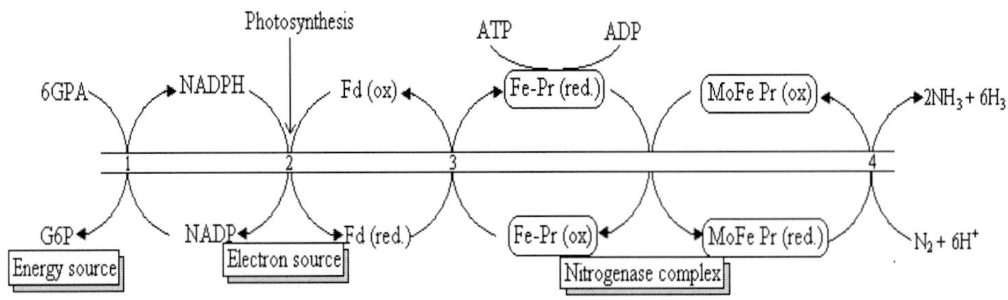

Fig 6.11 : Overall reaction in N_2-fixation involving ferridoxin (Fd), ATP and nitrogenase enzyme (1 = Glucose 6 phosphate dehydrogenase; 2 = NADPH ferridoxin oxido-reductase; 3 = Fe-Pr nitrogenase; 4 = Mo-Fe-Pr nitrogenase); red = reduced, ox = oxidized); Fe-Pr = iron protein

The reductase protein is imperfect and reduction of N_2 is always accompanied by the formation of one molecule of H_2. Thus there is additional consumption of two electrons. The associated reduction of H^+ in enzyme complex involves the reaction as follow :

$$N_2 + 6e^- + 12ATP + 8H^+ \xrightarrow{Mg^{2+}} 2NH_4^+ + 12ADP + 12Pi$$

$$2H^+ + 2e^- + 4ATP \xrightarrow{Mg^{2+}} H_2 + 4ADP + 4Pi$$

Overall reaction : $N_2 + 8\,e^- + 16ATP + 10H^+ \longrightarrow 2NH_4^+ + 16ADP + 16Pi + H_2$

Nitrogen fixation – Energy intensive process : The 8 high potential electrons come from the reduced ferrodoxin that is produced either in chloroplast by the action of photosystem I or in oxidative electron transport. The overall reaction of biological N_2 fixation is energetically very costly, with at least 16 ATP molecules being hydrolyzed. The actual requirement for N_2 fixation is considerably in excess of 16 molecules per N_2 fixed, as a rule between 25-30 molecules. Hydrogen evolution is a regular feature of nitrogenase activity. H_2 evolution precedes the binding of N_2 and a bound intermediate [like N_2H_2 complex] occurs enroute to NH_4^+ formation. Hydrogen exerts an inhibitory effect on nitrogen fixation by roots and by *Azotobacter*. The enzyme responsible for normal H_2 evolution is hydrogenase. The side reaction of H_2 evolution during N_2 fixation may account for low efficiency of N_2 fixing cells. In some plant-bacteria associations about 35% of their available ATP and reducing powers may be lost in H_2 evolution. In some organisms, like *Azotobacter*, the lost energy is recaptured by hydrogenase enzyme generating ATP. The presence of uptake hydrogenase (*Hup*$^+$) in specific rhizobia results in a significant increase in nitrogen content in plants. Hydrogenase is widespread in aerobic diazotrophs. It contains Fe-S and requires N_2 for its activity. The enzyme splits H_2 to $2H^+$ and 2 electrons, so recycles some electrons for subsequent use in N_2 reduction. In legumes, selection of *Rhizobium* strains with high *Hup* activity has been considered as an important factor in increasing the efficiency of N_2 fixation.

The NH_4^+ is first product of fixation. NH_4^+ itself may repress nitrogenase enzyme (Fig. 6.12). From NH_4^+, glutamine is formed through NH_4^+ assimilating enzyme 'glutamine synthase'.

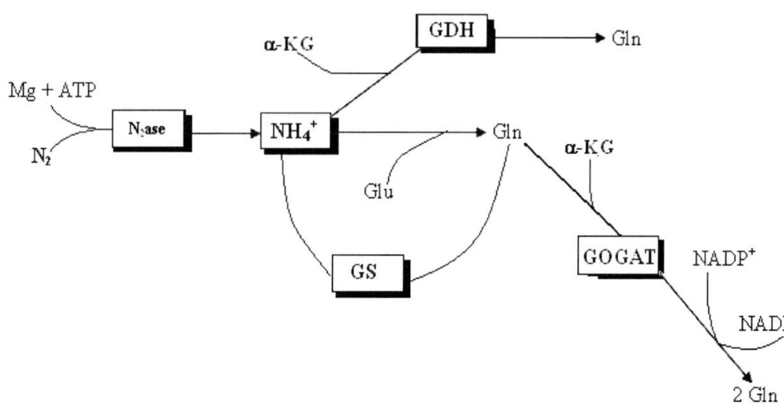

Fig 6.12 : Pathway of ammonia assimilation in legumes : N_2ase = nitrogenase; GDH = glutamate dehydrogenase; GS = glutamine synthetase; GOGAT = glutamine α-oxoglutarate amino-transferase (glutamine synthase); Glu = glutamic acid; Gln = glutamine; α-KG = α-keto-glutarate

Assimilation of fixed nitrogen : Ammonia (NH_4^+) is excreted by bacteriods in nodules and host cytosol by simple diffusion across the peribacteroid membrane. Cyanobacterial associations also excrete NH_4^+. Free living organotrophs are reported to produce NH_4^+ for their own growth so NH_4^+ fixed is available only upon the death and decomposition of microbes. However, ^{15}N experiments in *Azospirillum* and host plant sorghum have shown that plant leaves contained 40% of N_2 fixed over a 24 hours period. This depicts rapid release of NH_4^+-fixed in plant-microbe associations.

NH_4^+ produced by bacteriod is incorporated in organic forms into the associated plant cells. Ammonia assimilation occurs via two main reactions catalyzed by glutamate dehydrogenase (GDH) and glutamine synthetase-glutamine synthase (GS/GOGAT). It follows either GDH or GS/GOGAT enzyme pathway (Fig 6.12). GDH system provides major assimilatory pathway when ammonia is abundant. It requires a minimum of 1.5 mM NH_4^+ to operate. Such a high level depresses nitrogenase enzyme. Even high concentration of end product of NH_4^+ assimilation, glutamine, represses the nitrogenase activity. The combination of glutamine synthetase (GS) and glutamate synthase (GOGAT) maintains ammonium at low level and GS regulates nitrogenase and production of bacteriod protein.

Glutamine is transferred in secondary reaction to other products before leaving root via xylem. Temperate legumes transfer their nitrogen to amide asparagine and glutamine while some tropical legumes export ureides, allantoin and allantoic acids. These nitrogenous compounds have low C :N ratios (1 :1 to 1 :3). Actinorhizal association (*Frankia-Alnus* symbiosis) export ureide citrulline. The amides (asparagine and glutamine) are more soluble than ureides and better adopted to lower temperature conditions.

Generally glutamine is the form of nitrogen exported only in small amounts in nodules. In legumes having indeterminate nodules, such as *Mimosa pudica,* glutamine is mainly metabolized to asparagine in the nodules and delivered in the xylem to shoots. In contrast to this, legume species with determinate nodules, such as soybean, glutamine, etc. is metabolized in the nodule to ureides, allantoin and allantoic acid, and translocated in the xylem to shoot.

6.8.7 ENERGETICS OF NITROGEN FIXATION

The dinitrogen molecule is normally very unreactive due to very stable triple bond which links two atoms in the structure. Diazotrophs make N_2 molecule very reactive at ordinary temperature and pressure through enzyme nitrogenase. The enzyme is remarkably slow and takes 12.5 seconds for a molecule of enzyme to form two NH_3. This slowness arises in parts because the two proteins have to come together and separate 8 times to reduce one molecule of N_2 molecule. It is their separation that is the slowest step.

Malic acid acts as a substrate for the reductive synthesis of carboxylic acids, fumerate and succinate, which are transported to bacteroids to serve as an energy substrate for nitrogenase. Other carbon skeletons from glycolysis also serve substrate for NH_3 assimilation in the cytosol of legume nodule. In legumes plants, such as cowpea, nodules may consume as much as 50% of the photosynthates produced at certain stages. About 50% of these photosynthates are respired as CO_2. However, 25-30% of respired CO_2 can be re-assimilated by nodules via phosphoenolpyruvate (PEP) carboxylase providing about 25% of the carbon needed for malate and aspartate synthesis, required for the assimilation of NH_3 and export to host. On fresh weight basis, PEP carboxylase activity in nodules is between 20-fold (in pea) and 1000-fold (in soybean) higher than their roots. In *Medicago sativa* carbon costs per gram of nitrogen fixed varies between 5.1 and 8.1 g C, depending on the rhizobial strain, and about 50% of these costs are required for nodule growth and maintenance.

On an average, different N_2 fixing systems (e.g. differing in *Hup* activity or type of nitrogen compound exported) involve 36-39% carbon costs for nodulation, nodule growth and maintenance, 42-45% for nitrogenase activity and 16-22% for NH_3 assimilation and export. Biological reduction of nitrogen to NH_3 is a highly endergonic process with minimum energy requirement i.e. 960 KJ mol^{-1} N fixed. Reduction of one molecule of N_2 theoretically requires 0.22 molecules of glucose but measured efficiencies of utilization reveal a range of 0.66 to 1.38 moles of glucose per mole of N_2 fixed whereas its NH_4^+ assimilation into organic form requires 0.15 moles of glucose per mole of N_2. Transport of fixed N_2 and growth of organism requires additional 0.2-0.7 moles of glucose. Thus overall 1.13-2.37 moles of glucose are utilized per gram of N_2 fixed in legume symbiosis. Nitrogen fixation requires 2.9-6.1 g C g^{-1} N while as the use of fertilizer NO_3 involves 0.8-2.4 g C g^{-1} N in plants.

6.8.8 GENETICS OF NITROGEN FIXATION

The entire symbiosis process is controlled by several genes, located either on plasmids or chromosomal material. Gene transfer experiments have yielded the information about the location of genes on chromosomes. Extensive molecular biological work on *Rhizobium meliloti* has led to the development of molecular linkage map. These include genes for attachment, root hair curling, nodulation, host specificity, nitrogenase function, etc. The establishment of a legume symbiosis involves the coordinated expression of about 50 genes. The ability of a number of rhizobia to interact with plants is determined by *sym* plasmid borne genes. Plasmids are 0.1-5.0% of bacterial chromosomal size. The host specificity is controlled by several nodulation (*nod*) genes. Some *nod* genes induce plant to produce 'nodulin', a flavonoid chemical which may be required for root hair curling and cell division. The flavonoid actually participates in switching on other rhizobial genes. The infection, accomplished through bacterial surface components lipopolysaccharides or exopolysaccharides (lipo-chito-oligosaccharides), is also gene controlled process. There are set of nitrogen fixing genes called *nif* genes that enable bacteria to fix nitrogen. The *nif* genes are regulated positively in response to nitrogen limitation and symbiotic developmental pathway; and negatively in response to O_2 and intermediate levels of fixed nitrogen. Other genes present in symbiotic *Rhizobium* are i) nitrogen fixation gene (*fix*), ii) host range gene (*nod H* and *nod Q*), iii) hydrogenase (*hup*), iv) curl root hair genes (*hac*), v) nodulation genes (*nod A, B, C*) and vi) *sym* plasmids carry some *nod* genes in *R. meliloti.*

Nitrogen fixation genes in Klebsiella : In *Klebsiella pneumonia nif* lies fairly close to *his* on the chromosome but far away from *trp* (Fig. 6. 13). Finer details of map reveal that *nif* cluster has several genes, at least two for Mo-Fe-protein and one each for Fe-protein and enzyme regulation. Complete sequence of *nif* has been done and products for these genes identified (Table 6.12). The fine map of nitrogen fixation genes and their relative positioning in *K. pneumonia* is shown in Fig. 6.13 and 6.14. In *K. pneumonia* a cluster of 20 contiguous genes, grouped into 7 operons, code for nitrogenase enzyme. The *nif* gene lies next to the genes regulating the biosynthesis of histidine. *K. pneumonia nif* genes are subject to two levels of regulation in response to ammonia and oxygen, one at global level and the other is *nif* specific. *nif* D and *nif* K control the synthesis of Mo-Fe proteins whereas *nif* B is involved in the control of Fe-Mo protein cofactor. The global level of *nif* regulation is mediated by centralized *ntr* system which controls the expression of various nitrogen assimilatory genes in enteric bacteria. Genes *nif* A and L regulate the reading of *nif* operons. *nif* A is a regulatory gene whose product switches on the reading of other operons. *nif* A product is involved in the activation of *nif* genes and initiate transcription of genetic code. However, *nif* L product act as repressor for all genetic transcripts (except for its own operon) and switches off reading *nif* gene in response to O_2, NH_4^+, temperature, etc.

Regulation of gene function : The nitrogen regulation gene *ntr* A, renamed now as *rpo* N product, is essential for the activation of *nif* A product. In absence of *rpo* N the product *nif* A does not switch on *nif* genes. The product of *ntr* C is necessary for *nif* LA operon to be read (Fig.6.15). However, the formation of *ntr* C product is regulated by ammonia and repressed in presence of enough ammonium. *ntr* C product binds to the promoter region of *nif* LA DNA and enables the initiation of its transcription. *ntr* B product interferes with *ntr* C action. The *ntr* BC is actually contiguous with *gln* A, the gene for glutamine synthetase, in a cluster *gln*A.*ntr*BC. *ntr* genes control *nif* gene operation and termination whereas *nif* LA system provides fine tuning. *Rpo* N product in *Klebsiella* is also involved in nitrogen metabolism.

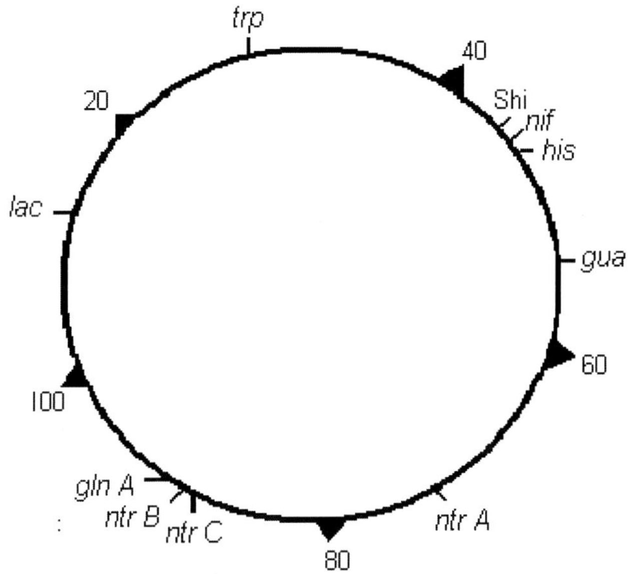

Fig 6.13 : Different nitrogen fixation genes in chromosomal map of *Klebsiella* : lac = genes for lactose utilization enzyme; trp = genes for synthesis of tryptophane; Shi A = genes for synthesis of shikimic acid uptake; nif = genes for N_2 fixation; his = genes for synthesis of histidine; gua = genes for guanine synthase; ntr A, B, C = genes for N_2 regulation; gln A = genes for synthesis of glutamine synthase

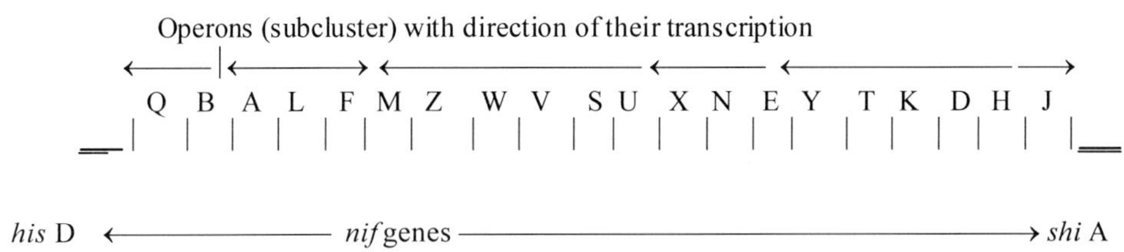

Fig. 6.14 : Fine map of nitrogen fixation genes of *Klebsiella pneumonia* (Arrows show the direction of transcription of operons)

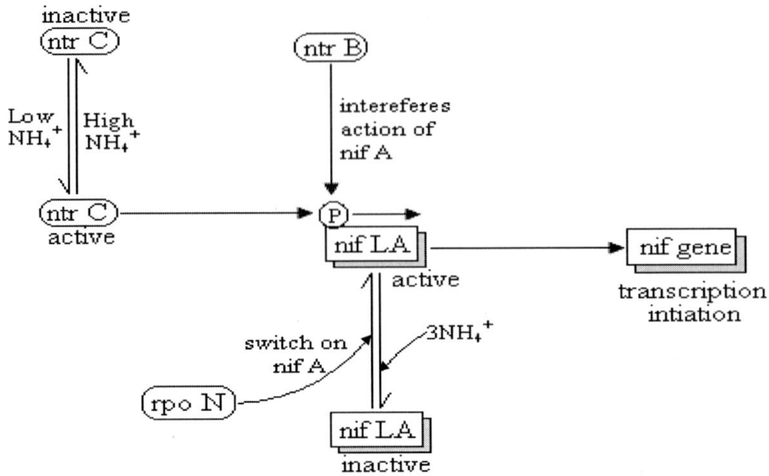

Fig. 6.15 : *Ntr* gene expression in *Klebsiella*

Table 6.12 : The various *nif* genes in *Klebsiella pneumonia* and their role

Genes	MW (KDa)	Code for
nif J	128	Enzyme pyruvate flavonoxin oxidoreductase involved in electron release for dinitrogen fixation
nif H	35	Peptide molecule of dinitrogen reductase (Fe protein)
nif D	56	α–subunit peptide of dnitrogenase (Mo-Fe protein)
nif K	60	β–subunit peptide of dnitrogenase (Mo-Fe protein)
nif T	80	Protein molecule of unknown function
nif Y	19	Protein molecule of unknown function
nif E	46	Peptide molecule forming *nif* E product
nif N	50	Peptide molecule forming *nif* N product
nif X	18	Regulatory protein
nif U	32	Proteins concerned with Fe-S centre of dinitrogen reductase
nif S	45	Protein concerned with Fe-S centre of dinitrogen reductase
nif V	41	Peptide molecule forming enzyme which make homo-citrate from 2-oxo-glutarate and acetyl coenzyme A
nif W	10	Protein involved in full activation of nitrogenase
nif Z	16	Protein involved in the insertion of Fe-Mo-Co into nitrogenase
nif M	28	Protein involved in the activation of nitrogenase reductase
nif F	19	Flavodoxin, electron acceptor from pyruvate
nif L	45	Regulatory protein, involved in *nif* cluster activation
nif A	57	Regulatory protein, involved *nif* cluster repression
nif B	49	Protein involved in processing of Fe-Mo-Co synthesis
nif Q	19	Protein involved in Mo uptake

The *Azotobacter vinelandii* and *A. chroococcum* have a partial *nif* cluster in its chromosome gene which is just like that of *K. pneumonia* with component genes in similar order but more widely separated. But some of its *nif* genes are outside the main cluster. It also has a second copy of *nif* H, identified as one of the *vnf* genes which codes for alternative nitrogenase in *A. chroococcum.*

Numerous transfers of *K. pneumonia* have now been made using plasmids pRD1, FN68 etc. Artificial recombinant plasmids such as pSA30 carrying *nif* operons have been made carrying entire alien genes such as *lac* gene, *nif*-promoter etc. Bacteria such as *Salmonella typhimurium, Serratia marcescens, Erwinia herbicola* or *Klebsiella aerogenes* attain perfect nitrogen fixing ability (anaerobically) when *nif* genes are inserted into them.

In *Rhizobium* (*R. meliloti*) the inactive protein product of *nod* D gene is activated through contact with plant signal by way of flavonoid compound. The flavonoids act as co-inducer for bacterial *nod D* proteins (nod factor) which in conjunction with flavonoids activate the transcription of nodulation genes. The composition of nod factors excreted by bacteria, as a set of structurally similar lipo-oligosaccharide molecules, is characteristic for the species or strains. The nod D factor then interacts on the promoter genes which control *nod A, B, C* gene action to induce nodule formation in host roots (Fig 6.16). Worthwhile to mention is that hydrogen flavones produced in infection-zone of emerging root hair stimulate the expression of nod genes within minutes whereas coumarins and isoflavones produced behind the root tips repress *nod* gene transcription. The success of infection depends on the ratio of these stimulatory and inhibitory compounds which provide the plant a mechanism for the regulation of nodulation. Once *nod* D has interacted with flavonoid, other *nod* genes such as *nod* F and E, involved in host range specificity, are induced and they, in turn, may regulate other non-sym plasmid genes such as those involved in polygalactourase production. Genetic mapping have revealed the presence of *fix* genes (in *R. meliloti*) on bacterial chromosome. The plasmids carry *nod* and *nif* genes. In *R. meliloti* a 14 kb *nif*/ *fix* gene cluster has been found. *vnf* genes code for vanadium nitrogenase and *anf* genes for iron-based nitrogenase. *vnf* D, K and H signify for structural genes of vanadium nitrogenase. However, in *A. chroococcum* a small gene *vnf* G, lying between *vnf* D and *vnf* K, codes for low molecular weight ferrodoxin like subunit of vanadium nitrogenase. Further, *vnf* H is not contiguous with them as *nif* H in *Klebsiella*. The third nitrogenase does have contiguous structural genes, arranged as *anf* HDGK. Both *vnf* and *anf* systems have activator genes *vnf* A and *anf* A similar to *nif* A in function. These share certain regular *nif* genes such as B, N, V, etc. Genetics and regulation of alternative nitrogenase is not yet fully clear.

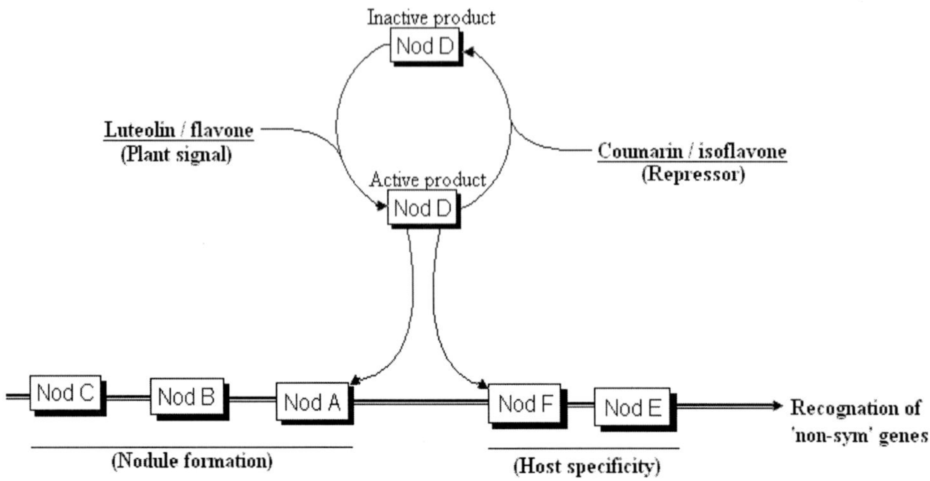

Fig 6.16 : Events at molecular level in the root region during *Rhizobium*-legume interaction (P = promoter)

In *Bradyrhizobium,* the functional analogues of *nod, nif* and *fix* genes are carried on chromosomes (as against the *sym* plasmid in *Rhizobium*). Four of the nodulation genes *nod* D, A, B and C are contiguous and highly conserved. These genes control root hair curling. *nod* D is a regulatory gene which is constitutive in *R. leguminosarum* bv. *trifolia.*

The *nod* gene inducing activity of root exudates and adsorption of rhizobia at the root surface is strongly affected by calcium and pH. Low pH and low Ca concentration negatively affect nodulation of *R. meliloti* in alfalfa in the very first steps of nodulation. Depending upon the plant species and environmental conditions, N_2 fixation starts at the earliest between 10-21 days after infection. During the period bacteria rely on host for all mineral supplies, photosynthates and amino acids, etc.

6.8.9 FACTORS AFFECTING NODULATION AND N_2 FIXATION

Several factors influence the process and development of nodulation in the roots and nitrogen fixation by symbiotic microbe. These include :

a. **Nitrate-nitrogen :** It has inhibitory effect on both nodulation & nitrogen fixation. High NO_3 level may inhibit nitrogenase activity. The mechanism of inhibition is not clear. It has been hypothesized that NO_3 in converted to NO_2 which destroys auxins and indole acetic acid. The critical balance between available carbon and available nitrogen in root system is offset by extraneous nitrate. Moreover, NO_2 formed could inactivate leghaemoglobin by the formation of nitrosylhaemoglobin complex and destroy its oxygen regulating capabilities. The nitrogen source such as NO_3, urea, amino acids, etc. suppress nitrogenase synthesis, probably these are converted transiently to ammonia during the cell metabolism.

b. **Ammonical-nitrogen :** Mineral nitrogen beyond a certain level interferes with the infection process of bacteria in root hair and the number and structure of nodulation in plants. High NH_4^+ immediately switch off the nitrogenase enzyme activity, a process which presumably conserves ATP.

c. **Oxygen :** The rate of oxygen diffusion into nodules of legume affects the extent of N_2 fixation. Water logging, even for a shorter period, is detrimental to the nodule functioning. Atmospheric O_2 concentration >20% enhances the process of N_2 fixation but levels beyond 50% O_2 causes inhibition. Import and export of water in nodule indirectly affects the rate of N_2 fixation. Restriction on import of water from phloem decreases the turgor pressure in nodule thus indirectly affects the nodule permeability for O_2. The rate of N_2 fixation might be limited by the availability of water for the export of reduced nitrogen in xylem from nodule.

d. **Carbon :** In general there is a favourable correlation between N_2 fixation rate and continuous supply of newly fixed carbohydrates to root nodules. The supply of available carbon and O_2 has major effect on symbiotic N_2 fixation. The addition of sucrose enhances nodulation in plants. ATP generation is greater when carbohydrates are oxidized than H_2O oxidation.

e. **Minerals :** Rhizobial cells have specific requirement for Ca or Mg and Mo. Ca stimulates nodule formation. Mo is indispensable for symbiotic activity because of being a part of prosthetic site of nitrogenase. Co stimulates symbiotic N_2 fixation and is also required for leghaemoglobin synthesis. The concentration of vitamin B_{12} in soybean nodules and chlorophyll content in leaves are positively correlated with Co. Similarly, PO_4 as a component of ATP is a source of energy supply to activate the reactions involved in nitrogen fixation. The presence of adequate phosphates enhances nodulation, nitrogen content and growth of plant. The phosphorus requirement by nodules may be higher than those of host root or shoot. The minimum concentration of phosphorus required for nodulation is about 0.5 μg P L^{-1} in external solution. Increase in concentration from 200 to 500 μg P L^{-1} causes greater increase in dry nodule weight relative to shoot. Seed pelleting with 100 g Mo ha^{-1} increases nitrogenase activity and nodule weight. Iron is required by several enzymes of nitrogenase complex such as in ferrodoxin electron carrier, some hydrogenases etc. Ni is constituent of hydrogenase so its deficiency affects nitrogen fixation. Vanadium is long known to decrease the need for Mo at least in *Azotobacter.* Elements like Fe, Cu, and Co have been implicated in leghaemoglobin synthesis. Deficiency of Cu causes depression in bacteriod content, accumulation of starch in invaded cells and reduction in the rate of nodule apex development.

f. **Soil reaction :** Legumes grow less luxuriantly under acid soil conditions. Low pH causes poor colonization in rhizosphere. Size and number of nodules also gets reduced. Besides, soil acidity reduces Mo uptake.

g. **Temperature :** Root temperature affects infection and nodulation processes. A temperature <10°C retards root hair growth whereas temperature >24°C promotes it. The temperature of 20-30°C is optimum for bacteroid tissue formation. The distribution of nitrogen between root and shoot is greatly affected by root temperature.

h. **Host genotype :** Varieties of some legume plant species are known to respond differently to a given *Rhizobium* sp. especially with respect to nodulation. Usually the plants sensitive to nodulation have high C : N ratio.

i. **Effectiveness of rhizobial strains :** Host plant is dominant factor in assessing the effectiveness or ineffectiveness of a rhizobial strain.

j. **Growth substances :** Gibberellins and IAA are found in fairly good amounts in root nodules. Lower levels of IAA are conducive for early nodulation whereas higher levels inhibit root morphogenetics. Similarly, NAA inhibits nodulation in clovers whereas PCPB (para chlorophenoxy isobutyrate) stimulates it. Colchicine, which induces polyploidy, increases the nodule number in plant roots.

k. **Pesticides :** Recommended levels of fungicides and insecticides, in general, exhibit no effect on nodulation. However, herbicides influence the host plants so influence the root nodulation and N_2 fixation.

l. **Saprophytic competition in soil :** The associated microflora may have beneficial or detrimental effect on *Rhizobium*. For example, increase in the number of protozoa and other predators, viruses, *Bdellovibrio,* etc., may decrease the *Rhizobium* population in soil. The alfalfa sickness or clover sickness are considered due to viral infestation. The rhizobial strains should also be able to endure antagonistic microflora. The aerobic spore forming bacteria and *Streptomyces* are the most active antagonists of rhizobia. In general, slow growing rhizobia appear to be most susceptible to the actinomycetes.

m. **Salinity and alkalinity :** The saline or alkaline conditions in soil decrease root hair formation as well as the mucilage deposited around roots.

n. **Light :** Light is involved in photosynthesis. Photoperiod influence the formation, size and number of nodules on root system. If nodulated plants are kept for several days under low light, not only nitrogen fixation rate decreases but also nodule conditions deteriorate.

Selected References

Addiscott, T.M., Whitemore, A.P. and Powlson, D.F.S. 1991. *Farming, Fertilizers and the Nitrate Problem.* CAB International, Willford, England.

Alazard, D., Ndoye, I. and Dreyfus, B. 1988. *Sesbania rostrata* and other stem nodulated legumes. pp. 765-769. In : *Nitrogen Fixation : Hundred Years After* (eds. H. Bothe, F. De Bruijin and W.E. Newton) Fischer, Stuttgart., Germany.

Alexander, M. (ed.). 1984. *Biological Nitrogen Fixation.* Plenum Press, New York, USA.

Atkins, C.A. 1987. Metabolism and translocation of fixed nitrogen in the nodulated legumes. *Pl. Soil,* 100 : 157-169.

Barak, R., Nur, T., Okon, Y. and Henes, Y. 1982. Aerotactic response of *Azospirillum brassilense. J. Bacteriol.,* 152 : 643-649.

Bartholomew, M.V. and Clark, F.E. (eds.) 1965. *Soil Nitrogen.* American Society of Agronomy, Madison, USA.

Benson, D.R. and Silvester, W.B. 1993. Biology of *Frankia* strains, actinomycete symbionts of actinorhizal plants. *Microbiol. Rev.,* 57 : 293-319.

Bergersen, F.J. (ed.). 1980. *Methods for Evaluating Biological Nitrogen Fixation.* John Wiley & sons, New York, USA.

Bergersen, F.J. 1991. Physiological control of nitrogenase and uptake hydrogenase. pp. 76-101. In : *Biology and Biochemistry of Nitrogen Fixation* (eds. M.J. Dilworth and A.R. Glenn). Elsevier, Amsterdam, the Netherlands.

Belser, L.W. 1982. Inhibition of nitrification. pp. 267-293. In : *Advances in Agricultural Microbiology* (ed. N. S. Subba Roa). Oxford & IBH Publ. Co., New Delhi, India.

Bothe, H., deBruijn, F.D. and Newton, K.E. (eds.). 1988. *Nitrogen Fixation : Hundred Years After.* Gustav. Fischer, Stuttgart, Germany.

Brady, N.C. and Weil, R.R. 1996. *The Nature and Properties of Soils* (11th edn.). Prentice-Hall, New York, USA.

Broughton, W.J. 1981. *Nitrogen Fixation, Vol.1. Ecology.* Clarendon Press, Oxford, UK.

Bundy, L.G. and Meisinger, J.J. 1994. Nitrogen availability indices. pp. 951-964. In : *Methods of Soil Analysis, Part 2 : Microbiological and Biochemical Properties* (ed. R.W. Weaver) Soil Science Society of America, Madison, USA.

Burns, R.F. and Hardy, R.W. 1975. *Nitrogen Fixation in Bacteria and Higher Plants, Vol. I &II.* Springer-Verlag, Berlin, Germany.

Cole, J.A. 1987. Assimilatory and dissimilatory reduction of nitrate to ammonia. pp. 281-329. In : *The Nitrogen and Sulphur Cycles* (eds. J.A. Cole and S.J. Ferguson). Society for General Microbiology, Cambridge Univ. Press, Cambridge, UK.

Deroche, M.E. and Carrayol, E. 1988. Nodule phosphoenol pyruvate carboxylase : A review. *Physiol. Plant,* 74 : 775-782.

Dilworth, M.J. and Glenn, A. (eds.) 1991. *Biology and Biochemistry of Nitrogen Fixation.* Elsevier, Amsterdam, the Netherlands.

Djordgevic, M.A. and Weinman, J.J. 1991. Factors determining host recognition in the clover-*Rhizobium* symbiosis. *Aust. J. Pl. Physiol.,* 18 : 543-557.

Dobereiner, J., Burris, R.H. and Hollaender, A. (eds.). 1978. *Limitations and Potential for Biological Nitrogen Fixation.* Plenum Press, New York, USA.

Evans, H.J. and Barber, L.E. 1977. Biological nitrogen fixation for food and fiber production. *Science,* 177 :332-339.

Favilli, F. and Messini, A. 1990. Nitrogen fixation at phyllospheric level in coniferous plants in Italy. *Pl. Soil,* 128 : 91-95.

Fischer, H.M. 1994. Genetic regulation of nitrogen fixation in rhizobia. *Microbiol. Rev.,* 58 : 351-358.

Focht, D.D. and Verstraete, W. 1977. Biochemical ecology of nitrification and denitrification. *Adv. Micro. Eco.,* 1 : 135-214.

Freney, J.R. and Simpson, J.R. 1983. *Gaseous Loss of Nitrogen from Plant Soil System : Developments in Plant and Soil Sciences, Vol. 9.* Nijhoff-Junk, The Hague, Netherlands.

Graedel, T.E. and Crutzen, P.J. 1993. *Atmospheric Change : An Earth System Perspective.* W.H. Freeman, New York, USA.

Greenwood, D.J. 1962. Nitrification and nitrate dissimilation in soil II. Effect of oxygen concentration. *Pl. Sc.,* 17 : 365.

Greenwood, D.J. 1975. *Ministry of Agric. Fish. Fd. Tech. Bull. No. 29,* HMSO, London, UK

Gresshoff, P.M., Roth, L.E., Stacey, G. and Newton, G. (eds.). 1990. *Nitrogen Fixation : Achievements and Objectives.* Chapman & Hall, New York, USA.

Hart, S.C., Stark, J.M., Davidson, E.A. and Firestone, M.K. 1994. Nitrogen mineralization, immobilization and nitrification. pp. 968-1018. In : *Methods of Soil Analysis, Part 2 : Microbiological and Biochemical Properties* (ed. R.W. Weaver) Soil Science Society of America, Madison, USA.

Havelka, U.D., Boyle, M.G. and Hardy, R.W.F. 1982. Biological nitrogen fixation. pp. 365-422. In : *Nitrogen in Agricultural Soils. Agronomy Vol.22* (ed. F.J. Stevenson) Soil Sci. Soc. Am./Am. Soc. Agron., Madison, USA.

Hollaender, A. 1979. *Genetic Engineering of Nitrogen Fixation*. Plenum Press, New York, USA.

Jackson, A.R.W. and Jackson, J.M. 1996. *Environmental Sciences : Natural Environment and Human Impact*. Addison Wesley Longman Ltd, Harlow, Essex, England.

Knowles, R. and Barraquio, W.L. 1994. Free living dinitrogen-fixing bacteria. In : *Methods of Soil Analysis, Part 2 : Microbiological and Biochemical Properties* (ed. R.W. Weaver) Soil Science Society of America, Madison, USA.

Layzell, D.B., Gaito, S.T. and Hunt, S. 1988. Model of gas exchange and diffusion in legume nodule. I. Calculation of gas exchange rates and the energy costs of N_2 fixation. *Planta*, 173 : 117-127.

Lui, W.C., Lund, L.J. and Page, A.L. 1989. Acidity produced by leguminous plants through symbiotic dinitrogen fixation. *J. Envirn. Qual.*, 18 : 529-534.

Malhi, S.S. and McGill, W.B. 1982. Nutrification in three Alberta soils : Effect of temperature, moisture and substrate concentration. Soil Biol., Mac Gill, 14: 393-399.

McKay, I.R., Dilworth, M.J. and Glenn, A.R. 1988. C_4-dicarboxylate metabolism in free living and bacteroid forms of *Rhizobium leguminosarum*. MNF 3841.`*J. Gen. Microbiol.*, 133 : 1433-1440.

Merrick, M.J. and Edwards, R.A. 1995. Nitrogen control in bacteria. *Microbiol. Rev.*, 59 : 604-622.

Neilsen, F.E. and MacDonald, J.G. (eds.). 1978. *Nitrogen in the Environment, Vol. 1&2* Academic Press, New York, USA.

Nutman, P.S. 1978. *Symbiotic Nitrogen Fixation in Plants*. Cambridge Press, Cambridge, UK.

Nye, P.H. and Tinker, P.B. 1977. *Solute Movement in the Soil-Root System*. Blackwell Scientific, Oxford, UK.

Payne, J.W. 1981. *Denitrification*. Wleg-Interscience Publ., London, UK.

People, M.B. and Crasswell, E.T. 1992. Biological nitrogen fixation : Investment, expectations and actual contributions to agriculture. *Pl. Soil*, 141 : 13-39.

Peters, J.W., Fisher, K. and Dean, D.R. 1995. Nitrogenase structure and function : A biochemical genetic perspective. *Ann. Rev. Microbiol.*, 49 : 335-366.

Philips, D.A. 1992. Flavonoids : Plant signals to soil microbes. *Recent Adv. Phytochem.*, 26 : 201-231.

Postgate, J.R. 1998. *Nitrogen Fixation* (3rd edn.). Cambridge Univ. Press, Edinburg, UK

Postgate, J. R. 1989. Trends and prospectives in nitrogen fixation research. *Adv. Microbial Physiol.*, 30 : 1-22.

Postgate, J.R. (ed.). 1982. *Fundamentals of Nitrogen Fixation.* Cambridge Univ. Press, Cambridge, UK.

Professor, J.I.1986. *Nitrification.* IRL Press, Oxford, UK.

Professor, J.I. and Cox, D. 1982. Techniques for estimating nitrifiers and nitrification. pp. 178. In : *Experimental Microbial Ecology* (eds. G.R. Burns and J.H. Slater) Blackwell, USA.

Quispel, A. (ed.). 1988. *Biology of Nitrogen Fixation.* North-Holland Publishing Co., Amesterdam, Holland.

Randhawa, G.S. and Raad Hassani 2002. Role of rhizobial biosynthetic pathways of amino acids, nucleotides bases and vitamins in symbiosis. *Indian J. Exptl. Biol.,* 40 : 755-764.

Richards, R.A. 1991. The chemistry of dinitrogen reduction. pp. 58-77. In : *Biology and Biochemistry of Nitrogen Fixation* (eds. M.J. Dilworth and A.R. Glenn) Elseviers, Amesterdam, Netherlands.

Rosendahl, L., Glenn, A.R. and Dilworth, M.J. 1991. Organic and inorganic inputs into legumes root nodule nitrogen fixation. pp. 259-29. In : *Biology and Biochemistry of Nitrogen Fixation* (eds. M.J. Dilworth and A.R. Glenn) Elsevier, Amsterdam, the Netherlands.

Schwintzer, C.R. and Tjepkema, J.D. (eds.). 1990. *The Biology of Frankia and Actinorhizal Plants.* Academic Press Inc., San Diego, USA.

Silverster, W.B., Harris, S.L. and Tjepkema, J.D. 1990. Oxygen regulation and hemoglobin. pp. 157-176. In : *The Biology of Frankia and Actinorhizal Plants* (eds. S.C. Schwintzer and J.D. Tjepkema). Academic Press, San Diego, USA.

Sprent, J.I. and Raven, J.A. 1985. Evolution of nitrogen fixing symbiosis. *Proc. R. Soc. Edin,* 85 B : 215-237.

Sprent, J.R. *The Ecology of the Nitrogen Cycles.* Cambridge Studies in Ecology, Cambridge Univ. Press, Cambridge, UK.

Stacey, G., Burris, R.H. and Evans, H.J. (eds.) 1991. *Biological Nitrogen Fixation.* Routledge, Chapman & Hall, London, UK.

Stevenson, F.J. (ed.) 1982. *Nitrogen in Agricultural Soils. Agronomy, Vol. 22.* Soil Sci. Soc. Am./ Am. Soc. Agron., Madison, USA.

Stewart, W.D.P. (ed.). 1975. *Nitrogen Fixation by Free Living Organisms.* Cambridge Univ. Press, Cambridge, UK.

Subba Roa, N.S. 1982. *Advances in Agricultural Microbiology.* Oxford & IBH Publ. Co., New Delhi, India.

Subba Roa, N.S. (ed.) 1988. *Biological Nitrogen Fixation - Recent Developments.* Oxford & IBH Publ. Co., New Delhi, India.

Subba Roa, N.S. and Rodriguez-Barueco, C. (eds.). 1993. *Symbioses in Nitrogen Fixing Trees.* Oxford & IBH Publ. Co., New Delhi, India.

Subba Roa, N. S. and Dommergues, Y. R. (eds.). 1998. *Microbial Interaction in Agriculture and Forestry. Vol. 1.* Oxford & IBH Publ. Co., New Delhi, India.

Subba Rao, N.S. and Yatazawa, M. 1984. Stem nodules. pp. 101-110. In : *Developments in Biological Nitrogen Fixation* (ed. N.S. subba Rao). Oxford & IBH Publ. Co., New Delhi, India.

Tate, R.L. III. 2000. *Soil Microbiology.* John Wiley & sons, New York, USA.

Tiedje, J.M. 1988. Ecology of denitrification and dissimilatory nitrate reduction to ammonium. pp. 179-244. In : *Biology of Anaerobic Microorganisms* (ed. A.J.B. Zehnder). John Wiley & sons, New York, USA.

Triplett, E.W. and Sadowsky, M.J. 1992. Genetics of competition for nodulation of legumes. *Ann. Rev. Microbiol.,* 46 : 399-428.

Torry, J.G. 1978. Nitrogen fixation by actinomycete-nodulated angiosperms. *Bioscience,* 28: 586-592.

Tsai, S.M., da Silva, P.M. Cabezas, W.L. and Bonetti, R. 1993. Variability in nitrogen fixation of common Bean (*Phaseolus vulgaris* L.) intercropped with maize. *Pl. Soil,* 152 : 93-101.

Venkataraman, G.S. 1972. *Algal Biofertilizers and Rice Cultivation.* Today and Tomorrow's Printers and Publishers, New Delhi, India.

Vessey, J.K. and Waterer, J. 1992. In search of the mechanism of nitrate inhibition of nitrogenase activity in legumes. *Physiol. Plant,* 84 : 71-76.

Walters, D.T., Aulokh, M.S. and Doran, J.W. 1992. Effect of soil aeration, legume residues and soil texture on transformation of macro- and micro-nutrients in soil. *Soil Sci.,* 153 : 100-107.

Weaver, R.W. and Graham, P.H. 1994. Legume nodule symbionts. pp. 199-222. In : *Methods of Soil Analysis, Part 2 : Microbiological and Biochemical Properties* (ed. R.W. Weaver). Soil Science Society of America, Madison, USA.

— *** —

Chapter 7

Transformation of Phosphorus in Soil

"Change is not made without inconvenience, even from worse to better"
(Richard Hooker)

Phosphorus, an essential element, plays a central role in cell metabolism and reproduction. It is a structural component of energy transferring molecules (ATP, ADP and AMP), nucleic acids, coenzymes, phosphoproteins, phospholipids and sugar phosphates. In soils, the phosphorus does not occur as abundantly as nitrogen or potassium. The total phosphorus in an average arable soil is estimated to be about 0.05% (by weight) with an average of approximately 1100 kg P ha^{-1} (range 200-2000 kg P ha^{-1}). But only a meager proportion (only ~0.1-1.0%) of it is available for use by plants and microorganisms because phosphorus has poor solubility and gets fixed in soil. Not only the native inorganic phosphates but even the added soluble phosphates of fertilizers and manure readily combine with the cations present in soil solution to form low solubility products. Fixation reactions in soils allow only a small fraction (10-15%) of phosphorus fertilizer to be taken up by the plants in the year of its application.

Phosphorus sources on earth : Phosphorus constitutes about 0.1% of the earth's crust. The main natural source of phosphorus in soil is rock phosphate (apatite) $3[Ca_3(PO_4)_2]CaX_2$, strengite ($FePO_4.2H_2O$) and variscite ($AlPO_4.2H_2O$). The other phosphorus minerals are wavellite ($Al_3(PO_4)_2.(OH)_3$], dufrenite [$FePO_4.Fe(OH)_3$] and vivianite [$Fe_3(PO_4)_2.8H_2O$]. Apatites account 95% of phosphorus in igneous rocks. It has a general formula $Ca_2X_2(PO_4)_6$, though instead of Ca either Mg, Al or Fe may also be present. The X represents an anion which may be F^-, Cl^-, OH^- or CO_3^{2-}. Mineral deposits of fluoroapatite, chloroapatite and carbonate apatite are usually dominant, except in calcareous soils where hydroxyapatites are predominant. Different combinations of cationic Ca, Al or Fe and anionic X may result in about 200 different types of phosphorus occurring in nature. Rock phosphates, high in carbonate apatite, are commonly mined for fertilizers.

The total amount of phosphorus on earth's crust is 10^{15} Mg, mainly in the form of apatite, whereas in oceans the largest reserviour are sediments. Sediments are major sink for phosphorus as almost all the water-eroded phosphorus reach oceans. The major reserve of phosphrus on earth is oceanic sediments. Compared to it, only a meager amount of phosphorus is reserved in soil, water (dissolved organics), mineral rocks and biota.

7.1 PHOSPHORUS CYCLE IN NATURE

Phosphorus cycle in nature appears to be simple than carbon, nitrogen or sulphur cycle as it does not involve any exchange with atmosphere (Fig. 7.1). The unique aspect of phosphorus cycle is that phosphorus does not undergo any valency change in the cycle. Phosphorus exists in nature in a variety of organic and inorganic forms but primarily in either insoluble or sparingly soluble inorganic form.

Plants take up phosphorus in orthrophosphate ionic form (either as HPO_4^{2-} or $H_2PO_4^-$ ions) from soil, though some soluble organic compounds may also be taken up by the plant roots. The chemical species of phosphorus, present in soil solution, is determined by the pH of soil solution. Monovalent anions $H_2PO_4^-$ dominate in strong acid soils whereas divalent anions HPO_4^{2-} dominate in alkaline soils. Once translocated into plant, phosphates become part of plant tissues. The phosphorus is returned to soil directly through plant residues and roots or indirectly via animal or their excreta. This phosphorus, both of plant and animal origin, is recycled by microbial degradation of organic matter. However, the amount of dissolved phosphorus in soil at any time is extremely small i.e. <0.1 to 1.0 kg HFS^{-1} (generally ranging from 0.001 mg L^{-1} soil solution in very infertile soil to about 1.0 mg L^{-1} in heavy fertilized soil). The free inorganic phosphorus ion in soil solution plays a central role in phosphorus cycling and plant nutrition.

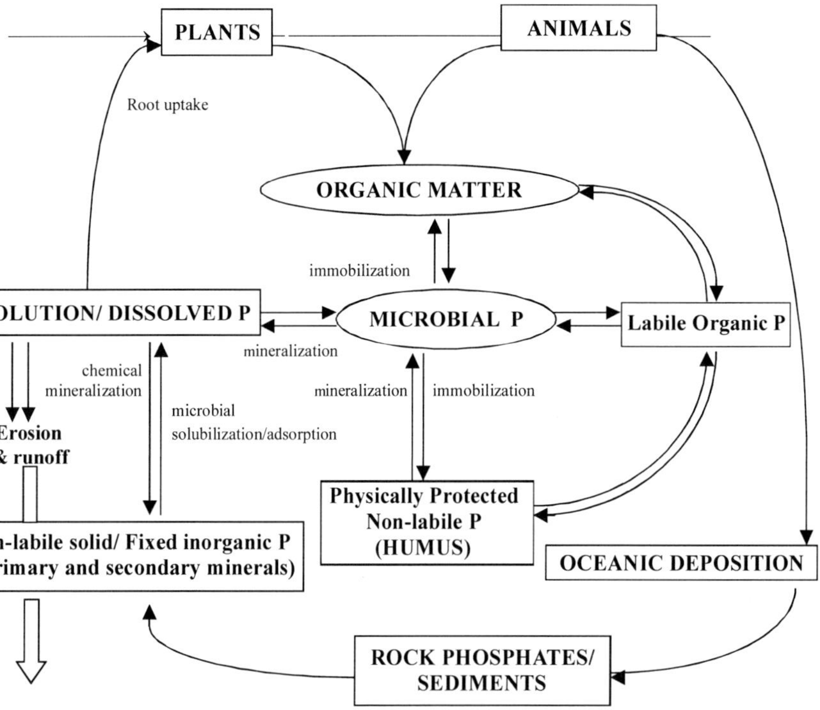

Fig 7.1 : Major soil phosphorus reserviours

Microbial immobilization-mineralization reactions in soil lead to the release of inorganic phosphorus which may then enter either into the soil solution or become part of labile organic P or humus. In soils, decrease in solution phosphorus due to microbial or plant absorption is buffered or compensated by both organic and inorganic phosphorus fractions either through mineralization of organic matter or through equilibrium reactions with dissolved phosphates and solid phosphorus compounds. Residual decomposition by microbial or enzymatic activities splits the bound phosphorus present in organic constituents and then hydrolyzes, immobilizes and mineralizes it.

Adsorbed phosphorus on clay and mineral surfaces, as HPO_4^{2-} or $H_2PO_4^-$, may be desorbed or released to balance/buffer any decrease in solution P. Primary and secondary minerals may also dissolve to resupply some phosphate ions in solution. However, phosphorus solubility in soil is controlled by the soil pH. Inorganic phosphates may be adsorbed by positively charged clay particles or soil organic matter. Microbial solubilization of rock phosphates or minerals may also contribute to plant phosphorus. Besides crop removal, some of the phosphorus is lost from soil system by the erosion of phosphorus carrying soil particles or organic and mineral particles and phosphorus dissolved in surface runoff water.

7.2 FORMS OF PHOSPHORUS IN SOIL

Phosphorus in soil exists mainly in two forms i.e. organic and inorganic form.

7.2.1 INORGANIC PHOSPHORUS

Inorganic forms of phosphorus occur as a simple calcium, iron, aluminium or fluorine phosphates or as adsorbed phosphates on iron and aluminium oxides, hydroxides hydrated oxides or simply as HPO_4^{2-} or $H_2PO_4^-$ ions in solution. Amongst the various inorganic forms of phosphorus, the calcium compounds predominate in most alkaline soils while Fe and Al forms are dominant in acid soils. Most of the soil phosphates are insoluble or sparingly soluble.

Phosphorus fixation : When soluble inorganic phosphates are added to soil, these react readily in soil to form insoluble phosphates. This loss of solubility, known as "phosphate fixation", reduces the phosphorus availability to plants and microorganisms. The phosphate fixation may be either due to the :

- Precipitation of phosphates as Ca/Fe or Al phosphates depending upon whether soil is alkaline or acidic in reaction, or
- Adsorption of phosphates to clay mineral (surfaces) of Fe, Mn and Al hydrous oxides or on clay lattice.

Fixation of phosphates is much more troublesome in both acid and alkaline soils. In acid soils (pH <5.5) phosphate ions react with Fe and Al ions and form insoluble Al and Fe hydroxy phosphates. Phosphate adsorption involves anion exchange and fixation by clay minerals. Phosphate ions become more or less firmly attached to soil-clay complex. This adsorption resembles that of NH_4^+ ion fixation except that NH_4^+ is a cation and PO_4^{2-} is an anion. Secondly, ammonium is released by microbial action as NO_3^- anion that is not fixed in soil. However, no such process operates in the release of PO_4^{2-} ions. It may be released by displacement with other anions as those of NO_3^- and SO_4^{2-}.

Under mild alkaline conditions, soluble phosphates revert to hydroxyapatite or other low solubility products/minerals because of abundant calcium in soil. In strongly alkaline conditions sodium phosphates are formed due to the abundance of sodium. Phosphorus availability at pH >9.0 is not a major problem but other associated adverse effects such as poor soil structure, low soil permeability and dissolution of organic matter are serious problems which do not allow the plants to grow properly.

Phosphorus availability : Phosphate fixation can not be avoided entirely, but it may be reduced by addition of competing ions for fixing sites. Organic anions from stable manure and silicates are effectively used for reducing phosphorus fixation in soil. Soils that remove more than 350 mg P kg^{-1} soil or have phosphorus fixing capacity of about 770 kg P ha^{-1} from soil solution are generally considered high phosphorus fixing soils. These soils tend to maintain low phosphorus concentration in soil solution. The only way to slow down phosphate fixation process is to somehow hold the soil pH at 5.8-7.0.

Phosphorus availability to biota in soil depends on the solubility of phosphorus, ionic species present, amount of phosphorus in solution and distance between phosphate ions from root or microbes. Solubility of phosphorus depends on the kind of ionic specie and concentration of other ions present. Both these factors are influenced by soil pH.

Phosphorus in soil solution : The soil solution phosphorus is often called 'intensity factor (I)' while organic and inorganic labile phosphorus is collectively known as 'quantity factor (Q)'. Maintenance of adequate solution phosphorus for nutrition in plants depends upon the ability of labile phosphorus to replace solution phosphorus taken up by plant roots and microbes. This is determined by the capacity factor (C) which is expressed as:

$$C = \frac{Q}{1}$$

The capacity factor provides idea about the relative ability of a soil to buffer the changes in its solution phosphorus. Larger the capacity factor greater will be the ability of a soil to buffer the solution phosphorus. Labile phosphorus is readily available portion of Q that exhibits higher dissociation rate and readily replenishes solution phosphorus. Depletion of labile phosphorus causes some non-labile phosphorus to become labile but at a slow rate.

The concentration of phosphorus in soil solution is 0.1 to 1.0 ppm and its diffusion rate is extremely low i.e. approximately 0.02-0.10 mm h^{-1} for HPO_4''. In most soils the amount of phosphorus available to plants from soil solution at any one time is very low which seldom exceeds 0.01% of the total phosphorus in soil. Plants can make satisfactory growth with a very small amount of dissolved phosphorus provided the supply is uniformly maintained. The rate of dissolution of inorganic phosphorus and mineralization of organic phosphorus generally seems to be fast enough to meet the plant requirement.

But transport or movement of $H_2PO_4^-$ or HPO_4^{2-} ion from the point of entry into solution to a plant root or microbe is a problem. Movement of PO_4^{2-} ions to root is mostly by diffusion process but roots cannot absorb or extract phosphorus in solution if greater than 2 mm away from the roots. Further, plants need metabolic energy to move phosphorus into their cytoplasm where phosphorus concentration may be 50-100 times higher than in soil solution. Therefore, a minimum level of phosphorus must be attained in roots before plant translocation can occur. The translocation of phosphorus into cytoplasm occurs at 0.05-0.5% level only.

Soil microorganisms and plant roots participate in the solubilization of soil phosphorus, mainly through their production of CO_2, organic acids and phosphorus solubilizing enzymes. Production of organic acids like citrate, malate, lactate, etc. and phosphatase enzymes by microbes solubilize soil phosphorus. The organisms involved include *Aspergillus, Bacillus, Penicillium, Pseudomonas,* mycorrhizae, etc. However, microbial activities are influenced by soil factors such as pH, moisture, aeration, etc.

7.2.2 ORGANIC PHOSPHORUS

In soils, the ratio of inorganically and organically bound phosphorus is around 1 :1. The ratio may vary from soil to soil and with the depth in a given soil. However, in mineral soils 30-50% phosphorus occurs in organic form whereas in organic soils organic phosphorus is usually dominant (>80%). The forms in which phosphorus is present in soil are usually in a state of dynamic equilibrium but differ from soil to soil. This dynamic equilibrium depends on a variety of factors like soil pH, $CaCO_3$ content, sesquioxides, clay, humus, moisture to air ratio, microbial activity, etc.

Major forms of organic phosphorus in soil : Organic phosphorus occurs in soil principally as phytates (inositol phosphates, $\{C_6H_6(OH)_6\}$), nucleic acids and their derivatives, phospholipids, phosphates-linked to sugars and derivatives of a phosphoric acid (containing C-P bonds), probably all occurring in organic complexes. Most organic phosphorus compounds identified in soil are esters of orthophosphoric acid and their contents in soil are as under :

i)	Inositol phosphate	10-50%
ii)	Phospholipids	1-5%
iii)	Nucleic acids	0.2-2.5%
iv)	Uncharacterized phosphates	50%

Phytin (mostly inositol phosphate) is synthesized both by microorganisms and plants and is the most stable form of organic phosphorus in soil. The phytins interact with higher molecular weight humic compounds. Inositol ring compounds carrying about 1-5 atoms of phosphorus also occur in soil. These possibly represent degradation products of inositol hexaphosphate. The inositol phosphates in soil include esters of myo-, chiro-, scyllo- and neo-inositols of which only first is a common plant constituent and other isomers might be of microbial origin. Nucleic acid P or phosphopyridine (NAD and NADP) occurs in soil organic matter as a result of plant litter decomposition and microbial immobilization-degradation process. Identified products are cytosine, adenine, guanine, uracil, xanthine, hypoxanthine etc. Nucleic acids are adsorbed by humic compounds as well as silicate clays. Much of this organic phosphorus appears to be associated with the fulvic acid fraction of soil organic matter. Adsorption on soil colloids probably helps in the protection of nucleic acid phosphorus from microbial attack. Organic phosphorus occurring in alcohol and ether extracts of soil is indicative of the presence of phospholipids. Sugar phosphates are also present in soil. These are easily decomposable. More than 50% organic phosphorus has not yet structurally been identified. It appears that organic phosphorus compounds found in soil are not simply the result of the accumulation of organic phosphorus compounds from plants that are resistant to decomposition, there must also be biochemical transformation which are not yet fully understood.

7.3 SOLUBILIZATION OF PHOSPHATES BY MICROBES

Microorganisms involved in phosphorus solubilization : A diverse group of microorganisms from autotrophs to heterotrophs, diazotrophs to phototrophs, certain fungi and actinomycetes solubilize mineral phosphates like apatite, rock phosphates and tri-calcium phosphates so are useful in enhancing crop productivity. At neutral pH maximum solubilization of phosphorus is carried out by bacteria whereas a pH of 4.0-5.0 is optimum for fungal activity. However, in soil fungi are considered to be the most efficient phosphorus solubilizers than bacteria, including actinomycetes or cyanobacteria. Bacteria that solubilize phosphorus represent about 10% of soil microbial biomass. The list of microorganisms involved in phosphorus solubilization, besides mycorrhizea, is given in Table 7.1.

Microbial inoculations and adding rock phosphates enhance organic waste decomposition. The mycorrhizal hyphae extend out into soil several centimeters from the root surface to absorb phosphate ions from soil solution and strongly bound phosphorus (Also see chapter 10). Various inoculants are nowadays used to solubilize bound phosphorus so as to maintain a regular supply of phosphorus to plant roots. The organisms used as inoculants are *Bacillus megaterium* var. *phosphaticum, B. polymyxa, Pseudomonas seratia,* mycorrhizae, etc. The use of "phosphobacterin", *B. megaterium* preparation, has been found useful in enhancing grain yield by 5-10% in soils.

Table 7.1 : The organisms involved in phosphorus solubilization

Organisms	Genus/species
Bacteria	*Alcaligenes phosphaticum, Agrobacterium* sp., *Aerobacter aerogenes, Achromobacter* sp., *Bacillus megaterium, B. subtilis, B. circulans, B. mycoides, B. mesentericus, B. pulvifaciens, B. striata, B. pumilus, B. polymyxa, B. cereus, B. coagulans, Bradyrhizobium* sp., *Brevibacterium* sp., *Curvularia lunata, Enterobacter asburiae, Erwinia* sp., *Escherichia fraundii, E. intermedia, Flavobacterium* sp., *Micrococcus* sp., *Nitrobacter* sp., *Nitrosomonas* sp., *Pseudomonas putida, P. liquifaciens, P. calcis, P. rathonis, P. fluorescence, P. straita, Rhizobium meliloti, Serratia phosphaticum, Thiobacillus ferroxidans, T. thiooxidans, Xanthomonas sp.*
Cyanobacteria	*Anabaena* sp., *Calothrix braunii, Nostoc* sp., *Scytonema* sp.
Actinomycetes	*Actinomyces* sp., *Streptomyces* sp.
Fungi	*Achrothecium* sp., *Alternaria teneus, Aspergillus awamori, A. carborum, A. niger, A. terreus, A. flavus, A. fumigatus, A. nidulans, A. foetidus, A. wentii, Candida* sp., *Cladosporium* sp., *Cephalosporium* sp., *Cunnighamella sp., Chaetomium globosum, Fusarium oxysporum, Glomus fasiculatum, Humicola inslens, H. lanuginose, Helminthosporium* sp., *Mucor* sp., *Mortiella* sp., *Oideodendron* sp., *Penicillium lilacinum, P. digitatum, P. balaji, P. funicolosum, Populospora mytilina, Pseudogymnoacsus* sp., *Rhizopus sp., Rhodotorula* sp., *Trichderma viride, Phoma* sp., *Pythium* sp., *Paecilomyces fusiporus, Populospora mytilina, Rhizoctonia solani, Micromonospora* sp., *Schwaniomyces oxydentalis, Sclerotium rolfsii, Torula thermophila.*

Mechanisms of solubilization : Microorganisms solubilize mineral phosphorus through the mechanisms of chelation, iron reduction or acidification. Phosphate dissolving bacteria reduce the pH of substrate by secretion of a number of organic acids such as formic acid, acetic acid, oxalic acid, propionic acid, glycolic acid, lactic acid, fumaric acid, succinic acids, etc. Some of these acids such as oxalic acid may form chelates with cations such as Ca, Mg and Fe so destabilize phosphate mineral and make phosphorus available.

7.4 IMMOBILIZATION OF PHOSPHORUS

The soil organic matter contains 1-3% phosphorus and usually organic phosphorus decreases with the soil depth. Microorganisms, in general, have greater requirement of phosphorus (1-2% present in cells) in comparison to the plants (0.05-0.1% in cells). Therefore, most of the organic phosphorus is first immobilized by microorganisms to build their cell constituents and does not ionize in early stages of decomposition. However, as a result of mineralization microbes also release inorganic phosphorus. Bacteria, on an average, possess 1-3% phosphorus, whereas in fungal mycelium phosphorus varies from 0.5 1.5%. In microbial cells the storage of phosphorus in the form of inorganic polyphosphates is widespread. Polyphosphate formation also occurs in the hyphae of mycorrhizal fungi. Hyphae take up inorganic phosphorus from soil solution and synthesize polyphosphates which act as transient energy and storage pool of phosphorus in hyphae. These are subsequently transferred to the host roots either as polyphosphates or as inorganic phosphates.

7.5 MINERALIZATION OF ORGANIC PHOSPHORUS

Numerous microbes through their enzymatic activities digest/decompose organic matter. Phosphorus mineralization is catalyzed by general soil heterotrophic microbial population. Mineralization of immobilized microbial P occurs at a slower rate. Phosphorus mineralization rate is controlled by physical and chemical factors influencing heterotrophic activity. Phosphorus mineralization and its assimilation into biomass is generally optimal under soil conditions suitable for aboveground plant community development.

Role of phosphatases : Phosphatase enzymes dephosphorylate all known organic phosphates of plant origin. Kirchner and his associates found phosphatase activity in soils ranging from 0.3-0.6 μmol PO_4^{3-} hydrolyzed g^{-1} soil h^{-1}. Phosphatase activity increases with organic C but is affected by soil temperature, pH and moisture.

C :P ratio : There are good evidences that phosphorus is the dominant element controlling C and N immobilization processes. The ratio of C :P in soil organic matter varies widely from 15 :1 to 75 :1. Organic matter, therefore, have indirect influence on phosphorus availability. It is assumed that at C : organic P ratio of 200 :1 phosphorus is readily mineralized as soluble phosphates ions into soil solution. At ratios greater than 300 :1 there will be immobilization of phosphorus into microbial biomass. The available HPO_4^{2-} or $H_2PO_4^-$ in soil solution will temporarily disappear. The C : organic P ratio seems to control the mineralization process in soil.

C :P <200	Mineralization will dominate
C :P 200 - 300 (0.2%P)	No gain no loss situation
C :P >300	Immobilization will dominate

Microbial decomposition of organic phosphorus also results in the formation of more recalcitrant phosphorus compounds which become part of humus. There may also be microbial synthesis of organic phosphorus compounds from inorganic PO_4 when an energy source is added to the soil. It has been shown that glucose usually causes immobilization of phosphates. In temperate regions mineralization of organic phosphorus in soil typically releases 5-20 kg P ha^{-1} yr^{-1}, most of which is absorbed by growing plants. Conversion of arable land to a grassland or pasture-land tends to accumulate organic phosphorus so increases organic matter content in soil. Even the addition of phosphorus fertilizer leads to the storage of organic fraction after being assimilated by microorganisms.

Organic matter and phosphorus fixation : Organic matter generally has little capacity to strongly fix phosphate ions. High soil organic matter exhibits relatively low level of phosphorus fixation. This may be owing to the :

- Interaction of humic-clay/hydroxide particles which render less fixation sites available for phosphorus;
- Organic acids produced during microbial decomposition compete with phosphorus for fixation sites; and
- Organic chelates of Al and Fe rendering Fe and Al less available for phosphorus fixation. Iron and aluminium ions can be complexed and tied up by organic matter. This leaves less iron and aluminum to precipitate insoluble phosphorus compounds.

Mild alkaline conditions favour mineralization. Much of the phosphorus used by plants other than applied phosphorus comes from organic phosphorus released by decomposition. Therefore, transfers and transformations are most important in crop production and ecosystem functioning. Mycorrhizae play important role in phosphorus transfer than other microbial groups in soil. Their significance arises because of solubilizing of phosphorus activities through CO_2 release, organic acid production and enzymatic action. These mobilize phosphorus from large distances through their hyphae to plants. The exploration of soil helps in the nutrient acquisition from distant places. Fungal hypae penetrate soil organic matter and macro-aggregates and acquire PO_4^- for newly developing roots. In soil, the transfer of soil organic matter phosphorus to organic phosphorus is primarily, but not exclusively, microbially mediated. The principal environmental problem related to soil phosphorus are land degradation caused by too little available phosphorus and accelerated eutrophication of streams and water bodies.

Selected References

Bolt, G.H. 1976. Adsorption of anions by soils. pp. 51-95. In : *Soil Chemistry : Basic Elements* (eds. G.H. Bolt and M.G.M. Bruggenwert). Elsevier Scientific, Amesterdam, the Netherlands.

Brady, N.C. and Weil, R.R. 1996. *The Nature and Properties of Soils* (11th edn.). Prentice-Hall, New York, USA.

Cosgrove, D.J. 1980. *Inositol Phosphates : Their Chemistry, Biochemistry and Physiology.* Elsevier, London, UK.

Foth H.D. and Ellis, B.G. 1997. *Soil Fertility* (2nd edn.). Lewis Publishers, CRC Press, Boca Ratoon, Florida, USA.

Gaur, A.C. 1990. *Phosphate Solubilizing Microorganisms as Biofertilizers.* Omega Scientific Publishers, New Delhi, India.

Illumer, P. and Schinner, F. 1995. Solubilization of inorganic calcium and phosphorus : Solubilization mechanisms. *Soil Biol. Biochem.*, 27 : 257-261.

Kirchner, M.J., Wollum, A.G. and King, L.D. 1993. Soil microbial populations and activities in reduced chemical input agroecosystems. *Soil Sci. Soc. Am. J.*, 20 : 264-270.

Miller, R.W. and Donahue, R.L. 1995. *Soils in our Environment* (7th edn.). Prentice Hall Inc., New York, USA.

Paul, E.A. and Clark, F.E. 1989. *Soil Microbiology and Biochemistry.* Academic Press, New York, USA.

Scheffer, F. and Schachtschabel, P. 1992. *Lehrbuch der Bondenkunde : Ferdinond Enke.* Verlag, Stuttgart, Germany.

Speir, T.W. and Ross, D.J. 1978. Soil phosphatase and sulphatase. pp. 197-250. In : *Soil Enzymes* (ed. R.G. Burns). Academic Press, New York, USA.

Stevenson, F.J. 1986. *Cycles of Soil Carbon, Nitrogen, Phosphorus, Sulphur, Micronutrients.* Wiley & sons, New York, USA.

Stewart, T.J. 1984. Interrelation of carbon, nitrogen, sulphur and phosphorus cycles during decomposition processes in soil. pp. 442-446. In : *Current Perspectives in Microbial Ecology* (eds. M.J. Klug and C.A. Reddy). Am. Soc. Microbiol., Washington, DC, USA.

Subba Rao, N.S. 1982. *Advances in Agricultural Microbiology.* Oxford & IBH Publishing Co., New Delhi, India.

Tan, K.H. 1993. *Principles of Soil Chemistry* (2nd edn.). Marcel Dekker, New York, USA.

Tisdale, S.L., Nelsa, W.L., Beaton, J.D. and Hartin, J.L. 1993. *Soil Fertility and Fertilizers.* MacMillan Publ. Co., New York, USA.

Wild, A. 1988. *Russell's Soil Conditions and Plant Growth.* Longman Scientific & Technical, Essex, UK.

— ★★★ —

Chapter 8

Sulphur Transformation in Soil

Death and life alternate, with each dependent upon the other

(S. E. Wedberg)

Sulphur in nature follows similar cycle as that of nitrogen and shares many common features with it. In soils, decomposition of organic matter releases major portion of nitrogen and sulphur bound in proteins, as cystine or methionine amino acids, or vitamins, etc. In proteins, on an average, for every 36 atoms of nitrogen one atom of sulphur is present. Mineralization supplies a large fraction of nitrogen and sulphur to the plants. Secondly, both nitrogen and sulphur have several chemical forms under different soil conditions (Table 8.1) and both are subject to microbial reduction. Thirdly, both can enter or leave soil in gaseous form and both are subject to some degree of leaching in the anionic form.

Table 8.1 : Forms of sulphur and nitrogen

Forms	Sulphur	Nitrogen
Oxidized negatively charged ion	$SO_4^{=}$	NO_3^-
Intermediate oxidation states	S^o, SO_2, SO_3^{2-}	N_2, NO_2^- & others
Reduced ionic forms	S^{2-}	NH_4^+
Amino acid form	-C-SH	$-C-NH_2$
Leachable form	$SO_4^{=}$	NO_3^-
Gaseous forms	CH_3-S-CH_3, H_2S, SO_2, SO_3	NH_3, N_2O, NO, N_2, NO_2

8.1 SULPHUR CYCLE

Sulphur is present in traces in air and forms about 0.1% of earth's crust. The mineral pyrite (FeS_2) is common to most soils. Moderately soluble gypsum ($CaSO_4.2H_2O$) is found in many arid soils. In industrial areas atmosphere contains high concentration of sulphur (sulphur oxides) owing to coal and fuel burning. The sulphur emitted may react with rain water and form sulphur oxide and sulphuric acids. Depositions of sulphur through green leaves, etc. may go up to 100 kg ha^{-1} yr^{-1}. Sulphur is present in volcanoes, rock sediments and igneous and metamorphic rocks as well as in some springs, etc. The sulphur cycle in nature is given in Fig 8.1. Sulphur is subject to atmospheric deposition (<5 kg S ha^{-1} yr^{-1}), leaching to ground waters, microbial reduction and oxidation and gaseous losses. Plants utilize sulphur as sulphate and, in general, require 10-40 kg S ha^{-1}; maximum in cruciferous plants. Sulphur may also enter plants through leaf as SO_2. Plants contain about 1% sulphur, and in their proteins about 2% sulphur is present. Sulphur enters the soil as organic compounds in plant and animal residues, elementary sulphur, H_2S through rain water or as S, SO_4 in fertilizers. These complex sulphur compounds are attacked by soil organisms and decomposed to liberate H_2S which is oxidized to SO_4 under aerobic conditions and to mercaptans under anaerobic conditions. Sulphur may enter into oceans through waters.

The most important mechanism of input for atmospherically derived soil sulphur is direct absorption of gaseous SO_2. Moisture status of soil is the most important soil property that governs the rate of direct absorption. The process increases with increasing moisture status.

In farming system pesticides and fertilizers have traditionally been the largest sulphur contributor to the soil-plant system. Acid rains also form important source for sulphur input in industrial countries. Burning fossil fuels releases considerable amount of SO_2 to air which acidify the soils and cause damages to buildings and other materials. In contrast to CO_2, SO_2 is highly water soluble (23 g 100 g^{-1} at 0°C; 11 g 100 g^{-1} at 20°C) and has a life time of only a few hours in the atmosphere. The SO_2 concentration of >0.0005% v/v in air is hazardous and toxic.

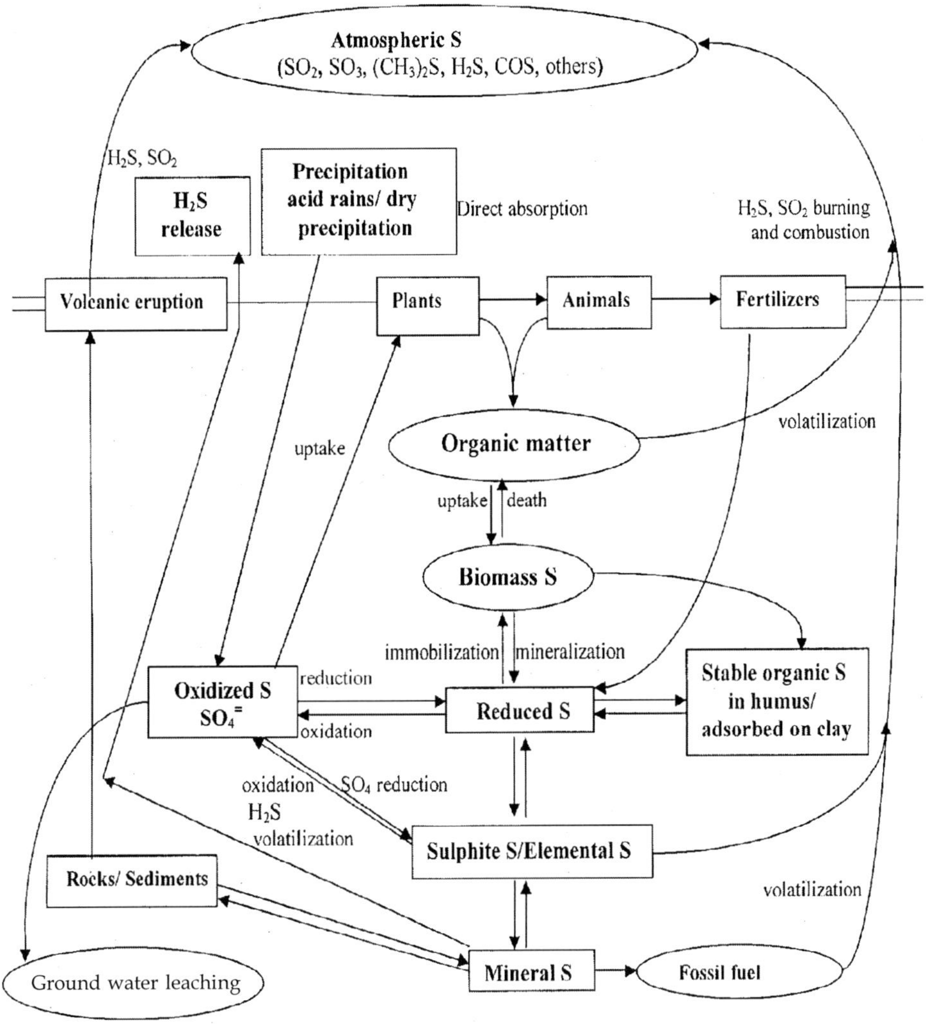

Figure 8.1 : Schematic diagram of sulphur cycle in nature

8.2 FORMS OF SULPHUR

Most agricultural soils contain sulphur in the range of 0.02-2.0 mg g^{-1} soil whereas volcanic ash and organic soils contain >3.0 mg S g^{-1} soil. Sulphur occurs in organic and inorganic forms. Over 90% of sulphur in most non-calcareous non-tropical surface soils is in organic forms, about half of which is in the form of sulphate esters and esters with C-O-S- linkage, about 20% in the form directly bonded to carbon such as amino acids and remainder in a variety of largely inert organic forms. The ratio between inorganic and organic sulphur in temperate soil varies between 0.17 and 0.44. In tropical weathered soil this ratio may be as high as 0.98. In soils, other than aridisols, the inorganic fraction is typically small compared to organic soils.

8.2.1 INORGANIC FORMS

Inorganic sulphur exists in nature in a number of oxidation states ranging from +6 in SO_4^{2-} and its derivatives to -2 in H_2S and its derivatives. Microbes are primary mediators for cycling the sulphur atoms among the various oxidation states (Table 8.2), although spontaneous chemical processes may contribute to these transformations to a limited degree. Weathering of sulphur containing minerals during soil formation occurs slowly. Primary minerals contribute non-significant amount of sulphur to plants. Inorganic sulphur exists primarily as SO_4^{2-} in calcareous soils of semi-arid and arid regions as well as in volcanic ashes and recently reclaimed soils. It may be present as gypsum ($CaSO_4.2H_2O$), as a basic aluminium sulphate and as a contaminant of $CaCO_3$ in concentrations ranging as high as 3000 ppm. Sulphur accumulates mainly as sulphides and polysulphides of iron in tidal marsh and saline lake conditions. SO_4^{2-} anion is not readily adsorbed by soil colloids and is held to some extent by clays and hydrous oxides of soil.

Table 8.2 : Oxidation states of major sulphur compounds in soil

Compounds	Oxidation state
Organic (R-SH)	-2
Sulphide (S^{2-})	-2
Elemental sulphur (S°)	0
Thiosulphate ($S_2O_3^{2-}$)	+2
Tetrathionate ($S_4O_6^{2-}$)	+2
Sulphur dioxide (SO_2^{2-})	+4
Sulphite (SO_3^{2-})	+4
Sulphate (SO_4^{2-})	+6

8.2.2 ORGANIC FORMS

Sulphur is essential nutrient for all living systems. In cell system it occurs is a variety forms. It is present in amino acids such as cyteine, cystine, glutathione and methionine, in cofactors as biotin, thiamine, coenzyme A, vitamin B_{12}, etc. and in iron-sulphur proteins like ferridoxin and lipoic acids (sulpholipids) so it is a structural constituent of these compounds. R-C-S-C-R acts as a functional group whereas R-SH group is directly involved in metabolic reactions. About 2% of organic-reduced sulphur in plants is present in water soluble thiol (-SH) fraction and tripeptide glutathione accounts >90% of this fraction. Other than these forms, sulphur in soil is attached to organic matrix via -O- or sometimes via -N- bonds i.e. as ester sulphate form (choline sulphate, phenolic sulphate, glucosinolates, glucose sulphate, sulfamates, adenyl sulphate, phosphoadenosine phosphosulphate, etc.). These are considered to be biologically more active or labile in soil. The sulphate esters often occur as storage products in plants and microorganisms. Glucosinolates constitute a significant portion of organic SO_4^{2-} in cruciferous plants.

Microbial biomass acts as a functional group whereas R-SH group is directly involved in metabolic reactions. Sulphur accounts about 2-3% of the total organic sulphur present in soil. Major fraction is in amino acid form and rest mostly as ester sulphates. Elemental sulphur has been identified in sporocarp of ectomycorrhizal fungi as well as in other self-inhibited and dormant structures. Elemental sulphur is deposited by some bacteria which utilize H_2S as a reducing agent during photosynthesis.

8.2.3 HUMUS

A considerable amount of sulphur is tied up in soil organic matter which is subsequently released on decay. The C :S ratio in soil organic matter is not as consistent as C :N ratio. Mollisol soils have C :N :S ratio of 90 :8 :1 while spodosol and luvisol soils have C :N :S ratio of 200 :12 :1. Worldwide this ratio is 100 :8 :1. The carbon-bound sulphur is difficult to analyze. Hydroiodic-reducible sulphur (most labile and biologically active organic form) accounts 25-75% of organic sulphur and generally increases with profile depth. It is mostly concentrated in clay-sized fraction and appears in fulvic acid when soil is extracted with NaOH or $Na_2P_4O_7$.

8.3 MINERALIZATION OF SULPHUR COMPOUNDS

In most soils inorganic sulphur pool is very small and biological uptake is dependent on the rate of mineralization. The rate of sulphur mineralization in soils, in turn, is influenced by environmental factors in the similar way like those controlling nitrogen mineralization. The mineralization of C-O-S, C-N-S and R-C-S compounds, originating from plants or microorganisms, involves both aerobic and anaerobic pathways. The hydrolysis of ester sulphates occurs by splitting of –O–S– bonds by enzyme sulphatase.

$$\underset{\text{Organic sulphur}}{R\text{-}O\text{-}SO_3^- + H_2O} \xrightarrow{\text{Sulphatase}} \underset{\text{decay products}}{ROH + H^+ + SO_4^{2-}}$$

The three sulphatases, well known for catalyzing enzymatic hydrolysis, are two aryl sulphatases and one choline sulphatase. The SO_4^{2-} hydrolyzed enzymatically is not available unless released extracellularly. Sulphur mineralization has been found less highly correlated with the degradation of carbon than that of nitrogen mineralization. The amino acids containing sulphur are degraded through anaerobic pathways via cysteine desulphahydrase or serine sulphydrase. Thus both inorganic sulphur and nitrogen are released as inorganic sulphur alone.

$$\underset{\text{cysteine}}{HSCH_2\text{-}\underset{NH_2}{CH}\text{-}COOH} + H_2O \xrightarrow{\text{Cysteine desulphahydrase}} \underset{\text{Hydrogen sulphide}}{H_2S} + NH_3 + \underset{\text{Pyruvic acid}}{CH_3COCOOH}$$

$$\underset{\text{Serine}}{HSCH_2\text{-}\underset{NH_2}{CH}\text{-}COOH} + H_2O \xrightarrow{\text{Serine sulphahydrase}} HOCH_2\text{-}\underset{NH_2}{CH}\text{-}COOH + H_2S$$

Serine sulphahydrase is reversible so can participate in sulphur assimilation as well. Methionine degradation may result in mercaptan (CH_3SH) and NH_3 formation. Mercaptions are odoriferous volatile sulphur compounds in nature. Mineralization releases approximately 3 kg S 100 kg^{-1} organic matter in soil in a year.

8.4 ASSIMILATION OF SULPHUR COMPOUNDS

Plants and microorganisms utilize SO_4^{2-} to form amino acids and other reducible substances. The process of assimilatory SO_4^{2-} reduction may occur in microbes and plants but not in animals.

Sulphur assimilation pathways : Once the SO_4^{2-} ion is taken across the membrane, the first step of its assimilation is the activation of SO_4^{2-} ion by ATP. Activation is a two stage process leading to the production of energy-rich sulphate nucleotide (esters) APS (adenosine 5′-sulphato-phosphate) and PAPS (adenosine 3′-phosphate, 5′-sulphato-phosphate).

i) $ATP + SO_4^{2-} \xrightarrow{\text{ATP sulphurylase}} APS + PPi$

ii) $APS + ATP \xrightarrow{\text{APS-kinase}} PAPS + ADP$

So there is involvement of two ATP per PAPS mole formed and approximately 6 kcal energy $mole^{-1}$. The PAPS can be then utilized to build either organic sulphate esters or may be reduced and added to serine to produce cysteine.

iii) $\text{PAPS} \xrightarrow{\text{PAPS reductase, ATP}} \text{Active sulphite } (SO_3^{2-}) \xrightarrow{H_2O} S_2^{2-}$

iv) $S_2^{2-} + \text{Serine} \xrightarrow[H_2O]{\text{Serine sulphydrase}} \text{cysteine}$

For the synthesis of sulphate esters, such as sulpholipids, the enzyme APS kinase catalyzes the formation of phospho-adenine phospho-sulphate (PAPS) in an ATP dependent reaction. From PAPS the activated sulphate can be transferred to an hydroxyl group forming a sulphate ester. In another mechanism, inorganic pyrophosphatase cleaves the pyrophosphates as under:

$$2\ \text{ATP} + SO_2^{2-} \xrightarrow{\text{ATP pyrophosphatase}} \text{ADP} + \text{PAPS} + \text{PPi}$$

PAPS can be incorporated through above reactions (iii) and (iv).

In SO_4 reduction pathway, activated sulphate of APS and PAPS are transferred by APS or PAPS sulpho-transferases to a thiol (R-SH) group of a suitable carrier. It is not clear whether glutathione acts *in vivo* as a carrier. This transfer of SO_4^{2-} to thiol group, mediated by sulpho-transferases, is associated with the reduction of sulphate (SO_4) to sulphite (SO_3^{2-}). The subsequent reduction of a carrier-bound sulphite to sulphide (S^{2-}) may involve sulphite reductase or organic thiosulphate reductase. The newly formed –SH group is then transferred to acetylserine, which is split into acetate and amino acid cysteine. Cysteine, the first product of assimilatory SO_4^{2-} reduction, acts as a precursor for the synthesis of all other organic compounds containing reduced sulphur as well as for other biosynthetic pathways. The mechanism for the synthesis of methionine in plants involves trans-sulphuration and direct suphuration leading to homocystein formation.

C : S ratio : About 1-3% of the organic sulphur becomes available each year through mineralization of organic matter. The balance between sulphur mineralization and immobilization in soil largely depends on the concentration of readily available/utilizable sulphur in organic residues relative to the available concentrations of carbon and nitrogen. The C : S ratio in organic matter determines the shifts in the process of mineralization-immobilization. At C : S ratio of <200 mineralization process will dominate so sulphates are released into soil solution. However, at C : S ratio of 200-400 mineralization and immobilization are balanced so there is no gain or loss. At C : S ratio >400, immobilization process dominates and inorganic sulphates tend to disappear from the soil solution. The immobilization of sulphur is encouraged by wide N : S ratio. During microbial breakdown of organic materials several sulphur containing gases are formed which include hydrogen sulphide (H_2S), carbon disulphide (CS_2), carbonyl sulphide (COS) and methly mercaptan (CH_3SH). All are more prominent in anaerobic soil or escape to atmosphere.

8.5 OXIDATION OF INORGANIC SULPHUR COMPOUNDS IN SOIL

Reduced sulphur is oxidized under aerobic conditions through a variety of intermediates by both chemical and biological pathways. The oxidation states of sulphur in soil are +6 in SO_4^{2+}, +4 in SO_3^{2-}, +2 in S^{2-}, 0 in S^o and –2 in H_2S and its derivatives. However, because the intermediates can be chemically oxidized, only H_2S, S^o (elemental sulphur) and SO_4^{2-} accumulate during microbial growth. Oxidation of sulphur causes acidity in soil. Sulphur oxidation occurs over a wide range of soil pH, from less than 2 to higher than 9, thus depicting greater flexibility than nitrogen oxidation process which is highly sensitive to pH change. Sulphur oxidation is controlled by soil moisture, pH and temperature. Elemental sulphur oxidation occurs between 4 and 55°C. Microbial oxidations proceed rapidly and such oxidations are dominated by thiobacilli which is physiologically and morphologically diverse microbial group capable of carrying inorganic sulphur oxidation in soil, water and sediments (Table 8.3). They are G –ve, aerobic to anaerobic and long rods to short filamentous in form. Thiobacilli are chemoautotrophic or facultative chemoautotrophic non-sporulating bacteria, widely distributed in soil ecosystem. They oxidize H_2S, elemental S, thiosulphate and polythionates to derive energy. Some thiobacillus species derive energy from oxidation of fixed C and ferrous iron.

8.5.1 ORGANISMS & PROCESSES INVOLVED IN SULPHUR OXIDATION

A wide range of heterotrophic bacteria, actinomycetes and fungi catalyze sulphur oxidation (Table 8.3). The process appears similar to the heterotrophic nitrification in utilization of both organic and inorganic compounds, and either process has shown to be energy-linked. The microorganisms utilize carbonaceous substrates as their carbon and energy sources and generally accumulate oxidized sulphur after certain period of growth. Heterotrophic sulphur oxidizers dominate a specific soil environment and the microbes responsible for sulphur oxidation vary from one soil to another depending up on the factors controlling sulphur oxidation. These may be primary oxidizers of sulphur in neutral and alkaline soil until pH is lowered sufficiently to permit oxidation by chemotrophic thiobacilli. The amount of available sulphur, soil moisture and temperature exhibit great influence on sulphur oxidation. Heterotrophic sulphur oxidation by some *Thiobacillus* is also becoming significant in microbial studies.

Photolithophilic purple and green sulphur bacteria are commonly found in mud and stagnant waters containing H_2S. Oxygen is not produced as an end product as observed in green plants, but sulphide is required and oxidized to sulphur by green bacteria. The photosynthetic purple bacteria oxidize sulphide directly to sulphate. They use sunlight as energy source and H_2S as electron source as under :

$$CO_2 + 2H_2S \xrightarrow{\text{Green bacteria, light energy}} \underset{\text{Carbohydrates}}{(CH_2O)_n} + 2S \text{ (elemental S)} + H_2O$$

$$H_2S + 2CO_2 + 2H_2O \xrightarrow{\text{Purple bacteria, light energy}} 2(CH_2O)_n + H_2SO_4$$

$$Na_2S_2O_3 + 2CO_2 + 3H_2O \xrightarrow{\text{Purple bacteria, light energy}} 2(CH_2O) + Na_2SO_4 + H_2SO_4$$

Table 8.3 : Various organisms involved in sulphur oxidation

Organisms	Characteristics
Thiobacillus thiooxidans	Motile, obligate aerobe, strict autotroph, grows at pH 2.0 to 5.0; acid tolerant, uses S as electron acceptor.
Thiobacillus ferroxidans	Strict aerobe, acid tolerant (pH 1.5-5.0); oxidizes sulphur and ferrous iron (Fe^{2+}) for energy.
Thiobacillus thioparus	Obligate chaemoautotroph grows in the pH range of 5.0 to 8.0; acid sensitive.
T. denitrificans	Autotrophic facultative anaerobe, acid sensitive, pH range 5.0 to 8.0; utilizes O_2 or NO_3 as electron acceptor; induces N_2 loss to atmosphere.
T. novalis	Facultative autotroph, oxidizes organic compound as well as sulphur salts but not elemental S, grows in pH range of 4.5 to 8.5.
T. intermedius	Facultative lithotroph, grows at pH range of 3.0 to 7.0; capable to use $S_2O_3^{2-}$ as electron donor; growth stimulated in presence of organic matter.

Contd...

Table 8.3. Contd...

T. thiocyanoxidans	Similar to *T. thioparus*; uses S or $S_2O_3^{2-}$ for growth; also found on thiocyanate in sewage effluents.
T. concretivorus	Strict autotroph, oxidizes thiosulphate to free sulphur and sulphuric acid, causes corrosion of pipes, cement degradation, acid tolerant, grows in pH range of 1.5 to 6.0.
T. coprotiticus	Found on dung, oxidizes S for energy.
T. versutus	Chemolithotrophs; grow on elemental suplhur; S-oxidizers.
T. acidophilus	Chemolithotrophs; grow on elemental suplhur; S-oxidizers.
T. prosperus	Facultative organotrophs; grow on elemental suplhur; S-oxidizers.
Ferrobacillus ferrooxidans	Autotroph, acid tolerant (pH range 1.5 to 5.5), oxidizes iron also.
F. sulfooxidans	Autotroph, acid tolerant (pH range 1.5 to 5.5), oxidizes iron also.
Sulfolobus spp.	S-oxidizing thermophilic organisms, grow in pH range of 2.0 to 5.0; temperature range 60 to 80°C; significant in geothermal sulphur rich springs; oxidize H_2S, cells also adhere tightly to sulphur crystals; organisms are spherical with distinct lobes, chemo-organotroph.
Thiomicrospora spp.	Small spiral bacteria (< 0.22 mm), one species grows anaerobically using NO_3^- as electron acceptor.
Thermothrix spp.	Filamentous bacteria, grows in hot springs, neutral pH, utilize O_2 or NO_3^- as electron acceptor; chemo-organotrophic in nature
Biggiatoa spp.	Colourless gliding bacteria, grow in presence of H_2S containing S granules, organotrophic, S also metabolized.
Thiothrix spp.	Gliding bacteria, metabolize H_2S and S, organotrophic
Leucothric spp.	Filamentous bacteria, organotrophic
Thioplaca spp.	Filamentous, oxidizes S
Green and purple bacteria (Rhodospirillaceae, Chromatiaceae, Chlorobaceae; Chloroflexiceae)	Accomplish photosynthetic sulphur oxidation under anaerobic conditions
Arthrobacter sp. *Bacillus* sp. *Chlorobium limicola, C. thiosulfatophilum, Chromatium okenii, Micrococcus* sp. *Rhodomicrobium vannielii.*	Capable of oxidizing inorganic S compounds

Contd...

Mycobacterium sp. *Pseudomonas (P. aeroginosa, P. fluorescens), Achromobacter stutzeri* some actinomycetes, some fungi like *Aspergillus (A. oryzae, A. ruba, A. repens, A. niger), Mucor flavus; Trichoderma harzianum; Aureobasidium pullulans; Microsporium* sp., *Penicillium* sp. etc. are heterotrophic sulphur oxidizers capable of oxidizing elemental S.

8.5.2 REACTIONS INVOLVED IN SULPHUR OXIDATION

The various oxidation reactions initiated by thiobacilli are given in Table 8.4. *Thiobacillus* has been used to produce suphuric acid (H_2SO_4) through action on elemental sulphur added to soil. It has a possible role in the management of plant diseases like potato scab caused by *Streptomyces scabies* and potato rot caused by *Streptomyces ipomeae*. Thus sulphur addition in conjunction with *Thiobacillus* inoculants can effectively control rot pathogen. Also, the sulphur-inoculants help in making alkali soils fit for cultivation. The H_2SO_4 formation augments nutrient-mobilization by increasing the levels of mobile PO_4, K, Ca, Mn, Al and Mg is alkali soils. In fact, Mo deficiency can be corrected by sulphur addition in such soils. *Thiobacillus* has been used as organic fertilizer, trade named as "Biospur", containing a mixture of rock phosphate, elemental sulphur and inoculant *Thiobacillus thiooxidans*. The H_2SO_4 produced as a result of microbial action dissolves phosphates and enhance phosphorus nutrient availability to plants. Also, inoculant can be added to manure along with soil, elemental sulphur and rock phosphate during compost making (Lipman process) to effectively enhance available nutrients. Because of acidifying reactions of sulphur, it may be applied along with suitable microbial inoculants to extremely sodic soils of arid region to bring soil pH to the favourable levels of plant growth. Soils with significant amount of sulphides or flooded for appreciable period with brackish or seawater when drained expose sulphides to air so cause its oxidation to sulphuric acid.

Cat-clays : The acidification is a rapid process and leads to pH decrease to as much as 1 or 2. Soils of this type are known as cat-clays. These cover significant areas of land in South-East Asia and in tidal areas of other parts of the world. Unless these soils are kept submerged, these soon become too acidic to sustain high plant growth. Sulphuric acid acidification can also be of significant problem in mining areas where reduced sulphur species such as pyrites are oxidized in spoil heap. Acidification prevents vegetation on affected areas and makes it difficult to restore certain lands disturbed by mining operation.

Table 8.4 : Various chemical reactions induced by thiobacilli

	Reactants	Organisms	Products
1)	$2S + 3O_2 + 2H_2O$	To, Td, Tf, Tt, Tco →	$2H_2SO_4$
2)	$Na_2S_2O_3 + 2O_2 + H_2O$	To, Td, Tf, Tt, Tn →	$Na_2SO_4 + H_2SO_4$
3)	$2Na_2S_4O_6 + 7O_2 + 6H_2O$	To, Td, Tt, Tn →	$2Na_2SO_4 + 6H_2SO_4$
4)	$2KSCN + 4O_2 + 4H_2O$	Td, Tt →	$(NH_4)_2SO_4 + K_2SO_4 + 2CO_2$
5)	$5S + 6KNO_3 + 2H_2O$	Td →	$K_2SO_4 + 4KHSO_4 + 3N_2$
6)	$5Na_2S_2O_3 + 8KNO_3 + 2NaHCO_3$	Td →	$6Na_2SO_4 + 4K_2SO_4 + H_2O + 2CO_2 + 4N_2$
7)	$12FeSO_4 + 3O_2 + 6H_2O$	Tf →	$4Fe_2(SO_4)_3 + 4Fe(OH)_3$
8)	$5Na_2\,S_2O_3 + 4O_2 + H_2O$	Tt, Tcy, Ti →	$5Na_2SO_4 + H_2SO_4 + 4S$
9)	$2Na_2S_4O_6 + 2Na_2CO_3 + O_2$	Tt →	$4Na_2SO_4 + 4S + 2CO_2$
10)	$2S + 2NO_3^- + H_2O + CaCO_3$	Td →	$CaSO_4 + N_2 + CO_2 + H_2SO_4$
11)	$H_2S + 2O_2$	Tco →	H_2SO_4
12)	$2Fe_2(SO_4)_3 + 12H_2O$	T_f →	$4Fe(OH)_3 + 6H_2SO_4$
13)	$NH_4CNS + 2O_2 + 2H_2O$	To →	$(NH_4)_2SO_4 + CO_2$
14)	$4FeS_2 + 15O_2 + 2H_2O$	T_f →	$2Fe_2(SO_4)_3 + 2H_2SO_4$
15)	$2FeS_2 + 7O_2 + 8H_2O$	Bacteria →	$10H + 4SO_4 + 2Fe(OH)_3$

To = *Thiobacillus thiooxidans*; Td = *T. denitrificans;* Tf = *T. ferroxidans;* Tt = *T. thioparus;* Tco = *T. concretivorus;* Tn = *T. novalis;* Tcy = *T. thiocyanoxidans;* Ti = *T. intermedius*

8.6 REDUCTION OF INORGANIC SULHPUR COMPOUNDS IN SOIL

Sulphate is reduced during its assimilation by bacteria. The assimilatory pathway is closely linked to meet the requirement for the biosynthesis of cystein. A few bacterial genera are also able to oxidize other forms of sulphur as alternative electron acceptors for anaerobic oxidation of H_2 and organic compounds (dissimilatory SO_4 reduction). No significant regulation of the enzymes involved in dissimilatory SO_4 reduction has been observed. Respiratory sulphate reduction is carried out by anaerobic organotrophic organisms using $SO_4^{=}$ as terminal electron acceptor in their anaerobic respiration. These organisms use various organic compounds as hydrogen donors viz. organic acids (lactate, malate, pyruvate, formate, etc.), alcohols, cholines, etc. These organisms are responsible for sulphide formation in water-logged soils and sediments. The sulphide accumulation in soil minerals has been defined as 'sulphidization' whereas exposure of sulphide bearing minerals to oxidizing conditions leading to sulphate production is termed as 'sulphurization'.

8.6.1 ORGANISMS INVOLVED IN SULPHATE REDUCTION

The sulphate reducing bacteria are obligate anaerobes. The main genera involved in this type of S transformation are *Desulfotomaculatum* (3 species), *Desulfovibrio* (5 species), *Desulfobacterium, Desulfococcus, Desulfosarcina, Desulfuromonas* and *Desulfomonas*. Other heterotrophic organisms include some bacteria, actinomycetes, fungi and algae (*Clostridium nitrificans, Aspergillus niger, A. nidulans, Penicillium chyrsogenum, Chlorella pyrenecidosa*, etc.) but these are of little significance. However, it may be borne in mind that sulphate to sulphide reduction by carbohydrates does not liberate as much energy as does the oxidation of carbohydrates with nitrate to form ammonium. Consequently in anaerobic environments with appreciable nitrate concentration, sulphate reducing microorganisms are out-competed by those that reduce nitrate. Under these conditions NH_4^+, and not sulphide, is the principal reduced species. The mechanisms of reduction are not yet fully resolved. Some possible steps involved are shown in Fig. 8.2.

The electrons are transferred through ferrodoxin or flavodoxin to cytochromes; and SO_4^{2-} is activated by sulphate adenyltransferase in presence of ATP and is subsequently reduced by APS reductases to sulphite (SO_3^{2-}) and AMP. The SO_3^{2-} is then protonated by a chemical reaction to form bisulphate (HSO_3^{2-}) and subsequent reduction reactions are catalyzed by bisulphate reductase, trithionate ($S_3O_6^{2-}$) reductase and thiosulphate ($S_2O_3^{2-}$) reductase to produce sulphide (S_2^-). The overall process of anaerobic SO_4 reduction is as under :

$$SO_4^{2-} \longrightarrow SO_3^{2-} \longrightarrow S_3O_6^{2-} \longrightarrow S_2O_3^{2-} \longrightarrow S_2^-$$

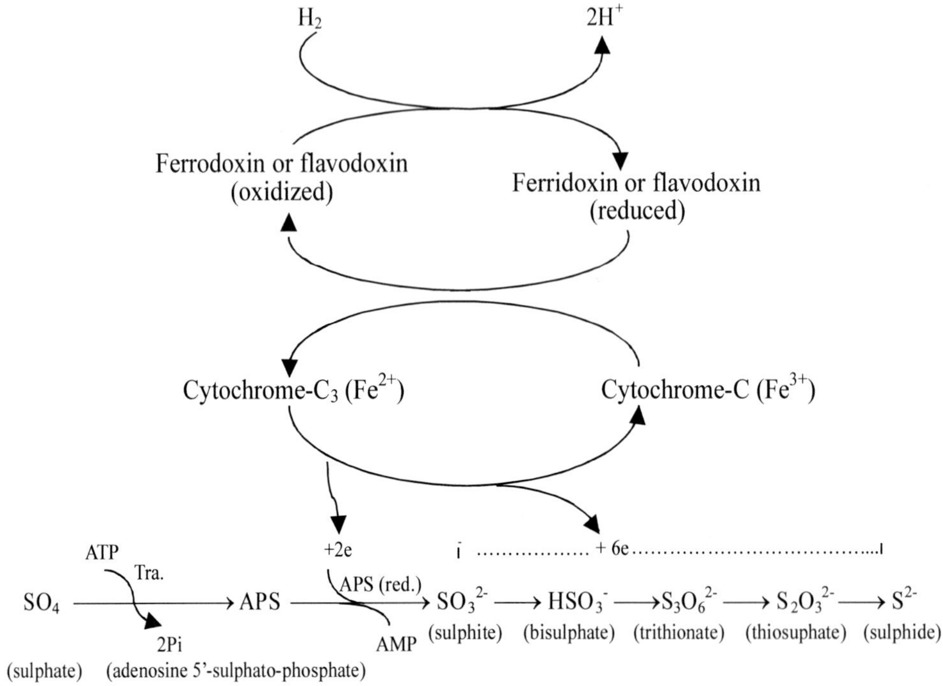

Fig 8.2 : Pathway of respiratory sulphate reduction in *Desulfovibrio* (Tra = ATP adenyl transferase; APS (red.) =APS reductase)

Most SO_4-reducers contain electron carrier desulfovividin which fluorescence red under alkaline conditions. Microbes which only reduce enough sulphates for self-assimilation utilize a different sulphate reduction pathway in which 3′-phosphoadenosine 5′-phosphosulphate is the first intermediate formed. Arylsulphates concerned with the hydrolysis of aromatic sulphate are involved in the formation of active forms of sulphate.

8.6.2 FACTORS FAVOURING SULPHATE REDUCTION

Sulphate reduction is generally restricted to water saturated soils because strong anaerobic conditions must prevail. A redox potential of approximately -220 mV prevails when SO_4^{2-} is dominant electron acceptor at neutral pH. Sulphate reduction is enhanced by organic matter additions because sulphur reducers are heterotrophs which need adequate supply of SO_4 in soil solution to activate sulphur reduction. Even sulphites (SO_3^{2-}), thiosulphates ($S_2O_3^{2-}$) and elemental sulphur (S^o) are readily reduced to sulphide form by microbial action. Sulphate reducing bacteria are found over a wide pH range. They can tolerate metals and dissolved sulphide up to the concentrations of 2%. Low molecular weight fatty acids (butyrate, acetate, propionate, etc.) are inhibitory to their growth. Either organic substrates or H_2 are used as electron donors.

$$2CH_3CHOHCOOH + SO_4^{2-} \longrightarrow 2CH_3COOH + HS^- + H_2CO_3 + HCO_3^-$$
Lactic acid

$$4CH_3COCOOH + SO_4^{2-} \longrightarrow 3CH_3COOH + CH_3COO^- + 4CO_2 + HS^-$$
Pyruvic acid

$$R2RCH_2OH + SO_4^{2-} \longrightarrow 2RCOOH + 2H_2O + S^{2-}$$
Organic alcohol — Organic acid

$$S_2O_3^{2-} + 4H_2 \longrightarrow 2HS^- + 3H_2O$$
$$S_4O_6^{2-} + 9H_2 \longrightarrow 2HS^- + 2H_2S + 6H_2O$$
$$CaSO_4^{2-} + 8H \longrightarrow H_2S + Ca(OH)_2 + 2H_2O$$

Although most of the SO_4^{2+} reducers require NH_3 as nitrogen source, some isolates reduce atmospheric nitrogen (N_2) to ammonia (NH_3). The sulphur reducers diminish the availability of sulphur to plants so affect crop production. In paddy fields sulphate reduction occurs in root zone. In poorly drained soils or paddy soils H_2S produced is, however, bound by Fe or Mn present in the soil. By tying up the soluble reduced Fe, the formation of FeS or MnS helps in the prevention of Fe/Mn toxicity in paddy marshy lands. Secondly in paddy soil inner rhizosphere is aerobic (being continually supplied with O_2 by aerenchyma of the plant), therefore any H_2S produced will be rapidly reoxidized by *Thiobacillus*. Sulphur reducers can be used to raise the soil pH.

8.7 ENVIRONMENTAL CONSEQUENCES OF SO_4 REDUCTION

The environmental consequences of SO_4 reduction are as under :

- Formation of H_2S is a serious problem in sewage treatment system as it causes corrosion of stones, concretes, pumps, pipes, etc. and emits foul odour (rotten egg smell). The process of cathodic depolarization by sulphur reducing bacteria accelerates spontaneous oxidation of iron containing compounds which leads to iron loss. However, if NO_3 is added to septic tanks H_2S production is eliminated as NO_3 acts as a more readily available electron acceptor.
- SO_4 reduction results in the disappearance of O_2 in environment.
- Under anaerobic conditions Fe^{3+} and SO_4^{2-} are reduced to Fe^{2+} and H_2S, respectively. These ions form a black precipitate of ferrous sulphide which can further react to form a number of sulphide minerals. The sulphides are corrosive, toxic and smelty.

8.8 VOLATILIZATION/ LEACHING OF SULPHUR FROM SOIL

A number of volatile sulphur compounds are released from the soils which are either rich in partially decomposing organic matter or waterlogged. The gaseous products are various oxides of sulphur (SO_x). Sulphur is released in the form of mercaptans, methyl and ethyl sulphides, thiols, carbon disulphide, carbonyl sulphide, dimethyl disulphide, diethyl disulphide, methyl thioacetate, etc. These contribute to global sulphur cycle. Various microbes such as lichens are extremely sensitive to SO_2 so they serve as good indicators of SO_2 pollution.

Sulphate ions are readily leached from soil especially in humid region, if not adsorbed by soil colloids, Fe/Al oxides and clays. Most soils retain some sulphate although the quantity held is generally small and its strength of retention is low as compared to that of phosphate.

Selected References

Castellano, S.D. and Dick, R.P. 1991. Cropping and sulphur fertilization influence on sulphur transformation in soil. *Soil Sci. Soc. Am. J.*, 54 : 114-121.

Cole, J.A. and Ferguson, S.J. 1988. *The Nitrogen and Sulphur Cycles.* Cambridge Univ. Press, London, UK.

Ehrlich, H.L. 1981. *Geomicrobiology.* Marcel Dekker, New York, USA.

Kelly, D.P. and Smith, N.A. 1990. Organic sulphur compounds in the environment : Biogeochemistry, microbiology and ecological aspects. *Adv. Microb. Ecol.*, 11 : 345-385.

Nriagu, J.O. (ed.) 1978. *Sulphur in the Environment. Part II Ecological Impacts.* John Wiley & sons, New York, USA.

Paul, E.A. and Clark, F E. 1989. *Soil Microbiology and Biochemistry.* Academic Press, New York, USA.

Roy, A.B. and Trudinger, P.A. 1970. *The Biochemistry of Inorganic Compounds of Sulphur.* Cambridge Univ. Press, Cambridge, UK.

Schiff, J.A. 1983. Reduction and other metabolic reactions of sulphates. pp. 401-421. In : *Encyclopedia of Plant Physiology, New Series Volume 15* (eds. A. Lauchli and R.L. Bieleski). Springer Verlag, Berlin, Germany.

Schmidt, A. and Jager, K. 1992. Open questions about sulphur metabolism in plants. *Ann. Rev. Pl. Physiol. Plant Mol. Biol.*, 43 : 325-349.

Schlesinger, W.H. 1991. *Biogeochemistry : An Analysis of Global Change.* Academic Press, San Diago, USA.

Smith, M. 1991. In : *Recycling Derelict Land* (ed. G. Fleming). Thomas Telford, London, UK.

Stevenson, F.J. 1986. *Cycles of Soil Carbon, Nitrogen, Phosphorus, Sulphur, Micronutrients.* John Wiley & sons, New York, USA.

Subba Rao, N.S. 1982. *Advances in Agricultural Microbiology.* Oxford & IBH Publishing Co., New Delhi, India.

Tabatabai, S.J. (ed.) 1986. *Sulphur in Agriculture, Agronomy Series No. 27.* Am. Soc. Agron., Crop Sci. Soc. Am., Soil Sci. Soc. Am., Madison, Wisconsin, USA.

Tisdale, S.L., Nelsa, W.L., Beaton, J.D. and Hartin, J.L. 1993. *Soil Fertility and Fertilizers.* MacMillan Publ. Co., New York, USA.

Vishniac, W. and Santer, M. 1957. The Thiobacilli. *Bact. Rev.*, 21 : 195-213.

Wainwright, M. 1984. Sulphur oxidation in soil. *Adv. Agron.*, 37 : 349-396.

Whitehead, D.C. 1964. Soil and plant nutrition aspects of the sulphur cycle. *Soil Fertil.*, 27 : 1-8.

— ★★★ —

Chapter 9

Microbial Transformation of Metals

Everything is everywhere, the environment selects.
(N. Beijerinck)

Soil microorganisms immobilize and mineralize various elements to satisfy their growth requirements. The subsequent turnover of microbial biomass also releases several mineral elements. The elements may be released as a result of organic matter decomposition (e.g. during the process of mineralization, ammonification, etc.) or through biological weathering. Other transformations may involve oxidation and reduction of various inorganic ions and compounds which are mostly catalyzed by chemoautotrophic bacteria during oxidation of ammonium, nitrite, sulphur, iron, etc. Such oxidations may also release energy for microbial growth. Some heterotrophs oxidize inorganic ions but do not derive any energy from the process. Reduction of oxidized state of elements, serving as electron acceptors, provides energy under reduced state or from more reduced substrate. Conversion of metals to organic forms or vice versa may also occur as a result of soil microbial activities though directly not involving any transformation of that specific metal. Also, transformation may be catalyzed by the end products formed by the microbes which induce changes in solubility, alkalinity, oxygen content, etc. and release elements in soil. Many metals undergo microbially-mediated transformations. There are evidences for direct and indirect biological alterations in the availability and solubility of K, Mg, Fe, Cu, Zn, Ca, Ni, Mn, Co, Cl, Se, As, V, Si, U, Pb, Sn, Cd etc. These transformations are important for the release of macro- and micro-nutrients for plants, soil formation, reclamation of metals of low grade ore, acidification of water supplies, production of metallic ores during geological times etc. Microbial activity contributes to the chemical, physical and mineralogical changes which are involved in weathering of a range of processes such as acid hydrolysis, complex formation, oxidation and reduction, and a chain of reactions that causes solubilization of minerals.

9.1 TRANSFORMATION OF IRON

Iron is a structural component of cytochromes, hemes and numerous electron transfer systems including nitrogenase enzyme.

Iron solubilization and precipitation : Soil iron has very low solubility. Even added soluble iron chemicals readily precipitate as low-solubility minerals. In most soils the maintenance of sufficient iron in soluble form for plant absorption is a major problem. Iron exists in oxidation states of Fe°, Fe^{2+} and Fe^{3+}. In strongly acid solutions below pH 5.0, metallic iron (Fe°) is readily oxidized to stable ferrous state (Fe^{2+}), becomes increasingly soluble and is seldom deficient. However, at pH >5.0 Fe^{2+} gets chemically oxidized to Fe^{3+}. In turn, Fe^{3+} is readily reduced under acidic conditions but precipitated in alkaline solution. The oxidation of Fe^{2+} is spontaneous at pH >5.0, so it is difficult to prove that reaction is enzyme mediated at high pH values.

$$\underset{\substack{\text{Simple cation}\\ \text{(Soluble)}}}{Fe^{3+}} \xrightarrow{OH^-} \underset{\substack{\text{Hydroxy metal cation}\\ \text{(Soluble)}}}{Fe(OH)^{2+}} \xrightarrow{OH^-} \underset{\substack{\text{Hydroxy metal cation}\\ \text{(Soluble)}}}{Fe(OH)_2^{+}} \xrightarrow{O^-} \underset{\substack{\text{Hydroxide}\\ \text{(Insoluble)}}}{Fe(OH)_3}$$

The reduced ferrous iron (Fe^{2+}) is also formed under anaerobic conditions and is more soluble than ferric iron. The ferrous iron is easily and rapidly oxidized in aerated soil and its solubility becomes equal to that of hydrous iron oxides. Alternate aerobic-anaerobic conditions cause some solubilization and precipitation cycles (Fig 9.1). Iron solubilization occurs under anaerobic (water-logged) conditions whereas precipitation occurs when soil becomes aerated (drained). The solubility of Fe^{3+} iron decreases about 1000-fold with each unit increase in soil pH.

Organisms involved in iron transformation : Various bacteria oxidize ferrous iron to ferric state which precipitates as ferric hydroxide around cells.

The iron bacteria may be filamentous (*Leptothrix, Crenothrix, Colnothrix, Cladothrix, Sphaerothrix* and *Toxothrix*) or non-filamentous spherical or rod shaped bacteria (*Gallionella, Siderophacus, Siderocapsa, Ferribacterium, Naumanniela, Siderosphaera, Thiobacillus, Ochrobium, Sideromonas, Sideronema, Ferrobacillus, Sphaerotilus, Siderobacter* and *Siderococcus*). Besides, some algae belonging to cyanophyceae, chlorococcales, euglenineae, conjugales, volvocales and ultrichales also transform iron salts to ferric form and deposit precipitates around their filaments. The classification of iron bacteria on the basis of nutritional behaviour is given in Table 9.1.

Table 9.1 : Nutritional groups of iron bacteria

Group	Energy source	Examples
a) Obligate chemoautotrophs	Utilize energy in the process of ferric hydroxide formation	*Gallionella, Thiobacillus, Ferrobacillus*
b) Facultative chemoautotrophs	Utilize energy derived in the process of ferric hydroxide formation or from organic matter	*Leptothrix, Sideromonas, Crenothrix, Hypomicrobium, Metallogenium*
c) Heterotrophs	No energy from iron compounds, derive energy from organic matter only	*Bacillus, Pseudomonas, Serratia, Clostridium, Micromonospora, Desulfovibrio, Alternaria, Ochrobium, Naumaniella, Fusarium*

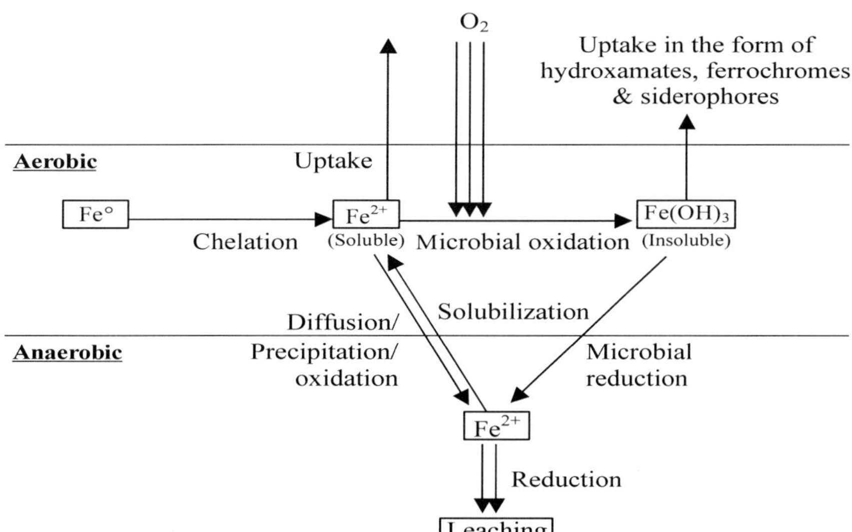

Fig 9.1 : Ferrolysis in soil

Microbial processes involved : Soluble organic substances like root exudates, humus decomposition products, microbial cell exudates, phyllospheric polyphenols, etc. are capable of reacting with iron and bind it into the soluble forms that are quite mobile in soil solution. Before Fe^{2+} is microbially oxidized, it is chelated by organic compounds (chelating ligands) so its bioavailability is increased. During oxidation electrons are moved through electron transport chain. Various organisms like *Thiobacillus, Leptothrix, Gallionella, Sulfolobus, Crenothrix, Hypomicrobium, Metallogenium, Pedomicrobium, Planctomyces, Seliberia* and *Sphaerotilus* either precipitate or oxidize iron. Under aerobic conditions, microbial iron oxidation dominates in acidic environment while chelation dominates in neutral environment.

Thiobacilllus ferrooxidans grows at pH 2.0-4.5 and can mediate oxidation at acidic pH as low as 0.8. It derives both energy and reducing power from iron oxidation reaction. *Leptothrix,* a filamentous sheathed bacterium, is present in well waters or iron pipes and precipitates iron around filamentous sheath of bacterium. *Crenothrix* grows in waters having low Fe^{2+} and dissolved O_2. It accumulates $Fe(OH)_3$ as mucilaginous sheath, about 0.2 μm thick, and causes clogging of pipes. Mineralization of this bacterium upon its death releases foul smell and causes red precipitation. *Gallionella,* a small rod shaped non-filamentous bacterium, forms mucilaginous encrustation of $Fe(OH)_3$ and contaminates well waters, pipes, river stream waters, etc. *Metallogonium, Pedomicrobium* and *Seliberia* deposit $Fe(OH)_3$ on the surface of filaments causing problems in drains. Some of the oxidation reactions involved are given as under :

$$2FeS_2 + 7O_2 + 2H_2O \xrightarrow{\textit{Thiobacillus ferrooxidans}} 2FeSO_4 + 2H_2SO_4$$

$$4FeSO_4 + O_2 + 2H_2SO_4 \xrightarrow{\textit{Thiobacillus ferrooxidans}} 2Fe_2(SO_4)_3 + 2H_2O$$

$$FeS + Fe_2(SO_4)_3 \xrightarrow{\textit{Thiobacillus ferrooxidans}} 3FeSO_4 + S$$

$$2FeCO_3 + O_2 + 3H_2O \xrightarrow{\textit{Leptothrix ochraceae}} 2Fe(OH)_3 + 2CO_2$$

$$CH_3COOH + 2Fe(OH)_3 \xrightarrow{\textit{Leptothrix ochraceae}} 2Fe(OH)_2 + 3H_2 + 2CO_2$$

$$CuFeS_2 + 4O_2 \xrightarrow{\textit{Thiobacillus ferrooxidans}} CuSO_4 + FeSO_4$$

$$4FeS_2 + 15O_2 + 2H_2O \xrightarrow{\textit{Thiobacilllus ferrooxidans}} 2Fe_2(SO_4)_3 + 2H_2SO_4$$

Oxidation-reduction processes in sediment result in the formation of minerals such as geothite ($Fe_2O_3.H_2O$), hematite (Fe_2O_3) and pyrite (FeS_2). In presence of sulphates and sulphate reducing bacteria, the reduced iron reacts with H_2S to form mineral hydrotrilite. This eventually forms pyrite (FeS_2).

Chelation of iron : Some microorganisms under iron stress conditions produce some chelating substances such as siderophores, iron chelate reductases etc. that solubilize or chelate soil iron. Production of such compounds makes these organisms more competitive to obtain iron and survive with an edge over other microbes. Ferric form (Fe^{3+}), common in most soils, is insoluble so its uptake is facilitated through exudations or reactions with organic compounds having strong affinity for Fe^{3+} (ferrichrome or siderophore) which often consist of hydroximates. Hydroximates have a hydroxyl group closely associated with a nitrogen portion of molecule. This results in a very strong affinity for multivalent iron through chelation. The siderophore (iron chelate reductase) producers have advantage in having easy access to Fe than non-producers in rhizosphere. Besides, these bacteria can help the plant in the control of root pathogens. The other iron specific chelators include heme, transferins, ferritin, etc.

Gleization : Iron can get reduced though a number of processes. For example Fe^{3+} can serve as respiratory electron acceptor or Fe^{3+} (red and insoluble) can be reduced with the formation of microbial end products like formate or H_2S. Many heterotrophic fungi (like *Alternaria, Fusarium,* etc.) and bacteria (like *Bacillus, Clostridium,* etc.) reduce Fe^{3+}. Iron oxidation-reduction plays a vital role in soil formation process known as 'gleization'. Gleying is a process involving changes in chemical speciation that result in changes in solubility. It is characterized by iron reduction and its subsequent partial solubilization and leaching. Gleization occurs in water-logged mineral soils that nonetheless contain some organic matter. Under anaerobic conditions any iron present in such soil may be reduced to Fe^{2+} (dull grey or green but insoluble). The light grey greenish colour, attributed to the reduced Fe, is produced. Gleying is predominantly a microbially mediated process. Fe^{3+} is reduced to Fe^{2+} as an alternative electron acceptor. A gleyed soil horizon may be identified by the presence of blue-green to blue-grey colours. Gleyed soils have usually mottled spot of Fe_2O_3 (hematite) where root channels or cracking cause localized oxidation. Gleying reactions result in sealing of soil under ponds.

Podzolization : It is the process in which insoluble ferric iron when solubilized in soil by acidification is subsequently complexed with organic matter. The process results in the formation of iron and organic rich sub-surface horizons in soil. Certain microorganisms produce organic acids and chelating acids, especially polycarboxylic acid, fulvic acid humic compounds and polyphenols, which form soluble chelates with aluminum cations and Fe^{2+}, respectively. The soluble Fe/Al-organic complex gets leached from 'A' horizon by percolating water during rainfall. As a result of the mobility of iron and organic matter, the bottom portion of 'A' horizon becomes characteristically distinct with pale colour, and is redesignated as an 'E horizon' (eluvial horizons that are less in organic matter). Iron complexes in 'B' horizon are broken down and the aluminum complexes precipitated. The deposition of Fe^{3+} hydroxides and organic matter causes darkening of this zone. This complex process is primarily because of microbial oxidation of organic chelating agents and the flocculating effect of relatively high concentration of polyvalent cations such as Ca^{2+}, Mg^{2+} and Al^{3+} found near weathering parent material. The end product of this phenomenon is podzol soil formation.

Corrosion of iron : Some microorganisms are also involved in the corrosion of iron pipes etc. under anaerobic conditions when pH is >5.5 and enough SO_4^{2-} are present.

$$4Fe + SO_4^{2-} + 4H_2O \longrightarrow FeS + 3Fe(OH)_2 + 2OH^-$$

Under these conditions a pipe of 3 mm wall thickness may be corroded within 5-7 years. To avoid any corrosion it is essential to carefully wrap the pipes with asphalt or plastic and maintain a small electric current. Various microbes involved in corrosion are *Desulfovibrio desulfuricans* (SO_4 reducing); *Thiobacillus thiooxidans* (S oxidizer); *Leptothrix* sp., *Crenothrix* sp., *Galionella* sp. (Fe-bacteria); *Thiobacillus denitrificans* (nitrate reducer) and *Hydrogenomonas* (H_2 reducer). Except *Thiobacillus,* all above organisms have optimum pH range of 6 to 9. In corrosion microbes contribute in many ways through :

- Sulphur oxidizers which form sulphuric acid,
- Heterotrophs which form organic acids, change Eh or electrode potential, create microgalvanic cells and depolarize the surfaces by oxidizing hydrogen, and
- Sulphate reducers which produce corrosive hydrogen sulphide.

9.2 TRANSFORMATION OF COPPER

Copper is essential cofactor in many enzymes like oxidases. It is also involved in many electron transfer reactions. Copper is bound by alanine and aspartic acid and may form complex with various enzymes (diphenol oxidase, galactose oxidase, tyrosinase, etc.) and proteins. Copper exists in soil mostly as cupric (Cu^{2+}) or cuprous (Cu^{+}) ions. Plants usually absorb it in cupric ion form. Copper exists as Cu^{2+} ionic species in high acidic soils, as $Cu(OH)_2^+$ in mild acidic soil and as $Cu(OH)_2$ in near neutral and alkaline soils. Most of the available Cu comes from exchangeable sites and soluble organic complexes or chelates. Copper is especially held by organic matter so its availability is usually very low in organic soils. This bound copper is readily released once organic matter is decomposed. Oxalic acid, for instance, is secreted by *Penicillium* which chelates and binds Cu and prevents its uptake.

Copper precipitation and leaching : Copper is precipitated by H_2S producing soil microorganisms. Apart from *Desulfovibrio desulfuricans* which reduces sulphates, bacteria such as *Clostridium lentoputrescens, Proteus vulgaris* and *Escherichia coli* produce H_2S from sulphur containing amino acids like cystein, methionine and glutathione under anaerobic conditions. Thus H_2S producing bacteria seem to be involved in the precipitation of copper rendering the element insoluble.

Copper sulphides, of both primary & secondary origin, are easily leached by microbial processes. The microbes involved are *Thiobacillus ferrooxidans, T. concretivorus, T. neopolitanus, Ferrobacillus ferrooxidans,* etc. which utilize either iron or sulphur or both occurring in sulphide minerals to produce appreciable quantity of ferric sulphate/sulphuric acid for dissolving copper.

$$Cu_2S + 2Fe_2(SO_4)_3 \longrightarrow 2CuSO_4 + 4FeSO_4 + S$$

$$3CuO + Fe_2(SO_4)_3 \longrightarrow 3CuSO_4 + Fe_2O_3$$

$$CuSO_4 + Fe \longrightarrow FeSO_4 + Cu$$

Copper toxicity to plant and microbes : Copper contamination of soil can cause phytotoxicity in crop plants. Vines are frequently sprayed with Cu-containing fungicides and this leads to the accumulation of copper in top soil. Further, copper sulphate is widely used as an algicide in ornamental ponds and even in water supply reserviours which are affected by bloom of toxic cyanobacteria and algae. Copper is highly toxic to soil microbial biomass thereby can affect the various aspects of soil fertility including mineralization process, and hence nutrient cycling and nitrogen fixation in legumes.

Copper deficiency : The worldwide mean range of soil copper is 0-80 ppm. The soils with less than 8 ppm of total copper may be deficient for proper crop growth. Deficiency of copper in higher plants, especially cereals, is dependent on the extent of its fixation in soil. The amount of copper fixed in soil depends on pH and such fixation is high in acidic peat soil. Black organic matter of soil is largely responsible for copper fixation in soil. In such soils copper is rendered unavailable by its complexation with humic materials.

9.3 TRANSFORMATION OF MANGANESE

Manganese occurs in acid soils in bivalent state (Mn^{2+}), the form most available to plants, whereas in neutral and alkaline soils it occurs in tri- and tetra-valent states (Mn^{3+} and Mn^{4+}), which are not easily available to plants. Therefore, oxidation state of manganese is the primary source affecting its absorption by plants. Inorganic forms of manganese in soil include oxides/hydroxides such as magnanite (MnOOH)), pyrolusite (MnO_2), carbonates ($MnCO_3$) and sulphides (MnS). Manganous (Mn^{2+}/MnO) oxidation to manganic ions (Mn^{3+} or Mn^{4+}/MnO_2) depends on pH, oxygen supply and organic matter content in soil. In soil solution, manganese occurs in Mn^{2+} ionic form. In oxidized soil, most of the manganese gets precipitated as insoluble MnO_2 (pyrolusite). Manganese solubility increases about 100-fold per unit decrease in pH. Organic matter decomposition aids in manganese solubility by furnishing electrons to reduce manganese as the decomposition proceeds.

Oxidation-reduction processes : Microbes are active in catalyzing both oxidation and reduction reactions in which manganese may be precipitated or solubilized (Fig. 9.2). Microbes precipitating Mn are wide-spread throughout the biosphere. Manganese cycling is dependent on carbon input which influences redox potential and microbial activity. Bacterial oxidation of Mn^{2+} occurs both in soils and sediments. Under acidic and neutral conditions Mn^{2+} oxidation is microbially mediated.

$$2MnO_2 + 2H_2 + 2S + 3O_2 \xrightarrow{\textit{Thiobacillus thiooxidans}} 2MnSO_4 + 2H_2O$$

$$2MnO_2 + 4S + 5O_2 + 2H_2O \xrightarrow{\textit{Thiobacillus thiooxidans}} 2MnSO_4 + 2H_2SO_4$$

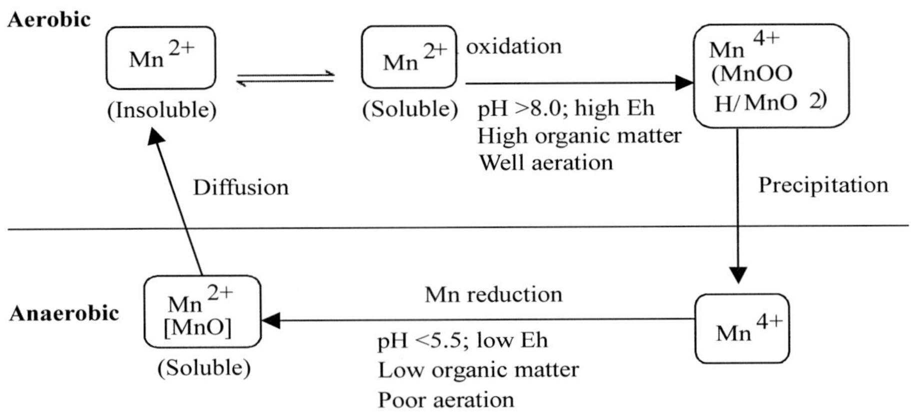

Fig 9.2 : Mn transformation in soil

Biochemical reduction of MnO_2 may also be catalyzed by certain bacteria e.g. *T. thiooxidans* but coupled with other elemental (S) oxidation. Mixed cultures of bacteria and fungi are more effective in the conversion of manganous to manganic ions than pure cultures. It has been observed that precipitation of Mn in the filaments of mycorrhizal fungi, before it enters the plant, allows the growth of manganese-sensitive plants under manganese toxic conditions. Easiest method to control Mn^{2+} oxidation is by genetic selection of host plants which, in turn, controls the rhizosphere populations responsible for Mn^{2+} oxidation. The number of manganese oxidizers in soil varies considerably but often accounts for 5-15% of the total viable microflora.

Organisms involved : Various types of microorganisms involved in Mn^{2+} oxidation include bacteria such as *Azotobacter chroococum, Pseudomonas fluorescens, P. trifolii, Erwinia* sp., *Bacterium herbicola, Chromobacterium* sp., *Arthrobacter* sp., *Bacillus* sp., *Aerobacter* sp., *Proteus* sp., *Corynebactrium* sp., *Leptothrix* sp., *Citrobacter* sp., *Klebsiella* sp., *Pedomicrobium* sp., *Hypomicrobium* sp., *Metallogenium* sp., *Gallionella* sp., *Sphaerotilus* sp., *Flavobacterium* sp. and *Streptomyces* sp., fungi such as *Fusarium* sp., *Trigschemia* sp., *Helminthosporium* sp., *Cladosporium* sp., *Pleospora* sp., *Curvularia* sp., *Periconia* sp., *Mycoplana dimorpha, Cephalosporium* sp., etc. and yeast (*Crytococcus albidus*). About 5-10% microbial population in soil is manganese oxidizer. *Hypomicrobium manganoxidans* derives energy from enzymatic oxidation of soluble Mn^{2+}. Under certain conditions rhizosphere microorganisms mobilize manganese i.e. oxidize Mn^{2+} to Mn^{4+} form (Mn_3O_4, Mn_2O_3 and MnO_2) and deposit MnO_2 outside the roots as deposits of black precipitate. The manganese inside soil may also be reduced through acid production or microbial mobilization in rhizosphere by root excretion of protons via decomposition of exudates and mucilages. In calcareous soils, mobilization of sparingly soluble manganese oxides in rhizosphere can be quite high as a result of root and microbial activity. Decrease in pH, reduction in oxidation potential and removal of oxygen as a result of microbial metabolism may increase the level of exchangeable manganese in soil. MnO_2 is reduced to Mn^{2+} under anaerobic conditions. In soils having pH <5.5, Mn^{2+} predominates because of chemical equilibrium whereas in soils having pH >5.5, biological forces play vital role in oxidation-reduction process.

$$\underset{(MnO)}{Mn^{2+}} \xrightleftharpoons[\text{biochemical reduction}]{\text{Biochemical or chemical oxidation}} \underset{(MnO_2)}{Mn^{4+}}$$

Chemical oxidation of Mn^{2+} occurs only at pH >8.0. But below pH 8.0 a number of soil bacteria and fungi oxidize Mn^{2+} ions. For instance, *Arthrobacter* does it through constitutive enzymes whereas *Leptothrix* oxidises Mn^{2+} to Mn^{4+} and accumulates MnO_2 in its sheath and filament during iron oxidation.

Manganese toxicity and deficiency : Manganese in plants is involved in many enzyme systems and in electron transport. Manganese is found in many peroxidases, superoxide dismutase and photosystem II. Manganese oxidizing bacteria in rhizosphere may decrease the manganese availability thereby induce manganese deficiency symptoms in plants in aerated and calcareous soils. However, in poorly aerated or submerged soil the risk of manganese toxicity is greatly decreased by the presence of manganese oxidizing bacteria. Susceptible varieties of oat develop 'gray speck' disease due to Mn^{2+} deficiency. The disease in pea is known as 'marsh spot' and in sugar beet as 'speckled yellow'. These diseases can be rectified by the addition of $MnSO_4$ as foliar spray or by acidifying or flooding the soil. Manganese toxicity in soils may occur either due to high manganese content and partly because strongly acid soils dissolve manganese to toxic concentrations.

9.4 TRANSFORMATION OF POTASSIUM

Potassium is a major plant nutrient. It activates over 80 different enzymes in plants, animals and microbes, especially during photosynthesis, respiration and protein formation. Potassium is essential for photosynthesis, protein synthesis, nitrogen fixation in legumes, translocation of sugars, etc. In plants, on dry weight basis, potassium content varies from 1 to 3% whereas in microbes it varies from 0.5 to 3.0%.

Available potassium : Potassium accounts for about 2.6% of earth's crust but is often inadequate to meet plant requirements. Though in one hectare furrow slice land (15 cm depth), on an average, about 90,000 kg potassium is present, the available (exchangeable and soluble) potassium in upper 15 cm soil is <100 kg ha^{-1}. Potassium is a mobile element and seldom complexes with any macromolecule. Many soil minerals such as mica, illite, montmorillonite, vermiculite, zeolite, feldspar, etc. contribute to K^+ fixation in soils. K^+ fixation is currently considered more a disturbing than a harmful reaction. Little is known about potassium transformation by microorganisms. Crop residues and animal remains partially replenish soil potassium.

Microbial solubilization of potassium : Potassium is not bound strongly in plant so microbial action is not as critical as for nitrogen or sulphur. Microorganisms immobilize a good amount of available potassium for their growth, which is later released on their death and turnover in soil. Organic and inorganic acid production is a major mechanism of potassium solubilization in soil. Carbonic acid, sulphuric acid, nitric acid and other acids, released by various heterotrophs as well as some chemotrophs, help in the weathering of minerals like muscovite, biotite, orthoclase, microcline (primary minerals), vermiculite, illite, chlorite (secondary mineral), etc. and release potassium in soil. Microbes also induce shift in equilibrium in clay minerals between soluble and insoluble forms of potassium (Fig. 9.3). Some potassium is released from organic residues during decomposition. Certain heterotrophic fungi and bacteria are capable of decomposing alumino-silicate minerals thereby release a good amount of potassium contained their in. Typical of this group include bacteria like *Bacillus, Clostridium, Pseudomonas,* etc., fungi like *Aspergillus, Penicillium, Mucor,* etc. and some actinomycetes. The presence of humic and fulvic acids in soils accelerates the fixation of potassium in soil. Many *Clostridium* and *Bacillus* species produce volatile and semi-volatile organic acids such as acetic acid, butyric acid, etc. which solubilize calcium and potassium from granite sand. *Aspergillus fumigatus* can use mica as a source of potassium, as potassium is immobilized. Sodium ion in solution exchanges for potassium ions in mica which result in the formation of vermiculite. Similarly, 2-ketogluconic acid has been identified during the microbial weathering of silicates and during the solubilization of $CaCO_3$, montmorillonite and hydroxyapatite.

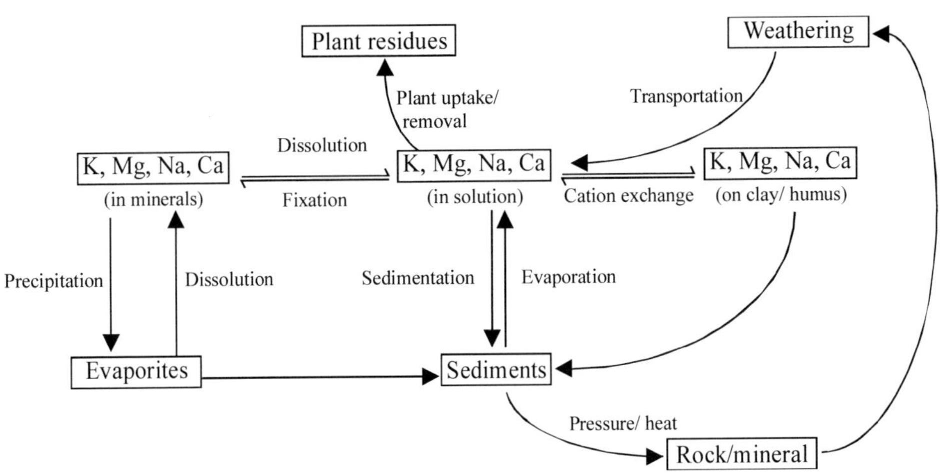

Fig 9.3 : Cycling of K, Na, Mg and Ca

9.5 TRANSFORMATION OF ZINC

Zinc is essential for numerous enzyme systems and is capable of forming stable bond with nitrogen and sulphur ligands. Zinc occurs in soil solution as Zn^{2+} and is quite immobile in soil. Above pH 7.7, it becomes $Zn(OH)^+$ and at pH 9.1 it precipitates as $Zn(OH)_2$. Generally, in soil zinc does not exist inorganic form but bonds strongly with sulphide forming ZnS (sphalerite) together with PbS (galena). Zinc, in contrast to iron or manganese, is less soluble in anaerobic conditions than in aerobic soil. Zinc cations generally become reduced when O_2 supply is low, as occurs in wet soils containing decomposing organic matter. Reduction can also be brought about by organic metabolic reducing agents such as NADPH or caffiec acid, produced by microorganisms and plants in the soil. Change from one valence to another state of zinc is brought about by microorganisms and organic matter. Sewage sludge, fertilizers, pesticides and some animal manure contribute to zinc accumulation in soils.

Zinc is required as a micronutrient by plants and its deficiency induces little leaf symptoms in tree plants. Several species of soil microorganisms are also susceptible to zinc toxicity. Nitrogen fixing bacteria e.g. *Rhizobium leguminosarum* bv. *trifoli,* in root nodules of legumes are adversely affected even at the zinc concentration within the permissible limits of 300 mg kg^{-1}. However, zinc is required or stimulatory to growth of a number of fungi, yeasts and bacteria. Cells contain about 100-400 ppm zinc. Microbes increase solubility of zinc through organic acid production. Fall in pH arising from the oxidation of ammonium salts by nitrifiers, decomposition of plant residues and oxidation of sulphides by *Thiobacillus* release zinc in a water soluble form.

9.6 TRANSFORMATION OF MERCURY

Discharge of metallic mercury (Hg^o) into soil and water environments has been a major concern for animal health including human being. Metallic mercury has high vapour pressure and is volatilized from soils treated with phenyl mercuric acetate, ethyl methyl acetate and mercuric and mercurous ions. Anthropogenic release of elementary mercury (volatile form) into atmosphere, soil and water has become a serious problem with extreme adverse toxic consequences on human health. Coal contains 1,200-21,000 ppb mercury. Burning of fuels releases enormous mercury and other metals. The metallic mercury (Hg^o) was produced upto seventies during electrolysis of brine to sodium hydroxide and bleach (Cl, OCl) and discharged into water courses.

Methylation of mercury : Inorganic mercury is not itself very toxic, however it does not remain in metallic form once released into environment. Microorganisms in sediments transform metallic mercury (Hg^o) to methylated forms (monomethyl and dimethyl mercury).

$$\underset{\text{(Elemental mercury)}}{Hg^o} \xrightarrow{\text{Microorganisms}} \underset{\text{(Methyl mercury)}}{CH_3Hg^+} \xrightarrow{\text{Microorganisms}} \underset{\text{(Dimethyl mercury)}}{CH_3)_2Hg}$$

Methylated forms are not only volatile and lipophilic but also toxic. Monomethyl mercury is soluble in water and readily absorbed and retained in animal tissues thus proving carcinogenic. Dimethyl mercury is insoluble in water and volatile, bioaccumulates in tissues and is difficult to eliminate. Methyl mercury is able to disrupt the blood brain barriers, acts as neurotoxin and causes behavioural disturbance in affected individuals.

In soil, mercury metal concentration is usually <1 ppm and mostly exists in cationic form of mercury (Hg^{2+}). Reduction to metallic elemental form (Hg^{o}) in soil is easily achieved by both biological and chemical reactions.

$$Hg^{2+} \longrightarrow (CH_3)_2Hg \xrightarrow{\text{Biological/photolysis}} Hg^{o}$$

Under anaerobic conditions soil microbes methylate mercury and form volatile organo-mercury compounds that are bioavailable and toxic to microorganisms. At the same time, anaerobic conditions can convert Hg^{2+} into exceedingly insoluble HgS (Fig. 9.4). Microbes involved in methylation are bacteria, fungi and yeasts such as *Bacillus, Clostridium, Mycobacterium, Pseudomonas, Aspergillus, Neurospora, Scopulariopsis,* etc. Besides methylation, these microorganisms release organically bound mercury during their heterotrophic activity. Methyl mercury may be demethylated by many organisms resulting in the release of mercury.

$$CH_3Hg^+ + 2H \longrightarrow Hg + CH_4 + H^+$$

Mercuric or mercurous ions may react with H_2S produced during microbial reduction of sulphates.

$$Hg^0 \text{ or } Hg^{2+} + H_2S \longrightarrow HgS \text{ (precipitate)}$$

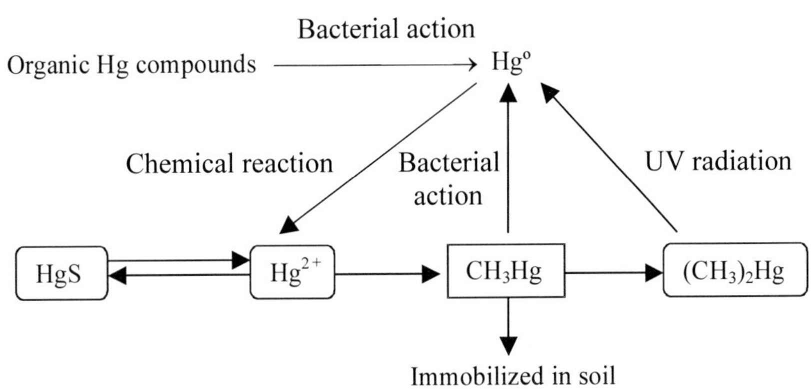

Fig 9.4 : Transformation of mercury in soil

The methyl sulphide, so generated, is insoluble but less toxic. Mercury accumulation in soils tends to correlate with organometallic level. Acidic oxidizing conditions in soil tend to stabilize Hg^{2+} which complexes with phenolic and carboxylic functional groups. Adsorption of Hg^{2+} on silicate clays and oxides is more favourable at higher pH. In reducing soil environment strong association of Hg with sulphide renders it less mobile.

9.7 TRANSFORMATION OF ARSENIC

The most probable oxidation states of arsenic (As) in soil environment are +3 and +5 although –3 and 0 oxidation states are at least possible in strongly reduced soils and sediments. Arsenite (+3) [AsO_2^-] is the reduced state of arsenic present in anaerobic soil and may occur as $As(OH)_3$, $As(OH)_4^-$, $AsO_2(OH)^{2-}$ and AsO_3^{3-}. Arsenate (+5) [AsO_3^{3-}] is oxidized state and most stable in aerobic soils.

Arsenate has chemical behavior similar to that of phosphate in soil. It is chemisorbed by iron and aluminum oxides, non-crystalline aluminosilicates and layer silicate clays. Arsenate is absorbed more effectively at low pH. It is fairly low in mobility in acid conditions but mobile in soluble sodium arsenate form. Soil microbes especially manganese oxidizers are able to promote the oxidation of arsenite to arsenate under aerobic conditions. In anaerobic soils both arsenate and arsenite are released into soil solution by dissolution of iron and manganese oxides. Subsequently desorbed arsenate is reduced to arsenite, which eventually is converted to insoluble form. Thus arsenic solubility may first increase and then decrease if anaerobic continuous flooding conditions continue. The sulphides formed under anaerobic conditions may co-precipitate arsenic in its lower oxidation state. However, volatile alkyl-arsene compounds may also be formed under strongly reduced conditions causing loss of some arsenic from soil. Various biological and chemical transformations interact to control solubility, mobility and availability of arsenic.

Arsenic is constituent of phospholipids in algae and fungi and its phytotoxicity level is >5 to 20 ppm. The fungal species participating in the conversion of arsenical compounds to volatile arsenic compounds include the species of *Aspergillus, Mucor, Scopulariopsis, Fusarium* and *Paecilomyces. Candida humicola* synthesizes some volatile metabolite from arsenite and arsenate. The gaseous arsenic containing compounds are released from soils treated with the salts of arsenic. Volatile arsenic compounds are often converted to methylated forms like those of mercury.

$$\underset{(As^{2+})}{\text{Arsenate}} \longrightarrow \underset{(As^{3+})}{\text{Methyl arsenate}} \longrightarrow \underset{(As^{+})}{\text{Dimethyl arsenate}} \longrightarrow \underset{(As^{3-})}{\text{Trimethyl arsine}}$$

Heterotrophic bacteria that convert arsenite to more oxidizing state have also been isolated. *Pseudomonas* and *Alcaligenes* reduce arsenate (AsO_4^{3-}). Though soils contain minor amounts of arsenic, its application through pesticides increases the arsenic concentration in the field. Several fungi are capable of making trimethyl-arsine $(CH_3)_3As$ from methyl-arsenate and dimethyl-arsenic acid.

9.8 TRANSFORMATION OF SELENIUM

Selenium and sulphur chemistry are almost similar except that selenates (SeO_4^{2+}) and selenites (SeO_3^{2-}) are weaker acids than their sulphur counterparts. Because of this weak acid character, SeO_4^{2+} and SeO_3^{2-} are retained more strongly by soils than their corresponding sulphur anions. SeO_4^{2+} (+6) and SeO_3^{2+} (+4) are mobile forms of selenium whereas elemental selenium Se^0 (0) is immobile and selenide HSe^- (-2) is volatile form. The selenite is stable under strongly oxidizing conditions over most of the ranges of soil redox conditions.

Soils contain, on an average, 0.05 to 1.27 ppm selenium. Plants can bioaccumulate selenium in their tissues so create a potential toxicity hazard to foraging animals. Organic forms of selenium are seleno-methionine and seleno-cystine. Organically complexed selenium is important fraction of soil selenium. Upto 40% of the total selenium in some soils is present in humus. Soluble organic selenium compounds are liberated through the decay of seleniferous plants. Organic selenium is more soluble under basic than acid soil conditions.

$$Se(IV) \longrightarrow \underset{\text{(Dimethyl selenide)}}{(CH_3)_2Se^+} \longrightarrow \underset{\text{(Dimethyl diselenide)}}{CH_3Se.SeCH_3}$$

A number of bacteria, actinomycetes and fungi reduce selenium salts. In filamentous microbes and bacteria the end product of the reaction in common media is elemental selenium. The species of *Candida, Clostridium, Corynebacterium, Rhizobium* and *Micrococcus* cause such transformation. The formation of elemental selenium can be detected microscopically because metal is deposited within the cell as a distinct red granule. Selenium mobility and bioavailability depends on its oxidation state. Selenium like mercury is also subject to methylation and the product formed is dimethyl-selenide. *Aspergillus, Candida, Cephalosporium, Corynebacterium, Fusarium, Alternaria, Ulocladium, Acremonium* and *Penicillium* are example of the micro-organisms capable of bringing such transformation.

9.9 TRANSFORMATION OF OTHER METALS

Soil serves as a natural repository for all the materials so cleans our environment. Increased anthropogenic activities have resulted in huge deposition of large amounts of heavy metals which pose a serious threat to the soil ecosystem and environment. The 'heavy metals' is a group of metals and metalloids which have an atomic density of >6 g cm^{-3}. Soils, sediments, water and living organisms possess a range of normal background concentration of these metals. Therefore, frequent application of sewage sludge, industrial effluents, fertilizers, etc. increase the concentration of metals like Cd, Cr, Co, Ni, Hg, Mn, Mo, Pb, Zn, Fe, Sn, etc. in soils to the toxic level. In England, for instance, about 5 x 10^5 t dry sewage sludge, rich in heavy metals (Cu, Zn, Cd, Ni, Pb and Cr), derived from industrial sources is applied to land each year. The effects of heavy metal on soil biological processes persist for decades even after the input of pollutant has ceased.

Bioavailability of heavy metal : The microorganisms are the first to be affected by any heavy metal than higher organisms. The bioavailability of heavy metals to terrestrial organisms is determined by the solubilization, adsorption and precipitation reactions occurring in soils and sediments. The extent to which these metals are adsorbed depends on the properties of metal concerned (valency, radius, degree of hydration and coordination with O_2), the physico-chemical properties of the environment (pH, redox status, soil water content, temperature, etc.), the nature of adsorbent (permanent and pH dependent charges, complex forming ligands, etc.), other metals present and their concentration, and the presence of soluble ligands in the surrounding fluids.

Complexation/chelation of metals : All metals, except Mo, occur in soil primarily in cationic form. Cr'' and CrO_4^-, though initially anionic, are immediately converted to cationic Cr^{3+} form in soil. The metals in soil either form complexes or are precipitated or chelated. The cationic micronutrients react with certain organic molecules to form organo-metallic complexes called 'chelates'. If these complexes are soluble, these are carried in soil solution so increase availability of micronutrients and protect them from precipitation reactions with inorganic soil constituents. However, insoluble complex formation decreases micronutrient availability. These may occupy spaces on organic matter complexes and are rendered less available for microbial action, mineralization and other processes. However, various microbes synthesize numerous metal binding compounds, which may be grouped as given in Table. 9.2.

Table 9.2 : Various types of chelates synthesized by microbes for metal chelation

Types of metal chelate	Metals chelated
Amino acids	Co, Cu, Zn
Polypeptides/proteins/enzymes	Cu, Fe, Mn, Ni, Zn
Metallophenolic complexes	Al, Fe, other metals
Metallophosphate complexes	Cu, Na, Zn, other metals
Polyhydroxy complexes	B, Ca, Cu
Metalloporphyrins	Cu, Fe, Mg

Glycine, histidine and glutamic acid readily bind cobalt while copper is bound by alanine and aspartic acid. Zinc is weakly bound by sulphur containing amino acids. Protein polypeptide 'bactracin' chelates Zn, Co, Mn and Ni. Colloidal organic matter has strong affinity for heavy metal cations through chelation on the sites with functional groups as –COOH, phenolic, alcoholic, enolic, hydroxyl, carboxylic, etc.

Microbial oxidation-reduction of metals : The oxidation-reduction processes initiated by microbes cause release of various metals or through their end products the solubilization of metals in soil may occur. For example, mineral sulphides may be oxidized by *Thiobacillus* spp., which attack either or both metallic and sulphide components of minerals. Some bacteria and fungi also synthesize a variety of chelating agents capable of causing liberation of Ca, Si, Mg, Al, Na and other elements from minerals. Acid production is a major mechanism for the solubilization of insoluble minerals. Biological production of sulphuric acid, nitric acid and carbonic acid from S, NH_4^+ and other salts can cause solubilization of Ca and Al. Organic acid liberated by heterotrophs solubilize K, Si, Al, Mg, Ca, etc. through biological weathering of rocks as well as in soil formation. The various such reactions involved are given below :

i. ZnS (sphalerite) + $4H_2O \longrightarrow Zn^{2+} + SO_4^{2-} + 8H^- + 2e^-$

ii. NiS (millerite) + $4H_2O \longrightarrow Ni^{2+} + SO_4^{2-} + 8H^- + 2e^-$

iii. $2MoS_2$ (molybdnite) $+9O_2 + 6H_2O \longrightarrow 2H_2MoO_4 + 4H_2SO_4$

iv. $2ZnS + Fe(SO_4)_3 + 2H_2O + 3O_2 \longrightarrow 2ZnSO_4 + FeSO_4 + 2H_2SO_4$

v. $2Cu_5FeS_4$ (bornite) + $18O_2 + 3H_2O \longrightarrow 8CuSO_4 + Cu_2O + 2Fe(OH)_3$

vi. $6(Cu_2.S).Sb_2S_3$ (tetrahedrite) + $27O_2 \longrightarrow 12CuSO_4 + 2SbO_3$

vii. $4Cu_2S$ (chalcocite) + $9O_2 \longrightarrow 4CuSO_4 + 2Cu_2O$

viii. $2CoS_2$ (cattierite) + $9H_2O + 7O_2 \longrightarrow 2CoSO_4.7H_2O + 2H_2SO_4$

Except for the iron containing mineral sulphide, the sulphide is the preferred locus of attack. When iron is involved, ferrous forms may be oxidized to ferric ions, which may either precipitate as an hydroxide, if pH is sufficiently high, or possibly oxidize other metals and become reduced again to supply additional substrate for microbial growth. *Thiobacillus ferroxidans* oxidize various Cu, Ni, Mo and Zn containing minerals while *Hypomicrobium* is capable of decomposing silver containing minerals in soil (Table 9.3).

Soil microbes also utilize various mineral elements either during the synthesis of their cellular organic materials or for organic-metal complex formation or inorganic skeleton structure (deposit) or in inorganic compounds dissolved in their body fluids.

Methylation of various metals, like that of Hg, is well known for selenium, tellurium and arsenic. Various organisms involved in methylation are *Aspergillus, Candida, Cephalosporium, Fusarium, Penicillium, Corynebacterium* and methane bacteria.

Table 9.3 : Various minerals oxidized by soil bacteria

Minerals	Organisms involved
Arsinopyrite (Fe As S)	*Thiobacillus, Ferribacterium*
Chalcocite (Cu_2S)	*Thiobacillus thiooxidans, T. thioparus*
Chalcopyrite ($Cu\,Fe\,S_2$)	*Thiobacillus, Ferribacterium*
Enargite ($Cu_3\,As\,S_4$)	*Thiobacillus, Ferribacterium*
Galena (Pb S)	*Thiobacillus, Ferribacterium, Arthrobacter*
Millerite (Ni S)	*Thiobacillus ferrooxidans*
Molybdnite ($Mo\,S_2$)	*Thiobacillus ferrooxidans*
Pyrite ($Fe\,S_2$)	*Thiobacillus, Ferribacterium*
Pyrrholite ($Fe_{1-x}\,S$)	*Thiobacillus ferrooxidans*
Realgar (Ag S)	*Arthrobacter, Hypomicrobium*
Sphalerite (Zn S)	*Arthrobacter*
Tetrahedrite $(Cu, Fe, Zn, Ag)_{12}Sb_4S_{13}$	*Thiobacillus, Ferribacterium*

Tolerance of soil microbes to heavy metal : Certain microbial species have developed some kind of tolerance under constant metal stress situations. Such metal resistant bacteria are usually determined by the genes carried on plasmids that alter the membrane permeability of cell to a metal. *Micrococcus* and *Azotobacter* rapidly immobilize lead (Pb) in cell wall so transfer Pb across the cell membrane. The excessive amounts of metals in soil reduce the diversity and distribution of microbial species in soil. More tolerant selective species appear in soil under constant metal stress. Microbial biomass and their enzyme activities usually get altered due to the presence of metal in organic matter. Under certain circumstances, microorganisms are capable of converting bound element to an easily assimiliable form in soil thereby help in purifying the soil.

Molybdenum : Molybdenum, as molybdate, is the heaviest essential element required in low concentrations by the microorganisms. In soil readily available MoO_4^{2-} anion is dominant at pH >5.2. The ionic MoO_4^{2-} is absorbed by oxides of iron and aluminum, just as in PO_4^{2-}. But calcium molybdate is much more soluble than its phosphorus counterpart. At low pH relatively unavailable forms (H_2MoO_4 and $HMoO_4^-$) are prevalent. Molybdenum is involved in nitrogen fixation process (in nitrogenase) and nirate reductase enzyme, which catalyze conversion of NO_3^- to NO_2^-. It is essential for nitrogen fixation and nitrogen assimilation. Molybdenum concentration in nodules of legume crops has been observed to be 5 to 10 times higher than those of leaves. A good correlation between organic matter and molybdenum has also been noted.

$$H_4MoO_4 \underset{+H^+}{\overset{+OH^-}{\rightleftharpoons}} HMoO_4^- + H_2O \underset{+H^+}{\overset{+OH^-}{\rightleftharpoons}} MoO_4^{-2} + H_2O$$

Various microbes during transformation of sulphur minerals also release molybdate ions. Molybdenum oxidation is increased several fold in the presence of *Thiobacillus ferroxidans*.

$$4MoO_4 + 2FeS_2 + 13O_2 + 6H_2O \xrightarrow{\textit{Thiobacillus ferroxidans}} 4H_2MoO_4 + 2H_2SO_4 + 2Fe_2(SO_4)_3$$

Vanadium : Low concentration of vanadium is beneficial for the growth of microorganisms and plants. Evidences show that vanadium can partially substitute for molybdenum in nitrogen fixation by microorganisms. It also functions in biological oxidation-reduction reactions.

Cobalt : Cobalt is essential for the microorganisms having potential to fix atmospheric nitrogen. Cobalt is also required in the nodules of both legumes and alder tree and in nitrogen fixing cyanobacteria and actinomycetes. It has role in vitamin B_{12} formation, the co-factor required for the production of heme compounds. Cobalt forms a complex with nitrogen atoms in the porphyrin ring structure that provides a prosthetic group for association with a nucleotide in the vitamin B_{12} coenzyme. This co-complex is called 'cobamide coenzyme'. Cobalt also functions in leghaemoglobin metabolism and ribonucleotide reduction in *Rhizobium*. It also activates enolase and succinic kinase enzymes.

Boron : Boron is the only non-metal among the micro-nutrients essential for plant growth and its availability in soil is controlled by soil pH. It is mostly available in the pH range of 7.0 to 9.0. At lower and higher pH, boron is easily leached so soils show boron deficiency. Soluble boron is present in soil mostly in the form of boric acid (H_3BO_3) and a little as $H_2BO_3^-$. Boron activates certain dehydrogenase enzymes and facilitates sugar translocation and synthesis of nucleic acids. Boron is absorbed by humus, the strength of binding being even greater than for inorganic colloids. Moreover, boron is a component of soil organic matter that is released by microbial mineralization, consequently organic matter serves a major reserviour for boron in soil. The absorption of boron to organic and inorganic colloids increases with increasing pH.

9.10 BIOLEACHING

With the dwindling rich mineral deposits the extraction of various metals from ores has turned out to be a big problem for mining industry. The ores with low metal content are not suitable for direct smelting. This has necessitated the processing of low quality ores as well as developing better techniques for efficient extraction of metals from these ores. The extraction of metals from ores using microorganisms is known as 'bioleaching'. The earliest recorded metal bioleaching is by Liu-An (177-122 BC) from China. In bioleaching technique bacteria such as chemolithotrophs (*T. ferrooxidans* and *T. thiooxidans*) are utilized to oxidize metal (iron, copper and zinc, etc.) sulphides to soluble metal sulphates. The elementary metals can then be readily extracted from solution. Bacterial leaching is currently used mainly for copper and uranium. Nearly 5% of metal production and 20% of copper mined world-wide comes from this low cost technology application to low grade mineral ores such as pyrite, metal sulphides, etc. Somewhat similar techniques are applicable to enrich uranium in soil. There is possibility of its application for the extraction of other metals in future.

The microbial bioleaching agents are used to recover the metals from mined materials and in the up-gradation of ores (old mines dumps). *Leptospirillum* is used to recover low grade ores. Bioleaching can be used to extract metals directly from deep underground ore bodies. Removal of sulphur from coal, prior to its combustion, would help to solve the problem of SO_2 production to atmosphere. Microorganisms offer solution to desulphurize sulphur containing coal in an economic manner. Various species of bacteria can degrade sulphur compounds in coal to water extractable compounds. Sulphur occurs both in organic and inorganic form. Bacteria such as *Thiobacillus ferroxidans, T. thiooxidan* and *Sulfolobus acidocaldarius* can oxidize sulphides to sulphates which can be drained from the coal. *S. acidocaldarius* is obligate thermophile that is used in bioleaching of molybdenum as it is more tolerant of molybdenum than *Thiobacillus*. This *in situ* technique helps to minimize the adverse environmental impact caused by extensive excavation. Bioleaching method is reported to be helpful in leaching of copper, nickel, molybdenum, zinc, lead, arsenic, cobalt, antimony, selenium, titanium, germanium, etc. from their sulphates. However, leachate products produced by bioleaching are very toxic and must be rigorously controlled to avoid soil and water pollution.

Selected References

Alexander, M. 1977. *Introduction to Soil Microbiology* (2nd edn.). John Wiley & sons, New York, USA.

Alloway, B.J. 1995. *Heavy Metal in Soil.* Blackie Academia and Professionals, Glasgow, UK.

Alloway, B.J. and Ayres, D.C. 1997. *Chemical Principles of Environmental Pollution.* Chapman & Hall, London, UK.

Bath, E. 1989. Effects of heavy metals in soil on microbial processes and population (A review). *Water, Air Soil Pollu.*, 47 : 335-379.

Bohn, H.L., McNeal, B.L. and O'Connor, G.A. 1985. *Soil Chemistry.* John Wiley & sons, New York, USA.

Dar, G.H. 1995. Effect of heavy metals (Cd, Cr, Ni and Pb) on soil microbial biomass C, mineralization and enzyme activities. *International Journal of Ecology and Environmental Sciences*, 21 : 87-95.

Deepa, M.K., Murugesan, A.G. and Sasikala, K. 2003. Tackling the challenges of heavy metal pollution by microbial biosorbents. pp. 192-201. In : *Biotechnology, Emerging Trends* (eds. J. Selvin, A.S. Ninawe, V.S. Sugunan,, N. Sukumaran and A.P. Lipton), Biotech Books, New Delhi, India.

Doetsch, R.N. and Cook, T.M. 1973. *Introduction to Bacteria and their Ecologies.* University Park Press, Baltimore, Madison, USA.

Ehrlich, H.L. 1981. *Geomicrobiology.* Marcel Dekker, New York, USA.

Hopkin, S.P. 1989. *Encyclopedia of Metal in Terrestrial Invertebrates.* Elsevier Applied Sciences, London, UK.

Krumbein, W.E. (ed.) 1983. *Microbial Geochemistry.* Blackwell Scientific Publ., Oxford, England.

Manahan, S.E. 1991. *Environmental Chemistry.* Lewis Publishers, Chelsea, MC.

Marshall, K.C. 1979. Biogeochemistry of manganese minerals. pp. 253-292. In : *Biogeochemical Cycling of Mineral-forming Elements* (eds. P.A. Trudinger and D.J. Swaine). Elsiever, Amesterdam, Holland.

Murr, A.E., Torma, A.E. and Brierley, J.A. (eds.) 1978. *Metallurgical Applications of Bacterial Leaching and Related Microbiological Phenomenon.* Academic Press, New York, USA.

Nealson, K.H., Tebo, B.M. and Rosson, R.A. 1988. Occurrence and mechanisms of microbial oxidation of manganese. *Adv. Appl.Microb.,* 33 : 279-318.

Ross, S. 1989. *Soil Processes.* Routledge, London, UK.

Sajic, J.E. 1969. *Microbial Biogeochemistry.* Academic Press, New York, USA.

Stevenson, F.J. 1986. *Cycles of Carbon, Nitrogen, Sulphur and Micronutrients.* John Wiley & sons, New York, USA.

Summers, N.O. and Silver, S. 1978. Microbial transformation of metals. *Ann. Rev. Microbiol.,* 32 : 638-672.

Tan, K.H.1978. Effect of humic acid and fulvic acids on release of fixed potassium. *Geoderma,* 21 : 67-74.

Tiller, K.G. 1989. Heavy metal in soil and their environmental significance. *Adv. Soil Sci.,* 9 : 113-142.

Vishniac, W. and Santer, W. 1957. The Thiobacilli. *Bacteriol. Rev.,* 21 :195-213.

Weinberg, R.P. (ed.) 1977. *Microorganisms and Minerals.* Marcel Dekker, New York, USA.

Wild, A. (ed.). 1988. *Russell's Soil Conditions and Plant Growth* (11th edn.). Longman, Harlow, USA.

Zajic, J.E. 1969. *Microbial Biogeochemistry.* Academic Press, New York, USA.

— ★★★ —

Chapter 10

Mycorrhizae — Plant Relationships

Mycorrhizae is not a panacea for correcting the numerous problems faced by the people all over the world – but it is an additional biological weapon which has greater potential for combating the current and future problems of supplying enough food, fiber and fuel for the expanding populations

(Marx and Schenk)

Mycorrhizae denote the symbiotic associations between plant roots and fungal mycelia. Literally it means fungal root. In 1840, Vittandini proposed that 'tree rootlets are nourished by certain fungal mycelia which mantle them'. The first illustrations of mycorrhiza were published in the same year by Robert Hartig. However, in 1885 A.B. Frank coined the term "mykorrhizen" to the mutualistic symbiotic fungal association of roots ('*mykos*' fungus '*rhizen*' roots). He characterized the sheath-forming fungi on tree as 'ectotropisch' (ectotrophic) and those without such compact sheath as 'endotropisch' (endotrophic). Fossil records reveal that mycorrhizae are as old as terrestrial plants.

Mycorrhizae involve dynamic conditions of mutual exploitation of invaders (biotrophic fungi) and defenders (absorbing organs of plants) resulting in beneficial impact on both the partners in terms of growth and development. The infection induces plant root development and consequent physiological changes in a characteristic way for better absorption of nutrients from soil especially when the soils are deficient in phosphorus (Fig. 10.1). It is now documented that mycorrhizal symbionts constitute an integral part of any and almost every root system. In fact, mycorrhizae show a sort of controlled parasitism. The fungus-plant association is therefore defined as *'a mutualistic symbiosis between plant and fungus localized in a root or root-like structure in which energy flows primarily from plant to fungus and inorganic nutrients move from fungus to plants.'*

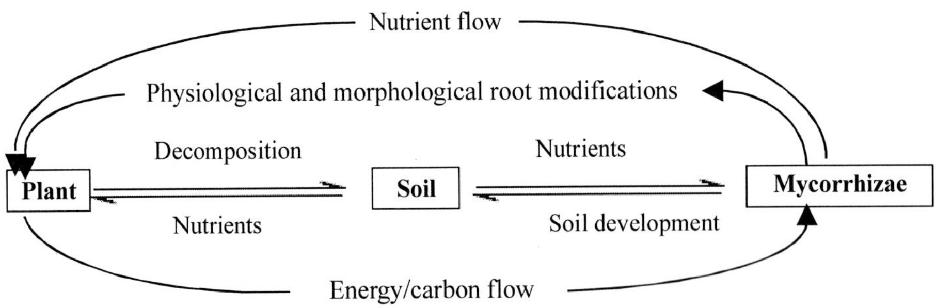

Fig. 10.1 : Plant-soil-mycorrhizae relationship

As per the conservative estimates about 70% of higher plants are obligately dependent on mycorrhizae, 12% facultatively and rest 18% remain uninfected. Mycorrhizae are ubiquitous soil microorganisms found almost in all the regions of world from tropical or temperate to arctic region. On global basis 83% dicotyledonous and 79% monocotyledonous plants and all gymnosperms are mycorrhizal. Mycorrhizae are absent in only 14 plant families including Cruciferaceae and Chinopodaceae.

10.1 MYCORRHIZAL MORPHOLOGY

On the basis of morphology and anatomy, the mycorrhizal associations are categorized into seven groups viz., endomycorrhizae (vesicular arbuscular mycorrhizae), ectomycorrhizae, ectendomycorrhizae, arbutoid mycorrhizae, monotropoid mycorrhizae, ericoid mycorrhizae and orchid mycorrhizae. Groups such as arbutoid, monotropoid, ericoid and orchids represent associations with specific host types of taxon Ericaceae, Monotropaceae, Ericaceae (arbutoideae) and Orchidaceae, respectively. The ectendotrophic association demonstrates the continuum between ecto- and endo-mycorrhizae. The characteristic features of three important kinds of mycorrhizae are summarized in Table 10.1.

Table 10.1 : Characteristic features of main mycorrhizal groups

Characteristics	Endomycorrhizae	Ectomycorrhizae	Ectendomycorrhizae
Mycelium			
- Septate	-	+	+
- Aseptate	+	-	-
Fungal sheath	-	+	+ or -
Hartig net	-	+	+
Intracellular hyphae	+	-	+
Hyphal coil in cells	+	-	-
Dichotomous haustoria	+	-	-
Vesicles in cells/tissues	+ or -	-	-
Fungal class	Zygomycota	Ascomycota and Basidiomycota	Ascomycota and Basidiomycota
Host plants	Bryophyta, Pteridophyta, Gymnosperm and Angiosperm	Gymnosperm and Angiosperm	Gymnosperm and Angiosperm

10.1.1 ENDOMYCORRHIZAE

Endomycorrhiza are obligate symbionts. About 80 fungal species form endomycorrhiza which are mainly the members of Zygomycota. Mostly they fall in a few genera in calss Zygomycota (order Mucorales; family Endogonaceae). The hyphae of endomycorrhiza enter the cortical cells of feeder roots of nearly all the cultivated plants and many forest trees, shrubs and wild herbaceous plants. Endomycorrhiza form large vesicles and arbuscules in their host cells and because of this they are also known as 'vesicular arbuscular mycorrhizae (VAM)'. Though VAM are not specie-specific yet some fungal species and even strains within a species greatly vary in their association with plants. Cross inoculation with any susceptible host is fairly easy.

Host diversity : VAM are geographically ubiquitous occurring in plants from arctic to tropical regions over a broad ecological range, from desert to aquatic environment. VAM occur practically in all angiosperm families (except Pinaceae, Betulaceae, Commenlinaceae, Ericaceae, Fumeriaceae, Orchidaceae and Urticaceae), gymnosperms and many pteridophytes and bryophytes. Most of the economically important grain, forage and vegetable crops and a few major commercial fruit, nut and berry plantations bear endomycorrhizae. Less commonly infected ones are Cyperaceae, Citraceae, Juncaceae and Caryophyllaceae.

Major genera : The taxonomy of VAM fungi is in a state of flux and purely based on spore morphology. The genera of VAM, known on the basis of spore morphology, colour, ramification, hyphal attachments in spores, etc., are *Glomus, Gigaspora, Acaulospora, Sclerocystis, Scutellospora, Entrophosphora* and *Glaziella.* The genus *Endogone* forms zygospore in the plants but is not known to produce VAM. About 150 species of VAM fungi have so far been identified and characterized. Characteristic features of important genera are given in Table 10.2. VAM fungi can be maintained only on live plants inoculated with spores of the species. VAM have the largest known resting spores of any fungi and may attain the size upto 600 μm.

Table 10.2 : Characteristic features of important genera of VAM

Genera	Characteristics
Glomus	Spores (chlamydospores) have characteristic straight or angular hyphal tips (stalks) and produce sporocarp. The spore germinate from the re-growth of attached hyphae
Gigaspora	Spores have bulbous spensors (stalks), rarely form vesicles. The spores germinate through wall of the spore.
Sclerocystis	They form sporocarp as a single crowded layer of erect spores that surround the sides and top of spore free central mass tightly interwoven with hyphae. The spores germinate from the regrowth of attached hyphae.
Acaulospora	Spores are sessile (without stalk) but laterally form funnel shaped inflated hyphal terminus. The spores germinate through wall of the spore, except the spores of *Acaulospora laevis* where spores germinate after 1-2 months.
Entrophosphora	They have azygospores within funnel shaped stalk of a large inflated hyphal terminus.
Glaziella	They have chlamydospores, found singly at hyphal tips scattered in the wall of large orange to red coloured sporocarp. *Glaziella* is confined to tropical lowlands only.

Infection process : VAM fungi do not produce any large aboveground fruiting bodies as ectomycorrhizae do. They usually exist as conspicuous thick-walled spores (chlamydospores) or sporocarps (aggregation of spores) or as vegetative propagules in roots and germinate in rhizosphere or in rhizoplane (Fig. 10.2). The external structures on the exterior of plant roots that have in past been identified as vesicles are now considered as chlamydospores in the formative stages. Therefore, the term vesicle is restricted to expanded tips within plant cortex.

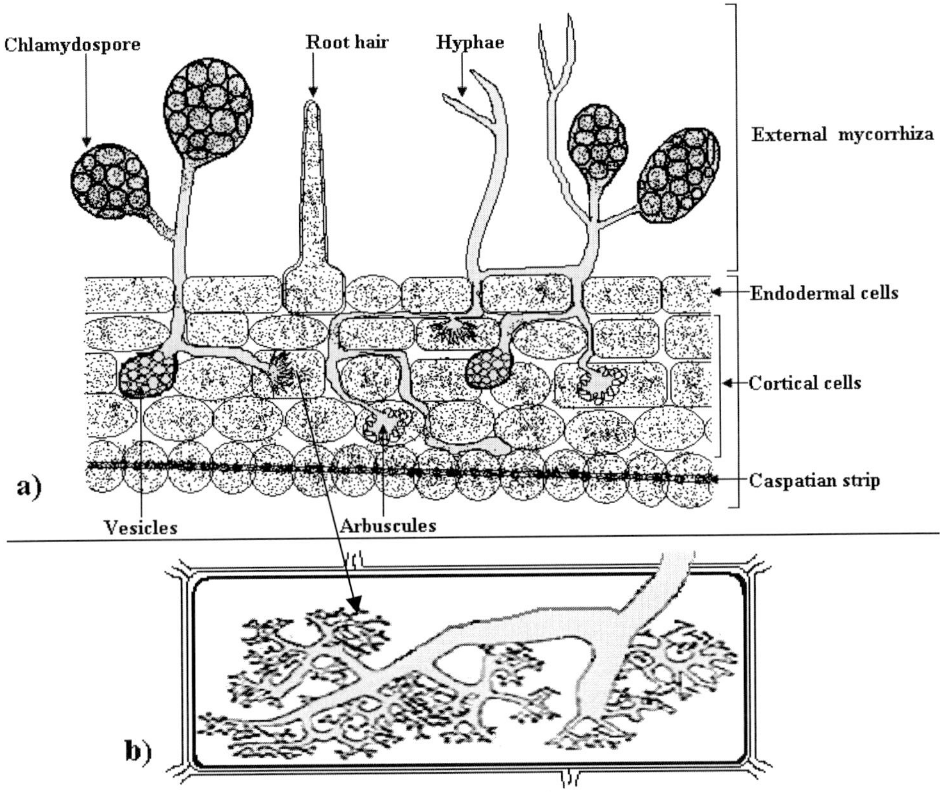

Fig 10.2 : a) A typical VAM infection in host; b) Magnified view of arbuscule development in a host cell

VAM spores germinate in soil under favourable conditions of growth. If a receptive root is not encountered by a germinating tube, the cytoplasm retracts from the germ tube and spore retains the capacity to germinate on some other occasion. This multiple germination reduces the vulnerability of fungus to grazing soil mycophagous fauna. There appears no exchange of message between growing hyphal tips and host until hyphae is just a few millimeters away from the root. VAM gets attracted by volatile substances emitted by the root that are active a few millimeters away from the roots. Infection process includes the arrival of fungus on root, penetration and development of infection and its spread to other parts of the root. Formation of appresorium, once in contact with roots, often occurs as preclude to infection. The fungal hyphae penetrate fine root cells between epidermis and produces fan-shaped complex of narrow (2-7 μm) branches (appresoria) that produce infection in the roots (Fig. 10.2). The hyphae progress into the cortical cells through dichotomously branched haustoria. Repeated dichotomous branching of invading hyphae within host cell gives rise to an arbuscule. Between plasma-lamella of host cell and hyphal wall is a polysaccharide matrix. The mycelium/hyphae ramify intercellularly and develop intracellularly within cortical cells in the form of specialized nutrient exchanging structures either as branched hyphae (arbuscules) or coiled hyphae. These structures are short-lived and persist only for 4-15 days in plant cell, then are digested by host cells and new branches are formed. Arbuscules are the main sites of solute exchange within the host. Following the arbuscular development, thin walled spherical to ovate irregularly lobed vesicles may be produced in between or inside cortical cells by these fungi as swelling along or at the tip of hypha. Vesicles serve as temporary storage organs of fungus. The vesicles are 10-100 μm expansions/swellings of fungal hyphae between cells, usually filled with lipids and occasionally with glycogen. The presence of oil globules in vesicles and branched arbuscules is the sign of infection. The nutrient transfer is accomplished through arbuscules i.e. between finely branched hyphae/mycelium and highly invaginative plant membrane. The infection does not cause any morphological change in roots but induces significant physiological change in the host. The presence of vesicles and arbuscules is a diagnostic criterion for identifying a VAM fungus in a root (Fig. 10.3).

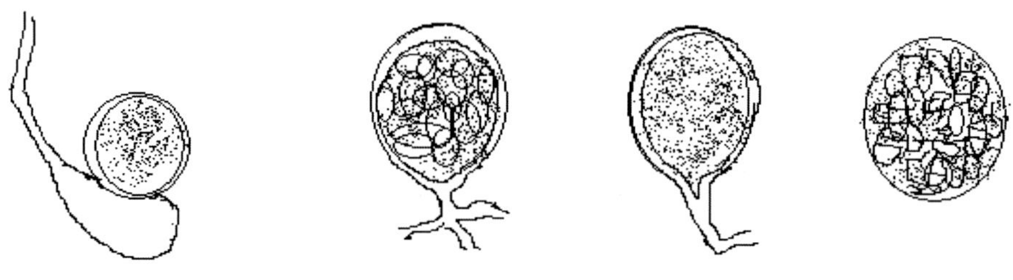

Fig 10.3 : Some important VAM fungi

Externally VAM fungi form a loose network in the soil around the roots. The hyphae are dimorphic and essentially aseptate. Extra-radical hyphae penetrate into the surrounding soil, ramify and produce resting spores. VAM spores are 10 to 400 μm in diameter and borne on external hyphae in soil singly (chlamydospores) or in aggregates (sporocarp). These resting spores are produced asexually on straight or subtended hyphae. The spores, being larger, are not wind-dispersed, although they may be disseminated with soil transported by wind or water. Earthworm, ants, birds, cricket, grass hoppers and wasps transport a small amount of spores with soil. Small mammals like rats, mice, lemming, voles, shrews, pikas and other sporocarp feeders seem to carry them through long distance. The broad range of these fungi seems to compensate for poor dispersal. The VAM infected roots, sometimes, show colouration e.g. yellowing in onion. However, there is not any noticeable change in root structure outside.

10.1.2 ECTOMYCORRHIZAE

Host specificity : Ectomycorrhizae (ECM) may show some sort of host specificity. ECM occurs mainly on woody plants and occasionally on herbaceous and graminaceous perennial plants. ECM are more abundant in borel and temperate forests whereas VAM in warmer climates with drier soils in pasture-lands and denuded forests. Many ECM fungi show broad host spectrum. Likewise, a host may be infected by more than one fungal species e.g. *Pinus sylvestris* may have as much as 40 ECM species present at a time.

Important genera : Ectomycorrhizae are formed by hundreds of different fungal species belong to class Basidiomycota (mushrooms and puffballs) and Ascomycota (cup fungi and truffles). Most of the fungi forming ECM with trees are basidiomycetous belonging to Hymenomycetes and Gasteromycetes (families : Agaricaceae, Amanitaceae, Boletaceae, Cortinariaceae, Geastraceae, Lycoperdaceae, Russulaceae, Tricholomataceae and Sclerodermataceae) and some belong to class Discomycetes (Pezizales), Plectomycetes (Eurotiales) and Zygomycota (Mucurales). The well-known genera in ECM are *Pisolithus, Thelephora, Scleroderma, Cenococcum, Boletus, Laccaria, Lactarius, Tricholoma, Suillus, Russula, Rhizopogon, Amanita, Hebeloma, Inocybe, Paxillus, Entoloma,* etc. They produce edible above-ground and under-ground fruiting bodies (Plate 10.1). Some are well-known mushrooms. More than 2100 species of ECM have been recorded and characterized so far, and over 5000 species are estimated to be present in the world. These fungi can be artificially grown on media containing simple sugars and vitamins but spore germination is often poor on these media.

Plate 10.1 : Fruiting bodies of ectomycorrhizae

Infection process : Ectomycorrhizae consist of septate fungal species infecting roots of trees and shrubs. Infection may arise from the existing mycorrhizal roots which act as point inoculum source. The feeder root surfaces are completely interwoven or covered by well-developed compact fungal hyphae known as "sheath" or "mantle". The sheaths are usually 20-100 µm thick and depending upon fungal associate, the host plant and environmental conditions as many as 1-2 to 40 hyphae form the network. The hyphae also penetrate roots between epidermal and cortical cells of feeder root and apparently replace middle lemella. Inside, fungus grows intercellularly forming inter-connecting complex network within intercellular spaces between cortical cells. Such a mycelial network is called "Hartig net". This provides a large surface area for interchange of nutrients between host and fungus. The extra-radical hyphae also develop and strands of hyphae or rhizomorphs may radiate from fungus and ramify into the soil and to the base of fruiting bodies. External modification of short roots (morphological changes) is accompanied with physiological changes in plant. ECM infected root system consists primarily of short stubby white rootlets with a characteristic 'Y' shape. These Y shaped rootlets provide visible evidence of mycorrhizal infection. Ectomycorrhizal roots are devoid of root hairs and are covered by a layer or sheath resembling a host tissue. This layer is called "pseudoparenchymatous sheath". The ectomycorrhizae remain active for period ranging from several months to three years.

The fungus survives in soil or roots as sporocarp, sclerotia or basidiocarp or as rhizomorph (hyphal strands) segment or as vegetative propagule. Rhizomorph are well-differentiated multi-hyphal organs with a diameter upto 200 µm and are important for solute transport over longer distances. The solute interchange between host and fungus takes place at hartig net. The colour of a symbiont may be brown, black, white, red, yellow or blend of these colours which represents the characteristic colour of fungus itself. Externally ECM may appear as single unforked root, bifurcate root, multi-forked (coralloid) root, nodular root or in any other configuration. This type of association is characteristic of many important forest and horticulture plant species, members of which are distributed among gymnosperms (especially the Pinaceae family) as well as in angiosperms (in families like Betulaceae, Fagaceae, Juglandaceae, Ceaslpinioideae, Myrtaceae, Dipterocarpaceae, Tiliaceae, Rosaceae, Salicaceae, etc.).

The surface area of mycorrhizal roots is almost 10 times greater than uninfected ones. Under normal conditions fine fungal hyphae extend up to 8 cm into the surrounding soil so increase nutrient absorption and mobilization.

Ectomycorrhizal fungi produce some auxins that are responsible for some morphological differences between mycorrhizal and uninfected roots. Mantle serves as a storage organ in ECM. ECM have poor competitive saprophytic ability. Many ECM fungi are facultative symbionts which can live independently in soil and can be cultured easily.

10.1.3 ECTENDOMYCORRHIZAE

Ectendomycorrhizae share some qualities both from ectomycorrhizae and endomycorrhizae. They are septate fungi belonging to class Ascomycota or Basidiomycota. Fungal hyphae enter the host cortical cells, coil and form Hartigs net around the root. Haustoria and vesicles are absent. Fungal sheath may or may not be present. Such mycorrhizae are found in both gymnospermic and angiospermic plants. It is believed that when defense reaction of higher symbionts decreases or diminishes, likely to happen in diseased or senescent trees, lower symbionts become endotrophic. Such instances are designated as ectendotrophic.

10.1.4 ERICACEOUS MYCORRHIZAE

Ericaceous mycorrhizae include arbutoid, ericoid and monotropoid. These fungal associates in all cases are septate and hyphae enter the host cell. However, fungal sheath and Hartig net are absent in ericoids and hyphal coils are not formed in monotropoids. Haustoria are absent in all cases. The fungal associates, in general, belong to class Basidiomycota, with some exceptional ericoids belonging to class Ascomycota.

Arbutoid mycorrhizae are formed as a mantle of a few to about 90 μm thickness. The mantle may serve as storage organ. The Hartig net penetrates the outer layer of cortical cells. The intracellular hyphae formed disintegrate within the cell. The short roots of *Arbutus* trees are infected extensively. Such mycorrhizal associates are found in *Rhododendron* and *Arctostaphylos* (manzanita).

Ericaceous plants like *Calluna* (heather) and *Vaccinium* (blue berries) are characteristics of acid or peat soils. Ericoid mycorrhizae are reported to increase the uptake of soil organic nitrogen in presence of labeled-NH_4^+. Plants in monotropaceae are infected by basidiomycetous fungus (*Boletus*) alongwith the roots of neighbouring trees. The *Monotropa* seeds are very small and germinate in soil, but without further development, until infected with *Boletus*. The carbohydrates and other nutrients are derived by the germinating plant from neighbouring plants through mycorrhizae.

10.1.5 ORCHIDACEOUS MYCORRHIZAE

Endotrophic mycorrhizal associations are found in all orchid family although a few genera are capable of growing with a fungal partner. The orchid species of genera *Epipogon, Coralliorhiza, Galeola, Neottia, Limodorum, Gastrodila, Didymoplexis* and *Vanilla* are entirely dependent on their fungal partners for nutritional requirement in juvenile stage of growth since most of these have little or no chlorophyll. Like *Monotropa,* when orchid seeds germinate they become infected by hyphae of mycorrhizae in soil and then entirely depend on fungal partners right from early seedling stage. After penetration fungus appears in cortical cells in coiled form. Intracellular hyphal coils have a limited life even in stable mycorrhizal associations of orchid. The hyphal disintegration within a cell (also known as tolypophagy or ptyophagy) occurs as early as 30-40 hours after infection and is usually completed within 11 days. This contributes to the nutritional requirement of the host. The same host cell may again get re-infected and the process continues. The infection is restricted to the cortical cells or root or to the portion of plant acting as root system. The fungal species involved are either phycomycetous or basidiomycetous or deuteromycetous in nature. Mycorrhizal associates of orchids belong to class Basidiomycota (*Armillaria, Fomes, Marasmium, Corticium,* etc.) or Deuteromycota (*Rhizoctonia*). Many of these fungi are lignin and cellulose degraders so are organic matter decomposers also. The orchids are dependent on fungus for its carbon requirements in its early stages of establishment.

10.2 MYCORRHIZOSPHERE

The changes in root exudates, excreted from a mycorrhizae infected root, lead to the development of a new microbial equilibrium which is distinct from those present in non-mycorrhizal roots. High level of bacterial and actinomycete populations have been observed throughout the mantle of mycorrhiza. Many amongst them are antagonistic to root pathogens. Increase in fungitoxic activity in plant rhizosphere is caused by the synthesis of antifungal compounds (as in case of *Paxillus involutus*). The synthesis of these compounds is stimulated by the presence of *Pinus resinosa* root exudates. The term 'Mycorrhizosphere' describes enhanced microbial population of selected microorganisms near fungal structure in soil. In comparison to general rhizosphere, in mycorrhizosphere all such cycles are controlled by both input and soil humus contents. Moreover, in mycorrhizosphere nutrients are incorporated into fungal biomass and then transported to plant tissues. In contrast, in non-mycorrhizal rhizosphere nutrients mineralized are directly available to the plants and microbes. Further, microbial biomass synthesis is controlled by plant productivity and the rate of root exudate production. In mycorrhizosphere microbial biomass production is controlled by plant productivity as well as by the transfer of photosynthate to the fungus. Nutrients may be accumulated from non-rhizosphere soil by the fungus and transported through rhizosphere into plant roots.

The extra-metrical hyphae of mycorrhizal fungi release substances that cause soil and organic fractions to aggregate. Mycorrhizae induce stabilization of sand dunes. The changes in mycorrhizosphere or mycosphere cause alteration in functional groups, which indirectly affect host-pathogen relationship. These functional groups include extracellular chitinase producers, phosphate solubilizers, siderophore producers, antibiotic or hormone secretors, pathogen suppressors, exopathogen, mycorrhizae suppressors, vitamin producers or hypersensitivity reaction inducers.

10.3 PHYSIOLOGY OF MYCORRHIZAE

Carbon flow : The ectomycorrhizal fungi generally do not secrete cellulolytic or lignolytic enzymes, therefore, fungal associates are unable to utilize complex carbohydrates from the soil. The fungus through symbiosis uses simple carbohydrates, amino acids and vitamins etc. from their host plants to complete life cycle. In turn, mycorrhizae provide some relief to the host plant. Experiments with ^{14}C labeled sucrose, fructose, glucose, etc. and $^{14}CO_2$ have shown that carbon requirement of fungus comes from host plant in the form of root excretions. Host plants (except orchids) supply carbon to mycorrhizal symbiont for its growth. The ectomycorrhizae obtain most of its carbon from host in the form of sucrose which is hydrolyzed extracellularly by the fungus to glucose and fructose. These sugars are translocated from the root via Hartig net to the fungal mantle, where they tend to accumulate in the form of typical fungal carbohydrates viz., trehalose, mannitol and glycogen. The host plant can not readily metabolize these carbohydrates. This maintains unidirectional flow of sucrose from host plant to fungus. The accumulated sink of carbohydrates, stored by mycobiont, is mobilized by the fungus to produce flushes of their large fleshy basidiomata near the tree. In mycorrhizal roots a good amount of net photosynthates allocated to the roots is required for fungal growth and maintenance. In mycorrhizal plants the fungal symbiont competes effectively for carbohydrates with host root and between 10-30% of the carbohydrates allocated to the roots might be used by the fungus.

It is not clear whether ectomycorrhiza are able to degrade cellulose or lignin, however, endomycorrhiza are reported to do so by converting plant carbon to glycolipids. In VAM plants root respiration may be 20-30% higher than non-mycorrhizal plants and 87% of higher respiration is attributed to fungus. In general, the proportion of photosynthate allocation to fungus in ectomycorrhiza associations is higher than VAM association.

Phosphate translocation : The translocation of phosphorus in mycorrhizae is through cytoplasmic streaming mechanism. The solute transfer in hyphae is bidirectional involving carbohydrate in exchange of phosphate and other minerals. In mycorrhizal network polyphosphates make up a significant proportion of organic phosphorus which are translocated by streaming action. Both ecto- and endo-mycorrhiza store phosphorus in their cells in the form of polyphosphates. High affectivity of VAM hyphae in phosphorus uptake is caused by their small diameter, large surface area and accumulation of polyphosphates within vacuoles. Polyphosphates are strongly negative-charged poly-anions, important in maintaining the cation/ anion balance of the fungus. These also bind cations (such as Mg and K) and amino acids (such as arginine and glutamine). The insolubility of polyphosphates maintains low internal phosphorus concentration thus allows continuous phosphate uptake from the soil. The transfer of nutrients to plants is across arbuscules in VAM, through coiled hyphae in orchids and ericoids, and through Hartig net in ectomycorrhizae. Close association of fine fungal structures and specially developed plasmalemma of plant cell serve as sites for nutrient transfer. Sheaths and Harting net in ectomycorrhizae persist for 9-14 days whereas VAM and fine coiled hyphae have turnover time of 4-15 days.

10.3.1 ROOT EXUDATES FOR ROOT INFECTION

Recognition of root by VAM usually occurs at the appresoria stage, but these structures are not always associated with infection. Root infection is initiated either by soil borne propagule or by the hyphae from a neighbouring root. ECM mycorrhization occurs when an ECM encounters a feeder root of a hospitable tree species. Colonization occurs only in a specific unsuberized root zone just behind the root tip but prior to the region in which primary cortex has begun to deteriorate. The root tip are not invaded but covered by fungal mantle. The mantle then spreads along the root, rapidly envelops any emerging short root and sends hyphae into their cortex to set up the vital Hartig net interface.

Root exudates of host plant have strong chemotactic action on mycorrhizae. The affectivity of flavonoids in exudates is enhanced by the elevated CO_2 concentration prevailing in the surrounding micro-environment. The stimulation of VAM infection has been obtained by *Azospirillum* inoculation and in case of ECM (e.g. *Laccaria laccata*) with 'mycorrhization' helper bacteria. The synergistic effect of dual inoculation between VAM and *Rhizobium* is well documented and generally there is an increase in nitrogen, sulphur and calcium uptake by plants. In non-host plants of VAM such as the members of chinopodiaceae and cruciferae incompatibility seems probably due to the composition of root exudate, toxins, host reaction to infection, etc. In host plant species, such as red clover, two superoxide dismutase isoenzymes are induced by VAM infection, presumably in response to infection and elevated O_2^- level. Large varietal differences, in some cases, may involve variable VAM response as observed in cowpea and wheat where it varies between 18 and 30%. At extremely low soil phosphorus, the root infections with VAM and ECM fungi are low because of phosphorus limitation for the fungal growth itself. High nitrogen depresses mycorrhizal infection. Enhancement of root infection of short leaf pine seedlings with ECM, *P. tinctorius,* has been achieved with boron supply at levels exceeding the host demand.

10.3.2 MYCORRHIZA INDUCED PHYSIOLOGICAL CHANGES IN HOST PLANTS

Changes in root morphology : The growth of both ecto- and endo-mycorrhizal fungi is stimulated by the excretion of host root exudates i.e. organic acids, sugars, vitamins, amino acids and other compounds into the soil. Upon germination and invasion of fungus into the root cell, the host root morphology and physiology changes. The mycorrhizal roots in comparison to non-mycorrhizal ones are distinct in colour (because of the presence of fungal mantle), thicker and branched (often pinnately and racemosely). The cell volume increases and weight of plant cytoplasm within a mycorrhizal plant cell is about 20 times greater than that of uninfected one.

In fibrous VAM-infected root system mycorrhizae contribute about 3% of root weight whereas the contribution in VAM infected legume is 12-16%. In highly infected plants, VAM fungal biomass may reach upto 20% of the root biomass (typically 10%). The mycorrhizal plants compensate the higher demand of photosynthates by increase in the rate of photosynthesis per unit leaf area. More than 50% forest and herbaceous photosynthates are directed to the below-ground parts to support root mycorrhizae and associated microbes and half of it is respired. About 4-14% carbon photosynthesized by plant is utilized by fungal symbiont alone. In turnover of fine roots, fungal components contribute about 50% of the total turnover.

Photosynthate partitioning: Root elongation is inhibited and lateral root/ feeder root formation stimulated by fungal auxins. The flow of growth regulating compounds and other chemical constituents in host root cell also changes. This induces increase in the photosynthetic rates *vis-à-vis* change in the partitioning of photosynthates (carbohydrates) between shoot and root. There is dramatic increase in the flow of carbohydrates from leaves to roots. These physiological alterations coupled with improved mineral uptake from the soil enhance the nutrient status of host tissue. Moreover, there is change in cell membrane permeability as well as improvement in the quality and quantity of root exudates. The changes in phosphorus nutrition due to mycorrhizal symbiosis alter membrane permeability in such a way that root exudates are also affected. As root phosphorus content increases the membrane permeability is reduced and more carbon is allocated to mycorrhizal fungi in the roots. This results in less exudation out of root particularly of amino acids.

Changes in microbial biomass : The changes in root exudates enhance the quality and composition of soil microbial biomass as well as the rate of organic matter decomposition in mycorrhizosphere. There is alteration in the activities of microorganisms that respond to qualitative changes in root metabolites in and around roots. The qualitative change in microbial composition favours greater nutrient mineralization in soil. The extra-radical hypha increases the potential of root system for better nutrient acquisition, water absorption and selective ion uptake and accumulation. It also improves plant survival, growth and development, nutrient cycling, feeder root health, longevity and tolerance to abiotic stresses like drought, soil toxins, extremes of pH, temperature, etc. Besides, soil deterioration is minimized through improvement in soil structure and aggregation. The net effect of these changes is a healthier plant, capable to withstand harsh environments, and better plant growth. Tripartite associations of mycorrhizae-*Rhizobium*-legume are synergistic than bipartite associations of either mycorrhizae or *Rhizobium*.

10.4 FUNCTIONS OF MYCORRHIZAE

Mycorrhizal symbionts have many roles in plants which can be categorized as :

a. Nutrient mobilization and uptake;
b. Protection from pathogens and nematodes;
c. Tolerance to abiotic stresses;
d. Production of growth hormones; and
e. Soil conservation

10.4.1 NUTRIENT MOBILIZATION AND UPTAKE

Nutrient mobilization, particularly those of phosphorus and zinc, in plants is accelerated by mycorrhizae through various mechanisms.

Nutrient extraction beyond depletion zone : As a result of morphological changes in roots, the area of physical contact between roots and soil is increased. There is almost ten-fold increase in the absorptive root surface area. Since plant root can extract nutrients from a limited distance particularly that of phosphorus only up to 2 mm away from root, therefore in rhizosphere nutrient depletion zones develop in the vicinity of roots. However, mycorrhizal roots under normal conditions are capable of extracting nutrients beyond such depletion zones up to 15 cm. In non- mycorrhizal plants depletion zone extends not more than 1 cm away from the rhizosphere and phosphorus is uniformly depleted in hyphal compartments (2 cm away from rhizoplane). At hyphae-soil interface new depletion zones develop, extending several millimeter into the bulk soil. Of the total phosphorus hyphae contribute 70-80% in mycorrhizal plants.

Improved proton-extrusion mechanism : The required minimum concentration of phosphorus in soil solution for uptake by root is >0.1 ppm whereas mycorrhizae are capable of mobilizing phosphorus uptake at extremely low phosphorus concentration of <0.001 ppm in soil. A concentration gradient is formed in such case resulting in the dissolution of phosphorus. The mechanism of uptake of nutrients is also greatly increased by the soil volume explored by the fungus. Reports indicate that mycelial network can acquire nutrient within soil over a distance of 40 cm away from the roots. The hyphal network of mycorrhiza extends the plant's working area and provides about 10 times as much absorption surface as the root system of an uninfected plant, thereby increase plant root efficiency. There is also an increase in nutrient uptake efficiency per unit surface area through improved proton-extrusion mechanism and organic acid (such as oxalate, acetate, fumerate, etc.) production. The average diameter of external VAM hyphae is in the range of 1-2 μm. Also, the total length of external hyphae may be in the range of 1-10 m cm^{-1} infected root length in VAM and 1-3 m cm^{-1} in ECM. Only a small proportion of external hypha is active.

Large physical exploration : Most of the organic residues are solubilized or protected within soil aggregates by both physical and chemical binding mechanisms. Mycorrhizae through their physical exploration via external hyphal network scavenge and retain nutrients for longer periods. Moreover, it composes a rather large carbon sink and nutrient pool in nutrient cycling within rhizosphere. Fungal biomass accounts 45-55% of the total soil biomass. External hypha composes 26% of the extra-radical organic carbon pool. Reports indicate VAM hyphae amount 12-19 m g^{-1} soil or 17-45 m cm^{-3} soil or 2.6-54.0 km kg^{-1} soil. Such type of extensive physical entanglement contributes towards soil macro-aggregate formation as well as influences the residence time of nutrients within the system. This also minimizes loss of soil (nutrients) via erosion. Moreover, the soil matrix around hyphal network gets chemically altered and this helps in greater access of roots to nutrients. As a rule, in mycorrhizal plants per unit root length is 2-3 times higher than non-mycorrhizal plants. The physical network links the roots of two or more plants thereby facilitates direct transfer of nutrients from one root system to another or one ecosystem to another.

Increase in growth promoting microorganisms in mycorrhizosphere : The mycorrhizal associations induce increase in plant growth promoting rhizobacteria and beneficial microbes, thus improve organic matter decomposition, mineralization processes and other enzyme activities in soil. There is also increase in siderophore production thus enhancing the chelation of iron and increasing its bioavailability. In addition, mycorrhizae synthesize plant growth promoting hormones such as auxins, indole acetic acid, ethylene, gibberellins and cytokinins. Increased cytokinin increases root nodulation in legumes by facilitating phosphorus utilization from soil since legumes are poor competitors for soil phosphates. Mycorrhizal plants exhibit higher nitrogenase and nitrate reductase activities and also influence stomatal regulation.

Phosphatase activity : In soils only 0.1% of total phosphorus is available and rest remains fixed in clay lattice, etc. The phosphorus is sparingly soluble and least mobile. It shows very limited diffusion in soil. Hardly 15% of the applied fertilizer phosphorus is utilized by crops. About 50-67% of the total phosphorus supply is organically bound and cannot be utilized without solubilization. The mycorrhizae through their enzymatic activity such as acid phosphatases, etc. solubilize the organic phosphorus into orthrophosphate and scavenge them as polyphosphates. Acid phosphatase is an ectoenzyme and its activity is high all along the external mycelium and at the surface of the sheathed roots. The polyphosphates are sequestered within fungal vacuoles and accumulate in plant roots even when plants are under severe phosphorus limited conditions. The phosphates are then transferred to cytoplasm from hyphae. The fungal hyphae have greater affinity for phosphorus than host roots. About 80-90% of absorbed phosphates remain accumulated in fungal sheaths. So fungal sheath serves as phosphate reserviour to release phosphorus at deficiency or crisis times and provides a steady supply of phosphorus to plant when transport to roots is reduced e.g. during the period of drought. Increased phosphatase activity of mycorrhizae induces greater hydrolysis of soil organic compounds. Organic phosphates in soil are also easily decomposed by mycorrhizal fungi. The phytate phosphorus is taken up by mycorrhizae. The phosphates have low diffusivity and the mycorrhizal association benefits the plant in getting phosphorus even at the lowest solution concentration of 0.1 $\mu g\ g^{-1}$ in case of soybean.

Selective ion absorption : Mycorrhizal plants show increased nutrient uptake through selective ion absorption and accumulation. Thus mycorrhizal association, besides phosphorus, not only enhances N, K, S, Fe, Mg, Ca, Mn, Cu, Zn and Mo uptake from soil but relatively rapid turnover of mycelial fungus also leads to the concentration of nutrients in root region. Absorption of immobile elements in plants is increased by as much as 60 times by the symbiosis. Uptake of NH_4^+ is facilitated by ectomycorrhizae in forest trees. Zinc and copper uptake is increased under deficient conditions. In contrast, manganese is often much low in VAM plants probably due to the VAM-induced changes in rhizosphere microorganisms in general and manganese-reducers in particular vis-a-vis increase in manganese oxidizers. Low calcium content in mycorrhizal plants is probably related to the changes in root morphology and differentiation e.g. enhanced lignification and suberization of endodermis upon VAM infection. Similarly, fungal mantle of ECM can store inorganic nutrients like ammonium, phosphates, etc. and release them during deficiency period or in active phase. Nutrient cycling activities like organic matter decomposition, mineralization and elemental fixation are enhanced through boosting the growth of beneficial micro-organisms in mycorrhizosphere.

Organic acid production : The production of organic acids solubilize iron, aluminum and phosphorus in soil and precipitate them with calcium, thereby accelerate their mobility as well as mineral weathering and increase mineralization process. Moreover, acid production (formic acid, acetic acid, oxalic acid, etc.) by mycorrhizae lowers the soil pH. The acidification thus created enhances the mobility of calcium-bound phosphates and elevates CO_2 production by roots.

10.4.2 PROTECTION FROM PLANT PATHOGENS AND NEMATODES

The soil borne pathogens and symbionts share common habitat and show differential influence on host plant. Evidences support the view that symbiotic association induces tolerance against pathogens and nematodes and contributes to the suppression of root diseases. Mycorrhizal symbionts offer increased resistance to certain wilt and root rot diseases caused by *Fusarium, Pythium, Phytophthora, Verticillium, Rhizoctonia, Thielaviopsis, Olpidium, Macrophomina, Sclerotium,* nematodes, etc. These feeder root pathogens infect immature and meristematic cortical tissues of roots and cause necrosis. Mycorrhizae confer some cross protection to plants against several pathogenic fungi. Well-formed mycorrhizal roots are resistant to infection and non-mycorrhizal ones are prone to diseases. A selective pressure is usually exerted by mycorrhizae on soil microorganisms, some of which can antagonize root pathogen. Sporangium and zoospore production by *Phytophthora cinnamomi* was reduced in presence of the leachates from VAM plants. Common root rot of wheat caused by *Bipolaris sorokiniana* is inversely related to the root colonization with VAM. *Pisolithus tinctorius* and *Suillus granulatus* increase the survival rate of pine seedlings exposed to the attack of *Rhizoctonia solani* and *Phytophthora cinnamomi*. Resistance against root pathogens is accomplished through various mechanisms :

a) **Mechanical barrier :** Pathogens find it difficult to penetrate fungal mantle for root infection due to physical protection of feeder roots by forming a protective barrier in the form of mantle, as in case of ECM. Further, mycorrhizal plants are more healthy and capable of resisting the pathogenic attack.

b) **Chemical barrier :** Synthesis of toxic substances, analogous to phytoalexins in host-pathogen interaction, antimicrobial and polyphenolic compounds, stimulants or phytotoxins is found associated with the root tissue colonized by fungal symbionts. The orchid mycorrhizae synthesize orcinol, coumarin, hircinol and other phenolic compounds as defense reaction to the presence of root pathogens. The increased synthesis of phenolic compounds and release of volatile terpenes and sesquiterpenes (ethanol, isobutanol, isoamyl-alcohol, acetoin, isobutyric acid, etc.) produced by *Boletus varigatus* inhibit root pathogens like *Phytophthora cinnamomi* and *F. mesannosus*.

c) **Physiological and morphological alteration :** Root physiological and morphological alteration due to mycorrhizal infection causes more tilt towards induced resistance in plants. Changes in root exudation and host defense reaction and excretion of phenolic substances etc. by ECM have inhibitory effects on various pathogenic fungi. Phenolic accumulations inhibit the growth and spread of *Sclerotium* and *Fusarium* in mycorrhizal plants. Decrease in root exudation of reducing sugars is related to decrease in disease severity. Alteration in root exudation pattern reduces the chemotactic response of *Phytophthora* zoospores to root and reduces the stimulus for germination of chlamydospores and oospores. High level of amino acid production, especially arginine, and chitinolytic activity is found in mycorrhizal roots of several plant species, the former seems probably due to blockage in orinthine cycle of mycorrhizal fungi. The chitinolytic enzymes are produced by the plants for digestion of arbuscules. However, these are effective against other fungal pathogens as well. Starch content in mycorrhizal roots is decreased.

d) **Antibiosis :** Some mycorrhizae like *Lactarius, Cortinarius, Hygrophorus, Leucopaxillus,* etc. are reported to produce antibiotic substances such as diatretyne nitrile, diatretyne-3 etc. Some of these antibiotics are antifungal on *Rhizoctonia solani, Pythium debaryanum, Phytophthora cinnamomi, Fusarium oxysporum,* etc. *Boletus variegatus* produces volatile fungistatic compounds like isobutanol, isobutyric acid, etc. in pure culture. *Leucopaxillus cerelis* var. *piceina* inhibited germination of zoospores of *P. cinnamomi* at 50-70 ppb and killed zoospores at 2 ppm level. ECM infected plant cells may partly digest the arbuscules of fungus thereby plants show some selected chitinolytic activity which restrains the pathogen to invade further. Unlike ECM, VAM fungi do not interact directly with the pathogens through antagonism, antibiosis or predation but the effect is indirect physical or physiological alteration in host plant.

e) **Competition for space :** Fungal symbionts show better competitive ability for rhizosphere nutrients as compared to soil pathogens.

f) **Organic acid production :** Mycorrhizae produce various kinds of organic acids, like oxalic acid by *Paxillus involutus* against *Fusarium oxysporum.* This enhances disease suppression in rhizosphere and also helps the plant to have more favourable environment for beneficial rhizosphere microbes than root pathogens. In many cases absence of specific root exudates required by pathogens (sugars, organic acids, etc.) have been observed. Root infections by pathogenic nematodes are generally less on mycorrhizal plants than on non-mycorrhizal ones.

g) **Competition for nutrients :** Improvement in phosphorus uptake increases plant tolerance to some soil borne diseases like Verticillium wilt of cotton, Phytophthora root rot of citrus, take-all disease of wheat, etc. The fertility status of host may influence the host susceptibility to pathogenic infection and disease resistance. Nematode infected plants frequently show deficiency symptoms of N, Fe, Mg, B and Zn, and VAM fungi seem to augment this deficiency thereby reduce the chances of attack by pathogens. Moreover, mycorrhizae utilize root carbohydrates and other chemicals so reduce the attractiveness of root to pathogens.

10.4.3 TOLERANCE TO ABIOTIC STRESSES

Mycorrhizal association helps the host plant to overcome various abiotic stresses like water scarcity, drought, extreme soil and weather conditions, heavy metal, etc. Mycorrhizal tree seedling can often withstand drought conditions better than non-mycorrhizal seedling. Mycorrhizal associations may influence plant water relations directly or indirectly. Rhizomorph or extended extra-radical mycelium increases water transport to host. Evidences reveal that ectomycorrhizal fungi can grow better in solutions of higher osmotic stress than that which may plasmolyze non-mycorrhizal root hairs. Mycorrhizae belonging to Basidiomycota tolerate water stress upto -70 bars (-7MPa). For instance, *Cenococcum graniforme* is able to withstand low water potential and grows best at water potential of -1.5 MPa. Growth rate and hence colonization potential is, however, substantially reduced at much lower water potential. It is argued that VAM and ectomycorrhizal infections strongly increase water transport from soil to host plant leaves. This has been attributed to the presence of large number of hyphal entry points on the roots, greater hyphal length, greater surface area linking the roots with other plants, large exploration of soil and greater resistance/protection offered by the fungus against natural vagaries. It has been estimated that a hyphal diameter of 5 μm and a root diameter of 500 μm will lead to 1 m of hyphal length, having surface area equivalent to 1 cm of root length. Exploitation of water reserviours is accomplished beyond the root zone by mycorrhizal hyphal net work. Further, stomatal regulation is another way of VAM seedling possessing drought tolerance as found in citrus. Stomatal regulation is known to be stimulated by increased cytokinin level in plants. Sterols are also present in higher concentration in mycorrhizal plants which influence membrane permeability. Further, the ability of mycorrhizae to conserve soil and water enhances plant's ability to tolerate biotic and abiotic stresses. However, it is to be clarified that mycorrhizae are of little benefit to plants at very low water level. But at intermediate water level in soil drying mycorrhizal plants consistently maintain higher turgor pressure than non-mycorrhizal plants.

Heavy metal toxicity is reduced in VAM infected plants. Heavy metals are sequestered either in fungal structures or extra-matrical mycelium in the sheath. Heavy metals such as Zn, Pb, Ni, Cd, Cu, etc. are bound by carboxylic group present in pectic substances (hemicelluloses) of interfacial matrices between fungus and host cell. Copper, for example, can become available at phytotoxic concentrations in mine spoils but the fungal tissues that form mycorrhizal association with root system exclude copper under these situations by complexing the metal. Mycorrhizae in plants are reported to offer resistance against salinity, besides overcoming the harsh environmental conditions in soil.

10.4.4 PRODUCTION OF GROWTH HORMONES

Mycorrhizal association is reported to induce hormonal production, especially growth auxins, in plants. In ECM-infected plants elongation of short lateral roots are inhibited by mycorrhizal IAA production. *Pisolithus tinctorius* inoculation in *Pinus taeda* enhanced shoot growth and increased cytokinin content in needles although fungus itself does not produce cytokinin. Higher cytokinin has been found in flax and citrus inoculated plants. In maize inoculation with VAM increased abscisic acid by two-fold in shoots and roots compared to non-mycorrhizal plants. The increased cytokinin synthesis is attributed to increased photosynthetic activity, high transpiration, greater phosphorus uptake, improved iron transfer and better nutrient acquisition. Hormone production is also accelerated in root zone due to the boosting of beneficial microorganisms in mycorrhizosphere. The mycorrhizae produce various phytohormones such as indole acetic acid, ethylene, gibberellin, cytokinin and auxins that induce feeder root proliferation.

10.4.5 SOIL CONSERVATION

The extra-radical hypha of mycorrhizal association is an important, and much overlooked, contribution to the process of creating a stable soil aggregate structure. The hyphae of VAM fungi alongwith other saprophytic fungi and fibrous roots bind soil particles and micro-aggregates into macro-aggregates through physical entanglement as well as by their secretion of polysaccharides and organic compounds. The mechanical entanglement by extra-radical hyphae physically brings mineral particles and soil debris together thus bringing more stability to micro- and macro-aggregates through cementation by their secretions. This allows improved soil aeration and water percolation, thus accelerating biological activities in soil.

The mycorrhizosphere influences physical, chemical and biological properties of soil. Eroded soils have been reported to contain reduced number of mycorrhizal propagules. Mycorrhizal colonization imparts drought resistance and tolerance towards abiotic stress in plants thus enhances the chances of plant survival under extreme conditions, thereby does not allow the soil to be exposed to water and wind erosion. Conservation of nutrients within various ecosystems is enhanced by the formation of various semi-permanent mycelial system which form absorptive networks, inter-linking plants both at intra- and inter-specie level (Table 10.2).

Mycorrhizae, especially ECM, show marked spatial and temporal diversity. Maintenance of mycorrizal fungal diversity helps in minimizing site degradation by increased unpredictive environmental fluctuations. In forests, 3 to 5 times more organic matter is returned to soil in the form of root and mycorrhizae than is returned by the decomposition of litter. Of this, mycorrhizae account 85-95% and rest is from soluble root exudations, root caps and mucilages. The higher input in mycorrhizal plants of organic carbon into the soil has important consequence on the number, activity and distribution of soil microorganisms. It alters soil-root interface as well as the quality of rhizosphere microorganisms and their composition and activity so forming a new 'mycorrhizosphere'. Ectomycorrhizal forests characteristically show a relatively low diversity of host species but a great diversity of fungal associates than do VAM ecosystem. Mycorrhiza influence vegetation pattern within the community through extra-radical hyphal bridge between individual plants of the same or different plant species in a mixed stand. Such transfer, for example, is possible for N nutrient between root system of legumes and non-legumes in a mixed stand or in intercrop. In forest system, nutrient transfer from one root system to another is accomplished through mycorrhizal links. Hyphal strands appear to link plants directly to form a continuous transport network between the plants of same or different species. This allows the transfer of water, nutrients and carbon between plants. Interplant transfer, for both ECM and VAM systems, has been demonstrated and it is suggested that these mycorrhizal pathways facilitate nutrient conservation at ecosystem level. Mycorrhizal associations appear to influence community structure via mutualistic and competitive processes; whereas at ecosystem level, the mycorrhizal functioning/processes and composition may shift depending upon mycorrhizal vegetative pattern within the community (e.g. succession).

Temporal succession of fungal species is associated with the aging of forest stands. With the passage of time 'early colonizers' in forests are succeeded by 'late colonizers'. Such succession occurs due to the changes in resource quality and because of selection pressure. Late stage colonizers exhibit well-developed proteolytic potential. Besides phosphorus cycling, ectomycorrhizal sheath of tree contributes not less than 50% of total nitrogen input to a Douglus fir ecosystem. There are some indications that diversity of mycorrhizal species decline from natural ecosystem to intensive agricultural system.

Several ECM fungi have capacity to excrete organic acids and siderophores. These acids presumably contribute in enhancing weathering micas. *Paxillus involutus*, for example, releases a large amount of oxalic acid which dissolves sparingly soluble calcium phosphate. Production of siderophores for iron acquisition, oxalic acid production to bring calcium phosphate into soil solution and precipitation of calcium oxalate to restrict calcium uptake play key role in adoption of some host plants to calcareous soils. Siderophore production is probably also involved in enhanced weathering of geothite by *Suillus granulatus*.

Table 10.2 : Summarized mycorrhizal role

Structural organization level		Processes and phenomena
Individual mycorrhizal roots	a) Soil	i. Improvement in nutriention mobilization and uptake ii. Greater organic acid production iii. Larger hyphal exploration beyond depletion zones iv. Better soil aggregation/improvement in soil structure v. Immobilize heavy metals
	b) Plant	i. Root morphological and physiological changes ii. Selective and increased plant nutrient translocation iii. More growth hormone production iv. Improved water-use efficiency v. Protection from pathogenic invasion vi. Increased tolerance to abiotic stresses
Community		i. Higher biodiversity and versatility ii. Environmental stability iii. Competitive ability
Ecosystem		i. Nutrient retention and conservation ii. Soil conservation against erosion iii. Biogeochemical recycling iv. Positive feed back v. Ecosystem restoration

10.5 FACTORS AFFECTING MYCORRHIZAL DEVELOPMENT

The mycorrhizal development in plants is influenced by various agro-climatic and edaphic factors besides plant host. Mycorrhizae generally prefer acidic condition, good aeration and mesophilic temperatures. Growth depression can occur under stress conditions of low light intensity, low soil temperature (leading to low photosynthesis) and high level of root colonization resulting either in more carbon being respired per unit time or more carbon being transported to extra-matrical hyphae.

a) **Soil conditions :** The soil type, pH, drainage, moisture, fertility status, organic matter content, etc. influence the mycorrhizal development in plants. Organic matter improves the aeration and drainage in soil so favours mycorrhizal proliferation and seedling root growth and longevity. Alkaline pH or high fertility negatively influences mycorrhizal functioning. Light or course textured soil have been found better suiting for mycorrhizal growth. Anaerobic conditions in soil may eliminate mycorrhizal fungi which are obligate aerobes. The spores are usually dispersed through flood water and via wind driven soil.

b) **Cultural practices :** The disturbance of soil reduces the effectiveness of mycorrhizae. Tillage reduces phosphorus uptake and root colonization due to the disruption of extra-radical hyphal network and causes shift in species diversity and composition. Therefore, minimum tillage may increase effectiveness of VAM. Nutrient status is one of the key soil physico-chemical parameter controlling inoculum success. Fertilizer application, particularly phosphorus, type, rate and timing affects the seedling and mycorrhizal symbionts. High phosphorus fertilizer rates usually inhibit mycorrhizal growth whereas nitrogen deficiency or sufficiency reduces mycorrhizal colonization and sporulation. Phosphorus deficiency causes increased exudation from roots and this probably stimulates mycorrhizal infection. Certain crop rotations, as opposed to mono-cropping and periods of bare fallow, may favour the buildup of effective mycorrhizae in soil.

Lateral and vertical pruning improves the development of feeder roots and their associated ectomycorrhiza by increasing root fibrosity. Root pruning produces wounds that cause redistribution of primary carbon supply from seedling foliage to the root so increases energy supply to the feeder roots. Agricultural practices may alter the mycorrhizal population, specie-composition and colonization. Clear-cut logging or rigorous soil mixing or defoliation severely depress and delay mycorrhizal infection.

c) **Pesticide application :** The effect of pesticides on mycorrhizae is similar as on its host. The pesticides may reduce root colonization, spore number and effectiveness of mycorrhizae or indirectly by changing the host physiology. Usually killing of soil borne pathogens through use of 33% chloropicrin or 67% methyl bromide fumigation prior to ectomycorrhizal inoculation is effective for development in succession. Certain systemic fungicides such as triadimefon (bayleton) suppressed *Pisolithus* and other mycorrhizae even at low dosages. However, captan and benomyl stimulated ectomycorrhizal development so these two fungicides are recommended for the control of soil borne diseases in nurseries inoculated with mycorrhiza.

d) **Light intensity :** High light intensity promotes ectomycorrhizal development whereas intensity <20% of full sunlight reduces ectomycorrhizal production. During active photosynthesis plant roots accumulate excess soluble carbohydrates which together with root exudates are necessary for fungal symbionts to perfect root and their nutrient uptake from soil. Any decrease in light influences mycorrhizal development through alteration in host physiology i.e. photosynthate supply. Shading of nurseries, if made, to protect seedlings from scorching sun should be done for a limited time at mid day.

e) **Host specificity :** Most mycorrhizae have broad host range; however some, particularly ECM, exhibit host specificity. Mycotrophy in pines is pre-requisite for successful establishment. Roots of some plants are known to be colonized by different myco-symbionts at a time. For example, pine has been found to associate with 40 different fungi at a time whereas Douglus fir associates with about 2000 fungal species. Some plants form both ecto- and endo-mycorrhizae together. Although species of most of the field crops are colonized by mycorrhizal (VAM) fungi, members of Brassicaceae and Chenopodiaceae are considered to be non-host. Prior cropping with non-host reduces VAM fungal inoculum potential. Root colonization of cowpea planted after mustard has been observed to be 14% lower than root colonization of cowpea planted after cowpea. Similarly there has been 45% less colonization in barley-kale against barley-barley cropping. Coarser fibrous root systems have more external VAM hyphae than do fine fibrous root system.

f) **Competitiveness :** Interactions of mycorrhizae with indigenous soil microbial biomass are highly complex. Interactions with growth promoting rhizosphere microbes are additive or synergistic. However, presence of deleterious organisms reduces population, diversity and activities of mycorrhizae in soil. For example, many chytrid fungi, bacteria and actinomycetes parasitize VAM forming fungi. Mycorrhizae are physically disrupted and grazed by soil fauna like mycophagous nematodes, springtails etc. Therefore, it is essential that competitiveness needs to be reduced for successful mycorrhization. The mycorrhiza are ingested by a variety of small animals such as nematodes, mites, etc.

The fruiting bodies and sclerotia of mycorrhizal fungi are important food for larger animals so they accelerate the turnover of mycelial tissues and can disseminate mycorrhizal fungal spores.

10.6 PLANT RESPONSE TO MYCORRHIZAL INOCULATIONS

Soil phosphorus is usually a major critical limiting factor for plant growth. Plants respond better to mycorrhizal inoculation under low soil phosphorus conditions. The host genotypes and fungal strains seem to influence the response of plants to inoculants. Experimental trials conducted all over the world have shown stimulated plant growth response to both ectomycorrhizae and VAM inoculations. The VAM inoculation, in general, causes a saving of 30-50% phosphatic fertilizer without affecting the yield components. The inoculation of various *Glomus* species have shown about 20-60% increase in the yield potential of maize, wheat, barley, potato, millets, groundnut, cowpea, soybean, chickpea, beans, lucerne, clovers, tomato, chili, etc. The growth enhancement is well correlated with fungal root colonization, increased nodulation, plant biomass and nutrient uptake, etc. Increased uptake of zinc in apple, peach, maize, wheat and potato in zinc-deficient soils has also been reported. In red clover increased uptake of sulphur and greater production of cytokinin and chlorophyll has been documented. The beneficial effect of mycorrhizae is of special importance for those plants which have a course and poorly branched root system. Legumes with both short roots and short root hair show high mycorrhizal responsiveness. Studies have revealed two to four fold increase in the rhizosphere microbial biomass and soil enzymes as well as increased uptake of nitrogen in beans inoculated with *Glomus mosseae* or *G. calidonius*. Tolerance to water stress in citrus and avocado seedlings has been found in mycorrhizal inoculated ones.

Trees forming ectomycorrhizal association include most of the commercially grown horticultural and forest species in temperate and about 70% species in tropics. Mycorrhization of tree plantation has proved effective in the successful vegetation of denuded and eroded areas. Plantation of exotic pine *Pinus radiata*, oak, etc. without proper mycorrhization initially showed normal growth but within a few months failed to grow in Australia, Rhodesia, Philippines, New Zealand, etc. Even added fertilizers have shown no effect in boosting the crop growth and establishment on such sites. However, mycorrhizal supplementation/inoculation helped the plants to recover. Inoculation with specific fungus can be done at nursery stage with the help of a pure culture isolated from fruiting bodies found under/ near tree canopy. Ectomycorrhizae can be cultured at ease in laboratory as compared to VAM which are obligatory in nature. ECM normally develops within 1-3 months after the tree seeds germinate. Satisfactory results have been achieved in fungal species *Pisolithus tinctorius, Laccaria laccata, Rhizopogon luteolus, Suillus granulatus, Cenococcum graniforme, Boletus edulis, Scleroderma flavulum,* etc. in pine, eucalyptus, fir, poplar, quercus, acacia, cedar, avocado, grapes, strawberries, walnut, apple, etc. Mycorrhizal plants showed about two-fold uptake of NPK than non-mycorrhizal ones, improving plant growth and survival. A good correlation of plant growth with ectomycorrhizal colonization, plant biomass, resistance to biotic and abiotic stress, nutrient uptake and rhizosphere microbial biomass has also been reported. *Pisolithus tinctorius* has been exploited much and so far found more effective for rehabilitation of degraded sites. Stunted plant growth can be reverted to good vigorous growth when VAM or ectomycorrhizae is introduced into the site. The use of a combination of mycorrhizae and nitrogen fixing bacteria has shown synergistic effects on plant growth, vigour and yield in both leguminous and non-leguminous plants.

10.7 MYCORRHIZAL INOCULANTS FOR PLANT USE

Efficacious fungal inocula in various forms are necessary to support the varied and specific needs of a mycorrhizal management programme. The most common method of introducing the required fungus has been to use soil or humus from an established forest or nursery to inoculate the soil in which the seedlings are to be grown. This provides a mixed inoculum. However, this method does ot rule out the transfer of diseases and pests. As compared to VAM, which are obligate fungi, ectomycorrhizae are easy to culture, maintain and mass-produce.

Inoculation methods : Various methods are used to mass produce fungus and inoculate soils. These include :

- **Substrate-based moculum :** The isolated ECM fungus is grown for 3-4 months in a 2 liter container/jar containing sterilized peat-moss-vermiculite substrate saturated with a suitable liquid nutrient medium. The ECM generally grows slowly and takes several months to colonize vermiculite-peat moss mixture. The inoculum is washed in tap water to remove unsterilized substrate and then spread on nursery bed and mixed with soil to a depth of 8 to 10 cm. The nursery bed should be fumigated before seeding or seedling transplantation and inoculation by methyl bromide-chloropicrin or any other sterilant to reduce the competition from other rhizosphere microorganisms. Air-dried inocula have proved more successful with uniform mixing of fungal propagules in soil. Special machinery is used in advanced countries to incorporate ECM inoculum in soil.

- **Pot culture moculation :** Pot culture inoculum can be produced in bulk by growing maize or tomato plants from well-sterilized seeds in a sterilized soil. The roots of germinating seeds are inoculated with mycorrhizae and colonization of roots monitored at various time intervals. Application of water and nutrient is done as and when required. The root-colonized plant shoots are cut at ground level. The soil and roots mixed well. This serves as inoculum for future use.

- **Hydroponic system :** In this method, plants are grown in peat blocks that are standing in shallow nutrient flow culture. After VAM spores are produced in peat blocks. The blocks along with roots are ground up to serve as inoculum.

- **Modified nutrient film technique :** Here colonized roots are bathed by a thin film of nutrient solution allowing for greater aeration during fungal growth on plants.

- **Aeroponic culture method :** A fine mist of nutrient solution is used to colonize the roots. Contamination-free spores or fragments are effective up to 23 months. Inoculum density kept is 3.5×10^4 fungal propagules g^{-1} of dried roots.

- **Root organ culture :** Root transformation by *Agrobacterium rhizogenes* is done. This increases rate of root production and quantities of VAM propagules.

The VAM inoculum may be applied to the plants by band placement (2.5 cm below seeds in layers), seed treatment (surface pelting or coating), planting hole (burrowing into the soil and putting inoculum near the root of the tree), or side dressing methods. About 50 to 500 spores are required to inoculate a single pot. Alternatively, VAM inoculum and pregerminated seeds may be suspended in 4% methyl cellulose and applied as slurry in furrows @ 200 ml m^{-1}. Another method is to inoculate fresh fields with 25-30 kg root pieces ha^{-1}.

The ECM may be applied either by broadcast method in nursery soil or by banding the inoculum below seeds or by applying in slurry form over container-grown seedlings. Puff ball fruiting bodies, if used as inoculum source, are applied @ 5×10^8 to 10^{10} spores m^{-2} nursery bed. Another method is to use a known mycorrhiza infected soil @ 10% by volume for inoculating a new nursery bed for healthy growth of seedlings. Planting nurse seedlings that carry established mycorrhiza has also been used wherein seedlings are planted in seed bed at 1-2 m spacing. At the time of transplantation some seedlings are left behind in the nursery fields to infect next crop. The commercially produced inocula of *Pisolithus tinctorius*, under the trade name of 'MycoriMix' in Canada or 'MycoRhiz' in USA and 'Dr. Kinkon' in Japan are available.

The system used for the growth and inoculation of ECM involve large scale production of inoculum on vermiculite/peat/moss (as nutrient carriers) and application of inoculum on container grown seedlings or in the soil nursery beds. Isolated mycorrhizal mycelium is grown on modified Melin Norkan's medium (MMN) under aerated conditions at 30±2°C for 1 to 2 weeks in large fermentors. Then steam-sterilized vermiculite-peat-moss-MMN substrate is inoculated with the starter mother culture @ 5-20% of the fermentor volume, mixed well and incubated with or without agitation for 1-4 weeks, depending upon the nature and type of culture. Final inoculum is harvested by floatation on water and dried in aeromatic fluid bed dryer to a moisture content of 20-25%.

Isolation : Arbuscular mycorrhizal spores are retrieved from soil near the roots of infected plants by wet-sieving and decantation procedure and density centrifugation method. Usually such soils on an average have 50-100 spores g^{-1} dry soil in upper 20 cm soil. Composite soil samples are mixed with water (1 :4 w/v) in a measuring cylinder and stirred well for 2-3 minutes so that soil aggregates are dispersed and uniform suspension is obtained. The suspension is then decanted through sieves of different grades or meshes (mesh size 710, 530, 250, 175, 105 and 53 μm) in a series, placed in funnel, respectively. Filtrate of each sieving is collected and passed on through next lower sieve. The process can be repeated twice or thrice. The sieving and washing help in quantitative retrieval of spores on the basis of size. Root pieces and residues from 710, 530 and 250 μm sieve are examined directly under dissecting microscope for hypha, spores, sporocarps, etc. and stained under light microscope for observation. VAM spores usually range from 50 to 600 μm and are thick walled (upto 30 μm), darkly pigmented and filled with storage lipids. In floatation method vigorously shaken aqueous soil sample, prepared as above, is kept as such to allow the spores to float and accumulate at the top. Then aqueous soil suspension is gently passed through separating funnels. The spores retained on filter paper are collected by a needle using binocular stereoscopic microscope. Another method is to suspend the material retained by 175 μm sieve in water and allow it to settle down or centrifuge for 3 minutes at 3000 rpm. The suspension of spores/debris is then added to 50 ml centrifuge tube containing sucrose gradient established by placing 10 ml of 50% sucrose followed by 15 ml 25% sucrose and 10 ml water at the top with the help of a hypodermic syringe. After centrifugation for 5 minute at 3000 rpm the spores in centrifuge tube are found in the mid layer while soil etc. settles down. Spore in mid layer are carefully pipetted out in a beaker containing sterile water. The solution is decanted and spores washed several times.

VAM spore staining : VAM being obligate fungi are mass produced on some host plant in sterilized sand-soil mixture. Identified isolates of VAM spores, obtained from wet sieving, are first surface sterilized by dipping in 2% chloromine T and 200 ppm of streptomycin for 25 minutes and subsequently washed several times in sterile water. About 20 identical spores are transferred to moist filter paper towel-lets and wrapped around the roots of aseptically-grown seedlings. The seedlings with wrappings are then individually planted in sterilized sand + soil mixture in small containers. After 60-75 days, a portion of rhizosphere is examined by wet sieving procedure to determine whether VAM fungus has established in root region or not. The bulk samples of spores can be obtained from wet sieving and transferred to suitable potting substrates in pots containing desired hosts. Usually grasses or millets like pearl millet, sorghum, maize, etc. have been used for mass production of VAM. The VAM-grown soil and host are examined for VAM and gently macerated in sterile container and used for inoculation purpose @ 5-10% of potting mixture to about 2-5 cm below soil level and watered periodically. Seed pelting/coating is also adopted.

For quantification, roots from four plants of each treatment are cut into 1 cm segments. About 100 segments are taken and total infected roots counted. VAM colonization of fine roots is estimated by gentle washing of roots in tap water followed by immersion in 2.5% KOH for 30 minutes at 90°C. The roots are then rinsed in tap water and bleached in alkaline H_2O_2 for 30 minutes followed by immersion in 1% HCl for 1-24 hours. Stained roots are then put in 0.05% acidic glycerol-trypan blue in lactophenol for 3 minutes at 121°C or 30 minutes at 90°C. The stained roots are immersed in lactophenol overnight to destain cortical cells. Stained specimens are examined under microscope in lactic acid-glycerol (1 :1) medium when only hyphal strands and spores retain stain.

In comparison to VAM, ectomycorrhizae easily grow on synthetic media. Many ectomycorrhizal fungi do not grow on routine fungus culture media so require some selective media.

Mycorrhizal research has great scope in crop production because of its multiple benefits even under adverse conditions. There are indications that mycorrhizal system has much to offer for enhanced growth of target hosts in wastelands, reclaimed desert-lands, deforested areas and low productive sites. Genetics of mycorrhizae is poorly understood. Genetic engineering of mycorrhizae may, of course, be helpful in improving the inoculum potential and plant host response. If mycorrhizal association is to be exploited, a better understanding of source-sink relationship, especially of the carbon costs, is needed. Moreover, development of mycorrhizae in culture is essential to economize its use. Greater understanding of basic soil ecology and mycorrhizal interactions alongwith genetic improvement may in long run prove a valuable asset for greater crop production.

Selected References

Azcon-Aguilar, C., Barea, J.M. 1997. Applying mycorrhizae biotechnology to horticulture - Significance and potentials. *Scientia Hort.*, 68 : 1-24.

Bethlenfalvay, G.J. and Linderman, R.G. (eds.). 1992. *Mycorrhizae in Sustainable Agriculture.* Am. Soc. Agron.Inc.,/ crop Sci. Am. Soc./ Soil Sci. Soc. Am. Inc., USA.

Charoo, G.M., Dar, G.H., Beigh, G.M. and Zargar, M.Y. 2003. Impact of co-inoculation of *Glomus mossae* and *Rhizobium leguminosarum* on plant growth, nutrient uptake, root colonization and rhizosphere microbial biomass in common beans. *Applied Biological Research,* 5 : 35-41.

Dar, G.H., Zargar, M.Y. and Beigh, G.M. 1997. Biocontrol of *Fusarium* root rot in common bean (*Phaseolus vulgaris* L.) by using symbiotic *Glomus mosseae* and *Rhizobium leguminosarum. Microbial Ecology,* 34 : 74-80.

Dar, G.H., Beig, M.A. and Ganai, N.A. 2007. Growth response of kail pine (*Pinus wallichiana*) to various ectomycorrhizal fungi. *Ind. J. For.,* 30: 263-266.

Davis, R.M. and Menge, J.A. 1982. Influence of *Golmus fasciculatus* and soil phosphorus on *Phytophthora* root rot of citrus. *Phytopathology,* 70; 447-452.

Fitter, A.H. 1991. Characteristyics and functions of root system. pp. 3-25. In : *Plant Roots : The Hidden Half* (eds. Y. Waisel, A. Eshal and U. Kafkafi). Marcel Dekker, New York, USA

Graham, J.H., Leonard, R.T. and Menge, J.A. 1981. Membrane mediated disease in root exudation responsible for phosphorus inhibition of vesicular arbuscular mycorrhiza formation. *Plant Physiol.,* 68 : 548-554.

Harley, J. L. 1969. *The Biology of Mycorrhizae* (2nd edn.). Leonard Hill, London, UK.

Harley, J.L. and Smith, S.E. 1983. *Mycorrhizal Symbiosis.* Academic Press, London, UK.

Harris, R.F. 1980. Effect of water potential on microbial growth and activity. pp. 23-95. In: *Water Potential Relation in Soil Microbiology* (eds. J.F. Parr, W.R. Gardner and L.F. Elliot). Soil Sci. Soc. Am., Madison, Wisconsin, USA.

Hung, L.L. and Sylvia, D.M. 1988. Production of vesicular arbuscular mycorrhizal fungus inoculum in aeroponic culture. *Appl. Environ. Microbiol.,* 54 : 353-357.

Hayman, D.S. 1982. Practical aspects of vascular-arbuscular mycorrhizae. pp. 325-373. In : *Advances in Agricultural Microbiology* (ed. N.S. Subba Rao). Oxford & IBH Publ. Co., New Delhi, India.

Hekstra, A. 1996. *Sustainable Nutrient Management in Agriculture : A Simulation Programme for Calculating Field Level Plant Nutrition Recommendation.* Stiching Mondiaal Alternatief, Amsterdam, the Netherlands.

Krishna, R.R. 1981. *Studies on the Mechanism of Improved Plant Growth due to Vesicular Arbuscular Mycorrhiza.* Ph. D. thesis, University of Agricultural Sciences, Banglore, Karnataka, India.

Li, X.L., George, C. and Marschner, H. 1991. Extension of the phosphorus depletion zone in VA mycorrhizal white clover in a calcarious soil. *Pl. Soil,* 136 : 41-48.

Malajcznk, N. 1983. Microbial antagonism in *Phytophthora* . pp. 197-218. In : *Phytophthora: Its Biology, Taxonomy, Ecology and Pathology* (eds. D.C. Erwin, S. Bartniks-Garcia and P.H. Tsae). The American Phytpathological Society, St. Paul, Minnesota, USA.

Marschner, H. 1995. *Mineral Nutrition of Higher Plants* (2nd edn.). Academic Press, New York, USA.

Marx, D.H. 1972. Ectomycorrhizae as biological deterrents to pathogenic root infections. *Ann. Rev. Phytopath.,* 10 : 429-454.

Mercy, M.A., Shivashankar, G. And Bagyaraj, D.J. 1990. Mycorrhizal colonization on cowpea is host dependent and heritable. *Pl. Sci.,* 121 : 292-294.

Mitchell, R.J., Garrett, H.E., Cox, G.S. and Atalay, A. 1990. Boron and ectomycorrhizal influence on mineral nutrition of container-grown *Pinus ehinata* Mill. *J. Pl. Nutr.,* 13: 1555-1574.

Molina, R. and Trappe, J.M. 1982. applied aspects of ectomycorrhizae. pp. 305-324. In : *Advances in Agricultural Microbiology* (ed. N.S. Subba Rao). Oxford & IBH Co., New Delhi, India.

Mosse, B., Stribley, D.P. and Le Tacon, F. 1981. Ecology of mycorrhizae and mycorrhizal fungi. *Adv. Microbial Ecol.,* 5 : 137-211.

Palma, J.M., Longa, M.A. del Rio, L.A. and Arienes, J. 1993. Superoxide dismutase in vesicular arbuscular mycorrhizal red clover plants. *Physiol. Plant.,* 87 : 73-83.

Perrin, R. 1990. Interaction between mycorrhizae and disease caused by soil borne fungi. *Soil Use Mgmt.,* 6 : 189-195.

Powell, C.L. and Bagyraj, D.J. (eds.) 1984. *V A Mycorrhizae.* CRC Press Inc., Boca Raton, Florida, USA.

Read, D.J. 1991. Mycorrhizas in ecosystem. *Experientia*, 47 : 376-391.

Sanders, F.E., Mosse, B. and Tinker, P.B. (eds.). 1975. *Endomycorrhizae*. Academic Press, London, UK.

Schenck, N.C. (ed.). 1982. *Methods and Principles of Mycorrhizal Research*. Am. Phytopath. Soc., St. Paul, Minnesota, USA.

Schenck, N.C. 1981. Can mycorrhizae control root diseases? *Plant Dis. Reptr.*, 6 : 230.

Shah, T.A., Zargar, M.Y. and Dar, G.H. 1999. Interaction of *Rhizoctonia solani* with ectomycorrhizal inocula on deodar (*Cedrus deodara*). *Applied Biological Research*, 1 : 103-107.

Sharma, Mamta and Mukerji, K.G. 1992. Mycorrhiza - Tool for biocontrol. pp.52-80. In : *Recent Developments in Biocontrol of Plant Diseases* (eds. K.G. Mukerji, J.P. Tewari, D.K. Arora and Geeta Saxena), Aditya Books Pvt. Ltd., New Delhi, India.

Sieverdind, E. 1991. *Vesicular- Arbuscular Mycorrhiza Management in Tropical Agrosystem.* Deutsche Gesellschaft fur Technische Zusammerarbeit, Eschborn, Germany.

Skinner, M.F. and Bowen, G.D. 1974. The penetration of soil by mycelial strands of ectomycorrhizal fungi. *Soil Biol. Biochem.*, 6 :57-61.

Thompson, J.P. and Wildermuth, G.B. 1989. Colonization of crop and pasture species with vasicular-arbuscular mycorrhizal fungi and a negative correlation with root infection by *Bipolaris sorokiniana. Canadian J. Bot.*, 69 : 687-693.

Tinker, P.B., Jones, M.D. and Durall, D.M. 1992. A functional comparison of ecto- and endo-mycorrhizas. pp. 303-310. In : *Mycorrhizas in Ecosystems* (eds. D.J. Read, D.H. Lewis, A.H. Fitter and I.J. Alexander). CAB International, Wellingford, UK.

Utkhede, R.S. 1991. Biological control of *Phytophthora* on fruit trees. pp. 1-16. In : *Recent Developments in Biocontrol of Plant Diseases* (eds. K.G. Mukerji, J.P. Tewari, D.K. Arora and Geeta Saxena). Aditya Books Pvt. Ltd., New Delhi, India.

Wlicox, H.L. 1991. Mycorrhizae. pp. 731-765. In : *Plant Roots : The Hidden Half* (eds. Y. Waisel, A. Eshal and U. Kafkafi). Marcel Dekker, New York, USA.

Zak, B. 1964. Role of mycorrhizae in root diseases. *Am Rev. Phytopathol.*, 2 : 377.

Zargar, M.Y., Dar, G.H. and Zargar, A.H. 2000. Mycorrhizae in relation to ecosystem development. pp. 465-478. In : *Environment, Biodiversity and Conservation* (eds. M.A. Khan), A.P.H. Publ., New Delhi, India.

— *** —

Chapter 11

Soil Microbial Interactions with Plants

Man does not know the tiny roster of microorganisms in a single tiny lump of soil, let alone their complex interactions with each other and with their physical and chemical environment.

(Baker and Cook)

Soil is a living entity having dynamic physico-chemical and biological properties. Besides supporting the growth of various biological communities, it also serves as a medium on which the organic residues, in the form of dead plant and animal or their excretions, are degraded and decomposed. The biological properties to soil are endowed through the interactive activities of plant root, soil microbes and soil animals. The soil organisms live in a constant state of competition and have to struggle hard for their survival. In this struggle some species become dominant and some remain sub-dominant and others perish or remain in very low profile. So a sort of dynamic equilibrium is the net result of wide range of interactions occurring both at organism and population levels amongst different components of soil biota. Many interactions are short-lived and sporadic and others are more stable.

11.1 TYPES OF INTERACTIONS

Most associations between soil organisms can be considered as symbiotic because the term describes the living together of various organisms in an intimate association and not any special situation where they are mutually benefited. Symbiosis in soil varies with respect to the degree of intimacy (ecto- and endo-mycorrhizae, lichens, etc.), the balance of advantage (mutualistic or parasitic) and the extent of dependence (facultative or obligate symbiosis). The various types of interactions existing in soil inhabitants are summarized in Table 11.1. The soil organisms living symbiotically interact in many ways ranging from mutualistic to parasitic ones because the relation between two organisms change with their development or environmental conditions. In many instances the interactions may be beneficial to one or more species involved whereas in others it may be inhibitory. For example, one species may inhibit other species by competing for available nutrients or by forming toxic metabolic products. On the other hand, metabolic end products of one species may serve as nutrient for the growth of other species.

Mutualism : Mutualism indicates positive dependence of two interacting organisms that may or may not be obligatory. Such association is found in lichens, mycorrhizae, rhizobia, etc. In lichens, for example, algae or cyanobacteria grow in close intimacy with a fungal partner. The photosynthetic activities of algae or cyanobacterium provide organic compounds to the fungus and algae/cyanobacterium, in turn, get protection from dissociation. Similarly, nitrogen fixing rhizobia and phosphorus solubilizing mycorrhizae form associations with plants which have been discussed in previous chapters. Even heterotrophic ant (*Acromyrmax*) grows fungal garden of a basidiomycetous fungus *Xylaria* wherein host ant get cellulose digested by *Xylaria* in lieu of the nutrients supplied to the partner. These interactions not only influence the plant growth but also soil health.

Table 11.1 : Various types of biological interactions existing in soil ecosystem

Types of interactions	Description	Effects on partners
a) Mutualism	Positive interaction between species where both partners are benefited e.g. mycorrhizae-plant interaction or legume-*Rhizobium* symbiosis, lichen, etc.	+/+
b) Protocooperation	Positive interaction between two species but without the co-operation being obligatory for their existence or for their performance of some reaction. Two organisms grow	+/+

Contd...

Table 11.1. Contd...

	independently under obligatory conditions e.g. nutritional protocooperation between *Proteus vulgaris* and *Bacillus polymyxa* for nicotinic acid and biotin, respectively, excreted by other partner on a medium lacking both the vitamins.	
c) Synergism	A particular type of mutual association where combined activity exceeds the sum of the individual activities e.g. mycorrhizae-*Rhizobium*-legume association.	+/+
d) Commensalism	One species derives benefit from other while other associate remains unaffected e.g. cellulolytic or proteolytic microbes degrade complex macro-molecules to simple monomeric units which are utilized by secondary heterotrophs.	+/0
e) Neutralism	Two species behave entirely independently, so no specie is benefited or harmed.	0/0
f) Antagonism	Negative interaction in which either partner may suffer or one is benefited at the expense of other e.g. *Trichoderma* checks the growth of *Aspergillus*.	- or +/0
i) Competition	Negative or harmful interaction in which species compete for the same limiting resource such as nutrition, oxygen, space, etc. so both suffer e.g. various heterotrophs competing nutrients or space in soil.	-/ -
ii) Amensalism	Harmful association in which one species is inhibited/repressed by toxic compound or antibiotic production or allochemicals by other e.g. antibiosis caused by *Streptomyces*, *Bacillus*, etc. in soil, the other organism remains unaffected.	0/-
iii) Parasitism	One specie derives nutrients for growth by parasitizing another live species e.g. bacteriophage virus parasitizing bacteria; *Gliocladium* parasitizing *Trichoderma*.	+/-
iv) Predation	One specie ingests or kills the other specie for food e.g. protozoa feed on bacteria; slime mold and *Myxobacteria* feed on soil bacteria.	+/-

Protocooperation : Protocooperation is a non-obligatory beneficial relationship that develops when two different organisms get associated. So the process involves adoption on cooperative basis. The fungus *Polystictus audus* is unable to synthesize its own thiamine, though it forms biotin required by it. Reverse is the case with *Nematophora gossypii* which forms thiamine but not biotin. The two fungi when grown separately are not able to survive but if grown together grow well. Similarly, *Desulfovibrio* supplies H_2S to *Chlorobium* and in return gets organic material and sulphate from *Chlorobium*.

Synergism : In synergism various mutually beneficial relations between microbial populations may add to the better effects on each other. The relationships are not obligatory. The decomposition of lignin or pesticides by specialized microorganisms may show such beneficial effect on rhizosphere inhabiting microorganisms. Similarly, mycorrhizae-*Rhizobium* present on the same host may enhance each others beneficial contribution to the host.

Commensalism : Commensalism is a relationship between two populations in which one is benefited and the other remains unaffected. Commensalistic associations are common but not obligatory. Such associations help a combination of species to attack a particular substrate which an individual can not do alone. Lignin, for example, is a major constituent of wood and resistant to degradation by individual microbes under laboratory conditions. However, in forest soils it is easily decomposed by mixed population in soil. Similarly, many nutritionally fastidious bacteria occur in soil suggesting that their growth and survival depends upon the synthesis and excretion of vitamins and amino acids by less fastidious species.

Neutralism : In neutralism the two organisms present in a same ecosystem entirely behave independent of one another. The non-interactive behaviour of organisms is more likely at low density than at high density.

Antagonism : Antagonism is the inhibition imposed by the metabolic product of one species on the other. Such effect may be due to the competition, ammensalism or parasitism. Antibiotic production by bacteria, cyanide production by some fungi and antibacterial fatty acid production by algae are classical examples of antagonism. Similarly, methane, sulphide and volatile sulphur containing compounds produced by some soil microbes can also inhibit the growth of many microbial species. Myxobacteria and *Streptomyces* species have the ability to secrete a powerful cell wall degrading lytic enzyme to derive nutrients for use.

i) **Competition :** Competition involves exclusion or suppression of the growth of an organism by another struggling for the same limited resources such as nutrient, space, oxygen, vitamins etc. from the environment. Therefore, competition exclusion precludes two populations from occupying the same ecological niche. Good competitive ability of an organism depends upon high growth rate, rapid spore germination, good enzyme and antibiotic producing capability, tolerance to antibiotic produced by other organisms, etc. Competition results in major fluctuation in the composition of microbial populations. Such competition may be intra- or inter specific.

ii) **Ammensalism :** An association in which one specie is suppressed and the other remains unaffected is known as ammensalism. In soil environment many organisms may show such ammensalistic relations with other organisms. For example, antibiotic production modifies the growth of others. Similarly, high production of ammonia by ammonifiers inhibits nitrite oxidizing microbes.

iii) **Parasitism :** Parasitism depicts the relation between two living organisms in which one is benefited at the expense of other partner. Parasites are dependent on host. Parasites feed on the host cell, tissues or body fluids in intimate physical contact and utilize metabolites of the host. The host is usually harmed, not necessarily killed. All major groups of plants, animals and microorganisms are susceptible to attack by microbial parasites. For example, bacteriophage in bacterial cell and chytrid fungus parasitize algae, other fungi and plants. A motile zoospore of a chytrid fungus attaches to an algal cell and forms protrusions that penetrate the host cell to obtain nutrients. Meanwhile the zoospore on the outside of the host cell grows in size and eventually becomes a zoosporangium that produces more zoospores.

iv) **Predation :** In predation the predators feed on prey and digest it. For example, protozoa feed on bacteria and some algae. Similarly some fungi attack and prey nematodes whereas *Bdellovibrio*, a widespread bacterium in soils, preys other bacteria. *Bdellovibrio* can penetrate and grow within periplasmic space of a host bacterium. In the process, they destroy the host and derive nutrients from later.

Mutualistic interactions lead to the beneficial influence whereas pathogenic ones suppress or deteriorate plant growth. The sum total phenomena occurring at or near the root-soil interface have great impact both on plant and soil biota. Microorganisms grow associatively in soil on plant right from seed germination till death. Even some microbes act later after the death of plants. Other terms frequently used in interaction are obligate and facultative. On temporal criterion and durability basis, interaction may be persistent or transient. On the basis of the nature of their specificity, these interactions may be specific or permissive. Nutritional classification categorizes organisms into autotrophs and heterotrophs. Chemoheterotrophs are dependent on organic compounds as both energy and carbon source and may show symbiotrophy or saprotrophy. In symbiotrophy organisms derive organic compounds directly from another organism by virtue of permanent intimate contact. Such interaction may be either biotrophic or necrotrophic depending whether the derivation of organic compound is from a living cell or killed cell of symbiotic host. Saprotrophy describes the situation when derivation of organic compound is from a non-living environment and it involves either osmotrophy or phagotrophy depending whether the assimilation of organic compounds occurs in a dissolved form or in particulate form.

11.2 SPERMOSPHERE

Seed is a microcosm of microbes with potential for carrying a wide variety of fungi, bacteria, viruses, etc. While most of the seed microflora is harmless, of course some are beneficial and a few are pathogenic under favourable conditions. It was in 1904 when M. Duggal found that healthy seeds possessed specific number and species of bacterial flora. Since then seeds of different plant species were found to carry, both internally and externally, a good quantity and quality of microorganisms.

Spermosphere : Spermosphere is the region around a germinating seed that has greatest potential for microbial effects to have lasting influence on plants. G. Picci in 1959 coined the term 'spermosphere' and phenomenon as 'spermosphere effect'.

Spermosphere as a zone of intense microbial interaction : Some organisms borne on seed exert beneficial influence on its germination through production of growth hormones etc., while others are deleterious. Germinating seeds also excrete some chemicals that influence the quality and quantity of microorganisms associated with seeds and/or in their immediate vicinities. Some microorganisms are present under the seed coat and these generally originate at flowering stage. Many saprophytes and weak pathogens live on seed surface and do not visibly affect seed. They may be harmful when environmental conditions for their growth and reproduction become favourable. A large number of fungal and bacterial species colonize seeds during their formation under humid weather. Microorganisms also colonize seeds during storage, especially when seed moisture content is increased as much as >16% (w/w). *Aspergillus* and *Penicillium* invade seeds in equilibrium with relative humidity of 65-90%. In starchy seeds (cereals, grains), such invasions occur at moisture level of 13-20%, in soybean at 12-19% and in oily seeds (sunflower, peanut, safflower, etc.) at 5-12%. Also, when the seeds are sown, certain interactions between seed borne microflora and soil microorganisms do occur under the influence of the chemicals exuded by a germinating seed. This all decides the final shape of microbial nature in the region and later these selective microorganisms move from spermosphere to the rhizosphere region within a period of 3 days depending upon the seed type, plant species, population and diversity of microorganisms, soil qualities and its associated microflora, environmental conditions for seed germination and microbial growth rate.

Seed coat offers physical barrier and immunity to infection as well as to germination. Moreover, seed coats restrict diffusion loss of seed leachates into the soil thereby reduce the buildup of parasites around the seeds and their subsequent parasitism of seedlings. Some soil microbes may colonize seed coat or embryo (Fig. 11.1). For example, cellulolytic fungi harbour seed coat whereas saprophytes colonize embryo region. Therefore, spermosphere region may become concentrated interaction zone for both pathogenic as well as beneficial microorganisms because of the chemicals exuded by germinating seeds. Such interactions vary with the type of seed and its resident microflora, the soil and its associated microflora and with the surrounding environmental conditions.

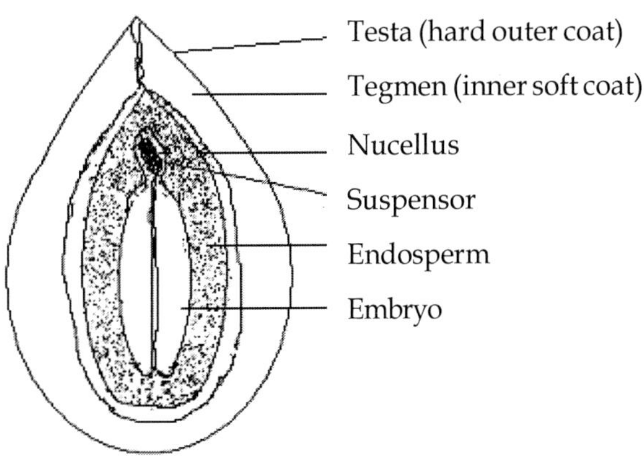

Fig. 11.1 : Parts of a seed

The organisms coated on seed get more support by virtue of their number and prior occupation of space so become dominant. *Rhizobium, Azotobacter, Azospirillum*, mycorrhizae, biocontrol agents, etc. are applied through pre-sowing treatment of seeds, therefore they get established to benefit the plant in the form of ready nutrient supply or growth promoting substances. Seed coating with any biocide influences the interaction between plant roots and microbes so alters the microbial population and diversity. When a seed is treated with an antagonist or organic material, competition between pathogen and non-pathogen sets in motion. Depending on the efficiency of either group, one is suppressed and eventually eliminated. Biocide treatments partially or completely eliminate this competition. Internally and externally seed borne pathogens of cotton *Xanthomonas campestris* pv. *malvacearum* is controlled when seeds to be sown are pretreated with cow dung slurry (rich in saprophytic microorganisms). The nature and level of organic matter in soil influences non-pathogenic organisms. Seed treatment with organic chemicals alter rhizosphere microflora of seedlings so alter plant-root-microorganism relationship in soil through seed.

11.3 RHIZOSPHERE

Most roots growing in soil bear dense cluster 0.1-1.0 mm long root hair numbering from 100 to 1000 cm^{-1}. The root hairs appear a few millimeters behind the root tip and increase the effective volume of soil being exploited by root for water and nutrients. Living plant roots have a profound effect on soil in their immediate vicinity.

Rhizosphere : The rhizosphere is a zone of microbial activity within and around the roots (Fig. 11.2). Operationally, rhizosphere may be defined as the region of soil extending a few millimeters from the surface of each root together with root surfaces (1-2 mm thick), in which microbial population of the soil is influenced by the chemical activities of the plant. It may be defined as the volume of interactions between a plant root system and its immediate soil environment. The chemical and biological characteristics of this zone usually vary from the bulk soil. Rhizosphere is sometimes divided into two regions.

1. *Endorhizosphere* : It is the root-epidermis-cortex zone (stele with conducting elements xylem, phloem, epidermis, endodermis and root cap) colonized by pathogens and non-pathogens.
2. *Ectorhizosphere* : It includes outer root region (plant and microbe-derived mucigel, root hairs, plant mucilages, sloughed off roots cap cells, etc.).

Biological interactions in rhizosphere : All normal roots under natural conditions support a large population of microorganisms in their rhizoplane and in rhizosphere. In rhizosphere there is proliferation of microbes than in bulk soil. This concentration gradient is as a result of interaction of rhizosphere microbial community and the diffusion rate of organic compounds through soil. The more active the microbes in decomposing the individual organic components, the lower will be the gradient with distance from the root. Apart from numerical preponderance of microorganisms in rhizosphere, the rhizosphere effect is also visible manifestation of the presence of amino acids, vitamins and specialized growth factors. The metabolic activities and their rates are quite different in this zone than non-rhizosphere soil. The soil acidity may be almost 10 times higher or lower in the rhizosphere than bulk soil. A wide range of enzymes catalyze decomposition of root exudates, organic residues, sloughed off roots, etc.

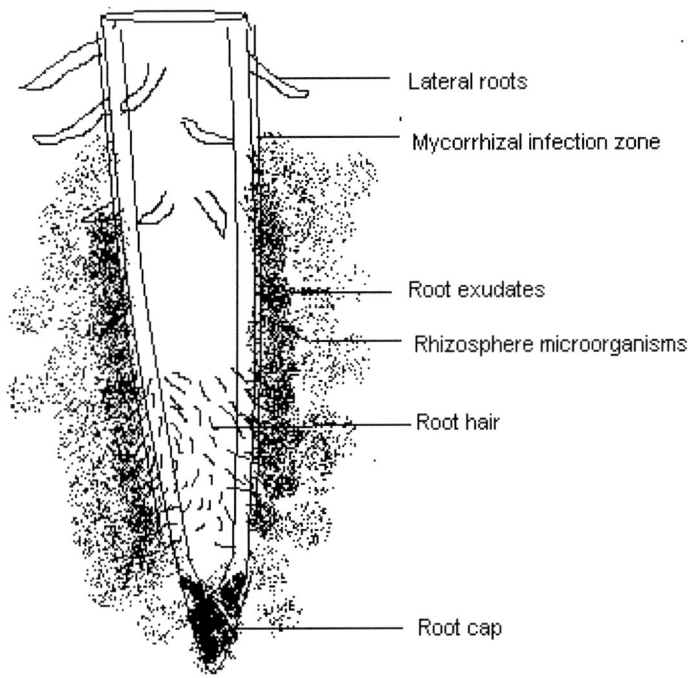

Fig. 11.2 : The diagrammatic presentation of a typical rhizosphere

Rhizoplane : The area where ecto- and endo-rhizosphere merge is probably the zone where most intimate plant and microbial associations do occur. This zone is termed as "rhizoplane" and is simply a zone of attachment of microorganisms to root surface (i.e. external root surface). Population densities in rhizosphere are highest on rhizoplane with fairly steep population gradient over a few millimeters away from root. Usually root hair zone tends to be most heavily populated with microbes. The root colonization by non-infecting microbes is not confined to the rhizosphere but takes place, to a varying degree also, in the apoplasm of the cortex. For such cases sometimes instead of term "endorhizosphere" the term 'bacterial endophyte' is used.

Characteristics of rhizoplane : Rhizoplane studies have revealed the presence of characteristic bacteria and fungi. The microbial count in the rhizosphere has been found to be ten to several hundred-folds as compared to non-rhizosphere soil (bulk soil) depending upon the plant age, plant species and nutritional status of the plant. Their composition would, however, also depend on the quality and amount of the nutrients available in the rhizosphere. For example, as the nitrogen supply increases both the number and proportion of diazotrophic bacteria would decrease at the rhizosphere whereas total bacterial count would increase. Structurally, the rhizosphere is known to harbour greater Gram negative bacteria (*Pseudomonas, Agrobacterium Alcaligenes, Achromobacter, Flavobacterium,* etc.) and denitrifiers and fewer Gram positive and Gram variable bacteria (*Bacillus, Azotobacter, Mycobacterium, Corynebacterium, Arthrobacter, Cellulomonas, Micrococcus,* etc.). The fungal genera encountered may include *Aspergillus, Pythium, Fusarium, Mucor, Phoma, Penicillium, Caphalosporium, Mortierella, Botrytis, Trichoderma, Gliocladium, Coniothyrium, Gliomastrix, Cylindrocarpon, Alternaria, Hypomicrobium,* etc. The population densities in the rhizosphere are generally highest on the rhizoplane with a fairly steep gradient (over a few millimeters) away from root. Maximum population and diversity is observed on rhizoplane; almost 3-4 fold than rhizosphere. Typically 4-10% space in the rhizosphere is covered by soil microbes. In fast growing roots there is usually a steep gradient of rhizoplane and rhizosphere microorganisms from apical to basal zone along the root axis. In maize, for example, of the total root surface area surface cover by bacteria is about 4% in apical zones, 7% in root hair zone and might raise upto 20% in basal zone. It is root hair zone that tends to be most heavily populated with microbes whereas growing root tip is typically free of microbes.

Microbial distribution in rhizosphere : Direct microscopic examination of plant root surfaces have revealed that over 5-10% of root surface appears to be covered by microorganisms. The distribution appears irregular with small clumps and larger aggregates often associated with cell junctions and microbial cells. Moreover, the zone of elongation behind the root tip carry bacterial population and older parts of roots carry a considerable microbial population especially fungi and bacteria. The proportion covered by fungi and bacteria is variable. Single bacterium is often associated with pits in root cell wall, and clusters of bacteria at cell junction. Within the mucigel layers of bacteria usually appear as single unit. Young roots show less colonization of epidermis and cortical tissues by non-pathogens but later non-pathogenic microbial growth becomes widespread. Though the distribution of microorganisms in the rhizosphere is not uniform, in some root parts bacteria may be observed even at 10 to 40 cell depth and many other root parts may be free of bacteria. Bacteria and fungi are often arranged in lines along the root axis. Some may traverse whole rhizosphere and enter several cells deep into root cortex. One gram of root can support around 36 mg of bacteria by weight equivalent to 2×10^{10} cells. Typical populations in rhizosphere are 10^9 bacteria, 10^7 actinomycetes, 10^6 fungi, 10^3 protozoa and 10^3 algae per gram of rhizosphere soil. Evidences show that microflora of surface layer of soil is carried to deeper layers by growing root system. Such movement helps in improving the soil fertility.

Rhizosphere effect : It depicts the overall influence of plant roots on soil microorganisms. There is qualitative and quantitative change in rhizosphere with respect to the population, activities and diversity of microorganisms than existing in non-rhizosphere soil. Rhizosphere effect can be demonstrated by removing a plant from soil, shaking off its loose adhering soil debris and then estimating the number of microorganisms per gram of firmly adhered soil. Greater rhizosphere effect is seen with bacteria than with actinomycetes or fungi. Negligible changes are noted with regards to protozoa and algae. Gram negative non-sporulating rod shaped bacteria dominate and form about 40% of total bacterial population in the rhizosphere. Among the Gram negative bacteria, fluorescent pseudomonads such as *Pseudomonas fluorescens* frequently dominate rhizosphere. Fungi tend to be relatively less important, probably because of dense packing of the minimum void zone around the roots, which effectively ensures that there are insufficient channels for some fungi to approach roots. Further, in soil 70-90% fungi exist as spores while in rhizosphere of several crops 55-95% are vegetative. Rhizosphere effect in soil under apple tree increased along the profile for all the studied groups of microorganisms that were concentrated in the rhizosphere.

The rhizosphere microbes are ecologically significant in mobilizing and immobilizing plant nutrients, producing growth promoting substances and antagonizing plant pathogens. If the ratio of activity per unit weight of rhizosphere soil to the activity in equivalent weight of non-rhizosphere soil (R :S ratio) is greater than one, it indicates selective stimulation, whereas value of less than one reveals inhibition activity in rhizosphere. Greater heterotrophic activities are attributed to the rhizodepositions made by plants and microbes. Excretion of organic nutrients by plant roots selectively favours certain nutritional types of bacteria, although no clear cut nutritional relationships have been discovered. Individual amino acid concentration is generally 2-3 fold greater in rhizosphere. Plant type has profound bearing on the nature of enzyme produced. For instance, invertase and urease activities are higher in the rhizosphere of barley, cotton and lucerne than in corresponding non-rhizosphere soil whereas phosphatases are higher in legume rhizosphere. Abundance of nitrogen fixing and phosphate solubilizing organisms in rhizosphere assumes a natural significance for plants. It is now recognized that nitrogen fixing bacteria occur in the rhizosphere, stalk and phyllosphere of plants and such bacteria have been also found inside root cells. Similarly, in rice rhizosphere *Rhodopseudomonas* and cyanobacteria are reported to fix 40-50 kg N ha^{-1}. Species of *Azotobacter, Beijerinkia, Methylomonas, Azospirillum, Clostridium, Desulfovibrio, Klebsiella, Enterobacter, Flavobacter, Pseudomonas,* etc. are the main rhizosphere nitrogen fixing bacteria. Apparently nitrogen fixing organisms together with other associated microbes that excrete growth factors or help in reducing oxygen tension are perpetuated through continuous practice or vegetative propagation. Atmospheric nitrogen may be fixed by about 20 genera of free living rhizosphere bacteria and a similar number of cyanobacteria. Rhizosphere N_2 fixation by cyanobacteria, photosynthetic free living diazotrophs and associative heterotrophic bacteria are of considerable significance in submerged rice soils. Mycorrhizal association with plant roots hair is a symbiotic process which helps in nutrient uptake, especially that of phosphorus, thus influences the plant growth.

In rhizosphere, denitrifying bacteria are also present which convert it to a zone of high nitrification potential. Denitrifying bacteria are generally more competitive as aerobic heterotrophs in soil than they are as denitrifiers. This means that sites of intense heterotrophic activity in soil will also tend to be the sites of high denitrification potential. In rhizosphere, there is generally adequate carbon for growth but a limited supply of nitrogen. Grazing protozoa may release approximately one-third of nitrogen (as ammonia) immobilized in bacterial biomass close to the roots. Therefore, increased protozoa population and their grazing activities in rhizosphere plays important role in the release of nutrients held by microbial cells. This ensures efficient nutrient cycling in this sphere.

Mycorrhizal infection considerably reduces the flow of available carbon from roots as mycobiont serves as a carbon sink. The mycorrhizosphere, therefore, generally has much lower microbial population than the rhizosphere adjacent to non-mycorrhizal root tip.

Some antagonistic rhizosphere microbial species such as *Fusarium, Pythium, Rhizoctonia, Aspergillus, Gaeumannomyces,* etc. can sometimes, under certain condition, become plant pathogenic and release various phytotoxins or other detrimental substances which are absorbed by roots producing abnormal metabolic reactions. Thus the situation in rhizosphere is always complicated involving various interactive processes. But in general overall net effect of rhizosphere microbes is stimulatory to plant growth.

Root exudations : On an average, 30-60% of net photosynthetic carbon is allocated to roots, of which an appreciable proportion is released as organic carbon into the rhizosphere. The release of this carbon, also termed as rhizodeposition, is highly variable in annual species which may be as much as 40% and in forest trees >70% are quite common. A great variety of organic compounds are released by roots in soil which vary from simple organic molecules to cells and tissues that are sloughed off in the process of growth. The zone of elongation behind the root cap is the area where excretions, particularly amino acid excretion, appear to be active. Organic materials released by roots serve as substrate for rizosphere microflora and occur in various forms as under :

i. **Exudates :** These are lower molecular weight compounds which leak from root cells into either intercellular spaces and then to soil via cell conjunction, or directly through epidermal cell wall into soil. These are not metabolically mediated. A wide range of substances found in exudates include soluble carbohydrates, amino acids, sugars, growth regulators (such as cytokinin, absicissic acids, etc.) organic acids and phenolics. Exudates may serve as nutrients, inhibitors or stimulants for various soil microorganisms. The role of exudates is significant in the turnover of organic matter in rhizosphere, in aggregate formation and in nutrient and water absorption.

ii. **Secretions :** These are the compounds of both low and high molecular weight; actively pumped out of the roots. These are secreted as a result of various metabolic processes.

iii. **Mucigels :** Mucigel is a layer of gelatinous material on the root surface which is visible under transmission electron microscope. Mucilages are poorly defined substances and may be (i) secreted by Golgi bodies of root cap and epidermal cells, (ii) hydrolysates of polysaccharides of primary cells between epidermal cells and sloughed off root cap cells; or (iii) produced by bacterial degradation of outer primary cell wall of old dead epidermal cells. Mucilages are hydrophilic polyglucuronate and polygalacuronate polymers across which the ions and uncharged soluble materials should be able to pass freely. Root mucilages are gelatinous materials, exhibit a double granular and fibrillar structure and contain fructose, glucose, galactose, xylose, arabinose, mannose, uronic acids, etc. Mucilages have high capacity to form complexes with heavy metal cations (like Pb, Cu, Zn, Cd, etc.) mainly through exchange with Ca^{2+}. Mucilages act as selective filters for these cations through modification of the flux of heavy cations to root, thereby complete the role of cell walls from which they are derived. Roots tend to excrete a layer of mucilage up to 5 µm thick, but on an average it is more likely to be 1 µm thick, which bridges the small gaps between root and soil.

iv. **Microbial mucilages :** Microbial growth around the roots, using exudates, secretions and sloughed off cells as substrates, lead to microbial mucilage which also form part of mucigel. It, besides above, contains colloidal mineral, organic matter from soil and bacterial cells.

v. **Lysates :** These are the compounds released either from older epidermal cells because of autolysis or through digestion of cells by bacteria. The amount and kind of exudates are influenced by the type of root system, age of plant, stage of development of plant and mechanical injuries. In general, roots can release 50-100 mg of organic C g^{-1} root day^{-1}.

Rhizodepositions decrease with plant age but increase with soil stress such as compaction and low nutrient supply. The rhizodeposition of carbon causes the microbial density in soil to be massively enhanced in the rhizosphere. As a rule, in soil grown-plants 75-85% of the total organic carbon supply for microbial activity in the rhizosphere is represented by sloughed off cells and tissues. About 11% of plant assimilates go to microbial respiration and 2% to soil organic matter with soil microbial growth efficiency of 27%. Bacteria influence the uptake of ions by plant roots growing under gnotobiotic conditions. The ion uptake by roots is stimulated by microbes possibly through chelating agents or growth regulators. The complex system of biological and biochemical relationship in rhizosphere continuously changes because of the changing physiological conditions of the plant. It has been established that foliar chemical sprays alter exudation pattern of roots. Foliar-applied C^{14} and P^{32} chemicals were detected in root exudates within 12-36 hours after spray and influenced rhizosphere microflora. This paves the way for manipulating relationship in such a manner that encourages most beneficial organisms which excrete auxins, mobilize plant nutrients in soil, inhibit plant pathogens and enrich soil with fast growing microbial biomass. Rhizosphere microorganisms may increase rhizodeposition, particularly those of low molecular weight fractions, and use them as a carbon source for the production of chelators or organic acids. These microorganisms enhance turnover rate of organic matter, especially carbon, nitrogen and phosphorus, and thus recycle organically bound nutrients.

11.4 PLANT GROWTH PROMOTING RHIZOSPHERE MICROORGANISMS (PGPR)

The beneficial symbiotic and non-symbiotic rhizobiota that influence plant growth positively, promote mineral nutrient uptake and availability, secrete plant growth promoting substances and suppress deleterious rhizosphere organisms through competition for ions and space are known as plant growth promoting rhizosphere microorganisms (PGPR). These microbes have the ability to colonize rhizosphere aggressively, benefit plant and inhibit minor pathogens. These, in broad sense, include symbionts (*Rhizobium, Bradyrhizobium,* certain actinomycetes, mycorrhizae, etc.) as well as free living saprophytes. The beneficial symbionts have been extensively used worldwide as legume inoculants for enhancing crop yield. Other non-symbiotic PGPR's improve plant health through direct effect on plant or by producing growth promoting substances. Some rhizosphere bacteria such as fluorescent pseudomonads (*Pseudomonas fluorescens, P. putida*) exhibit fluorescence under UV light. These produce siderophores under ion-limited conditions and so compete with minor pathogens for ferric iron in the rhizosphere. The siderophores of PGPR have high affinity for iron as compared to deleterious organisms which lack siderophore receptor system for iron assimilation. The application of *P. fluorescens* and *P. putida* to potato seeds improved root development and crop growth. The beneficial effects of PGPR are attributed to the ability of these organisms to compete for substrates, niche exclusion and production of siderophores and antibiotics.

Siderophores : The siderophores or ferrichromes are low molecular weight organic compounds consisting of hydroxamates (composed of a hydroxyl group closely associated with a nitrogenous portion of the molecules) which have high affinity for multivalent iron chelators that transport iron into bacterial cells. Fluorescent pseudomonads produce yellow green fluorescent siderophores, which specifically recognize and sequester the limited supply of iron in the rhizosphere and make iron less available to certain members of natural microflora, especially pathogens. Siderophore pseudobacterin, for example, is reported to inhibit growth of potato rot pathogen *Erwinia carotovora*. Soil solutions contain iron less than the optimal requirement for plant growth because the iron exists in insoluble ferric form. Iron availability in soil decreases with increase in soil pH. Therefore, PGPR function better in neutral and alkaline soils than acid soil.

Release of metabolites and phytohormones : About 10% of bacteria in rhizosphere appear to be aggressive in reducing the population of deleterious microorganisms. Some PGPR produce antibiotic, compete well with other microorganisms and inhibit pathogenic organisms. Agrocin-84 antibiotic produced by *Agrobacterium radiobacter* suppresses the growth of crown gall pathogen *Agrobacterium tumefaciens*. Similarly, *Pseudomonas fluorescens* produces phenazin type antibiotic to control take-all disease in wheat. PGPR's also induce systemic resistance in host plants conferring resistance against a broad spectrum of diseases. Various strains of these bacteria are very effective in reducing manganese oxide and suppressing the growth of *Gaeumannomyces graminis* var. *tritici* at least under *in vitro* conditions. However, beneficial effects depend on soil temperature, moisture, pH, clay content, etc. which influence the survival of PGPR's in rhizosphere.

Many rhizosphere diazotrophs, such as *Azospirillum brasilense*, enhance root growth directly by the production of phytohormones, especially IAA. However, stimulation of root growth by PGPR may much depends on the availability of specific substrate as precursor of phytohormones such as L-tryptophan for IAA, L-methionine for ethylene and adenine for cytokinin. Various flavonoids, released by roots act as signals for recognition of specific hosts e.g. luteolin by rhizobia, quercetin or biochanin for spore germination and hyphal growth of VAM fungi, palmatic acid by ECM and adenine by *Azotobacter chroococcum*. A relatively large number of phytohormones such as IAA, cytokinin, ABA, etc. are released several folds in rhizosphere than bulk soil by microorganisms. *A. chroococcum* and *Pisolithus tinctorius* enhance cytokinin and IAA production when applied with adenine or tryptophan. Although fungi prefer to grow along cell junction, the ability to translocate nutrients allows them to grow across the areas of low substrate supply. This allows some escape from antagonistic fungi and, therefore, lateral colonization from soil remains to be an important mean of infection under conditions favourable for root growth.

11.5 DELETERIOUS RHIZOSPHERE MICROORGANISMS (DRMO)

Deleterious rhizosphere microorganisms (DRMO) may be either i) true major pathogens, which penetrate plant, disrupt phloem and cause major disturbance/disease, or ii) minor pathogens, which basically are saprophytes but can become weak parasites to juvenile tissues when aggressively colonizing the root surface. They produce no visual symptoms and affect plants by their metabolites without parasitizing plant tissues. DRMO include deleterious rhizobacteria as well as deleterious rhizofungi. It must be remembered that some soils are conducive for soil borne diseases whereas others are suppressive because of soil structure or microbial composition. Many rhizosphere microbes produce auxins, ethylene, cytokinin, vitamins and other plant growth substances. However, their relative and total concentration determines the positive or negative effect of plant pathogens on plant growth. Some organisms produce cyanide (HCN) as a secondary metabolite directly from glycine, proline and cyanogenic glycosides found in root exudates. The susceptibility of roots to infection is more related with plant resistance to disease than soil-root interface. Approximately 4% of the aerobic bacteria isolated from tryptic soya agar from potato rhizosphere have exhibited potential to produce HCN and these mostly appeared to be pseudomonads. The DRMO probably survive on root residues that provide a source of inoculum for the root colonization of next crop. Cyanide production has also been found to depend on Fe^{3+} availability. PGPR compete well for Fe^{3+} required by DRMO for HCN production. However, beneficial effects of PGPR are dependent on soil temperature, pH, moisture and clay content. The ecological competence may diminish by repeated subculturing *in vitro*, possibly related to loss of cell surface structure or reduction in antibiotic and siderophore production.

11.6 ROOT-PATHOGEN INTERACTION

All plants grow in an environment in which they are continually exposed to a large number and variety of microorganisms, some of them capable of causing minor damage or competing for nutrients. Infection is more likely if root is damaged by abrasion of soil particles or by emergence of lateral roots. The minor pathogens cause interference in normal plant growth without producing well-defined disease symptoms. Minor pathogens are unspecialized weak parasites which can live as saprophytes on plant material and can also enter root under favourable conditions e.g. *Corticum solani, Macrophomina phaseoli, Furasium calmorum,* etc. They can enter and kill juvenile but not mature root. These pathogens inhibit root growth by producing toxins and cyanides. Harmful rhizosphere microorganisms often cause depression in growth and yield. In fruit trees like grapevine this situation is often described as 'soil sickness'.

Compared to minor pathogens which are basically saprophytes waiting for opportunities to switch over to pathogenic phase, major pathogens actively grow on and attack roots in soil. These are highly specialized parasites with a little or no ability to survive as free living saprophytes in soil, although these often survive in the form of resistant propagule on various roots e.g. *Ophiobolus graminis, Phytophthora infestans,* etc. In all, disease severity is individually or synergistically influenced by the survival and dissemination of soil-borne pathogens, prevailing soil environment viz., temperature, pH, moisture, electric conductivity and other physico-chemical properties, the composition of soil microflora, the aggressiveness of pathogen, susceptibility of host and cultural practices followed. This shapes the soil microflora and determines disease intensity. Hence, unlike the air-borne diseases, these complicate the pathosystem in soil-borne diseases. Majority of these pathogens are dormant or surviving in some other stage of their life cycle and grow when conditions become congenial. Growth may be stimulated or directed towards roots; and spores and sclerotia may germinate in response to exudates in general or specific components of them. The proximity of plant roots stimulates the germination of spores of many non-pathogenic fungal species lying dormant in soil.

PGPR's induce systemic resistance in host plants, conferring resistance against a broad spectrum of diseases. Root exudates influence the proliferation and survival of root infecting pathogens in soil either through soil fungistasis or inhibition of pathogens in the rhizosphere. Fungistasis means inability of non-dormant spore, sclerotia or propagules to germinate even under favourable conditions of pH, temperature and moisture in soil. The fungistasis could be released or undone by rhizosphere effect of plants, which create congenial conditions for spore germination. Sclerotia are perennating structures of certain plant pathogens and differ from spore in germination. Sclerotia are mass of interwoven hyphae (e.g. *Rhizoctonia solani, Sclerotinia* sp. etc.) or an organized structure (e.g. *Marcophomina phaseoli, Verticillium dahlia,* etc.). Spores or sclerotia of many pathogenic fungi (e.g. *Rhizoctonia, Fusarium, Sclerotium, Aphanomyces, Alternaria, Pythium, Colletotrichum, Verticillium, Phytophthora, Plasmodiophora,* etc.) as well as bacterial spores have shown germination by the stimulus provided by root exudates of susceptible cultivar of host plants. The fungistasis of sclerotia of *Sclerotium cepivorum* is released by certain volatile stimulators associated with the roots of *Allium*. The volatile compound alkyl sulphide is produced when alkyl cystein sulphoxide released from *Allium* roots are broken down by soil bacteria. Once spores are germinated, the mycelium or hyphae is less susceptible to fungistasis.

Root exudates also provide a good base for the growth of antagonists which suppress the growth of pathogenic bacteria and fungi. Evidences show that rhizosphere of resistant varieties harbour more antagonists like *Streptomyces* and *Trichoderma* than susceptible ones e.g. in pea 13-33% of rhizosphere isolates of *Streptomyces* are from resistant varieties and only 6% in susceptible ones. Similarly, *Phymatrotrichum* root rot-resistant cotton varieties had much higher number of blue green fluorescent bacteria of *Pseudomonas-Phytomonas* group in their rhizosphere than the roots of susceptible varieties.

Root exudates also influence the formation of infection structures like appresoria. Root exudates containing toxic substances such as glycosides and hydrocyanic acids may inhibit the growth of pathogens. The other mechanisms are exudates induced alteration in soil pH (neutralizing) and microclimate of rhizosphere through liberation of water and CO_2. Such changes influence the root infection by phytopathogens. Adoption of different practices of crop rotation i.e. mixed cropping, multiple cropping or green manuring, etc. in a given soil activates root exudates in various layers of soil in different ways. These help in enhancing the rhizosphere effect and often suppress pathogenic invasions as well. The rhizosphere environment can be altered by adding different types of decomposable organic matter to soil. The addition of 560 kg ha^{-1} of chitin to a soil 2 weeks before planting beans *(Phaseolus vulgaris)* markedly reduced the severity of root rot caused by *Fusarium solani* f. sp. *phaseoli*. On the other hand, the addition of decomposable rye or grass increased the severity of root rot by *Thielaviopsis basicola*.

11.7 PHYLLOSPHERE

The surfaces of leaves are usually hydrophobic due to the presence of cutin and wax. The amount of these substances varies with plant species/ cultivar and their environment. Plant parts exposed to dust, radiation, air current, etc. harbour a typical microbial population on their surfaces, aided by cuticle wax and appendages which help in adherence or anchorage of microorganisms. The leaf surface is called 'phylloplane' and the zone on the leaves/herbage inhabited by microorganisms is called 'phyllosphere'. This association develops because phyllosphere serves as a nutrient base in the form of leachates or exudates (such as amino acids, sugars, sugar alcohols, pectic substances, organic acids, etc.) released by leaf or as a support for the microorganisms. Exudates are produced by glandular structures or trichomes and deposited on plant surfaces. Sucrose esters and diterpenenes are exuded by trichome and alkanes, fatty alcohols and wax esters by epidermis. Colonization of phyllosphere begins with the emergence of new tissues such as leaf from a bud. There is a succession in colonization by bacteria, yeast and filamentous fungi in relation to host growth and environment. Residents are epiphytes if they inhabit exclusively the leaf surface only; and endophytes if they inhabit the internal structure of the leaf. However, epiphytes, sometimes, can move from leaf surface to an internal habitat or occupy both.

The phyllosphere microorganisms survive and proliferate on leaves depending on the extent of leaf diffusates and exudates. In turn, these resident microbes supply nutrients to their host plant e.g. nitrogen fixers supply the plant with nitrogen and growth promoting hormones/substances and impart resistance to plants through their secretions or toxin production. It is claimed that there is selective stimulation of biological nitrogen fixers and antagonistic microorganisms on phyllosphere. Biological nitrogen fixation has been observed in the phyllosphere of wheat, cotton, maize, etc. About 100 species have been found capable of fixing atmospheric nitrogen on phyllosphere. Dominant nitrogen fixing microbes encountered are *Beijerinkia, Azotobacter, Derexia,* etc. The other bacteria and fungi found in phyllosphere are *Pseudomonas, Phytomonas, Erwinia, Sarcina, Alternaria, Cladosporium, Cladosporoides, Epicoccum, Helminthosporium, Trichoderma, Torula, Penicillium, Saccharomyces,* etc. Algae and lichens are also present on leaf surfaces and include *Anabaena, Calothrix, Nostoc, Tolypothrix, Scytonema,* etc. About fifty fungal genera have been found on phyllosphere. Under temperate conditions most frequently encountered ones are *Aureobasidium pullans, Cladosporium* sp., *Cladosporoides* sp., *Alternaria alternata, Epicoccum purpurascens* and yeast like *Cryptococcus, Sporobolomycetes,* and bacteria like *Pseudomonas, Xanthomonas,* etc. The quality and population of microorganisms in phyllosphere vary with plant species and its morpho-physiological and environmental conditions. The age of plant, leaf spread, morphology of leaf surface, maturity level, nutrient content and atmospheric conditions under which it grows greatly influence the phyllosphere microflora. Both native or resident organisms and casual inhibitors are found on leaf surface. Phyllosphere microorganisms help the plant in controlling air borne pathogens through space occupation, besides epiphytotic microbes synthesize indole acetic acid which enhance plant growth and vigour.

Factors affecting phyllosphere colonization : Intensity of colonization is influenced by climatic and plant factors. High humidity favours and insulation disfavours the growth of most heterotrophs, however both factors favour phototrophy. Plant stem and bark are often encountered by algae and lichens. Broad leaved plants support more microorganisms on their leaf surfaces than do grasses. Gram -ve yellow pigmented bacteria dominate the bacterial groups and yeast among fungal flora. About 1% leaf assimilates are metabolized by phyllosphere biota. Non-pathogenic microorganisms occur in the tissues of fruits, stems and leaves also. Some studies have revealed that the entry of some pathogens into leaf is facilitated by the physiological activities of non-pathogens. The phyllosphere microflora is suppressed due to environmental pollution caused especially by cement and fertilizers. Any change in phyllosphere affects plant growth and so influences the physiological activities of root system as well.

Resistance is also imparted in plants by phyllosphere microbes against soil borne diseases. Antimicrobial compounds (a variety of secondary metabolites including aromatics, terpenoids and aliphatic substances produced by the plant), such as phytoalexins, accumulate rapidly when plants are challenged by incompatible or saprophytic microorganisms.

Phytoalexins : Phytoalexins are induced in plants once any foreign agent or spore interacts with it. Phytoalexins are produced by healthy cells adjacent to localized-damaged and necrotic cells in response to the materials diffusing from the damaged cells. Foreign organisms or spores cause injury or physiological stimuli e.g. release of malic acid in chickpea. Various types of phytoalexins produced by pathogens are given in Table 11.2.

Table 11.2 : Some phytoalexins incited by pathogens in their specific hosts

Plant	Organism	Phytoalexin
Potato (*Solanum tuberosum*)	*Phytophthora infestans*	Caffeic acid, chlorogenic acid, scopolin, α-solanine, α-chaconine, solanidine, Rishitin, phytuberin
Pea (*Pisum sativum*)	*Aschochyta pisi; Penicillium expansum*	Pisatin
Gram (*Cicer arietinum*)	*Aschochyta rabie*	Cicerin
Beans (*Phaseolus vulgaris*)	*Colletotrichum lindemuthianum Monillia fructicola; Fusarium solani* f. sp. *phaseoli; Rhizoctonia solani*	Phaseolin
Carrot (*Daucus carota*)	*Ceratocystis* sp	Chlorogenic acid; 6-methyl mullein
Soybean (*Glycine max*)	*Phytophthora sojae Phytophthora megaspermum* f. sp. *glycinea*	Hydroxy phaseolin; Glyceollin
Sweet potato (*Ipomeae batatas*)	Injury or infection	Caffeic acid, chlorogenic acid, scopoletin, Esculentin, umbelliferone, ipomaemarone
Cotton (*Gossypium* sp.)	*Verticillium albo-atrum*	Gossypol
Broad beans (*Vicia faba*)	*Botrytis fabae*	Weyerone
Alfalfa (*Medicago sativa*)	*Aschochyta imperfecta, Uromyces straitus, Helminthosporium turcian, Colletotrichum phomoides, Cylindrocladium scorparum*	Sativol, medicagol, medicarpin, coumesterol and other glycosides
Red clover (*Trifolium subterraneum*)	*Sclerotinia trifoliorum*	Formononetin, biochsnin A, trifolirhizin, medicarpin
Apple (*Malus sylvustris*)	*Venturia inaequalis*	Phloridzin, phloretin and other oxidation products
Tobacco (*Nicotiana tobaci*)	*Nectria galligena, Pseudomonas solanacearum*	Scopoletin, scopolin, benzoic acids

The compounds present on leaf prevent various pathogens to establish e.g. phenols produced in apple prevent disease development. Phytoalexin production is stimulated in a host by the presence of certain pathogenic substances called 'elicitors'. Most biotic elicitors are compounds of high molecular weight and contain complex polysaccharides and glycoproteins of fungal origin, plant cell wall, lipids, microbial enzymes and polyypeptides. Phytoalexins are not produced during biotrophic infection. Non-pathogenic organisms may also stimulate the accumulation of phytoalexins in non-host plants.

Any change in the phyllosphere influences the plant growth which, in turn, affects physiological activities of root system. Such changes in root result in the alteration in pH as well as in the spectrum of exudations. Therefore, there exists a link between phyllosphere microflora and rhizosphere microflora. The rhizosphere and phyllosphere studies and manipulations of these, therefore, can help in improving plant-microbe relations in soil.

Selected References

Agnihothrudu, V. 1955. States in which fungi occur in the rhizosphere. *Naturwissenschaften,* 42 : 515-516.

Alexander, M. 1971. Biochemical ecology of microorganisms. *Ann. Rev. Microbiol.* 25 : 361-392.

Baker, K.F. and Cook, R.J. 1974. *Biological Control of Plant Pathogens.* W.H. Freeman & Co., San Francisco, USA.

Balandreau, J. and Knowles, R. 1978. The rhizosphere. pp. 243-268. In : *Interactions Between Non-pathogenic Soil Microorganisms and Plants* (eds. Y.R. Dommergues and S.V. Kurpa), Elsevier, Amesterdam, Holland.

Barber, D.A. 1976. Nutrient uptake. pp. 131-162. In : *Interactions between Non-pathogenic Soil Microorganisms and Plants* (eds. Y.R. Dommergues and S.V. Kurpa). Elsevier, Amesterdam, Holland.

Bauer, W.D. and Cartano-Anolles, G. 1990. Chemotaxis induced gene expression and competitiveness in the rhizosphere. *Pl. Soil,* 129 : 45-52.

Bolton, H. Jr., Frdrikson, J.K. and Elliot, L.F. 1993. Microbiology of the rhizosphere. pp. 27-63. In : *Soil Microbial Ecology* (ed. F.B. Metting Jr.). Marcel Dekker, New York, USA.

Bowen, G.D. and Rovira, A.D.1991. The rhizosphere : The hidden half of the hidden half. pp. 641-699. In : *Plant Roots : The Hidden Half* (eds. Y. Waesel, E. Eshal and U. Kafkafi). Marcel Dekker, New York, USA.

Bowen, G.D. and Rovira, A.D.1976. Microbial colonization of plant roots. *Ann. Rev. Phytopathol.,* 14 :121-144.

Bruehl, G.W. 1987. *Soil-borne Plant Pathogens.* McMillan, New York, USA.

Curl, E.A. and Truelove, B. 1986. *The Rhizosphere.* Springer-Verlag, New York, USA.

Darvill, A.G. and Albersheim, P. 1984. Phtoalexin and their elicitors : A defense against microbial infection in plants. *Ann. Rev. Pl. Physiol.*, 35 : 243-275.

Dickinson, C.H. and Preece, T.F. (eds.). 1976. *Microbiology of Aerial Plant Surfaces.* Academic Press, London, UK.

Ebel, J. 1986. Phytoalexin synthesis : The biochemical analysis of the induction process. *Ann. Rev. Phytopathol.*, 24 : 235-264.

Foster, R.C., Rovira, A.D. and Cock, T.W. 1983. *Ultrastructiure of the Root Soil Interface.* Am. Phytopathol. Soc., St. Paul, USA.

Glinski, J. and Lipiec, J. 1990. *Soil Physical Conditions and Plant Roots.* CRC Press Inc., Boca Raton, Florida, USA.

Greaves, M.P. and Darbyshire, J.F. 1972. The ultrastructure of the mucilaginous layer on plant roots. *Soil Biol. Biochem.*, 4 : 443.

Hassal, K.A. 1990. *The Biochemistry and Use of Pesticides : Structure, Metabolism, Mode of Action and Uses in Crop Protection.* Macmillan, London, UK.

Hiltner, L. 1904. Iiber neuere Erfahrunggen und probleme aufdem Gebiet der Boden-backteriologie und unter besonderer Berucksichtingung der Grundunging und Brache. *Arb. Hsch. Landwirt Gis.*, 98 : 59.

Jackson, R.M. 1960. Soil fungistasis and rhizosphere. pp. 168-176. In : *The Ecology of Soil Fungi* (eds. D. Parkinson and J. S. Waid). Liverpool Univ. Press, Liverpool, UK.

Kerr, A. 1982. Biological control of soil borne microbial pathogens and nematodes. pp. 429-464. In : *Advances in Agricultural Microbiology* (ed. N.S. Subba Roa). Oxford & IBH Publ. Co., New Delhi, India.

Kleister, D.L. and Cregan, P.B. (eds.) 1991. *The Rhizosphere and Plant Growth.* Kluwer Academic Dordrecht, Germany.

Kosuge, T. and Nester, E.W. 1984. *Plant Microbe Interactions : Molecular and Genetic Perspectives. Vol. I.* Macmillan, New York, USA.

Kuc, J. 1972. Phytoalexins. *Ann. Rev, Phytopathol.* 10 :207-232.

Leong, J. 1986. Siderophores : Their biochemistry and possible role in the biocontrol of plant pathogens. *Ann. Rev. Phytopathol.*, 24 : 187-209.

Lynch, J.M.1982. The rhizosphere. pp. 395-410. In : *Experimental Microbial Ecology* (eds. R.G. Burns and J.H. Slater). Blackwell Scientific, Oxford, UK.

Lynch, J.M. and Whipps, J.M. 1990. Substrate flow in the rhizosphere. *Pl Sci.*, 129 : 1-10.

Marschner, H. 1995. *Mineral Nutrition of Higher Plants* (2nd edn.). Academic Press, London, UK.

Mench, M., Morel, L. And Guckert, A. 1987. Metal binding properties of high molecular weight soluble exudates from maize (*Zea mays* L.) roots. *Biol. Fertil. Soils*, 3 :165.

Metting, Jr. F.B. (ed.) 1993. *Soil Microbial Ecology*. Marcel Dekker, New York, USA.

Mukherji, K.G. and Subba Rao, N.S. 1982. Plant surface microflora and plant nutrition. In: *Advances in Agricultural Microbiology* (ed. N.S. Subba Roa). Oxford & IBH Publ. Co., New Delhi, India.

Nakas, J.P. and Hagedorn, C. (eds.). 1990. *Biotechnology of Plant Microbe Interactions*. McGraw-Hill Publishing Co., New York, USA.

Nutman, P.S. 1971. Perspectives in biological nitrogen fixation. *Sci. Prog., Oxford*, 59 : 55.

Nye, P.H. and Tinker, P.B. 1977. Solute movement in the soil-root system. In : *Studies in Ecology. Vol. 4*, Blackwell Scientific, Oxford, UK.

Pandey, K.K. and Upadhyay, J.P. 2000. Microbial population from rhizosphere and non-rhizosphere soil of pigean pea : Screening of resident antagonists and mode of parasitism. *J. Mycol.Pl. Pathol.*, 30 : 7-10.

Preece, T.F. and Dickinson, C.H. (eds.). 1971. *Ecology of Leaf Surface Microorganisms*. International Symposium, Academic Press, New York, India.

Rangaswami, G. 1986. Soil-plant-microbe interrelationships. *Indian Phytopath.*, 41 : 165-172.

Richards, R.A. 1991. The chemistry of denitrification. pp. 58-77. In : *Biology and Biochemistry of Nitrogen Fixation* (eds. M.J. Dilworth and A.R. Glenn). Elseviers, Amsterdam, the Netherlands.

Rovira, A.D. and Davey, C.B. 1974. Biology of the rhizosphere. pp. 153. In : *The Plant Root and its Environment* (ed. E.W. Carson). University Press of Virginia, Charlottesville, USA.

Ryder, M.H., Stevens, P.M. and Bowen, G.D. (eds.). 1994. *Plant Productivity with Rhizosphere Bacteria*. Division of Soils, CSIRO, Glen Osmond, South Australia.

Schippers, B., Bakker, A.W. and Bakker, P.A.H.M. 1987. Interactions of deleterious and beneficial rhizosphere microorganisms and the effect of cropping practices. *Ann. Rev. Phytopathol.*, 25 : 339-458.

Schonwitz, R. and Ziegler, H. 1986. Quantitative and qualitative aspects of a developing rhizosphere microflora and hydroponically grown maize seedlings. *Z. Pflanzenernahr. Bodenk.*, 149 : 623-634.

Spurr, H.W. 1990. The phyllosphere. pp. 271-278. In : *New Directions in Biological Control: Alternatives for Suppressing Agricultural Pests and Diseases* (eds. R.R. Baker and P.E. Dunn). Alan R. Liss., Inc, New York, USA.

Subba Roa, N.S. and Dommergues, Y.R. (eds.). 1998. *Microbial Interaction in Agriculture and Forestry. Vol. 1.* Oxford & IBH Publ. Co., New Delhi, India.

Tate, R. 1987. *Soil Organic Matter : Biological and Ecological Effects.* John Wiley & sons, New York, USA.

Tate, R.L. III. 2000. *Soil Microbiology.* John Wiley & sons, New York, USA.

Warembourg, F.R. and Billes, G. 1979. Estimation of carbon transfer in plant rhizosphere. pp. 183-196. In : *The Soil-Root Interface* (eds. J.L. Harley and R.S. Russel), Academic Press, London, UK.

Wiebe, C. and Winklemann, G. 1975. Kinetic studies on the specificity of chelate iron uptake by *Aspergillus. J. Bacteriol.,* 123 : 837-842.

Wood, M. 1995. *Environmental Soil Biology.* Chapman & Hall, Cambridge, UK.

— *** —

Chapter 12

Soil Microbial Products — Phytohormones, Phytotoxins and Antibiotics

Because of the microbe's eternal struggle to emerge as winner, man has been able to discover strange, new, useful and protective products.

Soil is inhabited by incredibly diverse living microorganisms that are significant to plant growth and development. Soil microorganisms produce a variety of metabolites, probably several thousands, which influence the plant growth either positively or negatively. Therefore, in such a complex situation, identification of any one metabolite or a compound as active phytohormone or phytotoxin is really an uphill task. Appropriate extraction procedure, knowledge about the proper site for its action, appropriate conducive physiological environment and the effective intensity or concentration is of prime importance. Compounds, in general, are inhibitory to growth at higher concentrations but usually stimulatory at lower concentrations. All growth regulators, irrespective of their provenance or structure, may be divided into two groups.

- Those which induce long term effects probably via nucleic acid metabolism; and
- Those inducing short term effects mediated mainly via membrane alteration or directly through influence on the existing enzyme systems.

In both cases the activities and concentration of certain enzymes may be changed by the activation of non-active or latent enzyme or by *de novo* synthesis of enzymes or by production of activators/inhibitors or through degradative processes. The microbial products that influence plant growth are discussed as under :

12.1 PHYTOHORMONES

The substances that trigger or initiate biochemical processes leading to organic formation or other aspects of plant growth are referred to as 'phytohormones' (Fig. 12.1). These induce changes in chemical composition within a plant. Such substances are effective in very low concentrations.

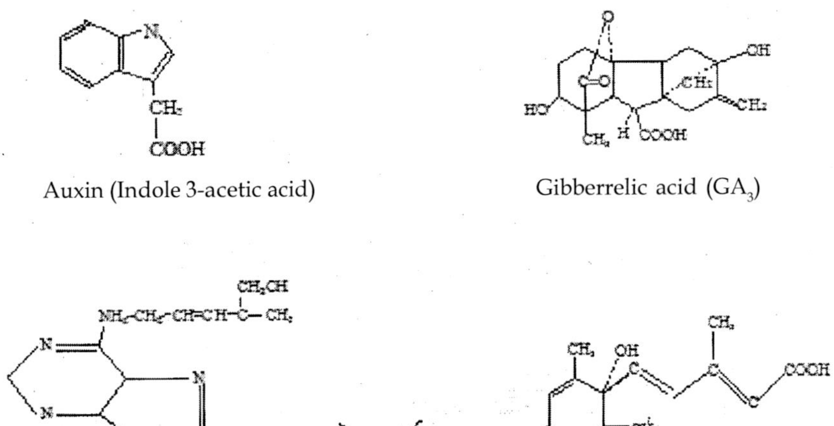

Auxin (Indole 3-acetic acid) Gibberrelic acid (GA_3)

Cytaokinin Ethylene Abscisic acid

Fig. 12.1 Some phytohosmona produced by microbes

12.1.1 AUXINS

Auxins, derivatives of amino acid tryptophan, are involved in natural regulation of plant growth. Indole 3-acetic acid (IAA) is one of the most common examples of this group. It is produced by a variety of microorganisms including fungi and bacteria such as *Agrobacterium tumefaciens, Pseudomonas fluorescens, P. syringae* pv. *savastonoi, P. cepacia, P. solanacearum, Rhizobium* sp., *Endophyllum sempervivi, Rhizopus* sp., *R. suinus, Ustilago maydis, Synchytrium endobioticum, Nectria galligena, Gymnosporangium juniperi-virginianea,* etc.

The balance between various phytohormones, especially auxins and cytokinins, is believed to influence plant cell division and morphogenesis. The morphological effects, induced by IAA, may be visualized in the form of stem elongation, gall formation, extensive branching pattern, etc. Auxins inhibit root elongation but simultaneously induce root thickening. Auxins promote plant growth along the longitudinal axis when applied in low concentration to the shoots of a plant. Stimulated lateral growth does not compensate for the dry matter loss. Microbially produced IAA by *Rhizobium* induces root hair curling whereas the one produced by *Agrobacterium tumefaciens* causes hypertrophy of plant cells forming crown gall tumors. Crown gall cells contain various types of auxins (IAA, IAN and other indole derivatives). Gall cells have about ten times more auxins than normal callus tissues. The accumulation of auxins in host tissues as a result of interaction between pathogenic microorganism and host plant is known as 'hyperauxy'. Several cases of hyperauxy have been reported e.g. wilt caused by *Verticillium*. IAA accumulation is required for gall formation, an expression of pathogenic virulence, during the infection of plants. Many effects of auxins are ascribed to indirect stimulation of ethylene formation in plants. However, most of these effects are because of endogenous IAA. Indole acetic acid is easily degraded by the microorganisms present in soil by enzyme IAA oxidase, therefore, it probably has significance to plants when formed symbiotically i.e. either in mycorrhizal association or in root nodule.

12.1.2 CYTOKININS

Cytokinins are N^6-substituted derivatives of adenine (6 furfural aminopurines) and as such are linked to phosphate esters of sugars to form nucleosides and nucleotides. The most common cytokinin found in plants is zeatin and in microorganisms is isopentyl adenine. Cytokinins promote cell division in plant tissues. These activate several enzymes probably via *de novo* synthesis and induce metabolism and transportation of various important components of cell nutrition concomitant with general stimulation of metabolic activities and cell division under favourable conditions. A high auxin to cytokinin ratio leads to more shoot formation whereas intermediate ratio leads to tissue growth as an unorganized callus.

Azotobacter and *Azospirillum* are known to produce cyptokinins in the rhizosphere alongwith other phytohormones. Various bacteria produce cytokinin or cytokinin-like substances. These include *Agrobacterium, Arthrobacter, Azotobacter, Azospirillum, Bacillus, Corynebacterium, Cytospora, Escherichia, Pseudomonas, Rhizobium,* etc. Besides, various fungi also produce cytokinin which include *Helminthosporium, Rhizopogon, Exobasidium, Monilia, Nectria, Saccharomyces, Taphrina,* etc. Normal *Rhizobium leguminosarum-Pisum sativum* nodules contain several types of cytokinins of both plant and bacterial origin. In uninfected plants bacterial cytokinins are absent and plant cytokinins occur in low amounts. Better recovery of VAM infected seedlings from drought stress compared to non-mycorrhizal ones has been linked to the increased cytokinin levels in infected plants which stimulate stomatal regulations so improve water relations of the host plant. Enhanced levels of cytokinins have also been observed in diseased plants supposedly one to deranged host metabolism.

12.1.3 GIBBERELLINS

Gibberellins are polycylic terpenoids with a basic structure of ent-gibberellane containing 19 or 20 carbon atoms. Isoprenoid is the active principle to the gibberellic acid which is synthesized from mevalonic acid. The compound gibberellin was first identified in 1926 from a fungal filtrate of *Gibberella fujikuroi,* imperfect state of *Fusarium moniliforme,* a pathogen of rice. The plants infected with this fungus are unusually taller popularly known to cause 'Bakanae syndrome', the foolish disease in rice.

More than 50 gibberellins (GA) are known so far. Gibberellins have been isolated from plants as well as from microorganisms. Amongst these, GA_1, GA_2, GA_3, GA_4, GA_7 and GA_9 were isolated from the culture filtrates of fungi and GA_5, GA_6 and GA_8 from the flowering plants. The role of gibberellins is to overcome the dormancy and dwarfism in plants, induce flowering in some photoperiodically-sensitive and other low temperature-dependent plants, cause alteration in sex ratio in flowers, improve fruit setting, stimulate stem growth and suppress lateral growth. The physiological changes induced by gibberellins include stimulation of cell division, increase in cell size, stem elongation, promotion of seed germination, induction of α-amylase synthesis (in barley seeds), release of bud dormancy, etc. Several species of bacteria, fungi and actinomycetes are known to produce gibberellin or gibberellin-like compounds. These include *Arthrobacter, Azotobacter, Azospirillum, Bacillus, Brevibacterium, Flavobacterium, Pseudomonas, Rhizobium, Actinomyces, Nocardia, Alternaria, Aspergillus, Cercospora, Elsinoe, Gibberella, Neurospora, Penicillium, Rhizopogon, Rhizopus, Sphaceloma* and *Suillus*. A glucose mineral salt medium, an incubation temperature of about 25°C, slightly acidic conditions and aerated submerged conditions are favourable for the production of GA_1. The production usually takes place within 2-3 days with the accumulation of GA, lagging behind the growth of the fungus.

12.1.4 ETHYLENE

Ethylene (C_2H_2) production has been detected in soil as well as in pure culture of microorganisms. The amino acid L-methionine is a precursor of ethylene. In soil, a number of other substances such as organic acids, carbohydrates, etc. are also known to serve as source of ethylene. The intensity of ethylene production in roots increases with the increase in organic matter content and also under the effect of organic amendments in the form of plant residues. Concentrations found under field conditions vary from 0.01 to 10 $cm^3\ m^{-3}$ soil. The effective hormonal ethylene concentration is 0.01 to 10.0 ppm.

Soil ethylene is of microbial origin. Ethylene production begins within a few hours after the O_2 concentration in soil falls below 1%, presumably brought about by the organisms that are also active in aerated soil i.e. facultative anaerobes. *Mucor humalis* has been found capable of synthesizing ethylene, a gas implicated in fungistatis. Anoxic and low oxygen conditions favour ethylene gas formation in soil. Various bacterial and fungal genera involved in ethylene production are : *Aeromonas, Arthrobacter, Citrobacter, Clostridium, Enterobacter, Erwinia, Escherichia, Klebsiella, Pseudomonas, Serratia, Streptomyces, Acremonium, Agaricus, Alternaria, Aschochyta, Aspergillus, Blastomyces, Botrytis, Candida, Fusarium, Laccaria, Mucor, Neurospora, Penicillium, Pythium, Rhizcotonia*, etc. The addition of L-methionine to soil leads to a marked increase in microbial ethylene production; thus release of ethylene from soil steeply increases. However, rhizosphere microorganisms may also affect root growth by enhanced decomposition of ethylene. Under field conditions in well-aerated soils the ethylene concentration, on an average, reportedly was only 38 nl L^{-1} in maize rhizosphere as compared to 207 nl L^{-1} in the bulk soil. In higher plants the gaseous phytochrome ethylene application induces several responses such as germination, root formation, fruit ripening, flowering, growth inhibition, senescence (pigment degradation) and defoliation depending upon individual plant specie and the age of plant. In addition, ethylene in plants is also formed after mechanical stress such as wounding, water loss, etc. or after pathogenic infection. Ethylene seems to be involved in wound healing reaction such as lignification and suberization.

12.1.5 ABSCISIC ACID AND XANTHOXINS

Abscisic acid (ABA) and xanthoxins are best known growth inhibitors. Both are terpenoids synthesized from mevalonate. ABA may also be derived from fragmentation of carotenoids (e.g. violaxanthins). ABA is involved in abscission, dormancy of seeds and ion uptake by roots. Probably these compounds are produced by microorganisms. Mucareous fungi are able to produce trisporic acids, similar to ABA. ABA is implicated in disease syndrome involving stunting, wilting and defoliation of diseased plants e.g. *Cercospora* produces ABA via mavalonate pathway.

12.2 PHYTOTOXINS

Phytotoxins are less commonly transported within plants and are sometimes active in greater concentration than growth regulators. 'Phytotoxin' is any metabolite, primary or secondary, which accumulates in toxic concentrations in soil. But for plant pathologists toxins are secondary metabolites. Phytotoxins are mostly low molecular weight compounds. All the toxins are not biochemically specific in action. Organic acids, fatty acid esters and phenols react non-specifically with a variety of plant constituents resulting in the rapid death of affected plants. These necrogenic toxins are produced by a variety of facultative organisms. Various chemical toxins released by bacteria and fungi cause many diseases in plants as well as influence the quality and composition of rhizosphere microflora. Phytotoxins also arise by the microbial degradations in soil.

Various organic acids like dihydroxystearic acid become active in soil when plants are under nutritional stress. Similarly, tyrosine, salicylic acid, α-carotonic acid, salicylic aldehyde, picolinic acid, cinnamic acid and aliphatic volatile acids are phytotoxic in nature. Acetic acid, butyric acid and propionic acids become phytotoxic in soil under anaerobic condition. Aliphatic acids are most phytotoxic at low pH whereas aromatic compounds are less toxic in soil. These organic acids vary with substrate composition and are released during decomposition of organic matter under different set of environmental conditions.

Plant pathogens produce a variety of chemicals during and after the infection of host. *Fusarium* sp. causing wilt produce fusaric acid (5-n butylpyridine 2-carboxylic acid) which interferes the permeability of host plasma membrane and causes water loss. Various toxins produced by several microorganisms involved in plant diseases are given in Table 12.1.

Table 12.1 : Various toxins involved in plant diseases

Toxins	Causal organism	Host
AAL toxin (dimethly hepta-decapentol esters of propanetricarboxylic acid)	*Alternria alternata* f. sp. *lycopersici*	Tomato
ACR toxin (hydrocarbon)	*Alternaria citri*	Rough lemon
ACT toxin	*Alternaria citri*	Tangerine
AK toxin (ester of epoxydecatrienoic acid)	*Alternaria fragariae*	Strawberry
	Alternaria kikuchiana	Pear
Alternaric acid (a dibasic acid)	*Alternaria solani*	Potato
Alternariol	*Alternaria* sp.	Various plants
AM toxin (cyclic tetrapeptidees)	*Alternaria mali*	Apple
Amylovorin (polysaccharide)	*Erwinia amylovora*	Apple and pear

Contd...

Table 12.1. Contd...

CC toxin	*Corynespora cassiicola*	Tomato
Cephalosporin P (terpenoid)	*Cephalosporium* sp.	....
Ceratoulmin	*Ceratocystis ulmi*	Elm
Cercosposin (polyketide)	*Cercospora* sp.	Beet
Coronatine	*Pseudomonas syringae* pv. *atropurpurea*	Grasses and soybean
Coumarin derivatives (shikimate)	*Phytophthora, Alternaria, Aspergillus*	---
Diaporthin (bianthraquinone)	*Endothia parasitica*	Chestnut
Enniatin A (a polypeptide)	*Fusarium oxysporum*	Potato
Funaric acid (butyl pyridine carboxylic acid)	*Rhizopus* sp.	Almond
Fusaric acid (pyridine-carboxylic acid)	*Fusarium lycopersici*	Tomato
Fusidic acid (terpenoid)	*Fusarium* sp.	---
Fusicoccin (a carbotricyclic terpene glycoside)	*Fusicoccum amygdale*	Almond and peach
Glycopeptide	*Corynebacterium insidiosum*	Alfalfa
HC toxin (cyclic tetrepeptides)	*Helminthosporiumcarbenum* race 1	Corn
HMT toxin (T toxin; linear polyketols)	*Helminthosporium maydis* race T	Corn
HS toxin (sesquiterpene glycosides)	*Helminthosporium sacchari*	Sugarcane
HV toxin (victorin)	*Helminthosporium victoriae*	Oats
Lycomarosamine (dipeptide)	*Fusarium lycopersici*	Tomato
Malformin A (cyclic pentapeptide)	*Penicillium* sp., *Aspergillus* sp.,	Various plants
Marticin (polyketide)	*Furarium* sp.	Various plants
Oxalic acid	*Sclerotium* sp., *Sclerotinia* sp.	--
Ophiobolins (terpenoid)	*Helminthosporium* sp.	Grain crop
Patulin (lactone)	*Penicillium patulum*	Potato
PC toxin	*Periconia circinata*	Sorghum
Phaseolotoxin (ornithine -alanine arginine tripeptide)	*Pseudomonas syringae*	Bean and legumes
PM toxin (linear polyketols)	*Phyllosticta maydis*	Corn
Pyricularin	*Pyricularia oryzae*	Rice
Rhizobiotoxin	*Rhizobium japonicum*	Soybean
Syringomycin (a peptide)	*P. syringae* pv. *syringae*	---
Syringotoxin	*P. syringae* pv. *tegetis*	Marigold
Tabtoxin (dipeptide)	*Pseudomonas tobaci; P. syringae* pv. *tobaci*	Tobacco
Tagetitoxin	*P. syringae* pv. *syringae*	Tobacco and citrus
Tentoxin (a cyclic tetrapeptide)	*Alternaria alternate* (syn. *A. tenius*)	Cotton
Trichothecenes (terpenoid)	*Fusarium* sp., *Trichoderma* sp. *Trichothecium* sp.	---
Zinniol	*Alternaria* sp.	Various plants

Source : Yoder (1980); Macko (1983)

12.3 ANTIBIOTICS

The phenomenon of antagonism is common in soil wherein an organism exerts a killing, damaging or inhibitory effect on other organism by the production of lytic enzyme or antibiotics or through competition. Various types of organic compounds are secreted by various microorganisms to have competitive advantage over other organisms. These include antibiotics, which operate at very low concentration of >10 ppm in killing or inhibiting the susceptible microorganisms. The minimum inhibitory concentration ranges from 0.005 to 1.0 μg ml^{-1}.

Antibiotics : Antibiotics are the secondary metabolic products; produced when cells virtually cease to grow actively. Antibiotics are organic compounds having anti-microbial properties. Useful antibiotics have higher therapeutic index. The 'therapeutic index' is the ratio between the minimum effective dose for the host and the minimum effective dose to kill the intended microorganism.

12.3.1 ANTIBIOTIC PRODUCING SOIL MICROBES

Antibiotics producing microorganisms are common in soil ecosystem than other systems. Most antibiotics (about 75%) known to date are the products of actinomycetes, especially the members of genus *Streptomyces* (around 50%). Other antibiotic producers include some fungi notably of genera *Penicillium* and *Cephalosporium* and a few species of bacterial genus *Bacillus* (Table 12.2). *Streptomyces griseus* produces more than 4% antibiotics; while *B. subtilis* yields around 60 antibiotics. These are mostly toxic to bacteria, including fastidious bacteria, mycoplasmas, etc. and fungi. More than 7000 antibiotics have been discovered and about 300 new ones are found each year. Most of these are not useful because of either their toxicity to animal or they are rendered inactive when used clinically. About 50 antibiotics are currently produced on larger scale for use.

Table 12.2 : Some soil microorganisms involved in the production of commercially important antibiotics

Microorganisms	Antibiotic produced
Streptomyces aureofaciens	Tetracycline, aureomycin
S. griseus	Cycloheximide, Streptomycin, Candicidin B
S. erythreus	Erythromycin
S. kanamyceticus	Kanamycin
S. linconensis	Lincomycin
S. fradiae	Neomycin
S. orchidaceus	Cycloserine
S. venezuela	Chloramphenicol, chloromycetin
S. noursei	Nystatin
S. rimosus	Oxytetracyclin, terramycin
S. antibioticus	Actinomycin D, Oleandomycin

Contd...

Table 12.2. Contd...

S. niveus	*Streptomyces niveus*
S. pimprii	Hamycin
S. nodosus	Amphitericin B
S. avermitilis	Avermectins
S. lygroscopious sub. sp. *aureoborimosus*	Milbemycins
S. aureus	Polynactins
S. ralstedii	Carbomycin
Nocardia lurida	Ristocetin
Micromonospora perpurea	Gentamycin
Fusidium coccineum	Fusidic acid
Paecilomyces variotin	Variotin
Penicillium crysogenum	Penicillin
P. griseofulvum	Griseofulvin
P. citrimun	Citrinum
P. notatum	Penicillin
P. brefeldianum	Brefeldin
Clitocybe gigaantea	Clitocybin
Cephalosporium acremonium	Cephalosporin C
Polstictus sanguineus	Polyporin
Aspergillus clavatus	Clavacin
A. fumigatus	Fumagillin
Gliocladium virens	Viridin
Lichen usnea; L. cladonia	Usnic acid
Bacillus polymyxa	Polymyxin B
B. colistinus	Colistin
B. subtilis	Bacitrcin, Attrimin
B. brevis	Gramicidin, Tyrothricin
B. lichiniformis	Bacitrcin

Microorganisms used for antibiotic production are all aerobes so are grown under vigorous aerated conditions. The antibiotics are produced by submerged cultivation method using steel fermentors that have continuous forced aeration and mechanical agitation devices fitted.

The chemical formulae of most antibiotics are variable and complex. Antimicrobial activities of antibiotics are measured by determining the smallest amount of the agent (minimum inhibitory concentration) required to inhibit the growth of a test organism. This may be accomplished by tube dilution technique and working out the changes in turbidity. Alternatively, agar diffusion method may be adopted by overlaying the test microorganism on culture media and using then antibiotic coated filter papers one may record the zone of inhibition.

12.3.2 ANTIBIOTICS IN DISEASE CONTROL

Various antibiotics are useful in overcoming many plant diseases. Antibiotics having antifungal or antibacterial action are used to control a number of plant and animal diseases e.g. griseofulvin (a matabolic product of *Penicillium griseofulvum*) and aureofungin (a metabolic product of *Streptoverticillum cinnammoneum* var. *terricolum*). Griseofulvin is a systemic fungicide and permeates plant tissue uniformly. It is used against the diseases caused by *Botrytis, Sclerotinia, Mycosphaerella, Esysiphe,* etc. Aureofungin is a broad spectrum antibiotic and has been effectively used against citrus gummosis (*Phytophthora* sp.), powdery mildew of apple (*Podosphaera leucotricha*), powdery midew and anthracnose of grapes, rice blight, maize blight, etc. Antibiotics used for plant disease control are generally absorbed and translocated systematically by the plant. Amongst the most important antibiotics in plant disease control are streptomycin, teteracyclin, cycloheximide, aureofungin and griseofulvin. Streptomycin, produced by *Streptomyces griseus,* is effective against broad range of bacterial phytopathogens. It binds to the bacterial ribosome and prevents protein synthesis. Sensitivity of microorganisms to antibiotics varies. G +ve are usually more sensitive to antibiotic than G -ve bacteria.

12.3.3 MODE OF ACTION

The antibiotics show selective toxicity to microbes because these act on peptidoglycan or ribosomes or nucleic acid synthesis. Broad spectrum antibiotics are effective against a wide range of G +ve and G -ve bacteria whereas some have narrow range, which act only on a specific group of microbes. Penicillin antibiotic, by inhibiting cross linking reactions, prevents cell wall formation and causes cell lysis in bacteria. Penicillin mostly acts on G +ve bacteria. It may cause allergic reactions in some individuals. Cell membrane inhibitors (polymyxins) act like detergents and destroy the membranes rich in specific lipid types found in G -ve bacteria. Polyenes are active against fungi and act on specific lipids found in the cell membrane of eukaryotes. Some antibiotics such as chloromphenicol, aureomycin, cycloheximde, erythromycin, neomycin, kanamycin, nystatin, etc. inhibit protein and nucleic acid synthesis. Many act as growth factor analogs owing to the resemblance of their structure with growth factors or metabolic intermediates of target organisms. Sulfa drugs or sulfonamides are analogs of growth factor p-amino benzoic acid (PABA), necessary part of vitamin folic acid, so inhibit folic acid synthesis. G-ve bacteria are often intrinsically resistant to antibiotics due to the presence of the outer membrane which acts as an additional barrier for the antibiotic to cross over and, therefore, protects peptidoglycan layer. In others the mechanism of production of such enzymes that degrade the antibiotic or cause reduction in the cell permeability to antibiotic or alteration in the target sites may be involved. The target sites and mode of action of various antibiotics in cells is summarized in Table 12.3.

Table 12.3 : Target sites and mode of action of antibiotics in microorganisms

Target sites in cell	Mode of action	Antibiotic groups involved
Cell wall peptidoglycan synthesis	a) Inhibition of cross-linking	i) β-lactam penicillins (penicillin G, methicillin, ampicillin) ii) Cephalosporins (cephalothin, cephamycin)
	b) Inhibition of polymerization	Glycopeptides (vancomycin)
Cell membrane (lipid metabolism)	Destruction of membrane, acts like detergent on specific lipids in eukaryotes	i) Polymyxins (polymyxin B) ii) Polyenes (amphotericin B, nystatin, filipin, candicidin
Protein synthesis	a) Inhibition of translocation/ ribosomal function	Aminoglycosides (neomycin, streptomycin, gentamycin), amillacin, kanamycin, tobramycin
	b) Inhibition of amino-acyl tRNA's and binding to ribosomes	Tetracyclin (chlortetracyclin, tetracycline, oxytetracyclin)
	c) Inhibition of translation	Chloromphanicol
Nucleic acid synthesis	a) Inhibition of tetrahydrofolic acid synthesis	i) Sulfonamide ii) Trimethoprin
	b) Inhibition of DNA gyrase	i) Quinolone ii) Ciprofloxacin

12.3.4 ANTIBIOTIC RESISTANCE

Pathogens often develop resistance to existing antibiotics so new antibiotics are required to be produced. The resistance is produced due to :

- Development of copious mucilages,
- Alteration of cell membrane so that antibiotics can not be recognized by the pathogen,

- Alteration of cell membrane which prevents antibiotic entry,
- Change to L-form by pathogen,
- Mutation in pathogen,
- Development of pathogenic enzymes capable of modifying antibiotics.

Antibiotic resistance may be inherent property of an organism or it may be because of the lack of site of action for antibiotics or impermeability of organism to antibiotics or because of antibiotic inactivation by microorganism. For example, enzyme penicillinases and β lactamases destroy β lactam ring of penicillin and render it active. Continuous exposure to antibiotics also leads to the development of acquired resistance in target organisms. The acquired resistance involves changes in genetic information. It may be the result of spontaneous chromosomal mutation or due to the entry of plasmids carrying resistance markers (R-factor) or transposons. Genes for antibiotic resistance are transferred through plasmids, etc. between bacteria by conjugation, transduction and transformation. For example, a recipient cell may acquire genes encoding penicillinase from a donor cell. Uncontrolled use of antibiotics has lead to the rapid development of antibiotic resistance in disease causing microorganisms. Antibiotics are extensively used in animal and plant production both as growth promoting substances and as prophylactic measures.

Microorganisms often produce more than one antibiotic, as many as five to six. Multiple antibiotic production presents serious problem. Another problem is the appearance of bacterial strains that are simultaneously resistant to several antibiotics, the so called 'multiple-resistant strains'. Multiple resistant bacterial strains owe their resistance to the presence of a resistance (R) factor. These factors are readily transferable between different bacterial genera because of their broad host range.

12.4 MYCOTOXICOSES

Fungi produce two categories of toxins. One those produced by some mushrooms (such as *Amanita phalloides, Cortinarius orellanus, Coprinus atrementarius, Inocybe* sp., *Gyromitra* sp., etc.) which are consumed directly as food. The poisoning caused by mushrooms is called 'mycetismus'. These cause gastrointestinal, kidney and neurological disorders in human sometimes leading to coma and death. Second type of toxins is those produced by certain molds, which grow on other food products. These are referred to as 'mycotoxins'. Mycotoxins may be mutagenic or carcinogenic and may display specific organ toxicity. Mycotoxins are produced by fungi as secondary metabolites at the end of exponential growth phase. Mycotoxins are generally synthesized via polypeptide route. Induction of mycotoxicoses i.e. disease of animals and humans caused by the consumption of feed and food invaded by toxin producing fungi, thus pose serious threat to human and animal health. Most mycotoxicoses are caused by common and wide spread fungi as *Aspergillus, Fusarium, Penicillium* and *Stachybotrys. Aspergillus* and *Penicillium* produce toxins mostly in stored seeds, hay or commercially processed food and feeds. *Fusarium* produces toxins on corn and other stored grains and *Stachbotrys* on straw, hay and other cellulose products used as fodder or bedding for animals. Main mycotoxins produced by these fungi are presented in Table 12.4.

Aflatoxin : Aflatoxins may be produced in infected cereal grains, most legumes, groundnut and other seeds. The moisture content in the seed or grains is the most important factor determining whether the aflatoxin producing fungus will grow or not. Moisture content >14% favours fungal growth. Aflatoxin, a polyketide toxin, is a metabolic product of *Aspergillus flavus, A. parasiticus* and other *Aspergillus* species. The polyketides are derived from acetyl CoA and melonyl Co A. Aflatoxins are known to possess carcinogenic properties. Aflatoxins consist of several chromatographically distinguishable components such as aflotoxin B_1, B_2, G_1, G_2, M_1, M_2, etc. The symptoms of mycotoxicosis caused by aflatoxin in animals, and presumably in human being also, vary widely with the concentration and type of toxin involved and type, dosage and age of animal. Young duckling, turkey and chicken fed with high dosage of aflatoxin showed serious illness and even caused death. Most of the ingested aflatoxin is taken up by liver. In some areas of Africa where climatic conditions are favourable for the growth of *A. flavus* on peanuts and grains, the incidence of liver cancer in population is moderately higher. Animal feeds containing aflatoxins greater than the permissible level (20 ppb) almost invariably develop liver cancer.

Table 12.4 : Various toxins produced by fungi

Toxins	Causal organisms
Mushroom toxins	
Psilocybin (tryptophan indole alkaloids)	*Psilocybe cubensis, Paneolus* sp., *Conocybe* sp.
Bufotenine (tryptophan indole alkaloids)	*Amanita* sp.
Amatoxins (cyclic polypeptide 8 aa)	*Amanita* sp., *Galerina* sp.
Amanitin	*Amanita phalloides* (death cap)

Contd...

Table 12.4. Contd...

Ibotenic acid, Muscimol, Muscazme	*Amanita mucaria*
Phallotoxins (cyclic polypeptide 7 aa)	*Amanita* sp., *Galerina* sp.
Orellanines	*Cortinarius* sp., *Orellanus* sp., *Galerina sulciceps*
Coprine	*Coprinus atrementarius*
Muscarine	*Clitocybe* sp., *Inocybe* sp.
Gyromitrin	*Gyromitra* sp. (inky cap)
Mycotoxins	
Citrinin	*Penicillium citrinum; P. viridicatum*
Lysergic acid (tryptophan)	*Claviceps* sp.
Ergotamine (tryptophan)	*Claviceps* sp.
Stergimatocystin	*Aspergillus versicolor; A. nidulans; A. rugulosus*
Patulin	*Aspergillus clavatus; A. giganteus*
Ochratoxin	*Aspergillus ochracens; A. mellus, A. ostianus; A. allioceus*
Stachybotryotoxins	*Stachybotrys* sp.
Penicillic acid	*Penicillium puberulum; P. cyclopium*
Sporodesmin (tryptophan, alanine)	*Aspergillus* sp., *Pithomyces* sp.
Gliotoxin (phenylalanine, serine)	*Myrothecium (Gliocladium)*
Pulcheriminic acid (leucine)	*Candida* sp.
Aflotoxin (polyketide)	*Aspergillus flavus*
Alteraria toxin	*Alternaria citri; A. alternate, A. solani; A. tenuissima*
Zearalenone (polyketide)	*Fusarium graminearum; F. tricinctum*
Cytochalasin T (polyketide)	*Phoma* sp., *Helminthosporium* sp., *Chaetomium* sp., *Metarhizinium* sp., *Zygosporium* sp.
Griseofulvin (polyketide)	*Penicillium* sp.
Ergochrome (polyketide)	*Aspergillus* sp., *Penicillium* sp., *Claviceps purpurea*
Xanthocillin (shikimate)	*Aspergillus* sp., *Penicillium* sp.

Aflatoxins affect higher plants and cause inhibition in seed germination, reduce seedling elongation, induce chlorophyll deficiency and mitochondrial injury, interfere with nucleic acids synthesis and functioning, especially mRNA, and inhibit various enzyme systems. Aflatoxins also inhibit *in vitro* growth of *Rhizobium* and root nodulation in legumes.

Fusarium toxins : Sevesal species of *Fusarium* produce zearalenone and trichothecene toxins in molded corn. Zearalenone (F-2-mycotoxin) is produced by *F. roseum, F. graminearum, F. oxysporum, F. moniliforme* and *F. tricinctum*. It causes 'estrogenic syndrome' in swine i.e. abnormalities and degeneration of genital system. Trichothecene toxins (T-2 mycotoxin) are produced in feeds infected with *Fusarium, Cephalosporium, Mycothecium, Trichoderma, Stachybotrys* species, etc. It affects swine, cows, chicks etc. causing vomiting, degeneration of cells of bone marrow, nodes and intestines, diarrhea, blistering, haemorrhagia and death.

Ergot alkaloids : These are produced by *Claviceps purpurea* and other *Claviceps* species which grow in the heads of cereal grains, produce sclerotinia and replace grains with the bodies containing ergot alkaloids. These fungi produce an extensive series of alkaloids, based on the amino acid tryptophan, dimethyl-allyl pyrophosphate and methyl methionine, and form ergoline ring structure. The alkaloids are structurally related to lysergic acid diethylamide (LSD), a well known psycho-active drug. Ergots cause abortion and smooth muscle contraction which sometimes may lead to the loss of limbs.

Patulin : It is a polyketide carinogenic toxin produced by *Penicillium urticae, P. patulum, P. griseofulvum, P. claviforme, Aspergillus clavatus, A. terreus, Byssochlamys nivea* and *B. fulva*. It is biosynthesized through acetate-malonate pathway. It is toxic to bacteria and some fungi, higher plants and animals, and causes repression of respiratory centers.

Stachybotrys toxins : These toxins are produced by the species of *Stachybotrys* on straw, hay, other fodders and animal feeds, etc. Stachybotrys toxins cause diseases like hemorrhage, necrosis and general cell damage in domesticated animals.

Alternaia toxins : Several species of *Alternaria* (*A. citri; A. alternata, A. solani; A. tenuissima*) produce toxic substances that have been found in apple, tomato, blue berry, etc. The toxins produced include alternariol, alternatiol monomethyl ether, alternuene, tenuazoic acid and altertoxin.

Aspergillus and *Penicillium* produce many other kind of toxins such as yellow rice toxins (citreviridine, citrinin, luteoskytrin, cyclochloroline, etc.) and tremorgenic toxins. Ochratoxin produced by *A. ochraceus* causes fatty liver disease; strigmatocystin produced by *A. versicolor* and citrinin produced by *P. citrinum* incite liver damage and, in some cases, liver cancer. Fungal allergins, endotoxins of zoopathogens and toxins that induce wilt in plants are usually of high molecular weight polymeric carbohydrates. Other toxins produced by fungi are primarily of low molecular weight (<1 kDa).

12.5 HYDROGEN SULPHIDE PRODUCTION

Hydrogen sulphide is produced by a number of obligate anaerobes in soil, if abundant sulphate and organic matter is present at a redox-potential below oxygen requirement level of microbes. H_2S production occurs in rhizosphere and spermosphere where O_2 is less. Plant respiration is inhibited under the soil conditions having $H_2S > 0.1$ μg ml^{-1} soil water. H_2S production can cause even greater agronomic and environmental problems than the products of denitrification because roots of many plants are sensitive to high H_2S concentration. Microbes responsible for H_2S production in soil are sulphate reducers like *Desufatomaculatum, Desulfovibrio* and *Desulfomonas*.

Selected References

Allen, M.F., Smith, W.K., Moore, Jr. T.S. and Christensen, M. 1981. *New Phytol.*, 88 : 683-693.

Arshad, M. and Frankenberger, Jr. W.T. 1991. Microbial production of plant hormones. *Pl. Sci.*, 133 : 1-8.

Griffin, D. H. 1994. *Fungal Physiology*. Wiley-Liss Inc., New York, USA.

Killham, K. 1994. *Soil Ecology*. Cambridge Univ. Press, Cambridge, UK.

Lynch, J.M. 1983. *Soil Biotechnology : Microbiological Factors in Crop Productivity*. Blackwell Sci. Publ., Oxford, London, UK.

Macko, V. 1983. Structural aspects of toxin. In : *Toxins in Plant Pathogenesis* (eds. J.M. Daly and B.J. Deverall). Academic Press, London, UK.

Misato, T. and Yoneyama, K. 1982. Agricultural antibiotics. pp. 465-506. In : *Advances in Agricultural Microbiology* (ed. N.S. Subba Roa). Oxford & IBH Publ. Co., New Delhi, India.

Mahadevan, A. 1984. *Growth Regulators, Microorganisms and Diseased Plants.* Oxford & IBH Publ. Co., New Delhi, India.

Mehrotra, R.S. 1980. *Plant Pathology*. Tata McGraw Hill, New Delhi, India.

Otani, T. and Ae, N. 1993. Ethylene and carbon dioxide concentrations of soils influenced by rhizosphere of crops under field and pot conditions. *Pl. Soil,* 150 : 255-262.

Rangaswami, N.S. and Bagayaraj, D.J. 1994. *Agricultural Microbiology*. Prentice Hall, New Delhi, India.

Subba Rao, N.S. 1999. *Soil Microbiology*. Oxford and IBH. New Delhi, India

Wolf, A.E. and Cruger, A. 1982. *Biochemistry : A Text Book of Industrial Microbiology*. Sinauer Assoc. Inc., Sunderland, USA.

Yoder, O.C. 1980. Toxins in pathogenesis. *Ann. Rev. Phytopath.* 18 : 103-129.

— ★★★ —

Chapter 13

Organic Waste Recycling

The soil is a natural medium for waste disposal and simultaneously is a major component of man's environment so needs utmost care and gentle treatment.

The soil microorganisms thrive on organic matter in soil by feeding readily on soluble substrates and rather slowly on insoluble substrates. Organisms derive carbon and energy during the process of organic matter decomposition and release organically-bound carbon, thereby decrease C :N ratio and utilize nutrients. Any decrease in organic matter content in soil is likely to affect the microbial biomass and their activities thereby influence the soil fertility and productivity. It is, therefore, imperative to recycle the organic matter in soil in whatsoever form it may be. The raw organic matter requires appropriate treatment to boost decomposition process. The objectives of organic waste recycling are to treat the waste and reclaim valuable substances present in wastes for possible reuses. These valuable substances include carbon, nitrogen, phosphorus and other macro- and micro-elements. The possible methods of organic waste recycling include agricultural use, bioenergy production and aquaculture use.

13.1 AGRICULTURAL USES

Organic wastes can be applied to the crops as fertilizers and/or as soil conditioners to improve the physico-chemical properties of soil. Direct application of raw organic waste does not yield the desired result as most of the inorganic nutrients and some organic matter is lost during the process of unscientific handling, processing and storage. We know that crops take up only inorganic form of nutrients like nitrate and phosphates. Microbial activities can be exploited to completely transform organic compounds into simple organic and then finally to inorganic compounds. During organic matter decomposition nitrogen, phosphorus, potassium and other nutrients are taken up by the microbial cells through immobilization processes and are later released on their death. This results in the accumulation of nutrients in inorganic form. Anaerobic conditions lead to the partial degradation of organic matter and accumulation of organic acids, methane, carbon dioxide, hydrogen, etc. as a result of microbial activities. The technologies of composting and anaerobic or aerobic digestion are examples in which organic wastes are stabilized and converted into products suitable for reuse in agriculture. Untreated wastes including sewage, sewerage and non-activated sludge may cause many undesirable public health problems and contamination. The wastewater that has been treated either by sedimentation and/or biological stabilization can be applied to the crops or grasslands through sprinkling or soil infiltration. Zeolite addition to extract heavy metals may be a key to the safe use of sludge as fertilizers and help to extend the life of sludge-disposal sites. Some of the methods of waste crop recycling through agriculture use are discussed as under :

13.1.1 COMPOSTING

Composting is a controlled aerobic decomposition process, carried out by successive microbial populations, combining both mesophilic and thermophilic activities which lead to the production of stabilized organic humus-like substance that have reduced bulk. Generally composting is applied to solid and semi-solid organic wastes such as night soils, sewage sludges, animal manures, crop residues and municipal refuse. In aerobic composting end products are CO_2, NH_3, organic acids, water and generation of heat. Anaerobic composting forms methane, CO_2, NH_3, trace amounts of other gases and other low molecular weight organic acids.

i) $\underset{\text{Organic matter}}{(COHNS)_x} + O_2 \xrightarrow{\text{Aerobic bacteria}} CO_2 + NH_3 + \text{energy} + \text{other end products}$

$$\text{ii)}\quad \underset{\text{Organic matter}}{(COHNS)_x} \xrightarrow{\text{Anaerobic bacteria}} \text{free sugars} + NH_3 \longrightarrow \text{energy} + \text{other end products}$$

Ammonia is further oxidized to nitrate (NO_3) by the nitrifying bacteria during the maturation or curing phase. Heat generation increases decomposition rate so favours the stabilization of large quantities of organic wastes. Anaerobic process is a slow process and can produce obnoxious odours originating from the intermediate metabolites such as mercaptans, sulphides, etc. Anaerobic conditions generate heat and raise temperature near to thermophilic levels (50-65°C).

a) Composting Process

Organic waste, highly heterogeneous in nature, is broken down during composting which according to temperature pattern has four phases viz., latent phase, growth phase, thermophilic phase and maturation phase (Table 13.1). In aerobic process degradation of organic matter depends on the presence of oxygen. During composting both qualitative and quantitative changes occur in the active microflora and their metabolic activities. This results in rise in temperature in composting heap. Organic wastes are initially decomposed by first level consumers. Mesophilic bacterial chemoheterotrophs appear first. As a result of their respiration activities, the temperature rises in the compost heap and thermophilic organisms begin to replace mesophilic population. Thermophilic fungi usually grow after 5 to 10 days of composting. Maximum thermophilic activity in compost is between 60 and 65°C. High temperature not only increases the microbial decomposition/activity but also kills the pathogens in compost material. If temperature is too high (>70°C) only spore forming bacteria, mostly *Bacillus* and actinomycetes, develop. Aeration, watering and turning compost material prevents its self-heating.

The temperature of compost declines as the readily available substrates are exhausted. As the temperature declines actinomycetes and fungi reinvade and dominate the scene. At this time heap surface appears white or gray in colour. Fungi and actinomycetes decompose cellulose, lignin and other resistant materials. *Streptomyces* sp., *Bacillus stearothermophilus, Clostridium thermocellum, Thermoactinomycetes* sp., *Thermomonospora* sp., *Geotrichum* sp., *Mucor* sp. and *Aspergillus* sp. are commonly encountered in compost heaps. The mesophilic macroflora and macrofauna flourish during final stage of composting and contribute to mixing of various degraded constituents. The population of various organisms in compost generally lies in the range given below :

Bacteria	10^8-10^9 g^{-1} compost
Actinomycetes	10^5-10^8 g^{-1} compost
Fungi, mold and yeast	10^4-10^8 g^{-1} compost
Protozoa	10^4-10^5 g^{-1} compost

Table 13.1 : Phases of organic matter decomposition

S. No.	Phases	Phenomenon
i.	Latent phase	• Microorganisms acclimatize and colonize in compost heap
ii.	Growth phase	• Mesophilic temperature of bacterial activity
iii.	Thermophilic phase	• Waste stabilization and pathogen destruction; • Temperature rises to 55-65°C; • Bacteria, fungi and actinomycete become active
iv.	Maturation phase	• Temperature decreases to mesophilic level so favours secondary fermentation; • Slow process favours humification and nitrification; • Nitrifiers inactivated at temperature 40°C; • Second and third level consumers (mites, nematodes, protozoa and beetles) grow and feed on first level consumers (bacteria, fungi and actinomycetes); • Nutrients returned as compost are in the form of microbial protoplasm and/or organic compounds that breakdown slowly.

A good quality compost of farm residues, containing about 1.5% nitrogen or more, may be prepared by amending the organic matter with phosphate fertilizers and inoculating with efficient cellulolytic culture. If organic matter is rich in lignin, the use of lignolytic microorganisms such as *Clavaria* sp., *Cephalosporium* sp., *Humicola* sp., etc. are advocated. The cellulolytic microorganisms such as *Trichoderma viride, Chaetomium abuanse, Myrothecium roridum, Aspergillus niger, A. terreus, Cellulomonas* sp., *Penicillium* sp., *Trichurus spiralis, Paecilomyces fussisporus, Nocardia* sp., *Streptomyces* sp. and *Cytophaga* sp. are often inoculated to accelerate decomposition process. *Paecilomyces fussisporus,* for example, is inoculated @ 300-500 g mycelial mass t^{-1} raw material. Maintenance of C :N ratio of 12-18, optimum moisture level of 50-60% and proper aeration is necessary for efficient conversion. Secondary inoculation, 28-42 days later, with nitrogen fixers (*Azotobacter chroococcum)* and phosphorus solubilizing fungi and bacteria (such as *Aspergillus awamori, Bacillus polmyxa,* etc.) may be done around ambient temperature after organic matter stabilization is achieved. Enriched compost @ 10 t ha^{-1} may be applied to the field 15 days before sowing, irrespective of recommended doze application of chemical fertilizers.

b) Environmental Factors Affecting Composting Process

Many environmental factors influencing the process of composting are as under :

1. **Nutrient balance :** A C :N ratio 25 :1 and C :P ratio of 100 :1 to 150 : 1 is desired for faster mineralization. Plant materials often have wide C :N ratio. Therefore, sometimes supplementation of inorganic nitrogen and phosphate fertilizer may be required. The exclusive use of high C :N ratio substrate does not allow microbial biomass to release nutrients in compost. Low C :N rates lead to nitrogen loss and production of foul odour. The inoculation of *Azotobacter* and phosphorus solubilizers under such a situation accelerates decomposition process. Optimum initial C :N ratio of compost may be 20-30.
2. **Particle size & structural support of compost pile :** The particle size of the material fed to composting pit should be small to allow more aeration for microbial action. Dried crop wastes need to be cut into small pieces (2-3 cm long). For agitated and aerated system particle size need to be maintained at 10 μm and for long heap and natural aeration at 50 μm.
3. **Moisture control :** Water is essential for nutrient stabilization and microbial activity. Further, water helps to cool the compost. The moisture content in compost pit if <20% severely inhibits the biological processes whereas high moisture leads to the nutrient leaching and creates anaerobic conditions. Moisture content of 50-60% in compost pit is most suitable during bacterial phase.
4. **Aeration :** Aeration allows sufficient oxygen for aerobic growth. Periodic turning of compost piles, insertion of perforated bamboo poles into compost piles allows more aeration. Forced aeration is done by pumping air via perforated pipes and orifice into the heap composts. Mostly aerobic conditions prevail at the outer surface of the compost heaps, while inside heap facultative or anaerobes dominate. Too much aeration leads to the loss of heat from piles. While anaerobic conditions lead to the release of odorous volatile compounds, it is recommended to maintain proper oxygen level around 10-18% for optimum activity during thermophilic stage.

5. **Temperature :** Microbial metabolism generates enormous heat which increases the temperature in compost pit, maximizes decomposition rate and produces a material which is microbiologically safe for use. The optimum temperature in compost piles may be controlled by proper aeration and moisture as well as by the use of screened compost in compost piles. Temperature pattern influences the type and species of microbial growth. In compost pits mesophiles (optimum; 25-45°C) grow first followed by thermophiles (optimum; 55-65°C). Frequent turning of the compost helps to cool the compost pile.

6. **pH :** Aerobic composting rarely causes extreme pH drop or rise, however a pH range of 6.5 to 7.0 is considered ideal.

7. **Type and nature of organic material :** The type and nature of substrate and the microflora associated influences the rate of humification. Highly heterogeneous substrates are fast decomposed. The process of decomposition and quality of compost is affected by the nature and composition of organic waste, surface area of the material, hydrophobicity and type and complexicity of chemical bonds in residues. A particle size of 0.75 to 2.5 cm is optimal for composting. Legumes are high in nitrogen so have low in C :N ratio and their decomposition process is faster than cereals. Young plants are easily metabolized as these have more soluble nutrient materials. Materials rich in lignin or heavy metal or toxicants offer greater resistance to decomposition. Inoculation of compost with materials from previous compost piles is beneficial.

8. **Activators and enhancers :** Efficient celluloytic fungi and biofertilizers (nitrogen fixers and phosphate-solubilizers) may be added to enhance decomposition process. Sometimes, limiting nutrients may also be supplemented to the composting process to proceed smoothly.

The compost is ready for stabilization or curing when turning no longer increases temperature. Curing is slow maturation of the compost after active composting is over. It allows maturity of compost which can be judged by decrease in organic content of composts [viz. chemical oxygen demand (COD), volatile solids, carbon, ash or C : N ratio], decline in moisture and temperature, presence of nitrates and absence of ammonia, lack of insect attraction, removal of obnoxious odour and presence of white or gray colour due to actinomycetous growth.

c) Composting Systems

Depending upon the location of the site of composting, the composting systems may be on-site (where composting of organic matter is done at the place of its generation) and off-site system (where organic matter is collected/ transported to a central treatment plant; and the process is controlled manually or mechanically).

i) Indian Composting Methods

Pit or Indore method : This is an aerobic method of composting wherein cattle dung and plant materials are dumped in circular or rectangular pits of 1.0 m depth and 1.5-3.0 m width near cattle shed. Aquatic weeds or green manure crop (0.7 t), rice straw (0.15 t), animal dung or excreta (1.0 t) and silt (7.5 t) are used in composting in pits. The pit sites should be at elevated places, often protected by a shed, to avoid rain-soaking of pits. Plant residues, cut into small shreds, are used as bedding material for cattle. Some 2-3 kg of well pulverized soil may also be spread in the evening time on the floor of cattle shed. This helps in soaking of urine which is rich in soluble nutrients so helps in their conservation. Wet beds alongwith excreta are removed from cattle shed and heaped in a layer about 3-5 cm in pits. On it a slurry of dung, made with 4.5 kg urine-earth and 4.5 kg inoculum taken from a 15 day old composting pit, is spread. Over this layer sufficient quantity of cattle dung is spread. A sufficient quantity of water is sprinkled to maintain optimum moisture. The pits are filled in layers, 15 cm thick, with chopped or shredded crop residues. The process of layering and spreading is repeated so that layers are formed one above the other till residues reach a height 30 cm above the ground level. Usually it takes 4-6 days for each pit to fill. The upper layer is plastered with mud and covered with a water column of 3 to 4 cm depth. After 15-20 days of pit filling, the material is turned over and superphosphate (0.02 t) added. The decomposing material is mixed thoroughly and moistened with water. This facilitates creation of anaerobic conditions and reduces nitrogen losses. The compaction of compost should be avoided. Residues are turned every fortnightly. Good quality compost is ready within 3 months. Manure (8 t) is ready after 12 weeks time with a C :N ratio of 15-20 :1 and organic matter content of 8 to 10%.

Bangalore method : In Bangalore method greater conversion of nutrients is attempted. In this method aerobic process is followed by anaerobic one. Organic residues and night soil are first put in alternate layers on a hard ground of pit. After filling, the pit is covered with 15-20 cm thick layer of refuse. The heaps are prepared in similar way as in Indore method, except that each heap in pit is sealed with mud plaster which increases temperature inside the plastered pit due to anaerobic fermentation. The material in pit is allowed to remain as such without turning or watering for 90 days. During this period the material gets settled due to the reduction in the volume of biomass. Additional night soil or refuse are put in alternate layers on top and pits once again plastered or covered with mud to prevent any moisture loss. The material once again undergoes anaerobic decomposition at a very slow rate. The mature compost, rich in nitrogen, is ready within 6 to 8 months period.

In certain locations liquid manure is kept in cemented pits to which weeds/aquatic plants are added. Alternatively, night soil and urine are kept in closed tanks and allowed to ferment for 15 to 20 days. Liquid manure may be used as top dressing.

Heap method : During rainy seasons or in the areas of heavy rainfall, composts may be prepared in heaps above the ground level. Small bunds are sometimes built around these piles to protect heaps from wind. The building of a heap is started with 20 cm layer of carbonaceous materials such as leaves, hay, straw, saw dust, dung, chopped agricultural residues, etc. This is then covered with 10 cm nitrogenous material such as fresh grass, weeds, garden plant residues, garbages, fresh or dry manure, digested sewage sludge, etc. The pattern of 15 cm carbonaceous residue and 10 cm nitrogenous material is followed until the pile attains a height of 1.5 m (Fig. 13.1). The piles are normally wetted so that they feel damp. Pile is sometimes covered with soil or hay to retain heat and is turned at 6 and 12 weeks interval.

ii) Chinese Composting Pile

The compost feed (human/animal manure and vegetable matter/crop materials in 1 :4 ratio) is piled up layer by layer into a heap of approximately 3.0 x 2.0 x 1.0 m (LWH). Each layer is about 15 cm thick. Bottom layer has green plants and aquatic weeds. The second and third layers have silt-straw mixture and animal excreta, respectively. Layering is repeated until pit is filled. The compost heap is then pierced by perforated bamboo poles or maize stalks to facilitate natural ventilation and provide a kind of structural support. Pile is covered with rice straw or plaster of mud. The poles may be removed after 1 or 2 days when mud hardens and temperature reaches 60-70°C. The holes are sealed with mud and compost pile becomes structurally stable. Sometimes the pits are plastered with mud and about 4 cm layer of water is maintained on the surface to create anaerobic conditions which helps to reduce nitrogen losses. However, some follow turning of compost material at regular intervals while others do not. Usually after 2 weeks the mud plaster is broken down and heap turned by adding fresh liquid excreta and superphosphate/ rock phosphate, followed by sprinkling of some water and resealing of pit with mud plaster.

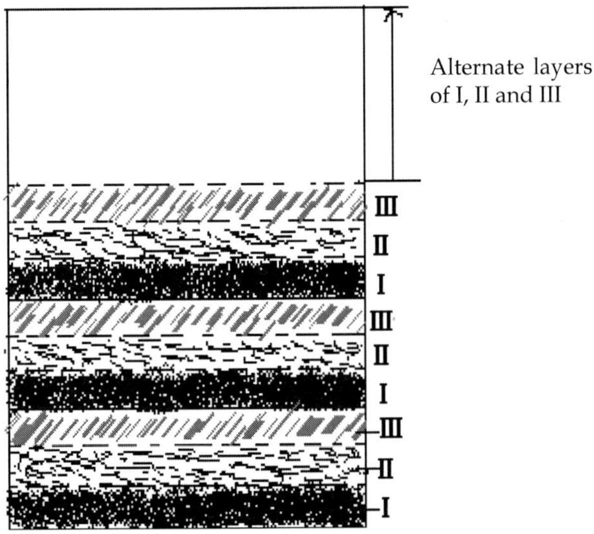

Fig 13.1 : Heap composting system : I = urine soaked stable manure; II = shredded straw; III = green manure + aquatic weeds/legume straw, etc.

Nitrogen fixing bacteria such as *Azotobacter* and *Azospirillum*, phosphate dissolving organisms such as *Psuedomonas, Bacillus, Aspergillus, Fusarium,* etc. may be inoculated to improve compost quality. City garbage (70-80%) and light soil (20-30%) may also be used. The time required for compost stabilization is 60 days. The compost formed has C :N ratio of 15-20 :1 and organic matter content of 8-10%.

iii) Windrow Composting

In windrow composting the approximate size of compost pile is 13.0 x 3.0 x 1.5 m. Periodic turning of compost is done once a week if done manually, otherwise daily if done mechanically so as to mix the materials and provide good aeration. The composting time is about 9 months.

iv) Aerated Static Pile Composting

Also known as "Beltsville forced air rapid composting", aerated static pile composting involves placing a mixture of sludge and wood chips over a base (consisting of compost and chips) and aerating it through a piping system.

The size of each pile is 12.0 x 6.0 x 2.5 m. The pile is connected to a centrifugal blower to draw air through sucking piles or to blow air through blowing piles. Entire pile is covered with a 30 cm layer of screened compost to avoid foul odours and maintain high temperature. It usually takes 30 days for composting the material through this method.

v) Dano System

The system consists of storage hoppers, rotary screen and magnetic separators to separate out non-combustible materials. Composting materials are fed into a biostabilizer which is a cylindrical chamber of 25-30 m in length and 3–4 m in diameter with slight tilt from horizontal. The cylinder rotates @ 1 rpm. The air is supplied at low pressure by fans through longitudinal ducts, each having several injector nozzles. This enhances effective aerobic decomposition of organic matter with a temperature of 60°C. Steam and waste gases are exhausted through extractor fan. The composting period required for ones reactor is 3-8 days.

vi) Jersey System

The system utilizes only municipal refuses. It is an aerobic process which involves aeration during movement of composting material from top (5th flours to a lower floor). Municipal waste is pre-processed to remove inorganic and non-compostable materials. Organic material is first carried to the 5th floor of the plant, kept as such there for one day and then released to next lower floor and so on. It takes five days for the compost material to reach lower floor. Composted products are further cured in an open lot for 30-60 days period before being used as a soil conditioner.

13.1.2 VERMICOMPOST

Vermicomposting involves the use of earthworms for composting the organic residues so as to accelerate the decomposition process and enhance the quality of manure. Earthworms also serve as biological indicators of soil fertility and health. The earthworm gut is an effective bioreactor and the organism releases many organic matter degrading enzymes such as protease, lipase, amylase, cellulase, chitinase, etc., besides growth hormones like cytokinins and auxins. Worldwide around 3,000 species of earthworm are present. Earthworms make tunnels in the soil, crush and mix it well and add rich casting materials thereby are useful in improving the soil quality. Physically, they work as aerator, crusher and mixer; chemically act as degrader and biologically serve as stimulator of organic matter degradation. The species suitable for composting are epigeic or surface living and reddish brown earthworms. The prominent species are *Lumbricus rubellus* (red worm), *Eisenia foetida* (red, brown or purple), *Eudrilus eugeniae* (animal flesh coloured) and *Periyonyx excavatus* (deep purple). One kg of earthworm consumes about one kg of organic matter per day. It is estimated that earthworm action can convert 1000 t moist organic matter into 300 t compost in a very short period. Earthworm excreta together with their cocoons and undigested feed make up vermin-castings which is rich in nitrates and available P, K, Ca, Mg, etc. Their casting also encourages the growth of bacteria and actinomycetes due to aeration in pore spaces. The actinomycetes population in worm-cast is extremely greater than the original soil.

The earthworm which have standardized culturing techniques show high consumption, digestion and assimilation, efficiently convert organic wastes, have fast growth rate, produce large number of cocoons, show good compatibility with other worms, disease tolerant and are adaptable to varied environment must be selected for rearing purpose. The earthworms are reared and multiplied in shallow perforated wooden boxes of 45 x 60 cm size and stored on shelves and tiers. Cemented tanks, earthen pots, earthen pits lined with stones and plastic are also used for the purpose. A vermicompost pit of 2.0 x 1.0 x 1.0 m size holds 20,000 to 40,000 worms, and yields one tonne manure per month. A vermin-bed of 15-20 cm is first made on a good loamy soil with a thin 5 cm layer of broken bricks and sand on it. About 100 earthworms are introduced into vermin-bed, followed by organic layering of fresh cattle dung and dry leaves (5 cm thickness each). Sometimes fungal and bacterial inoculants are also introduced to enhance composting process. A moisture level of 35-50% is maintained by regular watering. The bedding material is moistened for about 4 weeks and then chicken manure, green manure, water hyacinth, etc. are added to it. Other moist organic materials such as saw dust, crop wastes, cereal straw, rice husk, kitchen garbage, sheep dung, cattle dung, biogas slurries etc. may also be used. The moist green organic waste is spread over it to 5 cm thickness and the process is repeated after every 3-4 days. A pH of around 7.0 and temperature 20-30°C is maintained in pits. Mixing of waste without disturbing vermin-bed ensures proper vermin-composting. The process is repeated till the pit is full. A moist compost heap (35-40% moisture level) of 2.4 x 1.2 x 0.6 m at full activity period can support a population of more than 50,000 worms. Earthworms consume organic matter within a short period. At maturation regular watering is stopped for 3-4 days. This brings the moisture level down and ensures the migration of worms into vermin-bed. In some case boxes or pits of 4 x 3 x 1 m size, having sloping sides, constructed in shade are used. Bamboos are inserted in the filled pits as previously. The pits are lined with gunny bags to avoid the escape of worms. Worms are inoculated after the addition of substrate. Mature compost, a fine loose granular mass, is removed from the pit, sieved, dried and packed. Mature compost is applied to the crop @ 0.5 t ha^{-1}.

The advantages of vermicomposting are that the technology reduces environmental pollution, increases the soil water holding capacity, improves soil properties, provides fertilizer rich in vitamins, hormones and enzymes, increases the population of beneficial organism in rhizosphere so imparts pest and disease resistance and partially reduces the dependence on inorganic fertilizers.

13.1.3 COMPOSTING - BENEFITS AND RISKS

The application of compost to a soil influences its physical, chemical and biological properties. Soil physical conditions are improved through increase in water holding capacity, better aeration and drainage in clay soil, more compaction in sandy soil, better soil structure development and reduction in erosion of vital plant nutrients. Nutrients become available to the plants because of increased aggregation of humic substances, high cation exchange capacity and greater water retention. Stimulation of overall microbial biomass viz-a-viz specific beneficial microorganisms, especially those involved in the production of polysaccharides, nitrogen fixation and phosphorus solubilization, improves the biological activities in soil. The composts may be used in many ways which include :

- Fertilizer;
- Soil conditioner;
- Feed for fish in aquaculture;
- Land fill material;
- Horticultural medium on parkland, ornamentals and recreational areas.

Compost to be used as fertilizer or soil conditioner is usually fixed with chemical fertilizers to make its nutrient contents suitable for crop growth. The rate of application is 25-50 t ha^{-1}. Composted organic wastes serve as slow release-nitrogen fertilizers applied to soil. The average N, P and K contents in the compost are 2.58, 1.67 and 0.58%, respectively. The composting of organic wastes helps in waste stabilization, pathogenic inactivation and nutrient conservation and land reclamation. Compost application to soils reduces nutrient losses through leaching. Soil tilth is improved so permits better root growth and nutrient uptake.

Composting, especially when using municipal refuses, involves many health risks (Table 13.2). However, time and temperature are important parameters responsible for pathogen die-off during composting. Complete destruction/inactivation of pathogen is rarely achieved because of uneven temperature distribution, heterogeneous compost materials and partial inactivation of pathogen. The health-risk pathogens are *Salmonella typhosa, Salmonella* sp., *Shigella* sp., *Escherichia coli, Taenia saginata, Trichinella spiralis, Necator americans, Brucella abortussuis, Micrococcus pyogenes* var. *aureus, Mycobacterium tuberculosis* var. *hominis, Corynebacterium diphtheriae,* etc.

Table 13.2 : Pathogens associated during composting of sewage sludge

Microbial groups	Organisms	Disease caused
Primary pathogens		
- Bacteria	*Salmonella enteritidis*	Salmonellosis (food poisoning)
- Protozoa	*Entamoebae histolytica*	Amoeba dysentery (blood dysentery)
- Helminths	*Ascaris lumbricoides*	Ascariasis
- Viruses	Hepatitis virus	Jaundice (infectious hepatitis)
Secondary pathogens		
- Fungi	*Aspergillus fumigatus*	Aspergillosis (growth in lungs etc.)
- Actinomycetes	*Micromonospora* sp.	Farmer's lung (allergies)

13.1.4 MUSHROOM PRODUCTION

The mushroom industry is basically the utilizer of organic wastes and by-products but at the same time itself generates waste products in the form of spent mushroom substrate. The spent mushroom substrate can be used as additive to cattle feed or applied to soil as organic fertilizers or may serve as partial substrate for peat in artificial media for growing vegetables, flowers or tree saplings.

Several kinds of fungi serve as source of human food of which mushrooms have fascinated man since times immemorial because of their innate flavour, delicious taste and richness in quality proteins. About 10,000 species of macrofungi are known to mycologists and around 4,000 species are mushrooms, of which about 50% species are edible. Mushrooms are generally found in places where dead organic matter in the form of decaying leaves, straw, log, etc. are found in abundance. Presently, about 35 fungal species are cultivated in over 100 countries with a world-wide production of about 2.2 million tonnes. Many edible mushrooms appear to be mycorrhizal having symbiotic relations with the roots of certain trees. Mushrooms are fruiting bodies of basidiomycetous fungi formed during certain periods of year. Mushrooms are of variable colour, shape and size, and may be either edible or poisonous or plant parasitic. The poisonous mushrooms are popularly known as 'toadstools'. Of the cultivated mushrooms, five genera *Agaricus, Pleurotus, Volvariella, Lentinus* and *Flammulina* contribute more than 95% mushroom production. Both wild and commercial mushrooms are consumed. The mushroom commercially available in most temperate parts of the world is 'white button mushroom [*Agaricus brunnescens* (syn. A. *bisporus*)]' which is cultivated widely in mushroom farms. The other mushrooms such as 'Oyster mushroom (*Pleurotus sojar-caju, P. ostreatus*)' and 'Dhinghari mushroom (*Pleurotus* sp.)', Shiitake mushrooms (*Lentinus edodes*), paddy straw mushroom (*Volvariella volvaceae*) and honey mushroom (*Armillaria mellea*) are also cultivated on commercial scale. Commercial mushroom production technology is based on the utilization of agricultural wastes such as wheat or paddy straw, etc. Various methods of substrate preparation have been developed for different mushrooms. Steam pasteurization or hot water treatment (dipping substrate at 70-80°C for one hour) or chemical pasteurization (formalin application @ 500 ppm) are some of the effective methods widely adopted. Mushrooms convert nutritionally less valued materials into acceptable nutritious foods. From 3 kg of dry wheat straw 1 kg of fresh mushroom may be obtained.

Morphology : Mushrooms are fruiting bodies of fleshy fungi, majority of which belong to Basidiomycotina and some to Ascomycotina. The fruiting bodies are formed through growing together of a large number of individual hyphae (carpophores) which contain microscopic seed-like spores. The mushrooms first appear as white tiny balls consisting of stalk (short stem) and a cap (pileus), which later open up like an umbrella. The delicate membrane (viel) enveloping the cap tears off, if allowed to develop fully, and lamellae (gills) radiating from the stalk (stipe) into the cap become visible. A membranous annulus occurs on the upper part of the stipe in many mushrooms. The gills become darkened as the basidiospores (seeds) develop into millions and fall to the ground for starting their life cycle once again for a new generation of mushrooms.

If a mushroom fruit body is cut and examined under a microscope, the individual hyphae can be seen pressed tightly together. During most of its existence period, a mushroom lives as simple mycelium, growing within soil, leaf litter or decaying wood logs. When environmental conditions are favourable, the fruiting bodies develop; beginning first as a small button-shaped structure underground and then expanding into the full grown fruiting body that can be visualized aboveground. The nutrients for growth come from organic matter in the soil and are taken-up by the hyphal filaments which feed them to the growing fruiting bodies. Sexual spores called 'basidiospores' are formed. They are borne on the underside of the fruiting body, either on flat plates called gills or within deep pores. The spores are often coloured and impart a colour to the underside of fruiting body cap. Some mushrooms called 'puff balls' produce their spores within spherical fruiting bodies instead of the gills or pores. The spores puff out through cracks or holes that develop in the fruiting body as it dries. Spores are dispersed by wind and germinate when conditions become suitable. Classification is based on size, shape and colour of fruiting body, arrangement of gills or spores and the colour of spores.

Mushroom cultivation : Out door activities (pre-wetting and composting) can be done in any season but during indoor activity temperature requirement for spawn run (vegetative growth) is 22 to 25°C and for crop production (fructification) is 14 to 18°C. Indoor mushroom cultivation is done in trays. It requires safe handling and protection from the vagaries of nature. Mushrooms are grown on fermented composts which may be synthetic or natural, either using wheat or paddy straw as principal ingredient along with inorganic and organic fertilizers. Well decomposed compost, without any smell of ammonia, should have a C :N ratio of 17 :1 and moisture content of 65-70%. A synthetic compost, made up of wheat straw for growing *A. bisporus*, sufficient for one bed, contains calcium ammonium nitrate or ammonium sulphate 240 g, superphosphate 300 g, urea 96 g, potassium sulphate 120 g, gypsum 1200 g and saw dust 400 g. Mushrooms like *A. brunnescens, Volvariella volvacea, Pleurotus* sp. and *Armillaria mellea* are grown in special beds on wheat and paddy straw usually in buildings where temperature, humidity and ventilation are carefully regulated.

The growth requirements vary for different species of mushrooms. Generally, mushrooms require an atmosphere nearly saturated with moisture. Good aeration maintains congenial conditions and also removes toxic gases. A CO_2 level of 0.10 to 0.15 (volume) is necessary during production phase that can be achieved by providing 4 to 6 air changes per hour or by introducing 10 cu ft fresh air per sq. ft bed area per hour. Mushrooms grow independent of sunlight. Beds are prepared by mixing soil with a substrate very rich on organic matter, such as horse or chicken manure. These beds are inoculated with mushroom 'spawn'. The spawn is a pure culture of mushroom fungus that has been grown in large bottles on an organic rich medium (*A. bisporus* spawn is especially grown on wheat or sorghum grains in specialized spawn laboratories). The spawning rate for Oyster mushroom is 3-15% (dry weight basis) of substrate with 10% as optimal rate. A temperature of 22-28°C, with optimum of 23°C, and relative humidity of 90-95% is maintained in the initial stages of fungal growth. In the bed, spawn grows and mycelium proliferates into the compost. After several weeks it is ready for next step, the induction of mushroom formation. This is accomplished by adding to the surface of the bed a thin layer (4-5 cm deep) of well sterilized soil (peat or chalk or sphagnum-based) called "casing soil". Casing works as a reserviour of moisture for mycelial growth as well as regulates the flow of nutrients from substrate to the developing sporophore and also regulates water evaporation from bed surface. One week later temperature is lowered to 15-18°C. Pin-head mushrooms start emerging after about a week, develop to full size and continue to do so economically up to 2 months. A fine spray of water should be given over the casing and 70 to 80% humidity maintained during this stage. The appearance of mushroom on the surface of bed is called a "flush" and when flushing occurs the mushrooms are collected immediately while still fresh (Plate 13.1). After collection they are packed and stored in refrigerators at 4°C until marketed. The white coloured buttons turn brown and then black in a couple of hours if kept at high temperature. Several flushes can be taken in a single bed. After last flush the bed must be cleaned out and the process repeated once again.

Plate 13.1 : Mushroom raising

A productivity of about 18-22 kg fresh white button mushroom q^{-1} compost can be had in 60 days of crop duration. Mushrooms can be canned if market is not good. Other popular mushrooms such as *Lentinus edodes* (Shiitake mushroom) and *Grifola frondosa* can be grown on freshly cut wood of several tree species. They are medicinally important but costlier mushrooms.

Nutritive value of mushrooms : Mushrooms are prized delicacies at the dinner table. Although their food value is low (Table 13.3) but they add flavour to meat and vegetables. The protein content of dry mushrooms is 19-33%. It is twice as high as most vegetables, except pea, brussels sprout and pulses. They are rich in nine essential amino acids (isoleucine, leucine, lysine, cystine, methionine, phenylalanine, tyrosine, threonine, tryptophan and valine) and low in calories (<35 kcal 100 g^{-1}) with traces of sugar and free of cholestrol and fats. Mushroom fats contain 72% unsaturated fatty acids. Mushrooms are rich in minerals [P, K, Na, Ca, Mg and Cu], vitamin B_1, B_2, B_{12}, C, niacin and pantothenic acid than most of the vegetables. Mushrooms on dry weight basis contain 1.4-9.4 mg vitamin C (ascorbic acid), 0.35-1.14 mg thiamine, 54.9-108.7 mg niacin and 1.6-5.0 mg riboflavin 100 g^{-1}. Wild mushrooms are highly priced in Europe, America, etc. Most highly valued mushrooms are morels, chantrelles and truffles. Truffles produce a sex pheromone (found in male pigs) which is helpful in its hunting.

Table 13.3 : Food value of white button mushroom (*A. bisporus*) on fresh weight basis

Components	Content (%)
Total solids	9.00-11.00
Carbohydrates (as glucose)	2.45-5.25
Crude proteins	3.50-3.94
Actual proteins (amino acids)	3.15-3.30
Fats	0.20-0.41
Fibre	0.50-1.00
Ash	0.90-1.25

Mushrooms usually derive food by decomposing lignin, cellulose and hemicellulose present in agricultural wastes with the help of their extracellular enzymes secreted by the mycelium. Microbial protein available in compost is the chief source of organic nitrogen for their assimilation.

Diseases of mushrooms : Mushrooms are attacked by several fungi, bacteria and viruses like those of plants. The most common diseases encountered are soft mildew or cobweb (incited by *Dactylium dendroides*), brown plaster mould (*Papulospora byssina*), white plaster mould (*Scopulariopsis fumicola*), olive green mould (*Chaetomium olivacearum*), inky cap (*Coprinus lagopus* and *C. comatus*), green mould (*Trichoderma viride)*, Cinnamon mould (*Peziza* sp.), false truffle disease (*Pseudobalsamia microspora*), wet bubble disease (*Mycogone perniciosa*), dry bubble or brown spot disease (*Verticillium malthousei* or *V. psallistae*), button rot (*Sclerotium* sp.), brown spot (*Pseudomonas strutzi*), yellow blotch (*P. agarici*) and bacterial blotch (*P. tolaassi*). Besides, common viral diseases encountered are brown disease and watery stipe, X-disease and dieback disease. Sccarid, phorids and cecid flies and springtail also attack the mushrooms. To control fungal diseases application of carbendazim (50-100 ppm) or thirum or blitox or Dithane Z-78 (@ 0.2%) in beds is recommended while as for bacterial disease control tetracyclin or streptocyclin @ 0.025% may be applied. Often mismanagement or temperature and moisture variations contribute to the disease development. To minimize the use of chemical pesticides, integrated management with more emphasis on the use of resistant strains, manipulation of environmental conditions and use of biocontrol agents including botanicals.

13.2 BIOENERGY PRODUCTION

Biogas, a byproduct of anaerobic decomposition of organic matter, is considered as a low cost alternative renewable non-conventional energy source. This low cost technology besides energy production also minimizes environmental pollution and provides rich fertilizer slurry for use on farm. The process of anaerobic decomposition takes place in the absence of oxygen.

Biogas composition and energy value : Biogas contains 50-65% CH_4, 25-35% CO_2, 1-3% N, 0-2% each of H_2 and H_2S and trace amounts of ammonia and other gases. The energy of biogas originates mainly from methane whose calorific value is 9005 kcal m^{-3} at 15.5°C and one atmospheric pressure. However, the presence of CO_2 in biogas to reduces its thermal energy rendering the calorific value of biogas approximately 4,450-6,230 kcal m^{-3}. Each cubic meter of methane, at standard temperature and pressure, removes 3 kg of COD or BOD i.e. an overall reduction in about 65% BOD or COD by use of 40% total solids in a biogas plant.

The anaerobic digestion of agricultural wastes, sewage sludge, industrial effluents and city wastes in large waste treatment plants also yields biogas which is used as fuel for boilers and internal combustion of engines (for wastewater pumping, etc.). Based on the heat value of biogas, it is estimated that the combustion of 1 m^3 biogas is sufficient to (a) run a 1 HP engine for 2 hours; (b) provide 1.25 kW-hr of electricity; (c) provide heat for cooking three meals a day for 5 persons; (d) provide 6 hour light equivalent to a 60 W bulb and (e) run a refrigerator or an incubator of 1 m^3 capacity for 1 or 0.5 hour, respectively.

Biogas production technology : Biogas digesters of 2-3 m^3 size are used on small scale for individual households having 4-5 cattle. For big farms having 100-300 cattle a digester size of 300-600 m^3 is suitable. The gas generated may be used for cooking, heating and lightening the farm and households. Better biogas plant must have high conversion efficiency with optimum gas production and increased methane content, high utilization efficiency, operational simplicity, increased process stability and economy.

The anaerobic digestion serves as a source of energy, waste stabilization, nutrient reclamation and pathogen inactivation process. At present two most popular designs used are floating gas holder type (KVIC model) and Chinese drum-less or fixed dome type (Fig 13.2).

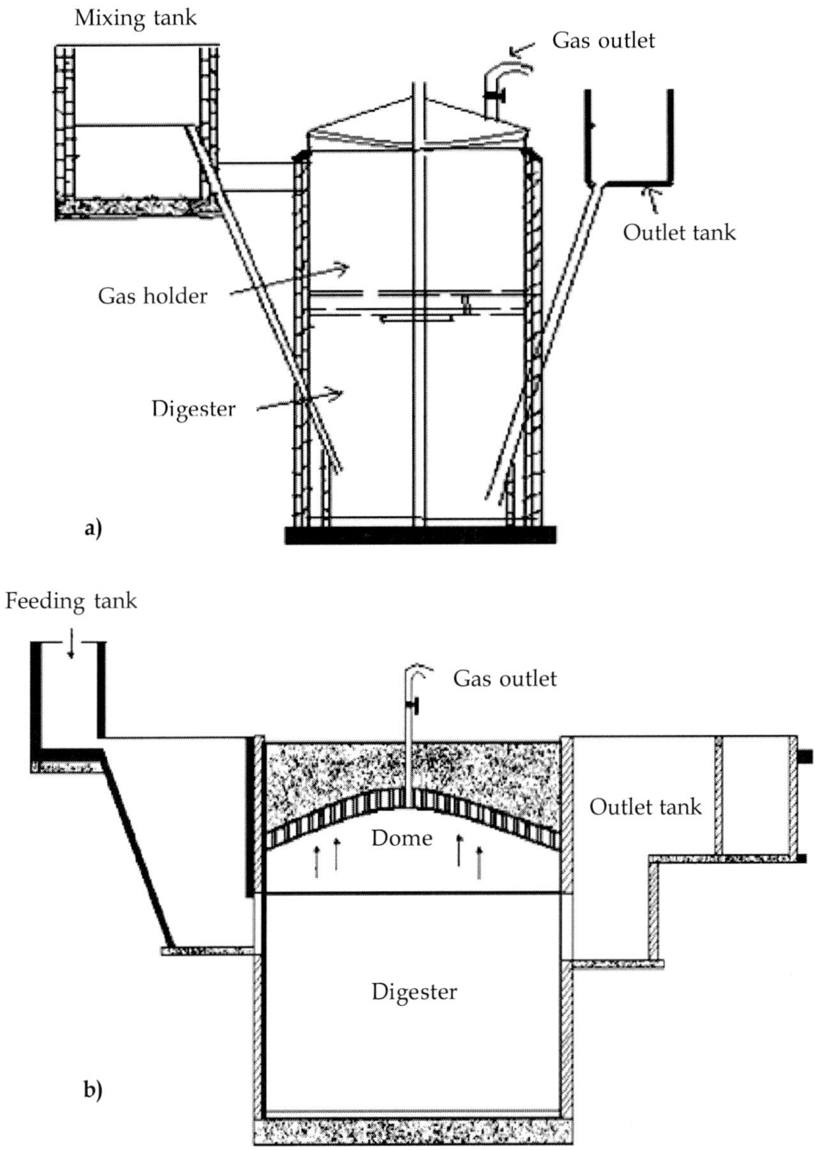

Fig 13.2 : Biogas plants a) Floating gas holder type (KVIC model) b) Fixed dome type (Janta model)

Depending upon the physical nature and periodicity of the availability of raw materials, digestion process is carried out in batch or semi-continuous type fermentors. Batch operations are useful for crop residues or seasonal industrial wastes or solid wastes which do not flow easily and mix properly. The proportion of mixing dung and water is 1 :1. Continuous type fermentors have a series of digesters and is suitable for free flowing wastes available without interruption. The dung feeding rate to run a biogas plant to produce biogas at optimal level varies as per the size of digester. In case of a community biogas plant the recommended feeding rate is 3500 kg day^{-1}. The more sophisticated way of methane production has been designed in USA which employs the use of algae and bacteria. The algae is harvested and put into digester. Bacteria consume algae to produce methane.

During anaerobic digestion about 60-70% conversion of solids to fuel occurs with leftover as rich organic manure. Biological reactions during biogas production reduce organic content of waste materials by 30-60% and produce a stabilized sludge which can be used as fertilizer or soil conditioner. After digestion at least 50% N present is in dissolved ammoniacal form which can be nitrified. The available nitrogen content is 30-60%, phosphorus 50% and potassium 80%. Further, the use of leftover improves physical properties of soil. The slurry or effluent from biogas digesters, though still polluted, is rich in nutrients so are used as fertilizers in either in dried slurry form and sprinkled on land or used for fish ponds. However, potential problems lie in biogas digester handing and reuse of slurry.

Stages in biogas production process : Anaerobic digestion is a three stage process :

i. Liquefaction or hydrolysis phase (involving solubilization and breakdown of polymers);

ii. Acid formation or hemiaceto- and acetogenic phase; and

iii. Methane formation or methanogenic phase.

$$\text{Organic matter} \xrightarrow{\text{Anaerobic digestion}} CH_4 + CO_2 + H_2 + NH_3 + H_2S$$

Four groups of organisms involved in above processes are :

Non-methanogenic Phase

a) Acid forming bacteria (hydrolytic and fermentative)

b) Acetogenic bacteria (acetate and H_2 producer)

Methanogenic Phase

c) Acetoclastic bacteria (methane forming)

d) Hydrogen utilizing methane bacteria.

Organisms involved : The biogas production process involves synergistic action of different types of microorganisms. During organic matter hydrolysis complex polymers are degraded to monomeric units by various microbes such as *Clostridium, Acetovibrio, Bacteriodes, Ruminobacter, Ruminococcus, Treponema, Eubacterium, Cellobacterium,* etc. Anaerobic cellulose degraders hydrolyze organic matter to ethanol and organic acids (like acetic acid, succinic acid, lactic acid, formic acid, etc.) and release a fair amount of CO_2 and water. Later, homoacteogenic and acidogenic organisms ferment organic compounds to acetate, CO_2 and H_2. Genus *Clostridium* is most predominant anaerobic cellulose degrader present in fermentor. Methanogenesis process, the conversion of acetate to methane and CO_2, is a rate limiting step and involves terminal reductive steps. Methane bacteria, which are secondary colonizers, include *Methanobacterium, Methanosarcina, Methanoautotropicus, Methanobrevibacter, Methanobacillus, Methanococcus,* etc. Bacterial population in biogas unit after stabilization is 10^6-10^7 ml^{-1} with cellulolytic population of 4 x 10^5 ml^{-1}. Biological activity in digesters is maximum at 35°C but slow at 6°C or below. Organic matter degradation is 45-55% in 25-30 days of fermentation. Methanogenic activity of 1 g volatile solids degraded produces around 300-500 ml methane g^{-1}. Under ideal conditions 10 kg organic matter produces 3 m^3 biogas. About 25% feed substrate is converted to gas and residues left become available as manure.

13.3 AQUACULTURAL REUSE

Three main types of aquaculture reuse of organic wastes in hot climate involve the production of micro-algae (single cell protein), aquatic macrophytes and fish. Micro-algae production normally utilizes wastewater in high rate photosynthetic ponds.

Fish production can often be linked to the existing waste treatment facilities to provide additional biological treatment. Waste recycling through aquaculture herbivorous species of fish like *Ctenopharyngodon, Cyrinus* or *Tilapia* feeding on phytoplankton is done. Feeding compost of C :N ratio 20 :70 @ 50 kg VS or COD ha^{-1} day^{-1} (in 20 x 10 x 1 m pond) increased fish yield two fold than feeding compost of C :N ratio 30 :1, at the same rate. The increase was six fold higher than control (no application). The organic matter content of digested effluents, including microbial biomass and reduced chemicals, is utilized by fish for their growth. However, these ponds need continuous aeration to have rapid growth of fish. A yield of around 50 t fish is harvested ha^{-1} yr^{-1} from the use of aerated digested effluents. *Tilapia nilotica* and *T. aurea* can be cultured in rice fields to convert insects and other aquatic organisms to edible fish protein. Over 250 kg ha^{-1} yield can be obtained this way.

13.4 LAND APPLICATION

The proper management of organic materials such as crop residues, animal manures and sewage sludge on land is one of the best means to stabilize soil, reduce water and waste erosion and conserve soil inputs against degradative processes to have a sustainable agriculture. It seems reasonable to expect that a one-inch layer of manure could be incorporated into soil every year without any difficulty, assuming that favourable soil moisture and temperature conditions prevail. The wastes added to soil involve many recalcitrants so extraordinary care is taken while allowing the land-fill practice in a productive soil. In landfill application, refuse is spread in thin compact layers, covered by a layer of clean earth. The pollution of surface- and groundwater is reduced by lining and contouring the fill, compacting and planting the uppermost cover layer, diverting drainage, and selecting the soil sites which are neither subject to flooding nor have high groundwater levels. The choice soil for landfill is clay due to its less permeability than other soil types. The leakage of the materials disposed of in landfill may avoided by solidifying them in materials such as cement, fly ash, asphalt or organic polymers. The non-biodegradable compounds should not be dumped in productive soil as it shall impair its quality.

Selected References

Alter, H. and Dunn, J.J. 1980. *Solid Waste Conversion to Energy*. Marcel Dekker, New York, USA.

Bahl, N. 1984. *Handbook of Mushroom*. Oxford and IBH Publ., New Delhi, India.

Bhardwaj, K.K. and Gaur, A.C. 1985. *Recycling of Organic Wastes*. ICAR, New Delhi, India.

Bhunavase, D.B., Jadhav, B.R., Rasal, P.H. and Patil, P.L. 1992. *25 Years of Research on Microbial Decomposition of Organic Wastes*. ICAR/MP Krishi Vidyapeeth, Pune, India.

Christensen T. and Payseur, T. 2003. *Anaerobic Digester Technology : Applications in Animal Agriculture - A National Summit*. Releigh, North Carolina, USA.

Chongrak Polprasert, 1996. *Organic Waste Recycling* (2nd edn). John Wiley & sons, Sussex, England.

Cook, D.B. and Kalbermathen, J.M. 1982. Prospects for resource recovery from urban solid wastes in developing countries. pp. 267-273. **In** : *Proceedings of Regional Seminar on Solid Waste Management* (ed. G. Tharun). Asian Institute of Technology, Bangkok, Thailand.

Dar, G.H. and Tandon, S.M. 1987. Recycling potential of waste plant residues for biogas and enriched manure production. *Science and Culture*, 53 : 93-95.

Dar, G.H. and Tandon, S.M. 1987. Biogas production from pretreated wheat straw, lantana residues, apple and peach leaf litter with cattle dung. *Biological Wastes*, 21 : 75-83.

Department of Environment (UK). 1995. *Waste Management Paper No. 266, Landfill Wastes.* HMSO, UK.

FAO. 1977. *China : Recycling of Organic Wastes, Soil Bulletin No. 40.* FAO, Rome, Italy.

Feachem, R.G., Bradley, D.J., Garelick, H. and Mara, D.D. 1983. *Sanitation and Disease – Health Aspects of Excreta and Waste Water Management.* John Wiley, Chichester, UK.

Ferguson, T. and Mah, R. 1987. Methanogenic bacteria. pp. 49-63. **In** : *Anaerobic Digestion of Biomass* (eds. D.P. Chynoweth and R. Isaacson). Elsevier Applied Sciences, London, UK.

Foth, H.D. and Turk, L.M. 1973. *Fundamentals of Soil Science* (5th edn.). Wiley Eastern Private Ltd., New Delhi, India.

Gaudy, A. and Gaudy, F. 1980. *Microbiology for Environmental Scientists & Engineers.* McGraw Hill, New York, USA.

Gaur, A.C. 1999. *Microbial Technology for Composting.* ICAR, New Delhi, India.

Gotaas, H.B. 1956. *Composting : Sanitary Disposal and Reclamation of Organic Wastes, WHO Monograph Series No. 31.* World Health Organization, Geneva, Switzerland.

Hagerty, D.J., Pavoni, J.L. and Heer J.E. (eds.) 1973. *Solid Waste Management.* Von Nostrand Reinhold, New York, USA.

Jadhav, H.V. 1997. Bioenergy. pp. 349-361. In : *Energy and Environment.* Himalaya Publishing House, New Delhi, India.

Loehr, R.C. Jewell, W.J., Vovak, J.D., Clarkson, W.W. and Friedman, G.S. 1979. *Land Application of Wastes, Volume I.* Van Nostrand-Reinhold, New York, USA.

Levanon, D. and Danai, O. 1997. Recycling agricultural wastes through mushroom production. pp. 305-308. In : *Advances in Mushroom Biology and Production : Proceedings of the Indian Mushroom Conference-1997* (eds. R.D. Rai, B.L. Dhar and R.N. Verma). Mushroom Society of India, National Research Centre for Mushroom, Solan, HP, India.

Nair, M.C., Gokulpalan, C. and Lalu Das (eds.) 1994. *Advances in Mushroom Biotechnology.* Scientific Publishers, Jodhpur, Rajasthan, India.

NAS. 1977. *Methane Generation from Human, Animal and Agricultural Wastes.* US National Academy of Sciences, Washington, USA.

National Academy of Science (USA). 1977. *Methane Generation from Human, Animal and Agricultural Wastes.* National Academy of Science, Washington, USA.

Nyne, E. J. 1990. Biogas the real fuel for developing countries. *Enzyme Microbial. Technol.,* 12 : 151-152.

Polprasert, C. 1996. *Organic Waste Recycling.* John Wiley & sons, Sussex, England.

Purkayastha, R.P. and Chandra, A. 1985. *Manual of Indian Edible Mushrooms.* Today and Tomorrow's Printers and Publishers, New Delhi, India.

Rai, R.D., Dhar, B.L. and Verma, R.N. (eds.) 1997. *Advances in Mushroom Biology and Production : Proceedings of the Indian Mushroom Conference-1997.* Mushroom Society of India, National Research Centre for Mushroom, Solan, HP, India.

Robinson, C.W. and Howell, J.A. (eds.). 1985. *Composting Comprehensive Biotechnology.* Pergamon Press, Oxford, UK.

Ruby, J., Murugesan, A.G. and Hussain, Z.M.I. 2003. Potential application of biotechnology in generating ecofriendly bioenergy. pp. 226-235. In : *Biotechnology, Emerging Trends* (eds. J. Selvin, A.S. Ninawe, V.S. Sugunan, N. Sukumaran and A.P. Lipton). Biotech Books, New Delhi, India.

Tam, D.M. and Thanh, N.C. 1952. *Biogas Technology in Developing Countries : An Overview of Perspectives, ENSIC Review No. 9.* Environmental Sanitation Information Centre, Asian Institute of Technology, Bangkok, Thailand.

Tapiador, D.D. 1981. *Vermiculture and its use in Thailand and other Asian Countries.* 1st National Earthworm Growers Convention. Metro, Manila, Philippines.

Tchobanoglous, G. Theisen, H. and Eliassen, R. 1993. *Integrated Solid Waste Management : Principles and Management Issues.* McGraw Hill, New York, USA.

TERI. 1982. *Biogas Handbook.* Documentation Centre, TATA Energy Research Institute, Bombay, India.

USEPA. 1983. *Process Design Manual for Land Application of Municipal Sludge.* EPA-625/1-83-016, US Environmental Protection Agency, Cincinnati, Ohio, USA.

USEPA. 1979. *Design Manual for Sludge Treatment and Disposal.* EPA-625/1-79-011, US Environmental Protection Agency, Washington, DC, USA.

USEPA. 1971. *Liquid Wastes from Canning and Freezing Fruits and Vegetables.* Program No. 12060 EDK, US Environmental Protection Agency, Washington, DC, USA.

Walsh, M.J. 1995. Sludge handling and disposal : An American perspective. *Water Qual. Int.,* 2 : 20-23.

— ★★★ —

Chapter 14

Pesticide — Microbe Interaction

I sometimes think that never blows so red
The Rose as where some buried Casesar bled
(Omar Khayyam)

In agricultural practices pesticide application is an inevitable factor for maintaining high crop productivity and is currently an integral part of modern farming. Pests (insects, microbes and weeds) cause approximately 20-30% loss in crop production so making pesticide use indispensable for crop protection.

Pesticide : A pesticide may be defined as '*any agent or chemical or a mixture of chemicals/ substances intended to prevent, destroy, repel or irritate one or more pest populations causing economic loss in crop production*'. Worldwide more than 450 active pesticide chemicals formulated into more than 10,000 commercial products are currently used by farmers. The most worrying aspect is that about two-third of the 800 chemicals tested so far have been found involved in causing or promoting tumor in rodents. Yet more than 50,000 synthetic chemicals/formulations remain untested so far.

Broader classification of pesticides : The first pesticide used for commercial purpose in agriculture was 'Bordeaux mixture' (a mixture of copper sulphate, lime and water prepared in 4 :4 :50 proportion) which was discovered by R. M. A. Millardt in 1882 and used against downy mildew of grape vine. Since then a long chain of pesticides have been discovered and used to serve specific purpose. These pesticides according to the target organism may be classified as insecticides, fungicides, bactericides, algicides, nematicides, acaricide, rodenticides and herbicides meant to suppress or control insects, fungi, bacteria, algae, nematodes, mites, rodents and weeds, respectively. Some of the main types of compounds currently in use are given in Table 14.1. As per their action, these may be systemic or contact/repellent or chemosterilent. Most pesticides frequently possess a benzene ring and also have heavy/ trace metals like As, Br, Cd, Cl, Cu, F, Pb, Hg, Ni, Mn, Sn, Zn, etc. Nearly all modern pesticides are organic neutral (uncharged) molecules. These chemicals are frequently applied to soil or plant in solid, liquid or gaseous form as emulsions, water soluble concentrate, pellet, wettable powder, water soluble powder, oil emulsion, suspension, dust, fumigant or injection. These are applied as such or along with some active or inert diluent, aerosol, granular, impregnate, volatile or non-volatile, fertilizer combination, bait, slow release material, etc. Because majorities of these compounds are not found naturally in soil-plant system, it is imperative to understand their behaviour and effects on soil ecosystem. The interaction between pesticides and microbes may be inhibitory for pathogenic ones but for other microbes such effect may vary from inhibition to stimulation.

Table 14.1 : Main types of compounds currently in use as pesticides

Pesticide class	Pesticides
Insecticides	
Organochlorines	BHC, DDT, Lindane, Aldrin, Heptachlor, Chlordane, Endosulphan, Aldrin, Dieldrin, Metoxychlor
Organophosphates	Parathion, Chlorpyrifos, Dichlorovos, Monocrotophos, Dimecron, Malathion, Metasystox, Methyl parathion
Carbamates	Aldicarb, Carbaryl, Carbofuran, Fenitrothion, Methomyl
Heterocyclic derivatives	Diazinon
Organosulphur	Tetradifon
Synthetic pyrethroids	Permethrin, Fenvalerate
Formamidines	Chlordimefom, Amitraz
Dinitrophenols	Dinoseb, Dinitrocresol

Contd...

Table 14.1 Contd...

Amidinohydrazones	Pyramdron
Botanicals	Nicotine, Retonone, Pyrethrum
Insect repellent	Dimethyl phthalate, Diethyl toluamide
Fumigants	Methyl bromide, Chlorothene
Herbicides	
Phenoxyacetic/ Phenoxy-alkanoic acids	2,4-D, 2,4,5-T, MCPA
Toluidines/ Nitroanilines	Trifluralin, Oryzalin, Ethalfuralin
Phenylureas	Fenuron, Isoproturon, Linuron, Diuron
S-triazines/ Heterocyclic-N	Simazine, Atrazine, Metribuzin
Bipyridylium	Diquat, Paraquat
Substituted amides	Propanil, Alachlor, Metachlor
Aliphatic acids	Dalapon, Dicanbe, DCPA
Glycines	Glyphosphates
Benzonitriles	Dichlobenil, Bromoxynil
Phenoxypropionates	Mecoprop
Translocated carbamates	Barban, Asulam, Chloropham, Propham
Thiocarbamates	EPTC
Hydroxyarylnitriles	Ioxynil, Bromoxydynil
Phenol derivatives	Dinoseb, DWOC, PCP
Chlorinated aliphatic acids	TCA, Dalapon
Fungicides	
Non-systemic	
Inorganic & heavy metal compounds	Bordeaux mixture-Cu, Kocide, $NiCl_2$, Fentin hydroxide (TPTH, Duter), Triphenyl tin chloride (Brestanol), Triphenyl tin acetate (Brestan), Lime-sulphur, Micronized sulphur
Dithiocarbamates	Maneb, Zineb, Mancozeb, Thiram, Nabam, Ferbam, Vapam, Ziram
Dicarboximides	Captan, Captafol, Dichlofluanid
Pthalimides	Folpet
Quinones/ Phenolics	Chloranil, Dichlone, Thiophenols

Contd...

Table 14.1 Contd...

Heterocyclic derivatives	Iprodione, Vinclozolin, Glyodin, Dyrene
Substituted aromatics	PCP, Quintozene (PCNB), Chloroneb, Dexon
Systemic	
Antibiotics	Agrimycin, Cycloheximides (Actidione), Blastidin S, Kasugamycin, Tetracycline, Streptomycin, Terramycin, Aureofungin, Giseofulvin
Triazoles	Triadimorph, Bitertanol, Propiconazole, Ectaconazole, Boutrizol, Triadimenol
Thiozoles	Ethazol (Terrazole; Truban)
Organophsophates	Iprobenfos, Fosetyl, Kitazin
Piperazines	Triforine
Carbamates	Propamocarb
Dinitrophenols	Dinocap (Kerathane)
Aliphatic nitrogens	Dodine
Thiourea	Thiophanate (Topsin, Cercobin), Thiophanate methyl (Cercobin-M, Topsin-M,)
Benzimidazoles	Carbendazim, Benomyl, MBC (Methyl benzimidazole carbamate), Thiobendazole, Terrazole
Pyrimidines	Ethirimol, Triforine, Diamethirimol, Fenarimol
Oxathiins	Carboxin, Oxycarboxin
Acylalanine	Metalyxl, Furalxyl
Morpholine	Tridemorph, Dodemorph
Rodenticides	
Inorganic & heavy metals	Zinc phosphide, Retafin, Fumarin
Coumarins,	Warfarin, Brodifscoum
Indandiones	Pindone, Chlorophacinone
Nematicides	
Halogenated hydrocarbons	BDCP, Methly bromide
Organophosphates	Phorate, Fenamiphos
Isothiocyanates	Vorlex, Dazomet
Carbamates	Aldicarb, Carbofuran
Others	Ethylene dibromide, Nemagon, Chloropicrin, Phenamiphos, Fensulphothion, Vapam

14.1 EFFECT ON NON-TARGET MICROORGANISMS

Non-target killing by pesticides : When a pesticide enters into a soil system it may either suppress the resident microbial population or may get metabolized by some resident microorganisms. Pesticides currently used range from those having low toxicity to microbes/soil animals to those highly toxic. Since the primary aim of a pesticide is to control the pests, which cause economic losses, therefore most of these pesticides are often applied several times in a cropping season. The lethal effect of pesticides is not usually confined to target pest alone but in the course the beneficial/ non-target organisms also get affected. The pesticides that tend to have non-target effect are inevitably those that are most mobile in soil or plant environment. For example, systemic fungicides are readily translocated by plants and affect the microbial population and activities in rhizosphere as well as target fungus or bacterium. Further, only 10-30% of soil applied pesticides do reach target/intended organism whereas hardly 0.1% sprayed pesticide reach target organism so rest ultimately enter soil or aquatic ecosystem, where if not readily metabolized or detoxified it may persist for longer periods and bioaccumulate. Soil serves as an ultimate sink for pollutants, which simultaneously influences soil biota.

Ecological imbalance : The soil biota is a complex entity of diverse organisms whose population always remains in a dynamic equilibrium influencing each other as per the soil fertility status. In soil, microorganisms account for 3-25 t biomass ha^{-1} whereas the biomass of insects and molluscs is about 20-190 g m^{-2}. Any interference in microbial population dynamics or microbe-mediated processes such as cycling of essential nutrients, organic matter decomposition and humus formation, soil biochemical processes and weathering, pathogenic and beneficial microbial survival may trigger ecological imbalance and thereby jeopardize soil fertility. For example, predatory animals account 2-10 g m^{-2} soil. It has been noticed that the recurring use of pesticide reduced spider population ($5x10^6$ ha^{-1}) by 12% each year. It is evident that different groups of organisms are affected in varied manner and this upsets the interaction involved between predators and prey. About 99% of micro-arthropod population is usually killed by fumigant DD and vapam and it take as long as two years for the population to fully recover. Fortunately the recovery time for microflora is generally much less from a few days to a few weeks. Whether such short term effects are ecologically significant needs thorough investigation. The pesticides, in general, do have negative impact on natural enemies of pests and those microbes involved in nutrient cycling and ecosystem restoration. Besides, pesticides may probably induce physiological disorders in soil biota.

Influence on microbial diversity : The heterotrophic populations in soil are usually altered by pesticide application. However, the response varies from one soil type to another. The dominant species may be drastically modified. As common genera are suppressed new groups emerge. This shift may be short-lived but the changed community may alternatively endure for longer periods. Mycorrhizae, root nodule forming nitrogen fixers, free living phosphate solubilizers, antibiotic producing organisms, antagonists, growth hormone producing rhizosphere and phyllosphere microbes are affected by fungicides, especially the systemic ones. The use of radiolabelled herbicides and insecticides have shown that many foliar applied compounds are readily translocated to roots and exuded into soil, so such chemicals have marked non-target effect on rhizosphere. Rhizosphere population may either increase or decrease depending upon the pesticide employed, its rate of application and changes caused in plant physiology, especially with respect to type and amount of root exudates released. Fungicides, especially those used as fumigants, may adversely reduce soil fungi and actinomycetes thereby slow down the process of humus formation in soil.

The diversity of resident microorganisms on phyllosphere and in rhizosphere is usually destabilized as pesticides reach these areas. Specific organisms, tolerant to pesticides, may remain and proliferate rapidly. Thus there is a shift towards less susceptible microbial growth in soil ecosystem. Pesticides, in some way applied to soil, change ecological balance through effects on non-target flora and fauna, by changing some soil physico-chemical properties and by influencing stabilization reached between antagonists and pathogens/pests in soil. For example, the insecticides that reduce the number of predatory mites may stimulate the number of springtails, which serve prey for the mites. Such inter-organism interaction is normal in most soils. Some insecticides in their formulations have carrier substances, which adversely influence microflora than insecticide itself. For example, petroleum distillates present in malathion is more toxic to bacterial growth than malathion itself. The soil-water pollution may arise through run off waters, pesticide drifts, leachates to ground waters, etc. which may adversely affect the soil-water biota.

Influence on soil processes : Biochemical reactions and enzyme activities occurring in soil may have interferences owing to the pesticide presence. Major types of biochemical effects of such pollutants on soil microbial and animal biomass are impairment of enzyme functions by the binding of toxicants to enzymes, co-enzymes, metal activators or enzyme substrates and alteration of cell membrane or carriers in cell membrane. There may be interference either in respiration activities or in the metabolism of carbohydrate and lipid or in protein synthesis or on various regulatory processes mediated by enzymes or hormones.

Ammonification process has been found relatively tolerant to insecticide and herbicide treatments. The fumigant and fungicide treatments generally lead to an increase in ammonium-nitrogen content in soil. The use of captan, thirum and verdasan leads to both qualitative and quantitative changes in amino acid composition of soil. Insecticides and herbicides have only marginal influence on microbial respiration. The supplementation of organic matter reduces medium pesticide toxicity and increases microbial degradation of pesticide in soil. However, fungicides and fumigants are of great concern. The breakdown of cellulolytic substances like straw is completely inhibited for at least 6 weeks following the application of captan or paraquat in soil. The litter decomposition is highly affected by pesticides as they require specific (cellulolytic, chitin degraders or ligninolytic) microorganisms. The soil enzymes such as dehydrogenase, phosphatase, saccharase, β-glucosidase and urease activities were reduced by $\geq$50% in soils where atrazine @ 4 kg ha^{-1} was applied to control weeds. In general, it has been observed that the activity of one enzyme is stimulated at the expense of other enzyme. Similarly, the elimination of one microbial group is often compensated for by an increase in another group. This type of ecological succession is even disturbed by pesticidal environment. Sensitive processes in soil like nitrification are suppressed even at lower concentrations of pesticides like phorate, permethrin, etc. The effect has also been found dependent on pesticide formulation. There may be temporary accumulation of ammonium especially in those soils where plants assimilate nitrate as nitrogen source preferentially. The nitrifying bacteria are the most susceptible, and inhibition of nitrification has occasionally been observed following a foliar application of a biocide. The *Nitrobacter* species are more sensitive to pesticides than *Nitrosomonas* species. Under such situation more accumulation of nitrite may occur. Denitrification is inhibited by herbicides and fungicides such as 2,4-D, phenyl urea, captan, maneb, nabam, etc. However, insecticides are non-toxic to denitrifying bacteria. Pre-emergence and foliar herbicides at normal rates have apparently no direct effect on rhizobia, but injury or abnormality in host unfavourably affects nodule number and atmospheric nitrogen fixation. Nodulation and nitrogen fixation processes are adversely reduced and even, sometimes, abolished by recurring pesticide use obviously because *Rhizobium* lives in close intimacy with root. The application of fungicides, such as captan, has detrimental effect on soil rhizobial population, root nodulation, etc. Recent evidences suggest that some pesticides may even enhance biological nitrogen fixation by reducing the activity of protozoa and other organisms that are competitors or predators of nitrogen fixing bacteria. These findings illustrate the complexity of life in soil. Algae are remarkably inhibited by herbicides. In soil, pesticides mostly kill nematodes, fungi and nitrifying bacteria whereas spore forming bacteria and actinomycetes are more tolerant.

Soil fumigants, nematicides and fungicides, in general, cause widespread non-target effect. There is commonly a depression in soil respiration for one to several weeks after pesticide application. Several strategies are now being adopted to reduce incompatibility between pesticides and inoculants and the recent ones involve the use of pesticide-resistant mutants for making inoculants.

Over last five decades the monitoring of pesticide behaviour in soil and water has generated enormous interest amongst the researchers. Nevertheless, uncertainty about the potential residual risks continue to prevail not only in human being, plants or animals but also in non-target beneficial soil flora and fauna which play vital role in maintaining soil ecological equilibrium and geochemical cycling.

14.2 PERSISTENCE, ACCUMULATION AND BIOMAGNIFICATION

Fate of pesticides in soil : Pesticides normally enter the soil at a level below the concentrations recommended for agricultural use (0.05 to 0.5 g m^{-2}) which, assuming a bulk density of 1.0 g cm^{-3} and an even distribution with the top 10 cm of soil, would give a concentration of 0.5-5.0 mg kg^{-1}. The fate and distribution of a pesticide in soil is governed mainly by the nature (chemical structure) and properties (such as adsorption to soil matrix, water solubility, persistence, etc.) of pesticide, the amount and frequency of pesticide sprayed/ applied and the prevailing meteorological, edaphic and environmental conditions. How long a pesticide persists in a soil is of greatest concern as it reflects environmental pollution (Fig.14.1). The escalating danger of pesticide pollution may arise owing to the :

- Persistent nature of a pesticide;
- Bioaccumulation of pesticides in plants, animals and microbes;
- Biomagnification due to chemical reactions;
- Adsorption into deep layers of soil and organic matter;
- Resistance of pesticides to biodegradation and chemical transformation; and
- Transportation by water.

Usually recommended doses of pesticides are not harmful but disastrous consequences become obvious when high or repeated doses of the same pesticide or recalcitrant pesticide are applied, especially under unfavourable environmental conditions.

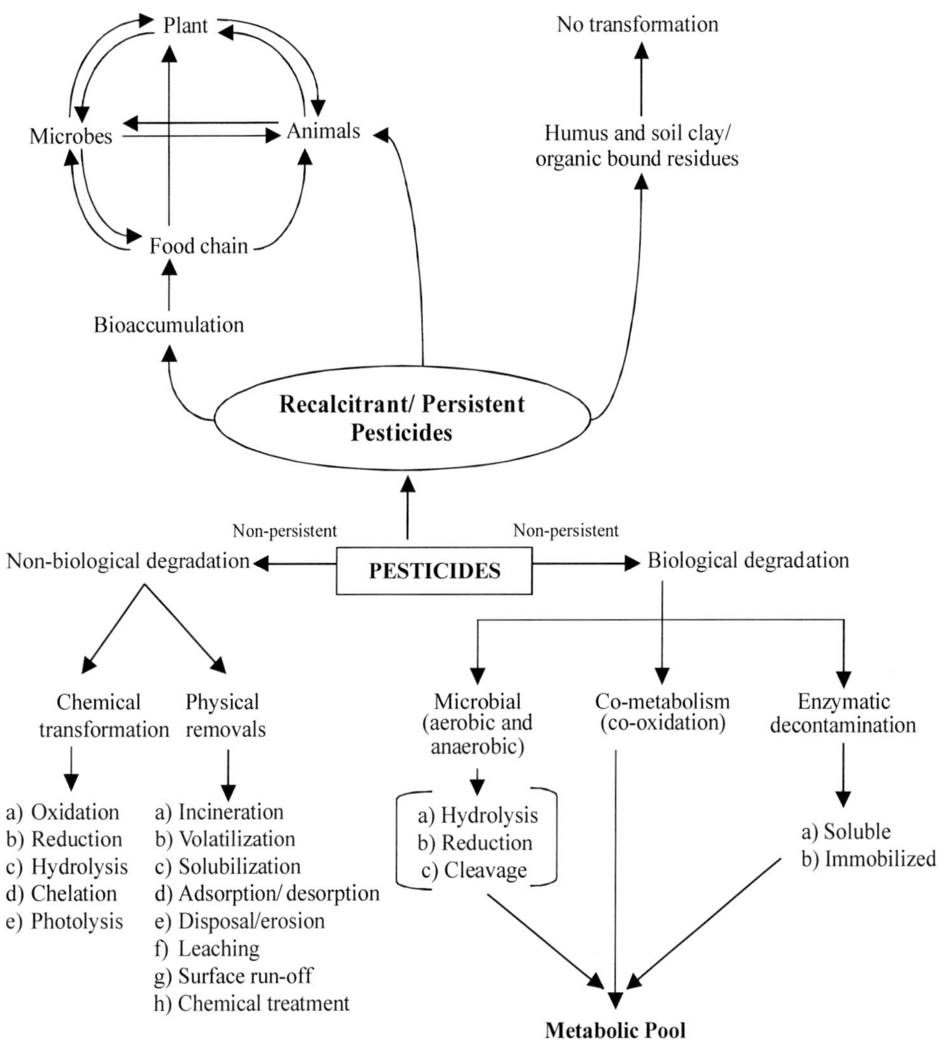

Figure 14.1 : Fate of pesticides in soil, aquatic and microbial systems

Pesticide movement in soil : Pesticides do possess some sorption coefficient, either low or high. Those with low sorption are more readily metabolized in soil. The pesticide movement in soil occurs through mass flow or diffusion. Mass flow is generally responsible for wide spread distribution or contamination of soil. Most of the organic pollutants are relatively insoluble and do not move down the profile of most soil types including peat. Some pesticides also tend to be relatively easily leached such as herbicide atrazine. Transport of non-sorbed pesticides or their residues by mass flow largely depends on the hydraulic properties of a soil and the extent to which rainfall or irrigation exceeds evapo-transpiration. Persistent pesticides have poor sorption properties so tend to move readily through soil into ground or surface waters especially after significant rainfall or irrigation. The longevity or residence time of a pesticide in soil/water is of greater concern because of their toxic nature or bioaccumulation and biomagnification. The maximum acceptable concentration of any pesticide in drinking water is 0.1 $\mu g\ L^{-1}$ (0.5 μg total pesticide) but this limit is often exceeded. The level during rains may even rise to 0.2-0.3 $\mu g\ L^{-1}$ in growing season. The pesticides most frequently causing problems with drinking water quality are herbicides such as atrazine, simazine, mecoprop, isoproturon, etc. and carbamate and choloropropionate insecticides. It is now established that most pesticides are decomposed by microbial mediation though volatilization and photo-oxidation. The persistence of a pesticide in soil is determined largely by the ability of soil microbial population to degrade the compound. Xenobiotic (the compounds foreign to biological system or those not degraded by microbial means) have greater persistence or remain as recalcitrant and react with concentrate by manifolds in their bodies. The birds and fish feeding on earthworm build up chemicals further to lethal level. Other pesticides may be either incorporated into various soil components or enter into food chain or may even migrate from the spill site. For example, earthworms are not generally susceptible to insecticides but since they ingest contaminated soil, therefore chemicals like organo-chlorine tend to accumulate in humus or get adsorbed by clay particles; so are never released as free products. Some may even react with other chemicals to form new toxic products.

Residual effect : Whether applied aerially to the plant or soil, some of the pesticides leave objectionable residues on crop, soil and surrounding waters. It has been observed that only 50% of the sprayed pesticides reach the intended orchard or farm. The remaining pesticide may be drifted away by wind or air currents to other non-target areas. Residual toxicity on plant surfaces and fruits is alarming. In Himachal Pradesh (India) about 1 t of fungicides has been found in 1,000 t of apple fruits collected from orchards frequently sprayed with pesticides. In India, only 3% food commodities have been found free from pesticide residues whereas 97% showed pesticide contamination out of which 25% contained residues above the permissible limits against the 2% in most developed countries. World-wide over 25 million cases of acute occupational pesticide poisoning are reported each year in developing countries and 14% of all known occupational injuries and 10% of all fatal injuries are caused by pesticides.

Detoxification of pesticides in soil : Most of the pesticides applied ultimately are broken down or degraded and detoxified by soil microbial biomass. Simple structural changes may convert a rapidly decomposable substrate into a persistent compound. For example, the addition of an extra chlorine atom to a molecule of 2,4 D converts it to 2,4,5 T. The effective persistence of pesticides varies from a few weeks to several years depending upon the nature (structure) and properties of a compound and, to some extent, on the availability of moisture and other soil factors such as organic matter, clay content, temperature, soil pH, decomposition process, plant uptake, etc. For example, 2,4 D, malathion and parathion have persistence value of 14 days and linuron, diazinon and 2,4,5-T about 3-4 months, whereas DDT and dieldrin have persistence value of >15 years. The highly toxic phosphates persist for more than 3 months at normal rate of application and in contrast chlorinated hydrocarbons may stay in soil for 4-5 years. Low pH usually increases the pesticide solubility in soil and higher pH lowers its adsorption. Similarly, at high temperature solubility is greater and usually adsorption increases upto 30-40°C and beyond this adsorption tends to decrease and volatilization increases with the rise in temperature.

Many genera of heterotrophic microbes use pesticides as substrate either co-metabolizing or directly using them as nutrients. Several compounds though non-toxic in nature are metabolized and converted to toxic form through process of activation, which is microbially-mediated. This results in the formation of carcinogenic or neurotoxic or phytotoxic compounds in soil. For instance, trichloroethylene is converted to vinyl chloride or chloral hydrate. Although the populations bringing about transformation of individual organic compounds are generally unknown, various bacterial and fungal species cause modification of one or more synthetic compounds in soil. These include *Arthrobacter, Acinetobacter, Agrobacterium, Achromobacter, Aerobacter, Bacillus, Clostridium, Corynebacterium, Escherichia, Erwinia, Flavobacterium, Klebsiella, Pseudomonas, Xanthomonas, Streptococcus, Kurthia, Torulopsis, Chlorella, Thiobacillus, Micrococcus, Rhodococcus, Flavobacterium, Alternaria, Aspergillus, Candida, Fusarium, Cladosporium, Chlorophenolicus, Chryosporium, Glomerella, Mucor, Penicillium, Phanerochaete, Trichoderma, Rhizoctonia, Sporocytophaga, Micromonospora, Nocardia, Streptomyces, Alcaligenes,* etc.

a) Trichloroethylene $\xrightarrow{\text{Bacterial action}}$ Vinyl chloride (a potent carcinogen)

Trichloroethylene $\xrightarrow{\text{Bacterial action}}$ Chloral hydrate (a mutagen/toxin)

b) Nitrate $\xrightarrow{\text{Bacterial action}}$ Nitrite $\xrightarrow{\text{Secondary amines}}$ Nitrosoamine (A potent carcinogen; mutagen even at low concentration)

c) Hg/Tin /As $\xrightarrow{\text{Methylation}}$ Mono-, di- and tri-methyl Hg/tin/As

d) Aldrin $\xrightarrow{\text{Bacterial action}}$ Dieldrin (persistence value increased to 15 years)

Pesticide persistence in soil : The persistence of a chemical in a soil is the summation of all the reactions, movements and degradation affecting these chemicals. Persistence of a compound is determined largely by the ability of soil microbial population to degrade the compound. Persistence depicts the duration of effectiveness of a pesticide and is a relative term. Greater the persistence of a pesticide in soil, greater will be the chances of its absorption/ adsorption by living organisms. Mineralization rate of persistent pesticides is almost zero, whereas residue formation and metabolite accumulation, in soil are high. The disappearance of a pesticide depends upon the initial concentration of a chemical used, soil properties, volatilization, photo-decomposition, erosion by wind and water, microbial degradation etc. The periodic application of even a biodegradable chemical may lead to an accumulation of pesticide in soil. Similarly, heavy metal containing pesticides result in the progressive accumulation of heavy metal such as Hg, As, Cd, Mn, etc. in soil even though the pesticide may have been degraded. The accumulated mercury or arsenic may enter food chain from seeds/crops treated with Hg or As-containing fungicides. In soils pesticides accumulate in earthworm and then accumulate in high levels in birds feeding on worm thereby enter food web of human being.

Persistence value : The persistence value of a pesticide is the period taken by a pesticide to cause >75% loss in normal arable soil. Rao and Davidson classified the pesticides on the basis of their half-life ($t_{0.5}$) value into three groups. The various pesticides falling under different categories are indicated in Table 14.2.

Table 14.2 : Pesticide categorized on the basis of $t_{0.5}$ value basis

Persistence level	$t_{0.5}$ (days)	Examples
Non-persistent	<20	2,4 D, Dalpan, Parathion, Methyl parathion, Malathion, Captan, Disulfotan, Thimet, Aminotiazole, Methyl demeton
Moderately persistent	20-100	Atrazine, Simazine, DVP, Carbofuron, Dichlobenil, 2,4,5 T, TCA, Diuron, Carbaryl, Heptachlor, Barban, Monuron, Aldrin, Linuron, Diuron, Glyphosphate, Ethirimol, Mecoprop, Calixin, PCP, Endrin, Chlorpyrifos, Diazinon, Tonofos
Persistent	>100	Trifluralin, Chlorobenzene, DDT, Picloram, Paraquat, BHC, Bromacil, Tetrachloroethane, Lindane, Dieldrin, Toxaphane, Chlordane, Chloroneb

Source : Nani and Thapliyal (1979); Rao and Davidson (1982); Rowell (1994)

14.3 TOLERANCE OR RESISTANCE DEVELOPMENT IN PESTS

As a result of repeated fungicide/insecticide sprays with the same chemical or application of increased dosages there is tendency in target organisms to develop resistance against that chemical. This occurs as a result of natural selection or biochemical changes or modification in genetic make up in targeted populations. There are reports of resistance development in target pests, and pests rebound over time. Therefore, the use of a pesticide repeatedly or its application in high concentration is not safe. The frequent pesticide-use adversely influences and reduces predator population in soil. This practice may also give rise to the emergence of new pests in the field, which otherwise are controlled by naturally through predators (non-target organisms).

14.4 BIODEGRADATION OF PESTICIDES IN SOIL

The amount and proportion of a pesticide applied that leaches to ground water is the result of a number of processes that occur in soil.

Biodegradation : The process of chemical breakdown of any substance into smaller products by the action of microorganisms or their enzymes is known as 'biodegradation'. All organic compounds including pesticides which reach soil are subject to microbial or chemical decomposition, provided the environmental conditions are favourable. In soil, most of the pesticides are usually deactivated but some may persist for longer periods.

14.4.1 FATE OF CHEMICALS IN SOIL

The fate of a pesticide in soil may be :

- Photochemical, chemical and microbial transformation,
- Volatilization,
- Plant uptake,
- Soil adsorption, and
- Movement through soil.

For many organic pesticides, adsorption onto soil colloids and presence of water are important catalysts for their degradation. The soil-water distribution co-efficient (k_d), octanol-water partition coefficient (k_{ow}), and its organic C coefficient give idea about the behaviour of pesticide in soil. 'k_d' is the proportion of a compound bound to solid relative to that remaining in soil solution at equilibrium. 'k_{ow}' indicates hydrophobicity of a compound whereas 'k_{oc}' is the amount of a compound adsorbed kg^{-1} organic carbon. These coefficients can be measured by :

$$k_d = \frac{X}{C}$$

where X = concentration of pesticide sorbed on solid surface
C = equilibrium concentration in soil solution

$$k_{ow} = \frac{Co}{C}$$

where Co = concentration of substrate in octanol

$$k_{oc} = \frac{k_d}{f_{oc}}$$

where f_{oc} = fraction of organic carbon in soil

k_{ow} less than 10 indicates low hydrophobicity with little likelihood of bonding on soil organic matter. Greater k_{ow} means greater ability of substance for bonding lipid and soil organic matter. Most organic compounds are relatively insoluble so show no movement down the soil profile.

Generally the amount of a pesticide leached from the soil decreases with increase in the rate of transformation. Therefore, persistent pesticides are likely to leach in higher amounts than degradable pesticides. In soil, morphologically and physiologically diverse organisms are present of which some are capable of degrading organic/synthetic molecules through mineralization processes.

When a foreign synthetic compound enters into a soil, certain reactions may be triggered. Recognition of these reactions often provides evidence that the process is in deed microbiological. Depending on its particular properties, the compound may be adsorbed to some extent on the clay or organic matter fraction of the soil. Only those pesticide molecules which are free in soil solution may be easily taken up by the plants and microbes. Therefore, measure of adsorption is a useful indicator of the availability of a chemical compound to non-target microorganisms. Soil organic matter and high surface area clays tend to be strongest absorbents for some compounds. The presence of certain functional groups like -OH, $-HN_2$, $-CONH_2$, -NHR, -COOR and NH_3 in chemical structure encourage adsorption, especially on soil humus. If the substance is in solution form, the compound will stay in solution without any change in its concentration until some microorganisms acquire the necessary enzyme systems to react with it.

14.4.2 MICROBIAL DECOMPOSITION OF PESTICIDES

The decomposition of a pollutant by microorganism is brought about by extracellular enzymes. Oxidation of organic pesticide occurs by the action of oxygenase enzyme excreted by microorganisms. In alkane hydrocarbons, the initial step in this oxidation is the conversion of a terminal CH_3 group to COOH group. Aromatic rings are cleaved by the addition of –OH group to adjacent carbon atom. The fungus *Cunninghamella elegans* is known to attack aromatic ring molecules such as naphthalene (Fig. 14.2). It is still not fully understood how enzymes are induced and whether genetic changes or mutations are involved. However, it seems clear that adopted organisms metabolize foreign compounds to derive carbon and energy for their growth and multiplication. The population of such organisms develops with a selective advantage. If all the conditions for microbial growth are favourable, then microbial growth occurs with consequent decline in the concentration of foreign molecule. Certain groups such as –OH, -COOH, $-HN_2$, etc., if present in a pesticide molecule, serve as the sites for microbial attack.

Microbial decomposition always shows an initial lag or acclimatization period during which microorganisms adjust themselves to the presence of organic compounds. This acclimatization process may take several days or some weeks depending upon the nature of molecule. During metabolism, heterotrophs catalyze a variety of reactions depending upon the species present. Such reactions include detoxification, hydrolysis, dehalogenation, demethylation, methylation, reduction, hydroxylation, deamination, ether-link cleavage, nitrite to amide conversion, conjugation, etc. These processes may result in the conversion of a toxic molecule to non-toxic product(s), or to simple products or form complexation with other groups etc. Most widely used organophosphate insecticides or herbicides such as 2,4 D, phenyl urea, aliphatic acids and carbamates are readily attacked by an array of microorganisms. However, other pesticides like triazine and most organic fungicides are slowly decomposed causing troublesome residue accumulation problems. Microbially synthesized surfactants may help in the acceleration of the rate of degradation of hydrocarbon oils in contaminated soils. In most cases, non-biological degradative processes such as photo-decomposition, volatilization, hydrolysis, oxidation-reduction, etc. occur at the same time when microbially catalyzed reactions initiate.

(i)

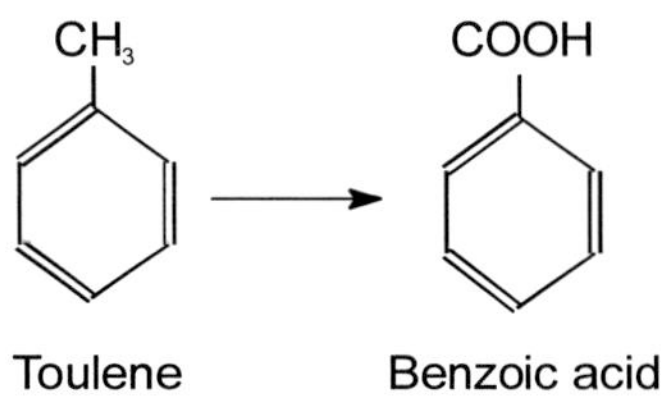

(ii)

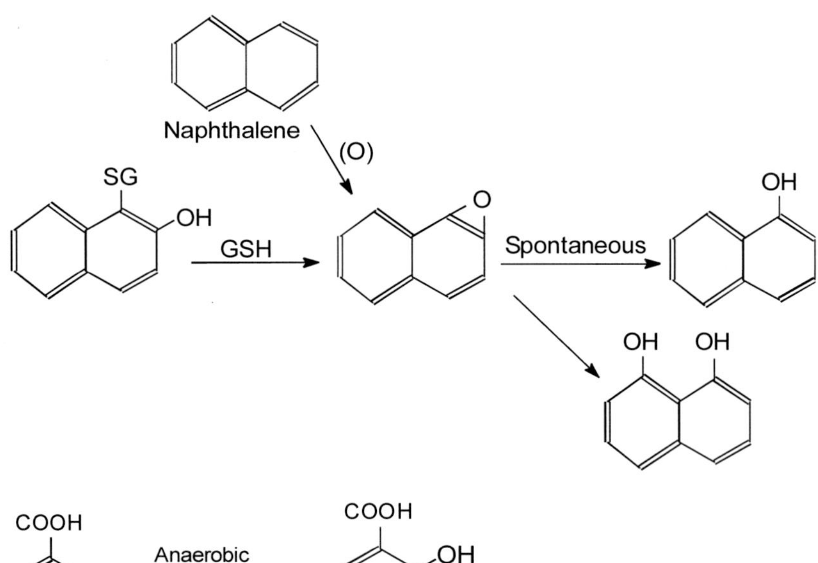

(iii)

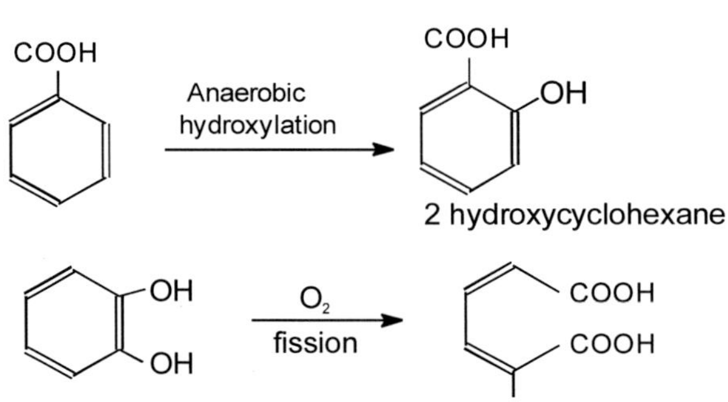

(iv)

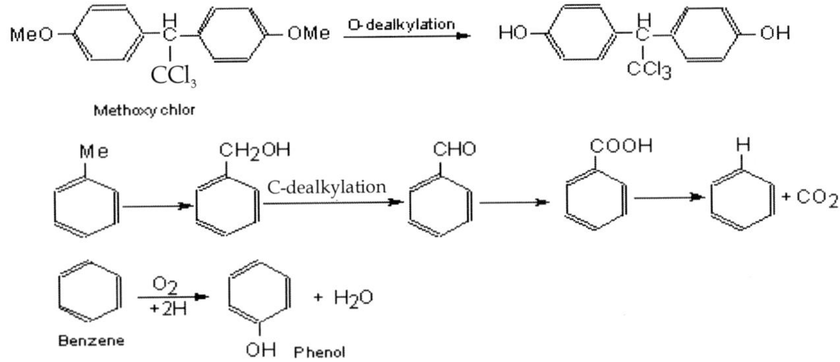

(v)

$$R=C(NHR_2)=C \longrightarrow R_1 - COOH + R_2 - NH_2$$

$$R_1O\text{-}P(=O)(OR_2) - OR_3 \longrightarrow R_1 - OH + R_2 - OH + R_3 - OH + H_3PO_4$$

Hydrolysis of carbamates/organophosphates and anilide herbicides

(vi)

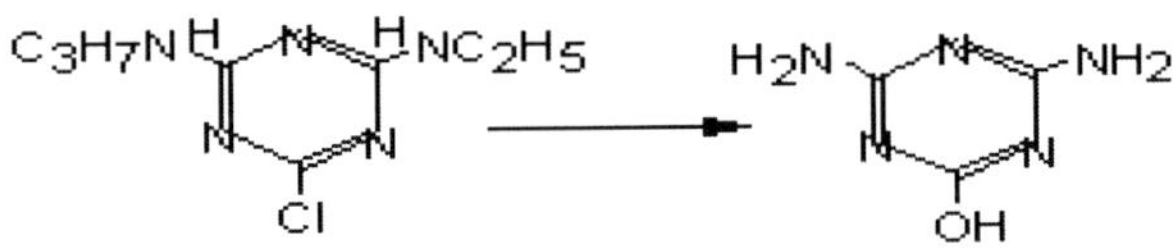

Side chain modification

Loss of NH_2 group as $NH\ 4^+$

Ring deavage ⟶ $CO_2 + H_2O$

Fig : 14.2 : Biodegradation of various organic compounds in soil

14.4.3 QUANTIFICATION OF PESTICIDE DEGRADATION

Quantification of pesticide degradation rate in soil may be done using rate models such as :

$$\frac{d_c}{d_t} = -k\,C^n \text{ or}$$

$$\frac{d_c}{d_t} = -V_{max}\left[\frac{C}{x + C}\right]$$

where, C = concentration of pesticide in mg L^{-1}

T = time in days

k = rate constant in days

n = reaction order

V_{max} = maximum rate

x = a constant

When n = 1, the first equation is reduced to first order kinetic equation. The degradation of simazine and linuron herbicides in field conditions follows the first order kinetics.

14.4.4 CO-METABOLISM

There are certain microorganisms that derive carbon and energy for their growth from the usual substrates without recourse to enzyme induction for degradation of a pesticide. During their process of metabolism these organisms cause partial degradation of pesticides through co-oxidation. The parent molecules are modified or detoxified or rendered harmless in the process by enzymatic reactions. This increases the susceptibility of chemical molecules to environment and other microorganisms. The phenomenon wherein a pesticide undergoes catabolism along with some other substance almost as an incidental process is called 'co-metabolism'. This phenomenon is now believed to play a significant role in the decomposition of many chemicals that are not readily metabolized. For example, aromatic hydrocarbons can be oxidized partially in *Nocardia* cultures growing on paraffin. Even the highly persistent and stable pesticides, such as DDT, reportedly are degraded by co-metabolic activities. Certain foliar-applied pesticides still reach the soil in small amounts due to their spray-dripping from foliage. It has been suggested that such compounds may be degraded incidentally by non-induced constitutive enzymes of the non-specific microbial cells. This process is known as 'incidental degradation'.

14.4.5 XENOBIOTIC

Xenobiotic are synthetic chemicals that are foreign to a soil ecosystem. Xenobiotic, sometimes referred to as recalcitrant, possess such structure and chemical bond arrangement that are not recognized by microbial enzymes responsible for degradation. Xenobiotic persist for longer periods in our ecosystem. Intrinsic properties of a xenobiotic, microbial population, extrinsic factors inherent in the environment and time determine the fate of a xenobiotic. Xenobiotic may serve as substrate for microbial growth and energy or may be co-metabolized. The microbes growing on xenobiotic may be isolated by enrichment techniques. Organo-chlorine molecules, such as DDT, PCP, PCDD, etc., are generally considered highly persistent in soil under natural conditions, with residence time >10 years. They have very slow decomposition rate because they contain carbon-chlorine bond, which is not found in nature. Most of the microbial species do not possess the enzyme capable of breaking this bond. Nevertheless, some species of bacteria and fungi have rapidly evolved the ability to decompose chlorinated organic compounds and these are being used in bioremediation of contaminated soils.

14.4.6 BIOREMEDIATION

Inoculation of specific microorganism, efficient in degrading a pesticide, into a soil is known as 'bioaugmentation'. This artificial inoculation reduces acclimatization period of indigenous organism as well. 'Bioremediation' is an accelerated process using micro-organisms (indigenous or introduced) and other manipulations to degrade and detoxify organic substances to harmless compounds, such as carbon dioxide and water, in a confined and controlled environment. Bioremediation is thus a specialized biodegradation process to remove or detoxify or cleanup man-made compounds from a contaminated environment, including soil. It involves the mitigation of chemical contamination of soil and is the method of enhancing the activity of native soil microorganisms, currently in vogue, using sophisticated engineering/biological agent to remove toxic pollutants. The process involves conversion of a pollutant to mineral forms such as CH_4, NH_4, H_2S, etc. Bioremediation is suitable for the treatment of a variety of organic chemicals, including volatile organic compounds, benzene, toluene, ethyl benzene, xylene compounds, phenolic compounds, polycyclic aromatic hydrocarbons and (particularly the simpler aromatic compounds), petroleum hydrocarbons, nitroaromatic compounds. Bioremediation is used to clean up leakage of petroleum or hydrocarbon in underground tanks and remove waste gases or metals from environment, sub-surface wastes, etc. Bioremediation is inexpensive method compared to physical methods of cleanup. Monsanto company is investigating the use of fluidized bed reactors wherein bacteria are immobilized on beads to prevent clogging, to process raw industrial waste waters for the removal of pollutants like glyphosates.

14.4.7 FACTORS AFFECTING BIODEGRADATION

The rate of biodegradation of certain pesticides has been found to be variable under different conditions and circumstances. The factors influencing the biodegradation process in soil include the nature (structure) and properties (the detoxicity of molecule and their intermediate products, water solubility etc.) of a pollutant, the genotype of microorganisms (whether they become adopted to decomposing pollutants), the soil type, temperature, pH, nutrient status, oxygen supply, moisture, clay and organic matter content, etc. The degradation rate of a pesticide also depends on the affinity of microorganism for the chemical, the susceptibility of chemical to attack by microbial enzymes and the presence of other chemical in soil. Many organic pollutants are more rapidly decomposed after they have been adsorbed onto soil. It has been seen that the level of degradation is fairly high under flooded conditions as compared to the upland conditions because of the anaerobic environment provided by the former. High levels of organic matter favour the metabolism of pesticides. Further, exposure of certain pesticides to sequential anaerobic and aerobic conditions results in faster biodegradation as compared to aerobic conditions alone. The general level of microbial activity in soil may be influenced in various ways so it is possible to assess pesticide degradation by either counting viable organisms/ measuring soil respiration or by determining biomass or enzymatic activities. The protoplasm on death of soil fauna becomes easy available nutrient for the surviving microorganisms. Similarly, soil drying may kill sensitive organisms such as protozoa and promote the growth of other microbes when such soils are re-moistened. It is difficult to assess the extent to which the degradation of foreign synthetic substances (pesticides) may be stimulated by these general nutritional effects, perhaps through co-metabolism, but it is certain that such processes do occur in soil.

14.5 GENETICS OF PESTICIDE BIODEGRADATION

Studies on genetics of pesticide biodegradation have revealed that specialized genes borne on degradative plasmids are responsible for inducing such reactions. For example, pseudomonads are characterized by their ability to catabolize a greater variety of unusual organic compounds through plasmid-controlled special metabolic pathways. In each case, a plasmid determines the enzymes of a single catabolic pathway or a portion thereof. For example, the enzymes of camphor degradation are determined by the '*Camphor* plasmid' of *Pseudomonas putida* and that of 2,4 D degraded on a conjugation plasmid in *alcaligenes paradoxus*.

A number of highly persistent pesticides such as lindane, MCPA, 2, 4-D, 2, 4, 5-T, etc. continue to be used extensively in under-developed countries with impunity, whereas most of these have been totally banned in developed countries. Most of the aerobic organisms that use aromatic compounds as respiratory substances attack them through either of the two branches of β-ketoadipate pathway. Through these reactions, six carbon atoms of aromatic nucleus in the primary substrate are converted to six carbon atom of an aliphatic acid, β-ketoadipic acid. This, in turn, is converted to acetyl CoA and succinic acid, both of which can immediately enter into the tricarboxylic acid cycle.

14.6 BIOAVAILABILITY OF PESTICIDE/POLLUTANTS

'Bioavailability of pesticide/pollutant' may be defined as the readiness with which a solute in a soil is within the reach of a microorganism or plant. The bioavailability of a molecule is dependent on its nature and properties or likely sorption (adsorption + absorption) and retention in a soil matrix and aquatic sediments composed of clay and organic matter. The organic fraction of soil and sediment has greater affinity for hydrophobic compounds such as polycyclic and aromatic hydrocarbons and other non-polar chemicals. Low bioavailability renders a pollutant or pesticide more resistant to microbial action. Physical pulverization provides greater aeration so increases the bioavailability and solubility of pollutants. Similar is the case with heat/rise in temperature or detergent application. The various methods by which biodegradation can be augmented include :

i) **Addition of surfactant like 'Surgee 2' or 'Corexit' in combination with bacteria to clean up oil spills :** Microbially synthesized surfactants help to accelerate the rate of degradation of hydrocarbon oils in contaminated soils.

ii) **Supplementation with inorganic nutrients such as N or P etc. :** This enhances the growth, population and metabolic rate of naturally occurring organisms capable of using the contaminant as a carbon source. For example, in the cleanup of crude oil following the Valdez oil spill in Alaska, microbe-enhanced fertilizers were applied to 79 miles shoreline to promote growth and metabolic activities of hydrocabon degraders already present there.

iii) **Inoculation with a known degrader :** About 30 or more different genera of bacteria and fungi are known to degrade hydrocarbons and pesticides are found in almost any soil or aquatic environment. Various enzymes present in them are responsible for inciting biodegradation of these compounds. These are mostly genetically coded in bacterial plasmids. Such plasmids occur in *Pseudomonas, Alcaligenes, Flavobacterium, Beijerinkia, Klebsiella* etc. *Pseudomonas* is often used to clean oil spills.

Selected References

Aizawa, K. 1982. Microbial control of insect pests. pp. 397-418. In : *Advances in Agricultural Microbiology*, (ed. N.S. Subba Rao). Oxford & IBH Publ. Co., New Delhi, India.

Alexander, M. 1994. *Biodegradation and Bioremediation*. Academic Press, London, UK.

Alloway, B.J. and Ayres, D.C. 1997. *Chemical Principles of Environmental Pollution* (2nd edn.). Chapman and Hall, London, UK.

Atwal, A.S. and Dhaliwal, G.S. 1997. *Agricultural Pests of South Asia and their Management.* Kalyani Publishers, New Delhi, India

Binulal, N.S., Zecharia, A. and Selvin, J. 2003. Bioremediation of oil pollution : Present status and future prospectus. pp. 236 - 243. In : *Biotechnology : Emerging Trends* (eds. J. Selvin, A.S. Ninawe, V.S. Sugunan, N. Sukumaran and A.P. Lipton). Biotech Books, New Delhi, India.

Chang, H.H. (ed.) 1990. *Pestisides in Soil Environment : Processes, Impacts and Modelling.* Soil Sci. Soc. Am., Madison, Wisconsin, USA.

Dinham, B. 1993. *The Pesticide Hazard : A Global Health and Environment Audit.* Zed Books, London, England.

Domsch, K.H. 1983. Principles of pesticide-microbe interactions in soil. pp. 179-194. In : *Soil Biology and Conservation of Biosphere, Vol. 1* (ed. J. Szegip). Academic Kiado, Budapest, Hungry.

Hassal, K.A. 1990. *The Biochemistry and Uses of Pesticides : Structure, Metabolism, Mode of Action and Uses in Crop Protection.* McMillan, London, UK.

Hayes, W.J. and Laws, E.R. 1991. *Hand book of Pesticide Toxicity.* Academic Press, San Diego, USA.

Hemond, H.F. and Fechner, E.J. 1994. *Chemical Fate and Transport in the Environment.* Academic Press, San Diego, USA.

Leisinger, T., Cook, A.M., Hutter, R. and Nuesch, (eds.) 1981. *Microbial Degradation of Xenobiotics and Recalcitrant Compounds.* Academic Press, London, UK.

Manahan, S.E. 1991. *Environmental Chemistry* (5th edn.). Lewis Publishers, Chelsea, MI, USA.

Martin, J.P. 1963. Influence of pesticide residues on soil microbiological and chemical properties. *Residue Rev.*, 4 : 96-129.

McBride, M.B. 1994. *Environmental Chemistry of Soils.* Oxford Univ. Press Inc., New York, USA.

Menn, J.J. 1996. Biopesticides : Has their time come? *J. Environ. Sci Hlth. B*, 31 :383-389.

Nani, Y.L. and Thapliyal, P.N. 1979, *Fungicides in Plant Disease Control*. Oxford & IBH Publ. Co., New Delhi, India.

Perry, A.S., Yammoto, I., Ishaaya, I. and Perry, R. 1998. *Insecticides in agriculture and Environment : Retrospect and Prospects*. Springer-Verlag, Berlin, Heidelberg, Germany.

Rao, P.S.C. and Davidson, J.M. (eds.). 1982. *Retention and Transformation of selected Pesticides and Phosphorus in Soil-Water System : A Critical Review*. EPA-6W/3-82-060, USEPA, Athens, Georgia, USA.

Ross, S. 1989. *Soil Processes*. Routledge, London, UK.

Rowell, D.L. 1994. *Soil Science : Methods and Applications*. Addison Wisley-Longman Ltd., Essex, England.

Schnoor, J.L. (ed.) 1992. *Fate of Chemicals and Pesticides in the Environment*. Wiley (Interscience), New York, USA.

Somerville, L. and Greaves, M.P. 1987. *Pesticide Effects on Soil Microflora*. Tylor & Francis, London, UK.

Sparks, D.L. 1989. *Kinetics of Soil Chemical Processes*. Academic Press, California, USA.

Steinhouse, E.A. 1977. *Insect Microbiology*. Academic Press, New York, USA.

Subba Rao, N.S. 1999. *Soil Microbiology*. Oxford & IBH Publ. Co., New Delhi, India.

Walker, N. 1982. Interactions of pesticides with soil microorganisms. pp. 377-395. In : *Advances in Agricultural Microbiology* (ed. N.S. Subba Roa). Oxford & IBH Publ. Co., New Delhi, India.

Ware, G.W. 1991. *Fundamentals of Pesticides : A Self Instruction Guide* (3rd edn.). Thomson Publication, Fresno, California, USA.

Watrud, L.S. and Seidler, R.J. 1998. Non-target effects of plant, microbial and chemical introductions to terrestrial systems. pp. 313-340. In : *Soil Chemistry and Ecosystem Health* (ed. P.M. Haung). Soil Science Society of America, Madison, USA.

— ★★★ —

Chapter 15

Biofertilizers and Biopesticides — Production and Application

One organisms's outhouse is another's castle

(Mark S. Coyne)

Boost to crop production has been the ultimate goal of researchers involved in agriculture so as to feed the ever-increasing human and domestic herbivorous animal populations. Crop production is not only constrained by the limited inherent potential of available varieties currently in use, but is also largely affected by the limitations of soil nutrient availability and damages by diseases, insects and weeds. To counter later challenges, augmentation of soil nutrients through use of biofertilizers in integrated nutrient management programmes and restricting the use of chemical pesticide by integrating biopesticides in plant protection programmes has received much attention in recent years because of consumer's preference for organic food.

15.1 BIOFERTILIZERS

The precious soil resource is presently facing extreme pressure from various quarters which has resulted in fast deterioration of its quality. Such pressures are on rise due to ever-increasing human demand for food, feed and fibre. Continuous tilling of a soil or bringing more forest areas under crop production has stripped the soil off the precious plant nutrients. The replenishment of these nutrients has been an uphill task. A wide gap exists between nutrient removal and its supply by the soil; and the gap, as per the practices in vogue, is mostly bridged by chemical fertilizers. However, continuous nutrient extraction and exclusive use of chemical fertilizers have posed serious threat to soil environment in the form of its quality degradation and chemical pollution. For instance, the application of larger amount of nitrogenous fertilizers has caused nitrate pollution in underground water and aquifers. Moreover, escalating fertilizer cost has been a great impediment in advocating chemical fertilizer use as a sole source for narrowing the gap between nutrient removal and supply. This has forced man to exploit and use other sources of plant nutrients.

Plant yield is considerably reduced by various abiotic and biotic stresses. The plants and microorganisms are closely inter-dependent and often enter into a mutual beneficial relationship that, among other things, enables specific stresses to be attenuated. The microbial inoculants have often been used in crop production to harness rich dividends from them (Table 15.1). The inoculants usually consist of bacteria or fungi suspended in or coated on some sterile carrier.

Table 15.1 : Classification of biofertilizers

A. Nitrogen fixers	
a) Symbiotic	- *Rhizobium* (fast growing)
	- *Azorhizobium* (medium growing)
	- *Bradyrhizobium* (slow growing)
b) Non-symbiotic	- Cyanobacteria (blue green bacteria)
	- Azolla (water fern)
	- *Azotobacter*
	- *Beijerinkia*
	- *Derxia*
	- *Azospirillum*
	- *Acetobacter*
	- *Frankia*
B. Phosphorus mobilizers/ solubilizers	
a) Mobilizers	-VAM fungi, ectomycorrhizae; ectendomycorrhizae
b) Solubilizers	- *Bacillus subtilis, Micrococcus* sp., *Pseudomonas* sp., *Flavobacterium* sp., *Xanthomonas* sp.
C. Compost inoculants	- *Trichurus spiralis*
	- *Aspergillus niger*
	- *Trichoderma viride*
	- *Chaeotomium* sp.

Microbial inoculants have been exploited either as biofertilizers or biocontrol agents. Some soil microorganisms are known to play key role in improving soil fertility and crop productivity through their ability to fix atmospheric nitrogen or solubilize insoluble phosphates or decompose farm waste or stimulate plant growth through synthesis of growth promoting substances. '*The artificially multiplied cultures of microorganisms helpful to achieve above described goals are referred as "Microbial inoculants or biofertilizers"*. Biofertilizers, in other words, are the carrier-based microbial inoculants containing cells of specific microorganism(s) mainly bacterium with ability to either fix atmospheric nitrogen or solubilize plant nutrients and render them available to the crops (Table 15.2). The extent of effectiveness depends on the inherent potential of these inoculants and their population in soil. Some organisms influence plant nutrient uptake and others directly benefit plant metabolism so they are also considered as biofertilizers. Besides, biofertilizers generally improve physical properties and health of soil. They also improve the soil humic acid status. Biofertilizers release a number of growth promoting substances and vitamins thereby help maintaining the soil fertility.

Table 15.2 : Some fertilizer equivalent of organic manures and biofertilizers

Components	Level of application	Nutrient equivalent fixed ($ha^{-1} yr^{-1}$)
Organic manure (FYM)	Soil application	36 kg "N + P_2O_5 + K_2O" (2 :1 :1)
Green manure (*Sesbania*)	Soil incorporation	44 kg N
Green manure (*Sesbania*)	Soil incorporation after 45 days of crop growth	50-60 kg N for HYV rice
Cowpea intercropped with castor	Legume buried after 6 weeks	30 kg N
Leucaenia loppings	Soil incorporation	25 kg N
Rhizobium	Seed inoculation/soil application	20-300 kg N
Azotobacter and *Azospirillum*	Seed inoculation	15-20 kg N
Cyanobacteria (blue green algae)	Soil application @ 10 kg ha^{-1}	20-30 kg N
Azolla	Soil incorporation @ 6-10 t ha^{-1}	40-80 kg N (3-4 kg N t^{-1})
Actinorrhizae (*Frankia* sp.)	Root inoculation	100-150 kg N
Sugarcane trash	Soil incorporation @ 5 t ha^{-1}	12 kg N t^{-1}
Rice straw + water hyacinth	Soil incorporation @ 5 t ha^{-1}	20 kg N t^{-1}
VA mycorrhiza	Seed pelting/ root inoculation	20-30 kg P
Ectomycorrhizae	Seed pelting/ root inoculation	20-30 kg P

The discovery of bacteria capable of reducing atmospheric nitrogen to ammonical form has been a major breakthrough in the field of agricultural research. The discovery of symbiotic N_2 fixer, *Rhizobium*, some non-symbiotic and associative nitrogen fixers such as *Azotobacter*, cyanobacteria, *Azospirillum*, etc. showing nitrogen fixing potential gave flip to search other such organisms present in nature. The vesicular arbuscular mycorrhizal and ectomycorrhizal fungi were found capable of solubilizing and mobilizing insoluble phosphates. Some organisms were capable of synthesizing either growth promoting substances or antibiotics/substances suppressing the pests. Some organisms possessed the above attributes in addition to their nitrogen fixation and phosphate solubilization capability. As such, these microorganisms play an important role in improving plant growth and combating plant diseases. Compost is the product of microbial action on complex lignocellulosic materials which is converted to easy soluble forms. Compost is an important biofertilizers that has been used world-wide by the farmers for centuries.

15.2 EMERGING NEEDS

Nitrogen is one of the major plant nutrients which profoundly influence crop production. The higher nitrogen requirement of crops and susceptibility of nitrogen to gaseous and leaching losses makes nitrogen the most demanding fertilizer material. Atmosphere has 80% nitrogen and the surface area of one hectare land has about 80,000 t atmospheric nitrogen over it. Unfortunately, plants can not tap this nitrogen because of it being in inert form. The present focus of research is to develop efficient microbial strains, suiting specific crop varieties, which can effectively tap atmospheric nitrogen under specific environmental conditions. Increasing attention is presently paid to harness maximum benefits from these sources due to reasons such as;

- Continuous depletion of soil fertility due to the widening gap between nutrient removal and replenishment resulting in decrease in crop production;
- Escalating costs of chemical fertilizers;
- Growing concern about environment protection and conservation; and
- Depleting crop productivity.

15.3 INTENSIFICATION OF BIOLOGICAL NITROGEN FIXATION IN SOIL

To derive maximum benefits from biological nitrogen fixation, attempts need to be intensified on :

- Horizontal expansion in area under legumes by introducing legumes in proper cropping pattern viz., inter-, multiple- and relay-cropping systems to enhance soil quality;
- Evolve efficient strains of N_2 fixing symbiotic and asymbiotic inoculants suitable for different crops;
- Overcome the preclusion or inhibition of chemical fertilizers on the functioning of biological nitrogen fixation;
- Refine agronomic practices for better on-farm fixation and conservation of nitrogen;
- Evolve practices that promote on-farm water/moisture conservation, as optimum moisture is essential for successful nodulation;
- Evolve appropriate technologies which may enable newly introduced N_2 fixers to successfully compete with the native strains of less effective N_2 fixing microorganisms;
- Expand the use of N_2 fixers to cereals by introducing N_2 fixation genes in their roots;
- Discover inexpensive nitrification inhibitors such as neem cake; and
- Evolve plant varieties responsive to both biologically and industrially fixed nitrogen, thereby enhance the plant's capability to absorb more nutrients by encouraging the growth of beneficial nitrogen fixers or phosphate mobilizers.

BNF in soil can be increased by implanting nitrogen fixation capacity factors present in N_2 fixing bacteria to non-legume crop plants through genetic manipulations or protoplast fusion or tissue culture and by overcoming the inter-generic hybridization between nodulating legumes and cereals.

15.4 NITROGEN FIXING BIOFERTILIZERS

15.4.1 SYMBIOTIC NITROGEN FIXERS

Rhizobium inoculants

Among the biofertilizers currently in use, *Rhizobium* inoculants are widely used on commercial scale. *Rhizobium* fixes atmospheric nitrogen in symbiotic association with legumes and sometimes with non-legumes such as *Parasponia*. *Rhizobium* bacteria are G-ve small rods measuring 0.5-1.0 x 1.0-3.0 μm in size. They colonize the roots of specific legumes, enter the root hair, multiply there and form tumor-like structures called 'nodules'. These nodules consist of central bacteroid zone surrounded by layers of cortical cells. Nodules are protective structures for bacterial nitrogen fixation enzyme, nitrogenase, against oxygen.

Nodules vary in size and shape according to the host and bacterial strain involved. They range from 2-6 mm in size in clovers to 3-10 mm in lucerne, cowpea and soybean. Usually nodules are round, club shaped, single or in clusters and translucent to white brown or pink in colour. Effective nodules are often fewer but large in size with pink central zone which is the active area for nitrogen fixation. Ineffective nodules are small, white, numerous and scattered all over the entire root system. The process of nitrogen fixation depends upon the enzyme nitrogenase, which is found in bacteroid zone. Bacteria can be cultured by streaking the nodule suspension on YEMA medium and incubating for 2-10 days at 28±1°C, depending upon the nature of organism i.e. whether fast or slow growing. Colonies are normally white, translucent, glistening and elevating with entire margin.

Rhizobium specificity : Both rhizobial inoculants and legume plants, to certain degree, show specificity. A particular species of rhizobial inoculant infects only a specific legume. However, they sometimes show cross infection due to mutations in the rhizobial inoculants. They are broadly classified in two groups on the basis of their growth behaviour i.e. whether fast or slow growing.

Method of production : *Rhizobium* culture is produced by either as agar broth culture or as carrier-based inoculant. In first case, the culture of *Rhizobium* is raised either in batch fermentors or in large flasks, containing agar broth medium, kept on rotary shakers. For the preparation of carrier-based inoculants, the rhizobial broth culture must have a cell population $>10^9$ ml^{-1}. The carrier is well-powdered (<100 mm mesh size), neutralized with calcium carbonate to raise pH upto 6.8 and then sterilized in autoclave at 15 lb per square inch for 3-4 hours. The broth culture is blended with sterilized carrier. The blending of broth with carrier is done in trays to attain 40% moisture holding capacity of the carrier. The blended culture is cured for 1 to 5 days at 25±1°C so that rhizobia get acclimatized with carrier. The blended culture is then filled in polythene bags and sealed, leaving two third space vacant for the aeration of bacteria. At the time of preparation, the inoculum should have a bacterial count of 10^8 cells g^{-1} soil or peat, and 15 days before the expiry date it should have a bacterial count of 10^7cell g^{-1} soil. Inoculant must be kept at a cool place. Optimum storage temperature is near freezing, though storage at 20±1°C can support the survival of a reasonable number of rhizobia. The maximum shelf life of the inoculant is 6 month from the date of its manufacture.

Characters of rhizobial strains and carriers : Rhizobial strains selected for biofertilizer use should possess high nodulation and nitrogen fixing capacity. The common carriers used are lignite, charcoal, peat, press-mud, soil (vermiculite clay <100 mesh size), compost mixtures, coir dust and farm yard manure. The other properties of the carriers should be as under :

- Carriers should be locally and easily available;
- It should be non-toxic to biofertilizers;
- It should have a neutral pH (6.0-7.5);
- It should contain high organic matter to sustain microbial growth;
- It should be finely grounded and pass through a sieve of 53 to 106 μm sieve; and
- It should have high moisture holding capacity (40%WHC).

Carrier-based inoculants should bear the information regarding the date of isolation, species contained, date of expiry, carrier used, crops for which to be used, net quantity, storage instructions, etc.

Method of inoculation : The usual inoculation technique is to apply the inoculant to the seed just before sowing either as dust or as slurry with water or adhesive solution. Adhesives like gum arabic or cellulose or jaggary not only ensure the adhesion of inoculum to the seed but also provide better environment for bacterial survival. First, 10% sugar or 40% gum arabic solution is prepared to serve as a sticker for rhizobial cells. The solution is then sprinkled over the seeds sufficient for one hectare. The seed is spread on polythene and solution mixed with the seed uniformly. The carrier-based inoculant is then thoroughly spread over coated seeds and mixed well. The seeds are spread uniformly for drying on polythene bags or floor in shade. After drying, seed are immediately sown in soil. Usually 400 g carrier-based inoculant is sufficient for inoculating the seeds sufficient for one hectare. The rhizobial count for effective nodulation under normal conditions should essentially be 10^6 per seed. The alternative method is soil treatment where seed treatment with non-compatible pesticide is required.

Crop response : The *Rhizobium*-legume association can fix high quantities of nitrogen from 20 to 300 kg ha^{-1}. The amount of nitrogen fixed by *Rhizobium* varies from crop to crop and with the agro-climatic conditions. Crops like groundnut, chickpea, cowpea, common bean, pea etc. leave behind about 20-30 kg of residual N ha^{-1} for the succeeding crop. The crops like danchia, sunhemp etc. add about 50 kg N ha^{-1}.

15.4.2 FREE LIVING NITROGEN FIXERS

i) Cyanobacterial (Blue Green Bacterial) Inoculant

The cyanobacteria [blue green bacteria (BGB), sometimes also called 'blue green algae'] are ubiquitous in distribution and capable of metabolizing elementary nitrogen from atmosphere. The cyanobacteria are photosynthetic nitrogen fixers. They occur either as a single-cell or in filamentous form which may be either branched or unbranched. Some cyanobacteria possess special structures called 'heterocysts' which are involved in atmospheric nitrogen fixation. Besides, cyanobacteria excrete vitamin B_{12}, auxins, ascorbic acids and other plant growth promoting hormones. Over 100 species of cyanobacteria are known to fix atmospheric nitrogen. The common genera which have most nitrogen fixing species include *Aulosira, Nostoc, Anabaena, Cylindrospermum, Calothrix, Tolypothrix, Lyngbya, Plectonema, Oscillatoria, Gloeocapsa, Scytonema,* etc. In wetlands, cyanobacteria cause ammonia accumulation in soils. In submerged paddy soils the biological nitrogen fixation is essentially a cyanobacterial process and contributes 20-30 kg N ha^{-1}.

Method of production : Mass production of BGB is done by open air shallow culture method or under artificially controlled field conditions. Under field scale production about 15 t wet BGB ha^{-1} can be obtained.

Open-air Shallow Culture

1. Shallow trays 3 m x 2 m x 23 cm of galvanized iron sheet or beds or pits lined with polythene sheets or bricks are prepared.
2. About 10 kg soil having a neutral pH is spread on trays and 250 g superphosphate and 0.2 g molybdenum (MoO_3) thoroughly mixed with soil. This promotes rapid multiplication of starter culture. If soil is acidic, lime @ 50 g per tray is added to neutralize soil pH.
3. The trays are filled with water upto 10 cm depth depending upon the rate of evaporation.
4. When the soil settles down, saw dust and starter culture @ 250 g per tray is sprinkled on the surface of standing water.
5. The trays along with contents are then exposed to sunlight. In hot summer the growth of BGB is rapid, so transpiration losses are high and more water needs to be added. In one to two week's time a thick blue green mass/scum floating on water surface is developed. When the growth becomes significantly thick, addition of water is discontinued and water is allowed to dry up. Pesticides such as parathion @ 0.05% or carbofuran 3G @ 3g m^{-2} or malathion 0.1%, etc. may be sprayed to prevent the

growth of pests like mosquitoes, snails, daphnids etc. For the control of green algal growth copper sulphate @ 4 g 10 m^{-3} may be added.

6. The dried BGB flakes are collected by scrapping the soil surface and stored in bags for future use. A tray yields 0.6-1.0 kg of inoculant under temperate conditions and 1.0-1.5 kg under tropical conditions.
7. The trays, after completion of one cycle growth, are filled up with water once again and multiplication process is repeated as above. After 3-4 harvests, the tray soil is replaced by fresh one.

Method of application : BGB flakes are broadcast @10 kg ha^{-1} over the standing water, one week after paddy transplantation. The field is kept waterlogged for a couple of days after inoculation. The BGB inoculation is done at least for three consecutive seasons to have long lasting effect. The use of multi-strain inoculants is more effective to offset the ecological and edaphic changes. Adverse conditions for multiplication and survival are caused due to direct contact with pesticides, chemical fertilizer, toxic materials (substance) and unfavourable soil conditions like low pH value, etc. The presence of predators affects the establishment of inoculants. Phosphorus and micro-nutrient deficiencies in soil need to be corrected for proper functioning of BGB.

Crop response : Long term field trails on BGB inoculation in Japan and India have shown that BGB applications compensates nitrogen equivalent to 60 kg ammonium sulphate ha^{-1}. In countries like Russia, Burma, Egypt and Philippines, BGB inoculation has resulted in 10-24% increase in grain yield. In India, it has been established that 1/3rd of nitrogen fertilizer can be conserved by BGB use without affecting the yield. Intensive field investigations have revealed that BGB inoculation causes 10-15% increase in grain yield. The response of BGB application is better in dry season as compared to wet season. Long duration varieties respond better to BGB than short duration varieties. Comparison of chemical nitrogen fertilizers and BGB has indicated that BGB applications are equivalent to 20-30 kg N ha^{-1} in rice. Further, the nitrogen fixed by cyanobacteria becomes rapidly available to the plants. BGB inoculations besides enhancing the paddy yield under temperate to tropical conditions produce certain growth promoting substances and vitamins. Superimposed inoculation of a field for 3-4 consecutive seasons helps in building up the population of useful cyanobacteria and afterwards there is no further need to apply more.

ii) Azolla Inoculants

Azolla, a floating fresh water fern, was recognized as organic manure for rice cultivation first time in North Vietnam in 1957. It is commonly found in ponds, pools, shallow ditches and rice fields. *Azolla* has branching stem, deeply bilobed leaves and true roots which penetrate the water body. Each leaf has dorsal and ventral lobes. The dorsal lobe is green due to the presence of chlorophyll, and exposed to air. The dorsal lobe also possesses a blue-green bacterial symbiont (*Anabaena azollae*) within a central cavity. The heterocyst of the symbiont (*Anabena*) is the site of nitrogen fixation. *Azolla* contains 3.0-3.5% nitrogen and produces rich organic matter. It is readily decomposed to ammonium which is readily taken up by rice plants. Since *Azolla* has high lignin content, it decomposes relatively at a slow rate (6 weeks or more) before all nutrients are released, a situation ideal for growing crop. The effectiveness of *Azolla* is ensured by its low C :N ratio (10 :1). The nitrogen fixed by *Azolla* is not tied up by the organisms involved in its decomposition.

There are six species of *Azolla* viz. *Azolla carolinia, A. nilotica, A. maxicana, A. filiculoides, A. microphylla* and *A. pinnata.* The initial inoculum requirement of *Azolla* is o.5-1.0 t ha^{-1}. *Azolla* grows very fast and forms green mat over water. *Azolla* applied @ 8-10 t ha^{-1} produces 10-20 t *Azolla* in 20 days. *Azolla* can accumulate up to 2 to 4 kg nitrogen ha^{-1} day^{-1} (equivalent to 10 to 20 kg ammonium sulphate). In 25 to 30 days *Azolla* can fix enough nitrogen sufficient for 4-6 t ha^{-1} rice crop during rainy season or 5-8 t ha^{-1} crop under irrigated conditions during dry season.

Method of production : A simple nursery method for large-scale multiplication of *Azolla* in the field has been developed for easy adoption by farmers. The field selected for its nursery raising is thoroughly prepared, levelled and divided into plots (2 x 20 m). Suitable bunds and irrigation channels are made. Water is maintained upto a depth of 10 cm. Fresh cattle dung slurry (10 kg mixed in 20 L water) is sprinkled over plots and *Azolla* inoculated @ 8 kg $plot^{-1}$. Superphosphate (100 g) is applied in three split dozes at four days interval as top dressing fertilizer. For insect control, Furadan granules @ 100 g $plot^{-1}$ or malathion or carbofuran 3G are applied 7 days after *Azolla* inoculation. After 15 days of inoculation, *Azolla* crop is harvested and introduced in main field as a source of primary inoculum or as green manure. About 40-55 kg fresh *Azolla* is harvested $plot^{-1}$.

Methods of application : *Azolla* is grown as monocrop before the transplantation of rice seedlings in fallow fields or as inter-crop with rice. Both these practices are used in India, although inter-cropping practice is more common. In intercrop, two rows of rice are planted about 10 cm apart with *Azolla* grown in 60 cm space on either side of the double rows. *Azolla* is incorporated either manually or by a rotary rice weeder and then killed by heavy shading and/or by high temperature.

Azolla as dual cropping : *Azolla* is the only green manure which can be cultivated as an inter-crop with rice under flooded condition. The fern is inoculated @ 0.5-1.0 t ha^{-1}, one week after rice transplantation. Superphosphate @ 10 kg P_2O_5 ha^{-1} and Furdan or Carbofuran 3G @ 2.5 kg ha^{-1} are applied for its rapid growth and control of insect pests, respectively. Split application of phosphorus produces greater biomass than single application. After the formation of mat in 20-25 days, the field is drained off and *Azolla* buried into the soil as above. *Azolla* gets decomposed and nitrogen is released into water and soil. *Azolla* nitrogen is available 20 days after fern incorporation.

Crop response : *Azolla* as a green manure produces 10-20 t of green *Azolla* per hectare containing 20-40 kg nitrogen. Although application of *Azolla* before planting is more effective, its dual cropping also increases the grain yield of rice. Fertilizer nitrogen and *Azolla,* inoculated as a dual crop @ 200 g m^{-2} in rice, are incorporated twice at minimum and maximum tillering stages to increase grain yield significantly. *Azolla* can be applied in fish culture ponds as well @ 25 kg ha^{-1} day^{-1} to increase crop production.

Limitations : Water is pre-requisite for fern multiplication so *Azolla* can not be used in uplands. The initial inoculum rate is very high. Storage value of fresh *Azolla* is very low. Moreover, *Azolla* is difficult to transport to longer distances.

iii) Azotobacter Inoculant

Bacteria of family Azotobacteriaceae have three major genera : *Azotobacter, Beijerinkia* and *Derexia.* Each genus has several species. *Azotobacter* is an aerobic free living nitrogen fixer and is one of the most important non-symbiotic N_2-fixing microorganisms. Based on morphological and physiological features, the genus *Azotobacter* has been classified into three species : *Azotobacter chroococcum, A. beijerinckii* and *A. vinelandii.* The first two species are most commonly found in India. *Azotobacter chroococcum* is dominant in acid soils while *A. beijerinckii* in neutral and alkali soils. They are found abundantly in the rhizosphere soil. The *Azotobacter* population in soil is usually $<10^4$-10^5 g^{-1}. Besides nitrogen fixation, *Azotobacter* exudes a number of growth promoting phytohormones as well as show antagonistic properties.

Mass production : Mass production of *Azotobacter* for preparing inoculants is done either by growing the culture in large flasks on a rotary shaker or in batch fermentors depending on the quantity needed. All the precautionary measures adopted during mass production of *Rhizobium* are also followed in mass production of *Azotobacter*.

Methods of inoculation : The methods of inoculation are :

a) *Seed treatment method :* For inoculating the seed sufficient for one hectare of land, first 10% jaggery solution is prepared or 50 g of sugar along with gum arabic is dissolved in 500 ml of water. The solution is boiled and then cooled. One packet of inoculant is mixed thoroughly in cooled solution. The seeds are heaped on a clean surface and the inoculant slurry is poured over it. The slurry is uniformly mixed with the seed. The inoculated seed are then dried in shade and sown in soil immediately.

b) *Seedling dipping methods :* In this method, 1.0 kg *Azotobacter* inoculant is mixed in 10 L water. Seedlings when ready for transplantation are dipped in the slurry for 20-30 minutes and then planted immediately.

c) *Soil inoculation : Azotobacter* inoculant is also directly applied to the soil. The inoculant (2 kg) is mixed with well-refined 50 kg soil or FYM. It is then applied to one hectare land 2 days after chemical fertilizers are applied or may be broadcast near the root zone at a depth of 7-10 cm when crop is 15-20 days old.

The use of *Azotobacter* is recommended due to following benefits :

1. Seed bacterization with *Azotobacter* helps in saving nitrogen fertilizers by 10-20 kg N ha^{-1}.
2. Besides its ability to fix nitrogen from air, *Azotobacter* secretes growth promoting substances which help in better germination and expand root system. Thus, inoculated plants can feed on soil nutrients in a better way.
3. *Azotobacter* suppresses the growth of saprophytic and pathogenic microorganisms near the root zone so reduces any crop damage by plant diseases. Thus *Azotobacter* plays nutritional, stimulatory and therapeutic role in plant growth.

Crop response : *Azotobacter* has been applied in a wide range of crops including vegetables, cereals, sugarcane etc. In case of cereals, it increases the yield from 8 to 25% and in cash crops from 10 to 20%. Vegetable crops, in general, respond better to *Azotobacter* inoculations than other crops.

iv) *Azospirillum* Inoculant

The ability of certain spiral bacteria to fix atmospheric nitrogen was first reported by A. Beijerinck in 1925. Later in 1978, Tarrand and his associates renamed the organism as *Azospirillum*. *Azospirillum,* more recently described as *Herbaspirillum,* is closely related to *Azotobacter* and is involved in associative nitrogen fixation. There are four major species of *Azospirillum* viz., *A. lipoferum, A. brasilense, A. amazonese* and *A. seropedicae*. Of the four species first two species are commonly found. These organisms increase the yield equivalent to 10-20 kg N ha^{-1}.

Method of mass production : Same as in *Azotobacter*

Methods of inoculation : Same as in *Azotobacter*

Crop response : *Azospirillum* inoculation has been found beneficial in a wide range of crops including cereal, forage and vegetable crops. The crop response, however, varies greatly with the type of crop, variety, location, season, soil fertility level, native microorganisms, etc. Crop response has been attributed to its ability to fix atmospheric nitrogen, increase root biomass and production of certain phytohormones. Bacterization with *Azospirillum* @ 2 kg ha^{-1} as soil inoculation significantly increases the vegetable yield.

15.5 PHOSPHATE SOLUBILIZING AND MOBILISING BIOFERTILIZERS

Next to nitrogen, phosphorus is the important nutrient required by the plants. Most of the soil phosphorus is in unavailable form and hardly 1-2% of it is incorporated into aboveground parts of the plant. Only about 15% of phosphatic fertilizer applied is used by plants. Phosphorus which gets fixed in soil is made available to crops by solubilizing action of various microorganisms. The insoluble and fixed forms of phosphates are converted into the forms readily taken up by the plants. Another factor which affects the effective utilization of phosphorus is its low mobility in soil. Fortunately, some filamentous organisms like phosphate solubilizing bacteria, vesicular arbuscular mycorrhizae (VAM) and ectomycorrhizae enhance the mobility of phosphorus in the soil.

Various bacteria able to solubilize insoluble phosphates in soil include the species of *Pseudomonas, Micrococcus, Bacillus* and *Flavobacterium*. Fungi like *Aspergillus, Fusarium, Sclerotium* and *Penicillium* also solubilize phosphorus. They are referred to as "phosphate solubilizing microorganisms (PSM)". The solubilizing effect PSM is due to production of organic acid such as formic acid, acetic acid, lactic acid, propionic acid, succinic acid and/or fumeric acid. These organisms are also known to produce amino acids, vitamins, IAA and gibberrelic acid, which help in better growth of plants. The PSM are found in the rhizosphere. Gerrtsen in 1948 was first to report that inoculating the seedlings with PSM increases phosphorus uptake and crop yield. The first phosphatic biofertilizer *Bacillus magaterium* var. *phosphaticum* was prepared in the then USSR under the name 'phosphobacterin' and extensively used in Eastern Europe. *B. magaterium* var. *phosphaticum, B. polymyxa, Pseudomonas striata, P. rathonis, Aspergillus awamori, A. niger, Penicillium digitatum* and *Schwanniomyces occidentalis* have been identified as the most efficient phosphate solubilizing microorganisms. Nowadays, *B. magaterium* var. *phosphaticum, B. polymyxa* and *Pseudomonas striata* are available for commercial use.

Method of mass production : Same as in *Rhizobium*

Method of inoculation : Seed/seedling treatment is given in a manner similar to *Azotobacter.*

Crop response : In India 'Microphos' culture containing *Pseudomonas striata* and *Bacillus polymyxa* was tested in 25 multi-locations field trails of potato, lentil, soybean, chickpea, rice and wheat. The results showed that 'microphos' increased grain yield by 5 to 10% through increased solublization of fixed soil-phosphorus. 'Microphos' cultures increased the efficiency of rock phosphates. The response to phosphate solubilizers was found to be positive only in soils having high organic matter but low in available phosphorus. Microphos culture saves 30-50 kg phosphate fertilizers.

15.6 MYCORRHIZAL INOCULANTS

Vesicular Arbuscular Mycorrhizae (VAM)

The role of fungi particularly those colonizing the root surface of growing plant roots in increasing the availability of phosphorus has received much attention. These organisms are known not only to mobilize phosphates but also transfer phosphorus from soil to roots. A. B. Frank in 1985 used the term 'mycorrhizae' to the beneficial association of fungi with the roots of higher plants. Mycorrhizae are indeed the major organisms involved in nutrient uptake of most of the land plants.

Mycorrhizae are ubiquitous in nature, found in tropics, temperate and arctic regions. *Glomus, Gigaspora, Acaulospora, Sclerocystis* and *Endogone* are some of the well-known VAM found in soils. VAM fungi inoculated to crop plants colonize plant root system, and increase growth and yield of crop plants. The improvement in plant growth is due to the increase in nutrient uptake particularly phosphorus, zinc and other micro-nutrients. Production of growth promoting substances, tolerance to drought and salinity and resistance to root pathogen induced by mycorrhizae also enhances the plant growth.

Methods of Inoculation

Planting holes : In this method soil inoculum is applied to planting holes where seedlings are aimed to be planted.

Placement method : In this method inoculum is placed near the seeds. First inoculum is placed in soil and then seeds are sown 2-3 cm above the inoculum layer.

Side dressing : This method is used when inoculum is not available in sufficient quantities. The inoculum is applied on the sides of fully established seedlings.

Mixed inoculum method : The method is adopted when sufficient quantities of inoculum are available and there is frequent variation in environment. This inoculation method leads to rapid infections.

Seed pelting : Here seeds are coated with pellet-inoculum before they are planted.

Crop response : The trails conducted with locally isolated *Glomus macrocarpum, G. mosseae, G. calidonius, G. fasiculatum,* etc. on chilli, chickpea, beans, tomato, potato, etc. have shown the possibility of saving 50% phosphatic fertilizers by using VAM without affecting the yields. Increase in nutrient uptake and yield of various crops in India has been common with VAM inoculation. The inoculums can be used along with *Azotobacter* or *Rhizobium* in non-legume and legume to synergistically benefit the crops.

15.7 COMPOST INOCULANTS

Crops remove large quantities of plant nutrients from soil, therefore recycling of farm wastes is essential for maintenance of soil health and replenishment of removed plant nutrients. One of the best methods of value addition of organic materials is 'composting'. Major portion of organic materials is made of cellulose, hemicellulose and lignin. Many cellulolytic microorganisms such as *Chaetomium, Trichoderma, Cellulomonas, Clostridium,* etc. decompose structurally resistant components of organic matter to easily available forms during composting. The ligninolytic microorganisms such as *Curuvlaria, Cephalosporium, Humicola,* etc. are capable of degrading lignin molecules. Inoculation with cellulolytic microbial culture reduces C/N ratio of organic residues earlier than uninoculated control. Further, the compost is ready within 6-8 weeks as compared to control which generally takes more than 12 weeks.

Cultures used for composting : The commonly used cellulolytic fungal cultures are *Trichurus spiralls, Aspergillus niger, A. terreus, Chaetomium abuanse, Myrothecium rosidum, Trichoderma viride, Cytophaga* sp. and *Cellulomonas* sp.

Method of application : Cultures are applied with water at pit filling stage. This helps in the quick reduction of C :N ratio to 20 :1. Besides rapid composting, microbial inoculation increases the quality of product. Paddy straw having a nitrogen content of 0.5% can be converted into humified organic material containing about 1.6% N within 6-9 weeks.

15.8 BIOPESTICIDES

Biopesticides form an essential component of integrated disease and pest management with focus onto avoid indiscriminate use of chemicals, reduce pollution and minimize any chance of resistance development in pests. Biopesticides are basically pest control agents of biological origin including fungi, bacteria, viruses, plant extracts, etc. that serve as a viable and ecofriendly alternative to chemical pesticides. US Environmental Protection Agency has officially defined a biopesticide as the one belonging to any one of the three groups :

- *Biochemical pesticides :* These pesticides are naturally occurring substances that control pests through nontoxic mechanisms. Biochemical pesticides include sex pheromones that interfere with insect mating, various scented plant extracts that attract insect pests to traps, hormones, natural insecticides, plant growth regulators, enzymes and repellents.

- *Microbial pesticides* : These comprize of microorganisms and their products, usually proteins as active ingredients. Microbial pesticides may be a naturally occurring or genetically controlled microorganism (e.g. a bacterium, fungus, virus or protozoan) as active ingredient. These pesticides can control many kinds of pests, although each separate active ingredient is relatively specific to its target pest(s).

- *Transgenic plant pesticides* : These are also known as 'Plant-Incorporated-Protectants (PIPs)'. Plants are transformed to produce pesticidal substances as a result of gene addition to the plant. For example, genes for Bt pesticide protein are introduced into the plant's own genetic material. Then the plant, instead of Bt bacterium, manufactures the substance that destroys the pest.

Presently biopesticides represent about 4.5% of world insecticide sale. The growth rate for biopesticides over next 10 years has been predicted to be 10-15% per annum in contrast to 2% for chemical insecticides. Over 90 species of bacteria, 400 species of fungi, 1200 species of insect-host associations and 50 species of botanicals reportedly possess pesticidal effect. The biopesticides have several advantages over chemical pesticides. Biopesticides are :

- Effective, biorational and create natural epizootics;
- Have limited field persistence and a short shelf life;
- Show relatively critical application times;
- Possess narrow target range and a very specific mode of action;
- Exhibit slow action and inherently less harmful;
- Suppress, rather than eliminate, a pest population;
- Often effective in small quantities; quickly biodegradable without residue problems; and
- Mostly self perpetuating and environmentally safe.

15.8.1 BIOINSECTICIDES

Insects damage a variety of plants as well as act as vectors of a number of animal pathogens. About 30% global agricultural production is destroyed annually by over 20,000 species of field and storage pests. Therefore, synthetic pesticides widely used for pest control are valued for their effectiveness, relatively long shell life and the ease with which they can be transported, stored and applied. Insect control greatly enhances crop yield and improves animal health. The development of pesticide resistance in insects necessitates the use of higher and repeated dozes of pesticides at increasing costs. For example, in South Africa, cotton is sprayed 10-12 times a season as compared to 5-6 times a decade ago. However, the associated problems with the pesticides use are soil-, water- and air-pollution and toxicity. As per the conservative estimates more than 3 million cases of pesticide poisoning are reported annually, of which 20,000 prove fatal. Due to the undesirable effects of chemical pesticides, alternative non-chemical means for disease and insect control have been tested. One potentially useful method is the use of microbial insecticides. These are microbial toxins which kill insects. The use of plant materials (botanicals) having biopesticidal properties and biological control agents are seriously being worked out. The integrated pest management, use of resistant varieties, clean cultivation, use of baits and traps, and planting of decay crops are other means of controlling the pests and diseases.

The microbial agents that have so far been used are essentially insecticides attacking the insects present at the time of application, but not multiplying in nature. Thus these microbial agents have little or no long term effect. They do not differ in principle from chemical insecticides, except in having more specificity and exhibiting least harmful effect on the environment. However, the ideal type of bioagent for the control of insect pests is an agent that multiplies after its application and is transmitted from desired target-insect to another desired target. Such an agent would spread through insect population and, if sufficiently effective, would eventually wipeout the insect in target area. To date, no microbial pathogens with these properties have been discovered.

Commercial production of biopesticides is restricted to a few bacteria, fungi and viruses (Table 15.3). Artificially formulated media with suitable carbon and nitrogen sources have served as substrate for some fungi and bacteria. A few bacteria and viruses need live insect larvae which are grown artificially in large fermentors to serve as substrate because of the obligate nature of these predators. For example, *Bacillus popilliae* is produced commercially in the form of a dust on Japanese beetle. However, *B. thuringeinsis* and *B. moritai* are produced by conventional fermentation techniques. Various biopesticides commercially in use are a mixture of *B. popilliae* and *B. lentimorbus* under the trade named Doom; *B. thuringeinsis* as Thuricide, *Heliothis* NPV as Elcar, *Lymantria* NPV as Gyp-check, *Ogyva* NPV as Biocontrol-1, *Beauveria bassiana* as Boverin and *Romanomermis* as Skeeter. The Bt based bioinsecticides account for >90% of the world biopesticide market. The use of natural enemies like parasitoid wasp *Epidinocarsis lopezi* to control cassava mealy bug in tropical Africa or *Bacillus thuringiensis* against diamond black moth (*Plutella xylostella*) pest of vegetables and maize stem borer have also been found effective. Similarly, pathogenic organisms are used to control weeds like *Lantana camera* in Hawaii.

15.8.1.1 Entomopathogenic Bacteria

Insects are susceptible to a number of microbial pathogens, especially bacteria. In almost all cases these insect-pathogens are quite specific and have no harmful effect on higher animals including man. Certain members of genus *Bacillus* are entomopathogenic in nature (Table 15.3). During sporulation, these bacilli produce certain proteins that are toxic to specific insects. Ishiwata in 1901 discovered *Bacillus thuringiensis* from a severe flacherie of silk worm, *Bombyx mori,* and named the disease as 'sotto bacillus' (sudden collapse). Beringer in 1915 proposed the name *Bacillus thuringiensis* when he isolated the organism from larvae of moth *Ephistia kukniella* in a flour mill in Thuringen (Germany). Later, *B. thuringiensis* has also been isolated from dead insects, litter of sericulture farms and soils.

Table 15.3 : Some microbial pesticides currently tested

Microbial agents	Target organisms
Insecticides	
Viruses	
Heliothis nuclear polyhedrosis virus (NPV) corn earworm	Caterpillar pests of cotton, soybean, tomato, corn and tobacco; cotton ball worm, corn ear worm and bud worm, tomato fruit worm and soybean pod borer
Orgyva pseudotsugata NPV of tussock moth	Caterpillar pests of forest and shade trees
Lymantria dispar NPV of gypsy moth	Caterpillar pest of forests especially hard woods
NPV of *Autographa california*	Caterpillar pest of vegetable and field crops
NPV of saw fly larvae	Caterpillar pests of forests
Bacteria	
Bacillus popilliae	Japanese beetle grub of lawn and pastures
Bacillus thuringiensis	Caterpillar pests of vegetable, forest tree and ornamentals
Bacillus thuringiensis sub sp. *isrealiensis*	Mosquito larvae and black flies
Bacillus thuringiensis sub sp. *kurstaki*	Lepidopteran larvae
Bacillus thuringiensis sub sp. *aizawi*	Wax moth caterpillar
Bacillus sphaericus	Mosquito larvae
Bacillus thuringiensis sub sp. *tenebrionis*	Beetles
Fungi	
Verticillium (Cephalosporium) lecanii	White fly, aphids and thrips
Hursutella thompsonii	Red mites on citrus
Phytophthora citropthora	Milk weed vine in citrus
Normuraea rileyi	Caterpillar pests of vegetable, field crops, forest and forages
Beauveria bassiana	Codling moth and Colorado potato beetle
Metarrhizium anisopliae	Beetle in wheat
Protozoa	
Nosema locusteae	Grasshoppers in rangelands
Nosema fumiferanae; N. pyravsta	Forest lepidopterons
Melaneba locustae	Grasshoppers and locusts
Vairimorpha necatrix	Black cutworm of maize
Nematodes	
Romanomermis culicivorax	Mosquito
Fungicides	
Trichoderma viride	Dutch elm disease pathogen
Bactericide	
Agrobacterium radiobacter	Crown gall pathogen
Herbicides	
Collrtotricum gloeosoroides	*Aeschynemone virginiea* (joint vetch)

B. thuringiensis is a Gram +ve rod shaped sporulating bacterium that produces a crystalline inclusion known as 'parasporal body' (a toxic crystal). The parasporal crystal contains crystal protein 'δ-endotoxin'. The classification of *B. thuringiensis* has been done on the basis of flagellar agglutination and variation in immunological properties of parasporal crystals. There are more than 30 sub-species of *B. thuringiensis* which are differentiated by their H (flagellar) antigens (H_1 to H_{19}). *B. thuringiensis* strain HD-1 var. *krustaki,* developed by USDA, and its derivatives are in commercial use since 1970 for the control of lepidopteron caterpillars such as cabbage loopers, cabbage worms, spruce and tobacco bud worm and gypsy moth larvae, etc. Another strain *B. thuringiensis* var. *israelensis* has been used for the control of mosquito and black fly.

Bacterial growth and toxin production : Crystalliferous bacilli which include *B. thuringiensis, B. entomocidus, B. popilliae* and *B. lentimorbis* produce endospores. The endospores germinate in insect gut to produce vegetative bacterial cells. They enter haemocoel where they cause septicemia, multiply rapidly to fill the haemocoel with bacteria and destroy tissues. Just before the death of host, thick walled highly refractile endospores are once again formed which are released on disintegration of host. In *Bacillus thuringiensis* infection death occurs within 3 days after ingestion.

The toxin producing organisms are cultured in laboratory and a preparation containing toxin and spores is obtained. This is then formulated into a dust or spray and distributed in the same manner as chemical insecticide. Because of high specificity of toxin, there is no danger to human or livestock. The most widely used *B. thuringeinsis* forms toxin as well as grows in various moth larvae and thus causes total disease.

Different strains of *Bacillus thuringiensis* have different crystal morphologies and even a single strain can form more than one type of crystal. *B. thuringiensis* produces several toxins such as α, β and γ exotoxins, δ-endotoxin and louse factor. However, β exotoxin and d endotoxin are important as microbial insecticides. α-exotoxin is a thermolabile toxin, produced and secreted by vegetative bacterial cells and not found in sporulating cultures. β-toxin is thermostable and produces malformation in susceptible insects, perhaps due to mutation. Moreover, β-toxin causes inhibition of pupation in housefly and affects bees, flies, grasshoppers and termites. The β exotoxin is a strain-specific property rather than a serotype-specific property. Usually one toxic bipyramid crystal is found in one spore, although the shape and size of toxic crystals may vary from strain to strain. For instance, *B. thuringiensis* var. *kurstaki* produces cuboidal crystal.

δ endotoxin is composed of glycoprotein subunits. The proteinaceous toxin shows toxicity by feeding and not by infection. However, some toxic peptides show toxicity both by infection and feeding. Lepidopterous insects and some selected species of dipterons and coleopterons show variable responses to δ endotoxin. The molecular weight of each subunit protein of toxic crystal varies depending on the method of dissolution of crystals (alkaline solution and the addition of reducing agents or enzymes) and the *B. thuringiensis* strains involved. The protein subunits have molecular weights of 230, 120-140 and 40-80 kDa. Bipyramidal crystal contains a protein of 120-135 kDa size (P_1 type) while cuboidal protein has protein of 65 kDa (P_2 type). P_1 protein appears to be active against lepidopteran insects and P_2 proteins against dipterans and lepidopterans. P_1 protein in biopyramidal crystal is a protoxin that is first inactive. On prolonged incubation inside larval gut, digestive enzymes (proteases, etc.) released in the mid gut cleave protoxin and convert it to 65-70 kDa active protein (toxin). *B. popilliae* toxin is more specific and affects Japanese beetle only.

Mechanism for action on host : The destruction of host occurs due to phospholipase C and lecithinase C present in α-exotoxins. In case of β-exotoxin there is inhibition of nucleotidase and prevention of RNA synthesis. δ-endotoxin, a glycoprotein unit, has silicon protein bond and high pH. It releases toxic component which break epithelial cells and causes gut paralysis thereby exposes gut to bacterial attack. Usually it causes increase in blood pH. β-exotoxin in lepidopteran insects shows three types of symptoms :

Type I :	Gut paralysis after injection of δ endotoxin, accompanied by alkalinity of hemolymph, fall in intestinal pH, destruction of epithelium e.g. in *Bombyx mori*
Type II :	Gut paralysis shortly after feeding δ endotoxin; no leakage from gut into hemolymph; no changes in blood pH; no general paralysis observed; insect larvae for want of food succumb to bacterial invasion and die within 2-4 days.
Type III :	Insects die due to infection by both spores and δ endotoxin e.g. Mediterranean flour moth *A.kuhniella*; gypsy moth *Porthetria dispar*

Endotoxin production is controlled by plasmids. The Cry^{+1} (δ-endotoxin crystal protein genes) from *B. thuringiensis* code for highly active linear polypeptide insecticidal proteins. These proteins (toxins) bind to the cell membrane receptors of epithelial cells of mid gut, insert itself into the membrane and create a pore through which other molecules pass freely resulting in osmotic imbalance. Punctured by many holes, the gut cells do not survive long. The insect larva is starved of nutrition and there is disruption in metabolism, cessation of insect feeding, paralysis and death within 24 hours. *Cry*$^{+1}$ *B. thuringiensis* strains can serve as plasmid donors to *B. thuringiensis* strains cured of *Cry*$^{+1}$ plasmid or to other bacterial species such as *B. cereus* grown in broth cultures where mating are allowed to occur. The genes controlling P_1 proteins have been cloned to *E. coli* and *B subtilis* host vector system. The *B. thuringiensis* var. *isrealensis* toxic protein is controlled by a 75 MDa transmissible plasmid. Other genes coding for insecticidal proteins from *B. thuringiensis* var. *tenebionis,* active against coleopteran insects such as Colorado potato beetle, have been cloned and characterized. Bt toxin genes have successfully been introduced into plant genomes or plant-associated microorganisms.

Production of transgenic insecticidal plants : A large number of Bt strains occur in nature and their actions towards different hosts also vary. Efforts are on to improve Bt gene expression in plants. Bt genes are rich in A-T while plant genes are rich in G-C. The partial or full synthesis of genes having higher G-C content has solved this problem and allowed the codon usage to be adopted for a particular crop. The resulting partial or full gene encoded exactly the same protein as the native one. This was followed by increasing the level of insecticidal protein in transgenic plant (hundreds of ng mg^{-1} protein) by about 100-folds through a combination of powerful promoters (i.e. CaMV35S with duplicate enhancers) and sequential modification in areas of gene with predicted mRNA secondary structure. The genes coding for Bt-toxicity have been engineered into cotton, tomato and tobacco as well as in root colonizing rhizosphere bacterium *Pseudomonas*. Transgenic cotton lines resistant to boll worm and pink boll worm have been evaluated in field and they performed well against these pests. A similar modified version of another δ-toxin encoding gene *cry*3A, isolated from coleopteron active Bt sub. sp. *tenebrionis*, has been transformed and expressed in potato plants. The gene conferred protection property to plants against Colorado beetle (*Leptinotarsa decemlineata*) under high natural field infestation conditions. Improvement in plant expression of *cry*3A gene has been obtained by increasing its overall G-C content from 36 to 49%. The expression of Bt genes using *cry*1A(b) gene has been achieved in monocotyledonous plants such as maize which provided protection against European corn borer (*Ostra nubilalis*) with several fold (>100 times) higher infestation than natural field conditions. The synthesis of Bt gene [*cry*1A(b)] completely replaced bacterial codons with maize-preferred codon and increased its G-C content from 37 to 65%. The first generation insecticidal plants were introduced to market in 1996. Transgenic cotton carrying Bt gene to protect against devastating boll weevil, developed by Monsanto Company, has been released for commercial use.

The bermuda grass endophyte *Clavibacter xyli* sub sp. *cyhodontis* has genetically been modified by cloning the gene coding for endotoxin protein of *B. thuringiensis* var *kurstaki* and later transferred to maize. In USA a transgenic *Pseudomonas* to which Bt toxic gene has been cloned is commercially distributed under brand name 'Dagger G' to control damping off of cotton. *B. popilliae* causes type A milky disease and *B. lentomorbis* type B milky disease in 70 insect species. Bt is commercially available as Bactimos, Tekner, Skaedal, Vectobec, etc., as wettable powder, flowable concentrate, granules, briquettes, etc. Mass production of *B. thuringiensis* insecticide is carried out in liquid broth fermentation tanks inoculated with selected strains. The broth is then concentrated and formulated into an aqueous oil based flowable product.

Infection dose : Infection dose is the number of bacteria to produce disease which is expressed as LD_{50}. Higher the infection dose, lower will be incubation period required. In most pupae and larvae the symptoms become apparent when bacterial population in host cells reaches 10^8 cells.

15.8.1.2 Insect Viruses

Several viruses have been found effective in the management of insect-pest menace. These viruses fall into five major groups :

1. Nucleopolyhedrosis viruses (NPV),
2. Cytoplasmic polyhedrosis viruses (CPU),
3. Granulosis viruses (GV),
4. Entomopox viruses (EPV), and
5. Non- inclusion viruses (NIV).

Insect viruses are highly host-specific obligate parasites (Table 15.4). They can be produced on live hosts (insects) and are safer to non-target microorganisms and man. *Baculovirus heliothis,* for instance, is raised on boll worm larvae on a semi-synthetic diet containing water-based mixture of sucrose, yeast, casein, salts, growth factors, etc. Alphacel is used as filler in the medium and agar for solidification. Growth of contaminants is inhibited by using formalin, sorbic acid, methyl parahydroxybenzote, aureomycin, etc. The hot diet is dispensed from a large tank by a filling machine into plastic trays and sealed. The diet is infested by caterpillars taken from insectaries, and incubated under controlled conditions to produce larval mass. A known volume of Baculovirus is sprayed on the diet which replicates 5,000-10,000 times on caterpillars in a week. The infected caterpillars are suctioned and then macerated with water, filtered, centrifuged, precipitated and spray-dried. Quality control testing is done before packaging. Various NPV, produced for different hosts, are *Lymantria, Mamestra, Neodiprion, Orgyva, Pieris, Prodenia* and *Trichoplusia.* NPV's are most promising because of their stability, safety and virulence as compared to CPU, EPV and NIV.

Table 15.4 : Some viruses successfully used as biocontrol agents

Type of viruses	Insect hosts	Host plants
NPV	*Anticarsia gemmatalis*	Soybean
	Cephalcia abielis, Chrostoneura fumiferana, Gilpinia hercyniae, Lymantria dispar, L. monarcha, Neodisprion lecontii, N. sertifer, N. swainei, Malacosoma disstria, Orgyia pseudotsugata; Spodoptera sp.	
Trees	*Heliothis armegera, Trichoplusia ni, Spodoptera sunnia*	Cotton
	Spodoptera exigua, S. sttoralis, Helicoverpa armigera	Vegetables and row crops
	Spodoptera litura	Grapes
	Wiseana sp.	Pastures
	Antographa california	Vegetables
GV	*Agrotis segetem, Pieris rapae*	Row crops
	Erinnyis ello	Cassava
	Hyphantria cunea	Trees
	Laspeyresia pomonela	Trees
BV (Baculovirus)	*Orycetes rhinoceros*	Oil and coconut palms

Nuclear polyhydrosis and granulosis viruses infect some of the same insects that are affected by *B. thuringeinsis.* However, these viruses have additional advantage of mass production in affected larvae under field conditions after their application; thereby show some degree of long term control. Repeated applications of viruses are necessary because viruses become inactive on foliage due to their exposure to sunlight. Moreover, continued crop growth leads to the production of new leaves that do not have any virus.

The first field trial of a genetically improved NPV that expressed insect-selective toxin gene (*Aa HIT*) was done in 1994. Experiments with cabbage looper *Trichoplusia ni* have shown 25% reduction in death time of target insects as compared to wild type virus; without alteration in pathogenicity or host range. Modified Baculovirus was very active and decreased crop damage and reduced secondary cycle of infection compared to the wild type viruses.

15.8.1.3 Entomopathogenic Fungi

Emtomopathogenic fungi show much taxonomic diversity. Fungal class Deuteromycota and Zygomycota contain most useful species having insect control properties. World-wide about 750 pathogenic fungal species spread over 100 genera reportedly infect insects. These pathogens cause mycosis in different insects belonging to arthropods, diptera, hemiptera, lepidoptera and coleoptera (Table 15.5). They infect all stages of host life cycle and are found in varied habitats like terrestrial, subterranean and aquatic habitats. About 25 fungal species that attack agriculturally and medically important pests have been identified as natural control. The fungi that produce spores or toxins are mass produced in culture media or on living hosts.

Table 15.5 : Entomopathogenic fungi used against different pest

Entomopathogens	Insect-pests
Metarrhizium anisopliae	Sugar cane spittle bug; sugarcane soldier fly; scarabaeiid grub
Metarrhizium flvovirdie	Brown plant hopper
Beauveria bassiana	Pine caterpillar; black vine weevil; Chlorado weevils; citrus root weevils; corn borer; eriophid mites; clear winged grasshopper; beetle; fire-ant
Beauveria brongnartii	European cock chaffer
Entomophaga grylli	White fly
Pandora neophides	Pea aphid
Entomophthora muscae	Muscoid flies; onion fly
Zoophthora sp.,	Clover weevil
Zoophthora phytonomi	Clover weevil
Zoophthora radicans	Spotted alfalfa aphid
Coleomomyces stregomyiae	Mosquito
Coleomomyces iliensis	Mosquito
Entomophaga anlicae	Brown tail moth
Entomophaga erupta	Green apple bug
Entomophaga maimaige	Gypsy moth larvae
Hirsutella thompsonii	Citrus rust mite
Hirsutella citriformis	Brown plant hopper
Paecilomyces fumosa-rosea	Sweet potato white fly
Nomuraea rileyi	Noctuid defoliators
Paecilomyces lilacinus	Brown plant hopper
Aschersonia aleyrodis	White fly larvae
Verticillium lecanii	Aphids, white fly
Lagenedium giganteum	Mosquito
Cuticinomyces clavisporus	Mosquito

Mostly entomopathogenic fungi infect their hosts through gut following consumption, but typically penetrate cuticle by mechanical or enzymatic means and rapidly proliferate within haemocoel resulting in the death of host within 1-8 days. Host death is followed by the production of infective spores on dead insects which either immediately repeat the cycle or produce dormant resting spores. Thus entomopathogenic fungi persist in environment for long time. Pathogens may consume the available nutrients in haemocoel and also produce toxins. *Beauveria bassiana,* for instance, is a beneficial soil fungus that secretes enzymes which dissolves insect cuticle, thus helps in fungal invasion into the body of insect. Environmental factors which control the spread and conidial production, survival and germination are critical for disease development.

Manipulation of entomopathogenic fungi for biological control in managed ecosystem is followed by using three basic strategies :

- ❒ Augmentation i.e. increasing the inoculum density through development of microbial insecticides;
- ❒ Enhancement of naturally occurring fungal pathogens by altering cultural practices (plant harvesting, irrigation, pesticide use, etc.); and
- ❒ Introduction of exotic fungal pathogens.

Amongst entomopathogenic fungi the successful ones are *Aschersonia aleyrodis, Beauvaria bassiana, B. globulifera, Metarrhizium anisopliae, Verticillium lecanii, Massopoa* sp., *Hisutella thompsonii, Coelomomyces* sp., *Aspergillus flavus, A. ochracens, Sphaerostilbe* sp., *Monosporella* sp., *Spicaria* sp., *Cordyceps* sp., *Entomopthora thaxteriana* and *Nomuraea rileyi.* Various entomopathogenic fungi are used on commercial scale. *Hirsutella thompsonii,* for instance, is used to control of citrus red mite and is commercially available under trade name "Mycar". North America produces 'Biotrol FMA' with *M. anisopliae* to control scale insects, white fly, plant hoppers, aphids etc. Spore suspension of *B. bassiana* is sprayed on the crop @ 5×10^7- 10^8 conidia ml^{-1} or @ 50 ml per pot or applied to soil @ 0.75-7.5 g conidia 200 $inch^3$ soil.

15.8.2 MICROBIAL HERBICIDES

The exotic or endemic pathogens have been exploited to kill weeds. To date over 100 weed pathogens have been reported. Some such pathogens used include :

i) *Puccinia chondrillina* against skeleton weed (*Chondrilla juncea*) in Australia;

ii) *Cercospora riparia* against *Ageratina riparia* in Jamica;

iii) *Cercospora rodmanii* against water hycinth (*Eichhrnia crassipes*) in USA;

iv) *Colletotrichum gloeosporioides* against joint vetch (*Aeschnemene virginia*) in USA;

v) *Phragmedium violaceam* against wild blackberry;

vi) *Phytophthora citrophora* against milk vine (*Morrenia odorata*) in USA.

In plants, glyphosphates act by inhibiting one of the enzymes necessary for the synthesis of amino acids in chloroplast. The genes coding for enzyme EPS synthase, involved in amino acid biosynthesis, have been isolated from resistant *E. coli* and modified in such a way that it is expressed in plant cytoplasm and transported to chloroplast. Thus transgenic tobacco, tomato and soybean plants fully resistant to normal glyphosphates have been developed. Phosphinotricin (Basta), a herbicide, inhibits enzyme glutamine synthase required for amino acid synthesis and nitrogen assimilation so causes ammonia accumulation to toxic level in plants. *Streptomyces hygroscopicus* encodes enzymes capable of converting phosphinotricin to non-herbicidal derivatives. Transgenic potato and tomato plants resistant to these herbicides have been developed.

15.8.3 BIOCONTROL OF PATHOGENS

Biological control of pathogens is reduction in the amount of inoculum or disease producing activity of a pathogen accomplished by or through one or more organisms including plants but excluding man. According to Garrett *'biological control may be defined as any condition or practice whereby survival or activity of a pathogen is reduced through agency of any other living organism, except man himself, so causing reduction in the incidence of a disease caused by a pathogen.* Antibiosis, competition, hyperparasitism and predation have long been recognized as the main ways by which biological control operates outside the host. Antibiosis includes any metabolic product(s) of antagonistic organism, which directly inhibits or destroys pathogenic organism. Competition for substrate (nutrient base) or space by non-pathogens at the site of infection court provides direct protection to the plant from pathogenic invasion. Competition for nutrient resource is particularly vigorous in soil which deprives pathogen of essential nutrients. The site competition principle is more prevalent on host plants. Hyperparasitism principle is based on exterior to host through which a parasitic microbe may parasitize or prey a pathogenic organism.

A biological control agent must have some special attributes such as it should be fast growing, capable to capture primary resources to colonize, have good survival in soil/plant near to the pathogen and capable to adapt varied environmental conditions including disturbed/cultivated environments. A biocontrol agent, therefore, should be able to reduce inoculum potential by lysis or parasitism of bacteria/fungi or their spores and should be stress-tolerant. Most work on biological control of plant diseases has been concentrated on the pathogens living in soil rather than on or inside the epigeal plant parts. Obviously, the chemical control of soil borne diseases is less effective and difficult to manage. Fungal biocontrol agents such as *Trichoderma viride, T. harzianum, Gliocladium virens,* and bacterial bioagents viz., *Agrobacterium radiobacter, Pseudomonas fluorescens, P. syringae* and *Bacillus subtilis* have received much attention. The use of *T. viride* against plum silver leaf disease, *A. radiobacter* against crown gall, *G. virens, T. viride* and *P. fluorescens* against root rot and damping off, and *P. syringae* to overcome frost injury are well-known examples. Some of the biocontrol agents currently in use against pathogens on commercial scale are given in Table 15.6. Commercially, *Trichoderma* species have considerable potential (Plate 15.1) and have been exploited to augment the native soil populations or used as seed dresser or seedling dip to control *Fusarium* spp., *Rhizoctonia solani* and *Pythium* spp. responsible for root rot, wilt and damping off diseases of nursery seedlings. These organisms thrive in/on soil organic matter and rhizosphere. These bioagents are fairly active and even show tolerance to some protectant fungicides. *Trichoderma* species are pathogenic to fungi either through secretion of fungal toxins or enzymes such as glucanases, cellobiases and chitinases, etc. which are responsible for fungal cell wall deterioration. Similarly, *Bacillus subtilis* has proved a beneficial soil microorganism that out-competes other invading pathogens. It is used to prevent and cure crown and root diseases, and some products claim a 2 year shelf life. Two strains of *Bacillus subtilis* are among the 187 biopesticidal active ingredients recognized by EPA. *Candida oleophila* is a yeast that out-competes many plant pathogens and has been exploited as commercial product to protect organically grown fruit. The use of *Rhizobium, Azotobacter* and mycorrhiza which augment the nutrient supply to the plants are also being tested for their plant protection potential. These organisms strengthen plant defense mechanism against the pathogenic attack or invasion by release of different kinds of hormones, cross protection etc.

Various saprophytes compete with disease causing pathogens. For instance, *Trichoderma* sp. competes with *Fusarium;* and *Verticillium* and *Peniophora gigantean* antagonize pine pathogen (*Heterobasidium annosum*) through competition exclusion, cross protection and antibiotic production.

Table 15.6 : Examples of some commercially available biocontrol agents

Trade name(s)	Active agent(s)	Manufacturer
Bioderm	*Trichoderma harzianum* + *T. viride*	Biotech International, New Delhi
Sppressivit	*T. harzianum* PV5736-89	FY Tovita, Czech Republic
Tri-002	*T. harzianum* KRLAG-2	Plant Support, the Netherlands
T 22, Planterbox & Root Shield	*T. hariozanum Rifai* strain KRL-AG2 (T-22)	Bioworks, Geneva, Switzerland
Ecofit	*T. viride*	Hoechst, Mumbai, India
Tricogaurd	*T. viride*	Annu Biotech international, Faridabad, India
Soil Guard 12G	*Gliocladium virens*	Margo Biocontrol Ltd., Bangalore, India
Biofox C	*Fusarium oxysporum* (non - pathogenic)	SIAPA, Italy
Aspire	*Candida oleophila* I-82	Ecogen, West, Langhorne
Kalisena SD/SL	*Aspergillus niger* AN-27	Cadila Pharmaceuticals Ltd. Ahmedabad, India
Paecil, Bioact	*Paecilomyces lilacinus*	Technological Innovation Corporation Pty. Ltd., Australia
Polyversun	*P. oligandrum*	Biopreparaty Ltd, Czech Republic
AQ 10	*Ampelomyces quisqualis* isolate M-10	Ecogen, West, Langhorne
Mycostop	*Streptomyces griseoviridis* K61	Kemira Agro Oy, Finland
Actinovate	*S. lydicus*	Natural Industries, Houston
Rhizo-Plus	*Bacillus subtilis* FZB24	KFZB Bitechnicv GnbH, Germany
Histick-N/T	*B. subtilis* MB1600 + *Rhizobium*	Becker Underwood
Companion	*B. subtilis* GB03	Growth Products USA
EcoGuard	*B. licheniformis*	Novozymes Biological Inc. Salem VA
No gall	*Agrobacterium radiobacter* K1026	Biocare Technologies, Australia
Galltrol	*I.radiobacter* strain 84	Ag BioChem Inc., USA
Bioshield	*Pseudomonas fluorescens*	Annu Biotech international, Faridabad, India
Intercept	*Berkholderia cepacia*	Soil Technologies Corporation, USA
Deny	*B. cepacia* type Wisconsin	Stine Microbial Products, USA

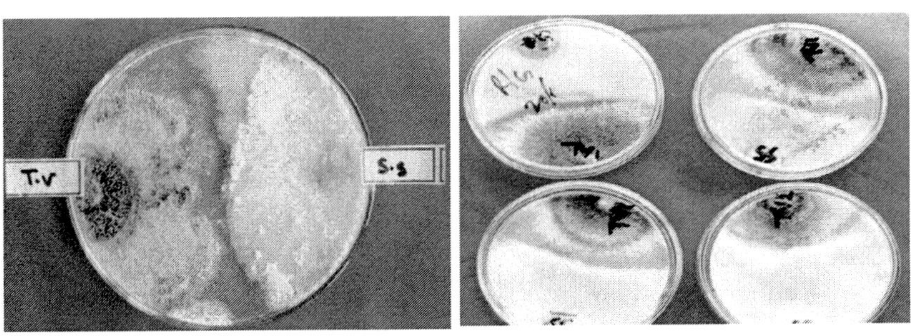

Plate 15.1 : *Trichoderma* inhibiting the growth of a pathogen

15.9 CONSTRAINTS IN BIOFERTILIZER AND BIOPESTICIDE USE

Biofertilizers and biopesticides have made major contributions to agriculture production in many countries, but despite these successes they have received less governmental attention. The acceptance and popularity of biofertilizers and biopesticides is limited due to the various constraints at production, resource, field and marketing levels.

Production constraints : Biofertilizer/biopestide production technology has greatly been refined over many decades. However, the progress in biofertilizer/biopesticide production is far below the satisfaction level due several limitations such as the unavailability of appropriate and efficient strains, mutation during fermentation/culture raising, lack of standard packaging, etc. Biofertilizers/biopesticides generally show specificity to a particular crop. Besides, they require specific edapho-agroclimatic conditions; hence lack of eco-specific or region-specific or specie-specific strains is one of the major constraints in their acceptance at large scale. Moreover, the strains selected for production should have various intended qualities. For instance, N_2 fixing inoculants should have the ability to fix nitrogen over a range of environmental conditions, ability to survive adverse physical conditions, ability to compete resident populations, ability to multiply in broth and ability to survive in inoculants carrier. Biofertilizers/biopesticides tend to mutate during fermentation. Repeated culture raising also enhances production and quality-control costs. It needs extensive research work to eliminate the chances of mutation. The inoculant development and use requires some technical training and it is not easy for the developing countries to produce high quality inoculants domestically. Further, there are no proper standards of packaging viz., material, type, size, etc. The claim of manufacturers that the projected demand for biopesticide has failed to become a reality, i.e. the lack of support, is proving to be the death knell of this industry which is still in infancy.

Marketing constraints : Biofertilizers/biopesticides are in use since late 50's in 20th century, however their consumption has not picked up to the desired level due to various market-related constraints which include lack of mass awareness in farming community, inadequate and inexperienced staff, lack of quality assurance, seasonal and un-assured demand and limited scope for marketing. In spite of the considerable efforts made in recent years, majority of farmers are not aware of biofertilizers/biopesticides and their usefulness in improving crop yields and soil health. Further, the acceptance of biofertilizers/biopesticides is poor due to inadequate and technically unqualified marketing staff. The sale of poor quality biofertilizers/ biopesticides has also resulted in loss of faith, and as a result product image has suffered. A small scale production unit, solely dependent on biofertilizer or biopesticide, is not economically viable because biofertilizer/biopesticide use is seasonal, and both production and distribution are seasonal. The private production unit holders are not sure of their demand. As biofertilizers are soil- and crop-specific, a large number of formulations of a microbial inoculants each having unique features are required to be produced by a manufacturer. This has limited the market size for each type of biofertilizer or biopesticide.

Resource constraints : In addition to technological and marketing constraints, resource constraint is also one of the main constraints in production of biofertilizers/biopesticides on commercial scale. The investments in biofertilizer or biopesticide production units are considerably low because of the risks involved, short life and no guarantee of off-take of biofertilizers or biopesticides. This has resulted in limited resource generation from private sector. Further, biofertilizers and biopesticides are considered as low cost input for poor farmers. Lack of trained personnel with essential combination of agronomic and microbiological skill for executing production-oriented research on inoculant technology has limited research and development efforts. The administrative structure has often thwarted the attempts to develop integrated innovative programmes. The crop response to biofertilizers and biopesticides is inconsistent; the poor farmers, therefore, find it difficult to take the risk of using them.

Field constraints : Many field level constraints too have limited the popularization of biofertilizers and biopesticides. The soil and climatic factors such as fertility status, unfavourable pH, high nitrate level, high temperature, drought or too much dampness, deficiency of P, Cu, Co, Mo or presence of certain elements in toxic amounts, etc. which significantly influence the crop response have hampered the wide acceptance of biofertilizers and biopesticides. Similarly, the presence of native antagonistic micro-organisms in soil competing with microbial inoculants many times does not allow the effective establishment of introduced inoculant. So proper soil management to avoid proliferation of less-effective antagonistic microbes in soil deserves attention. It is essential to understand interactions amongst the inoculant strain, specific crop and surrounding environment before making any recommendation for the use of inoculants. Inadequate understanding of these interactive forces has resulted in inability to predict whether a given crop will respond to inoculation in a particular region/season or not. Most of the farmers, stock-marketers and even agriculture extension workers do not have the proper knowledge about the inoculation techniques. Biofertilizers and biopesticides being living organisms require utmost care during handling, transport and storage; otherwise their shelf life and viability is likely to get greatly reduced. Delivery and acceptability of technology at farm level is low and no proper extension mechanism in this behalf needs to be developed.

The response to biofertilizers and biopesticides largely depends on balanced use of other inputs and better crop management. The poor crop management and use of poor agricultural practices has made biofertilizer or biopesticide use ineffective. Lastly, quality control of biofertilizers is essential. Quality standards have been setup for *Rhizobium, Azotobacter,* mycorrhiza and some biocontrol agents. But, its effectiveness in ensuring the supply of quality biofertilizers and biopesticides at field level is still doubtful. This has resulted in attrition of farmer's confidence in biofertilizers and biopesticides. Therefore, monitoring of product quality needs to be ensured at retail outlets. To feed the future generations without degrading our natural resource base that supports crop productivity, agriculture must become economically viable and ecologically sustainable.

Selected References

Aizawa, K. 1982. Microbial control of insect pests. pp. 397-418. In : *Advances in Agricultural Microbiology*, (ed. N.S. Subba Rao). Oxford & IBH Publ. Co., New Delhi, India.

Alison, F.E. 1973. *Soil Organic Matter and its Role in Crop Production*. Elsevier Scientific Publ. Co., London, UK.

Alloway, B.J. and Ayres, D.C. 1997. *Chemical Principles of Environmental Pollution* (2nd edn.). Chapman and Hall, London, UK.

Atwal, A.S. and Dhaliwal, G.S. 1997. *Agricultural Pests of South Asia and their Management*. Kalyani Publishers, New Delhi, India.

Burges, H.D. (ed.). 1988. *Formulation of Microbial Biopesticides, Beneficial Microorganisms and Nematodes*. Chapman and Hall, London, UK.

Campbell, R. and McDonald, R.M. 1989. *Microbial Inoculants of Crop Plants*. Soc. Gen. Microbiol., IRL Press, Oxford, UK.

Carruthers, R.I. and Hural, K. 1990. Fungi as naturally occurring entomopathogens. pp. 115-138. In : *New Directions in Biological Control : Alternatives for Suppressing Agricultural Pests and Diseases*. Alan R Liss inc., New York, USA.

Coppel, H.C. and Mertins, J.W. 1977. *Biological Insect Pest Suppression*. Springer Verlag, New York, USA.

Cory, J.S., Hirst, M.L., Williams, T., Hails, R.S., Goulson, D., Green, M.B., Carty, T..M., Posse, R.D., Cayley, P.J. and Bishop, D.H. L. 1994. Field trial of a genetically engineered baculovirus Insecticide. *Nature*, 370 : 138.

Dar, G.H. and Ganaie, M.R. 1990. Effect of cynobacterial inoculations on soil fertility and yield of paddy under temperate conditions. *Phykos*, 29 : 31-38.

Dar, G.H., Zargar, M.Y. and Beigh, G.M. 1997. Biocontrol of *Fusarium* root rot in common bean (*Phaseolus vulgaris* L.) by using symbiotic *Glomus mosseae* and *Rhizobium leguminosarum*. *Microbial Ecology*, 34 : 74-80.

FAO. 1977. *Recycling of Organic Wastes. FAO Soil Bulletin No. 40.* FAO, Rome, Italy.

Gaur, A.C. 1985. Phosphate solubilizing microorganisms and their role in plant growth and crop yield. pp 125-138. In : *Proceedings of Soil Biology Symposium*, HAU, Hissar, Haryana, India.

Hafte, H. and Whitley, H. 1989. Insecticide crystal proteins of *Bacillus thuringiensis. Microbiol. Rev.*, 33 : 242-255.

Jeevajothi, L., Mani, A.K. and Pappiah, C.M. 1993. Influence of NPK and *Azospirillum* on the yield of cabbage. *South Indian Hort.*, 41 : 270-272.

Joi, M.B. and Shinde, P.A. 1976. Response of onion to Azotobacterization. *J. Maharashtra Agric. Univ.*, 1 : 161-162.

Koziel, M.G., Beland, G.L., Dowman, C., Carrozi, W.B., Crenshaw, R., Crossland, L. Dowson, J., Desai, N., Hill, M., Kadwell, S., Launis, K., Lewis, K., Maddox, D., McPherson, K., Meghji, M.R., Merlin, E., Rhodes, E., Warren, G.W., Wright, M. and Evola, S.V. 1993. Field trial performance of elite transgenic maize plants expressing an insectidal protein derived from *Bacillus thuringiensis. Biotechnology*, 11 : 194- 200.

Lehri, L.K. and Mehrota, C.L. 1972. Effect of *Azotobacter* inoculations on the yield of vegetable crops. *Indian J. Agric. Res.*, 9 : 201-209.

Lysenko, O. 1985. Non-spore forming bacteria pathogenic to insects. *Annu. Rev. Microbiol.* 39 : 673- 695.

Maramorosch, K. (ed). 1991. *Biotechnology for Biological Control of Pests and Vectors.* CRC Press, Boca Raton, Florida, USA.

McEwen, F.L. 1978. Food production : The challenge of pesticides. *Bioscience*, 28 : 773-777.

Menn, J.J. 1996. Biopesticides : Has their time come? *J. Environ. Sci Hlth. B*, 31 :383-389.

Mishra, S. 1998. Baculoviruses as biopesticides. *Curr. Sci.*, 75 : 1015-1022.

Qxtoey, E. and Hieghes, M. A. 1990. Engineered herbicide tolerance into crops. *Trends Biotechnol.*, 8 : 61- 65.

Rewari, P.B. 1984. *Report of All India Co-ordinate Project on Pulse Improvement, 1974-1984.* Indian Council of Agricultural Research, New Delhi, India.

Schonbeck, F. 1979. Endomycorrhiza in relation to plant diseases. pp. 271-280. In : *Soil-borne Plant Pathogens* (eds. B. Schippers and W. Gams). Academic Press, New York, USA.

Singh, P.K. 1979. Use of *Azolla* in rice production in India. pp 407-418. In : *Nitrogen and Rice.* International Rice Research Institute, Manila, Philippines.

Steinhouse, E.A. 1977. *Insect Microbiology.* Academic Press, New York, USA.

Subba Rao, N.S. 1982. *Biofertilizers in Agriculture.* Oxford & IBH Co., New Delhi, India.

Subba Rao, N.S. 1999. *Soil Microbiology.* Oxford & IBH Publ. Co., New Delhi, India.

Sukhada M.D. 1987. Field response of tomato to inoculation with VAM *Glomus fasiculatum* and *Azotobacter vinelandii. Pl. Soil,* 98 : 295-297.

Venkataraman, G.S. 1972. *Algal Biofertilizers and Rice Cultivation*. Today & Tomorrow Printers & Publishers, New Delhi, India.

Venkataraman, G.S. 1981. *Blue-Green Algae for Rice Production, FAO Soils Bull. 46*. FAO, Rome, Italy.

Wenhua, T., Cook, R.J. and Rovira, A.D. (eds.) 1994. *Advances in the Biological Control of Plant Diseases*. China Agricultural University, Beijing, China.

— ★★★ —

Chapter 16

Soil Biotechnology — The Emerging Trends

Biotechnology, together with other technologies, could provide new solutions for some of the old problems hindering sustainable rural development and achievement of food security.

(FAO Report, 1999)

Biotechnological research started in early seventies with some impressive discoveries in the field of genetics, microbiology, biochemistry, biology and chemical bioengineering. The amalgamation of these sciences particularly integration of their basic principles and applied aspects gave rise to a new field, completely diverse and versatile in nature, what we call today 'biotechnology'. Originating from research laboratories in universities, biotechnology rapidly captured the attention of industries as the number of its potential products increased. This newly developed frontier science later found broad application in medical science, pharmaceutical industry, agriculture, veterinary sciences and chemical industry for the welfare of mankind. Biotechnology by definition means '*any technique that uses living organism or a part thereof to make or modify a product, to improve plant or animal or to develop microorganisms for specific purposes*'. The techniques of biotechnology are primarily directed to a single or a few genes rather than whole genome. Until 1981, there was no commercial application of genetically engineered processes, but by 1983 several commercial companies sprung up for using and exploiting the recombinant DNA technology industrially. There are more than 25,000 laboratories presently engaged world-wide in various genetic engineering programmes concerned with the development of new organisms with desired attributes, monitoring their behaviour in natural environment and analysis of their properties. For instance, in USA alone more than 1500 companies are at present engaged in insulin preparation. The first agricultural biotechnological product was swine vaccine. More than 2000 transgenic plants and microorganisms have been released in the world so far. Transgenic release procedures and regulations, generally well-defined, vary from country to country.

Biotechnological advances perpetuated even in complex heterogeneous soil ecosystem. The field of soil biotechnology has developed fast in recent years with the improvement in molecular and genetics studies of soil microorganisms, especially *Rhizobium, Xanthomonas, Agrobacterium, Azotobacter, Clostridium, Escherichia, Pseudomonas, Bacillus, Neurospora,* etc. The soil biotechnology involves '*the manipulation of soil organisms and their metabolic processes in such a way that leads to the optimization of crop productivity as well as yields maximum resource utilization or protects the environment for better human welfare*'. For instance, nitrogen fixation, desulphuration, oil recovery, bioremediation, soil development, pollution treatment, mineral extraction, biotransformation, bioenergy and biofuel production, etc. can be achieved in better way through biotechnological approaches. The artificial manipulations increasingly involve genetic changes through human interferences, although many simply involve selective addition, repression or deletion of genes in soil microorganisms. We know that gene exchanges in bacteria take place by the process of transformation, transduction or conjugation. So the apprehension of recombinant DNA transfer from a genetically engineered microorganism to other non-engineered soil microorganisms as a result of conjugative transfer of plasmids, phages etc. can not be ruled out.

16.1 BIOTECHNOLOGY IN IMPROVING THE CROP PRODUCTIVITY

In agriculture, genetically engineered microorganisms have been pressed into service to create the strains that may aid in enhancing the crop productivity, improving the soil fertility and/or protecting the environment. The genetic modification of plants to improve crop yields has increased the efficiency of agriculture and this represents one of the most rapidly developing areas of biotechnology. Great advances have been made in modifying the crop plants to induce or improve their resistance potential to pests (fungi, bacteria, virus, insects, etc.) and chemical pesticides. It seems now possible to modify the genes involved in photosynthetic mechanisms of plants to enable more efficient fixation of atmospheric CO_2 and enhance crop potential for primary productivity through genetic engineering of genes exclusively involved in the expression of ribulose bisphosphate carboxylase oxygenase (Rubisco). The gene responsible for Rubisco polypeptide synthesis present in photosynthetic bacteria has been cloned into *Escherichia coli* and introduced into a plant to synthesize large amounts of the enzyme.

Genetic modification of plants has certain advantages over microbial inocula with respect to their soil applications. Any microbial inoculum is basically exotic to the intended soil so it has to successfully cope with the resident soil population and then has to acclimatize to that heterogeneous environment. Further, the risks involved in release of genetically engineered organisms into the environment are more predictable and easily controllable for plants than microbes. However, manipulations in plants, sometimes, may not be easy or the conditions for the effective functioning of a genetically engineered soil organism may not be conducive. In such situations, inocula need to be prepared from the natural habitat through genetic engineering technology. The soil inocula involving genetically engineered organisms may be used in soil environment in a variety of potential ways (Table 16.1).

A naturally occurring conjugation phenomenon in *Agrobacterium tumefaciens* and *A. rhizogenes* which induce crown gall (tumor development) and hairy root diseases, respectively, in plants has been exploited by biotechnologists as vector to genetically engineer the foreign genes into plants. This action renders them transgenic in characters with imparted resistance against different diseases, insect-pests, herbicides or insecticides or any other transferred attribute. *A. tumefaciens* is a soil bacterium which infects crown region of dicotyledonous plants through mechanical wound to produce galls. *A. tumefaciens* has large extra-chromosomal 150 MDa mega-plasmid 'Ti plasmid' which contains the genes responsible for tumour formation. Ti plasmid possesses a tumour inducing segment (T-DNA) which carries genes for the synthesis of two growth hormones auxin (indole acetic acid) and cytokinin, and the genes controlling the synthesis of a group of amino acid derivatives known as 'opines' (nopaline and octopine).

Opines are used by the bacterium as carbon nitrogen and energy source. The T-DNA is presumed to be better vehicle in gene transfer from *A. tumefaciens* to plants because T-DNA is flanked by unique 25 base pair sequences on its left and right borders which can transfer any gene sequence to the plants placed between them. The expression of bacterial genes controlling growth hormones is not controlled by plasmid but is necessary for the development of crown gall symptoms.

Table 16.1: Some genetically engineered plants and microbes

Plants or microorganisms	Function introduced	Effects
Cotton, tobacco and tomato transformed with *Bacillus thurengenesis kurstaki*)	Toxin production	Control of lepidopterous pests; temporary rise in soil biomass and alteration in specie diversity of soil microbes.
Potato, tobacco, lettuce, sugar beet transformed with *Serratia marcescens*	Chitinase production	Control of fungal pathogens; plants become resistant to soil borne phytopathogens
Pseudomonas syringae ice variants	Frost resistance in plants	Site colonization, competitive exclusion and antagonism control ice-forming organisms.
Klebsiella planticola	Ethanol production	Decrease in arbuscular mycorrhizal infection in wheat; phytotoxicity to wheat
Pseudomonas fluorescens,	Degradation of 2,4 D herbicide	Decrease in soil fungal population
P. fluorescens	Biocontrol agent	Plasmid DNA transferred to several genera of indigenous bacteria
P. lividans	Cellulose degradation	Increase in CO_2 evolution on the addition of lignocellulose
P. fluorescens sp. B_{12}	3-cholorobenzoate degradation	Transferred gene for this character into alcaligens in a natural aquifer

The infection process in *A. tumefaciens* is initiated by a series of *vir* (virulence) genes at the wound sites on the host. In fact, the phenolic compounds released by wound and plant cells act as signal for gene induction. Formation of crown gall, due to the integration of T-DNA, results from the expression of genes from the plant growth hormones (auxin and cytokinin). This causes rapid proliferation of plant cells leading to the tumour development. The T-DNA transfer is mediated by the products encoded by 30-40 kb *vir* segment of Ti plasmid. This region is composed of at least six essential operons and two non-essential operons which are necessary for the recognition of susceptible cells in plants. To excise T-DNA from a plasmid and transform the T-DNA region into host cells, v*ir* gene is activated only by contact with cell metabolites (phenolics) released by plants. The methods available to transform plant cells include introduction of Ti plasmid encapsulated in liposome into protoplast or fusion of bacterial sphaeroplast with plant cell protoplast or infection of protoplast with partially regenerated cell wall with intact *A. tumefaciens.* This way one can generate a transgenic plant, able to synthesize opines. Inserting these genes into monocots using engineered *A. tumefaciens* as vector is now possible.

T-DNA mediated gene transfer system has stimulated research to generate transgenic plants by replacing tumour inducing genes with suitable selection marker (antibiotic- or resistance-marker) to select transgenic plant that have multiple cloning sites at which foreign gene can be introduced and have specific promoters for efficient expression of foreign genes. Improvement in plant growth can be achieved through use of improved/effective inocula, made through genetic manipulation, having better competitive ability or co-inoculation with antibiotic producing bacteria which may increase colonization and nodulation.

16.2 STRAIN IMPROVEMENT IN MICROBIAL SYMBIONTS

The study of rhizospheric processes, nodulation and effectiveness of bacterial symbionts and mycorhizal response falls under the realm of biotechnology. *Rhizobium, Azotobacter, Azospirillum, Clostridium* and mycorrhizal fungi seem obvious target soil organisms for research in DNA recombinant technology. The transfer of *nif* genes from *Klebsiella pneumoniae* MSa I to *E. coli* strain C 603 or from *Rhizobium trifolii* to *Klebsiella aerogenes* has enabled these microbes to fix nitrogen. The manipulation of *nif* gene into alien prokaryotes background is now a routine process. The extension of symbiotic host range of *Rhizobium* is quite promising and work in this direction is in progress. Similarly, enhancing biological nitrogen fixation in cereals is another important area of soil biotechnology which has received attention and success to some extent has been achieved. Molecular methods of introducing foreign genes into *Rhizobium* have enhanced the commercial value of these inoculants. The δ toxin genes (*cry II A*) of *Bacillus thirungenesis* sub sp. *terebrionis* have been introduced into *R. leguminosarum* for the control of larvae which feed on legume roots and nodules causing considerable damage. The transgenic *R. leguminosarum* has greater ability to compete for nodule occupancy than non-transgenic parent strain. Field tests of *Rhizobium meliloti* have yielded considerable improvement in nitrogen fixation by the bacteria as well as growth of the host plants. The use of mutagenized form of *Rhizobium* inoculants with better nitrogen fixing ability in soybeans has successfully been made. δ toxin genes *cry II A* has been used to produce Bt-toxin or transgenic cotton. Inability of endomycorrhiza to culture can be overcome through genetic engineering. Such studies need improvement and priority.

16.3 PLANT DISEASE CONTROL

Genetically engineered antagonists of soil borne plant pathogens are widely exploited as biocontrol agents of many diseases. Such antagonists include *Trichoderma viride, T. harzianum, Gliocladium virens* and *Pseudomonas fluorescens*. Antagonism is because of organic volatiles and gene technology increases inoculum resistance to fungicides (to be effective biocontrol).

Rhizosphere microflora has also been exploited for the control of soil borne pathogens. Soils generally harbour fluorescent pseudomonads, root rot pathogens such as *Pythium, Rhizoctonia, Phytophthora, Fusarium, Sclerotium,* etc. crown and collar rot pathogens like *Agrobacterium tumefaciens,* and many other fungal and bacterial pathogens. The use of recombinant DNA technology has permitted the identification and tailoring of genes involved in host-pathogen specificity, pathogenic virulence, function and symptom expression in host plant, toxin and enzyme synthesis and their regulation. A serious thought has been given to use monoclonal antibody technology to produce antibodies for use in diagnostic and epidemiological studies of specific plant diseases. Development of genetically modified organisms may prove valuable in tackling the disease at its initial stage of development. It is now possible to remove the plasmids and replace the disease-causing genes with genes that produce desirable characteristics. When the new plasmids were placed in *Agrobacterium tumefaciens,* the bacterium acted as a living delivery van, transferring the desirable genes directly into plant cells. Hence, it is sure that judicious application of soil biotechnology can facilitate development of plant genotypes which are tolerant to numerous biotic and abiotic stresses.

16.4 GENE CLONING FOR INSECTICIDE RESISTENCE/ TOXIN PRODUCTION

The potential application of transgenic techniques has facilitated the farming of crops with multiple durable resistances to insect, weeds and diseases, particularly in absence of pesticides. The use of *bxn* gene of soil bacterium *Klebsiella azaenae* which detoxifies bromoxynil by breaking down it into non-toxic constituent 3,5 dibromo-4-hydroxy benzoic acid has been used to produce transgenic tobacco. The cloned *bxn* gene was placed under the control of a light regulating tissue-specific promoter, Rubisco small sub-unit. The transgenic plants acquired resistance against the herbicide. Similarly, cloning of *Bacillus thuringenesis* genes (producing toxins into a strain of *Pseudomonas fluorescens* that lives in maize rhizosphere) can provide protection against soil insects. The use of toxin genes isolated from various strains of *B. thuringensis* has been the focus to develop plants resistant to insects etc. The δ endotoxin producing genes from *B. thuringenesis* have been transferred to plants through Ti plasmids to induce insecticide resistance in them. Different strains of *B. thuringenesis* produce toxins with different spectra of activities. Some of these genes have been introduced into rhizosphere bacteria like fluorescent pseudomonads, *Bacillus,* etc. Such recombinant strains are grown, killed and applied to crops which kill the feeding insects or provide crop protection. Similarly, genes for herbicide glyphosate tolerance have been transferred to some plants via Ti transfer system. These plants are not affected by herbicide applications, used for killing the weeds.

16.5 PHYLLOSPHERE PROTECTION

Frost damage is a major cause of crop loss in temperate and subtropical regions. Frost-sensitive plants like citrus, rosaceous plants (pear, apple, etc.), potato, strawberries, etc. are injured when temperature falls below 0°C because of ice formation within their tissues, especially in early spring and late autumn periods. Small volume of pure water can be supercooled to -10°C or below without ice formation, provided no catalytic centers or nuclei are present to induce ice formation, so super-cooled water will not damage plant cells even at low temperature. Various strains of epiphytic bacteria like *Pseudomonas syringae, P. fluorescens, P. putida, Xanthonmonas campestris* and *Erwinia herbicola* present on phyllosphere induce ice nucleation/ice crystallization at temperature as high as -1°C. Such bacteria make up a small proportion (0.1-10%) of phyllosphere population but cause severe damage to the host plant. Ice crystals are formed due to the presence of ice^+ genes in these microorganisms. These produce unique protein which acts as a nucleus for formation of ice crystals from supercooled water. These bacteria carry gene '*Ina*' (ice nucleation active gene). Genetic engineering of these microorganisms plays significant role in controlling frost injury in plants. *Pseudomonas syringae* is a known phyllosphere microorganism producing a protein causing ice nucleation. Deleting the ice^+ genes involved in ice nucleation make them ice^-. These ice^- organisms when sprayed on the plants prevent frost sensitive crops from ice crystal formation. The sprayed ice^- *P. syringae* on plant system out-compete the resident population of ice^+ bacteria on phyllosphere thereby restrict the growth of ice forming bacteria through site colonization, competitive exclusion and antagonistic effect. Near isogenic strains of *P. syringae* or genetically engineered strains of leaf residents serve as antagonist to ice nucleation bacteria. Antagonistic sprays facilitate early colonization of plant surfaces so avoid the risks of late spring frosts. During the spray of genetically engineered microorganisms on phyllosphere, a number of genetically engineered organisms are also carried to the soil. Intentionally introduced genetically engineered microorganisms may enter the soil while being applied on flora of the crop. Their better competitive ability and exclusion ensures their survival and dispersal.

16.6 MICROBIAL INOCULA FOR PLANT GROWTH REGULATION

Improvement in plant growth can be achieved through use of improved/effective microbial inocula, constructed through genetic manipulation, having better competitive ability or co-inoculated with antibiotic producing bacteria which may have increased colonization and nodulation capacity. In various rhizosphere bacteria such as *Azotobacter chroococcum* growth hormone production is activated and enhanced by root exudates.

16.7 GEM FOR DETOXIFICATION OF CONTAMINANTS/ BIOREMEDIATION

Genetic modification has enabled the development of genetically engineered microorganisms (GEM) that have enhanced capacity of degrading xenobiotic or aromatic hydrocarbon or recalcitrant compounds. For instance, the organisms which degrade chloro- and methyl-aromatic compounds have been constructed. Similarly, genetically modified oil-oxidizing microorganisms have great scope to restore soils contaminated by oil spills. Industrial and environmental biotechnologists aim at minimizing the pollution risks, enhancing the agro-product development and optimizing the energy use by maximizing production to recyclable or biodegradable products. GEMs are often sprayed on ores to accelerate metal leaching from ores in the mining industry. A number of chemoautotrophic bacteria of genus *Thiobacillus* have been engineered to increases ore leaching efficiency. Another example is that of citrate synthase gene of *Pseudomonas aeroginosa* that has been introduced into tobacco and papaya, making them able to secret citrate from their roots which chelates aluminum. These transgenic plants can withstand high Al stress without any yield loss. Many GEMs are used as indicators for detection of toxic or mutagenic compounds in soil. The selected bacterial strains are genetically engineered by insertion of reporter genes encoding for traits not present in natural soil bacterial community and easily detectable, like bioluminescence. The reporter genes are either constitutively expressed or inducible by exposure to different stresses. Reporter bacteria, capable of surviving in the soil environment, are then inoculated into soil samples and, immediately after re-extraction of the inoculated cells, the specific activity of reporter genes is determined. There is every chance that field-released genetically engineered microorganisms may enter soil. GEMs used for industrial purpose may contaminate soil as a result of their accidental release.

16.8 BIOFUEL AND BIOENERGY GENERATION

Living organisms are tiny chemical machines with enormous potential to transform organic compounds to novel chemicals. Microorganisms have evolved a variety of enzyme regulatory mechanisms to accommodate the changing needs of the microbial cell in a changing environment. This potential of microbes has successfully been exploited in bioprocessing engineering and chemical transformation. Bioprocessing (fermentation) technology including biofuel and bioenergy generation from biomass, industrial and agricultural wastes, etc. has become demanding and has played main role in yielding various valuable products. Interestingly, several soil microbes like *Saccharomyces cerviscea* have been identified to be efficient in converting wastes into valuable resources. The use of modified organisms has been made to catalyze fermentation processes under a defined set of environment. Biotechnological advancements have lead to engineering of recombinant forms which are highly efficient than the existing strains.

16.9 RISKS INVOLVED IN USE OF MODIFIED ORGANISMS AND BIOSAFETY OF SOIL ECOSYSTEM

Genetically modified plants/microorganisms on one hand offer immense potential to uplift agriculture and allied fields and, on the other hand, invite some serious environmental consequences. However, safe and effective use can be achieved only by developing better means of understanding these organisms, their interaction and ecological range. A broad approach to the engineering of soils and their resident flora can be valuable for agricultural economy of those countries where soils are of marginal fertility. The possibility of horizontal transfer of genes from one organism to another, ignoring specie barrier, when used in soil environment cannot be ignored. Such organisms may prove hazardous if they threaten soil health and cause damages than the expected goods. The fear of disturbances in ecological balance due to some accidental escape of foreign gene through pollen dispersal from transgenic plant to their wild relatives is another worry to the biotechnologists. This may lead to the creation of a super-weed or prove toxic to non-target beneficial organisms. This needs to be assessed for possible risks and tested for biosafety in field before any environmental introduction is made.

Biosafety assessment is done by monitoring the introduced genetically modified organisms for persistence, invasion, pathogenicity, toxicity to non-target microorganisms and dispersal behaviour beyond the intended targeted environment. Detection and monitoring of genetically modified microorganisms involves a number of extractive procedures and *in situ* techniques of great power, accuracy and sensitivity. Gene probing done in soil through PCR enables single copies of gene to be amplified and detected from the environmental samples. Genome tracking for detection of even a single cell *in situ* is also done. Direct microscopic *in situ* methods employed include immunoflourescence techniques, bioluminescence, CCD microscopy of bioluminescent-marked cells and other specific strain detection methods. For cell extraction, plate count colorimetric selection plus direct staining, most probable number estimation, immunological approaches, flow cytometry, affinity concentration, luminescence base techniques, etc. are currently in use. The genetic detection is made via gene probing, DNA analysis, amplification through PCR, gene sequencing and mapping/ RNA analysis, etc. In these cases reporter genes, screen-able markers, are used to identify a genetically engineered cells. Such cells are routinely incorporated into transgenic plant, animal or microbes. Antibiotic resistance markers are frequently used because they allow the selection of recombinant organism by plating onto a selective medium containing a specific antibiotic on which only resistant organisms grow. An important criterion in selection of such marker sequences is that these confer resistance only to antibiotic of limited pharmaceutical use. Most commonly used reporter genes for detection and monitoring are *GUS* (β-glucouraniside; *uid* A) and *Lac Z* (β-galactosidase) genes from *E. coli* and *lux* (luciferase) gene from firefly or bacteria that encode bioluminescence. The *lac Z* system encodes β-galactosidase production and has been used in tracing engineered *Pseudomonas* strains from soil. The strains incorporating this gene can be recognized by a characteristic blue colour produced on incubation with a specific substrate. Other genes encoding for chromogenic reaction potential such as production of catechol 2,3 oxygenase have also been used. Such assessments are easier for plants than microbes. For example, cloning of l*ux* gene for bioluminescence from marine *Vibrio* into a selected soil organism can be easily done. These cells can be imaged with some degree of bioluminescence (catalyzed by enzyme luciferase) proportional to their metabolic activity. The bioluminescence marker system has additional advantage that it does not give rise to environmental concerns associated with antibiotic resistance markers.

Survival and dispersal of released organisms are directly affected by their interaction with indigenous soil biota and physico-chemical environment. Inocula added to soil usually have poor survival. For instance, genetically modified organisms introduced in wet conditions may be wiped out from their target source. Cells which predominantly get dispersed in large soil pores usually are depleted by protozoal grazing. Similarly, extreme dry conditions may not allow such organisms to proliferate. So water potential has dominant control in such cases. The transportation risk of genetically modified organisms throughout the soil profile by earthworm and protozoal activities are more. Genetically modified organisms are unlikely to change the functional diversity of soil. The pesticides-resistant genetically modified organisms introduced into a new environment should have no marked effect on non-target microorganisms and/or on the processes they mediate like aggregation, nitrification, xenobiotic degradation, mycorrhizal infection, rhizosphere microfloral activities, etc. The chance of such interferences can be avoided by introducing suicidal genes in modified organisms wherein beyond a specified environment the suicidal genes may be activated thereby modified organisms may persist no longer in the system.

Many soil ecologists believe that the persistent production of Bt toxin by engineered plants could give insects greater opportunity to develop resistance.

This may deny organic farmers of the few remaining environmentally benign insecticides. Others believe that plants producing Bt may also kill eco-friendly insects and predators that help to keep insect pests under control. The extent of the phenomenon and its ecological impact remain uncertain.

Selected References

Alexander, V.S. and Slevin, J. 2003. *Agrobacterium tumefaciens* mediated gene transfer: An overview. pp. 93 - 101. In: *Biotechnology, Emerging Trends* (eds. J. Selvin, A.S. Ninawe, V.S. Sugunan, N. Sukumaran and A.P. Lipton). Biotech Books, New Delhi, India.

Altman, A. (ed). 1998. *Agricultural Biotechnology*. Marcel Dekker, New York, USA.

Bryant, J. A. 1985. Transformation of monocot plants by Ti plasmid of *Agrobacterium*. *Trends in Biotechnology*, 3:21.

Cruger, W. and Cruger, A. 1990. *Biotechnology: A Textbook of Industrial Microbiology* (2nd edn.). Thomas D. Brock, Sinauer Associates, Sunderland, USA.

Dale, J.W. 1998. *Molecular Genetics of Bacteria*, (3rd ed.). John Wiley & sons, New York, USA.

Dar, G.H., Anand, R.C. and Sharma, P.K. 1993. Genetically engineered microorganisms to rescue plants from frost injury. pp. 1-19. In: *Advances in Biochemical Engineering/ Biotechnology*, Vol. 50. (ed. F. Flechture), Springer Verlag, Berlin, Germany.

Davis, R.W., Botstem, D. and Roth, J.E. 1980. *Advanced Bacterial Genetics: A Manual for Genetic Engineering*. Cold Spring Harbor Laboratory, New York, USA.

Demain, A.L. 2000. Microbial Biotechnology. *Tibtech.*, 18: 26-31.

Ehrilich, H.A., Gibbs, R. and Kazazian, H.H. (eds.) 1989. *Polymerase Chain Reaction*. Cold Spring Harbor Laboratory, Cold Spring, Harbor, New York, USA.

Garret, S.D. 1965. Towards biological control of soil borne plant pathogens. pp. 4-17. In: *Ecology of Soil Borne Plant Pathogens* (eds. K.F. Baker and W.C. and Snyder.). Murry, London, UK.

Glazer, A.N. and Nikaido, H. 1995. *Microbial Biotechnology: Fundamentals of Applied Microbiology*. W. H. Freeman, USA.

Killham, K. 1994. *Soil Ecology*. Cambridge Univ. Press, Cambridge, UK.

Lindsay, K and Jones, M.G.K. 1990. *Plant Biotechnology in Agriculture*. Prentice-Hall, Eaglewood Cliffs, New Jersy, USA.

Nakas, J. P. and Hagedorn, C. (eds.) 1990. *Biotechnology of Plant Microbe Interactions*. McGraw-Hill Publishing Co., New York, USA.

Newbury, H. J. 1985. Some possibilities for genetic engineering of Rubisco. pp. 115-142. In: *Plant Genetic Engineering*. Cambridge Univ. Press, Cambridge, UK.

Postgate, J. R. , Dixon, R., Hill, S. and Kent, H. 1987. *Nif* gene in alien background. *Phil. Trans. Roy. Soc.* (London), 317B: 227-243.

Roizman, B. 1988. Molecular and genetic engineering: The principles, the power and the promise. *Science,* 239: 1534-1538.

Selvin, J., Ninawe, A.S., Sugunan, V.S., Sukumaran, N. and Lipton, A.P. (eds.) 2003. *Biotechnology: Emerging Trends.* Biotech Books, New Delhi, India.

Srivastava, S. and Srivastava, P.S. 2003. *Understanding Bacteria.* Kluwer Academic Press/ Springer (India) Pvt. Ltd., New Delhi, India.

Stalker, D.M., McBride, K.E. snd Malyi, L.D. 1998. Herbicidal resistance in transgenic plants expressing a bacterial detoxification gene. *Science,* 242: 419-423.

— ★★★ —

Glossary

A

A horizon : The surface horizon of a mineral soil having maximum organic matter accumulation, maximum biological activity, and/or eluviation of materials such as iron and aluminum oxides and silicate clays.

ABA : Abscisic acid, a terpenoid growth inhibitor synthesized from mevalonate as well as from the fragmentation of carotenoids.

Abiogenesis : Spontaneously generation of living organisms from non-living matter.

Absorption, active : Movement of ions and water into the plant root as a result of metabolic processes by the roots, frequently against an activity gradient.

Absorption, passive : Movement of ions and water into the plant roots as a result of diffusion along a gradient.

Acetylene-block method : Estimates denitrification by determining release of nitrous oxide (N_2O) from acetylene-treated soil.

Acetylene-reduction assay : Estimates nitrogenase activity by measuring the rate of acetylene reduced to ethylene

Acid phosphatase : An ectoenzyme involved in phosphorus solubilization

Acid rain : Atmospheric precipitation with pH values less than 5.6; the acidity being due to inorganic acids such as nitric and sulphuric acids that are formed when oxides of nitrogen and sulphur are emitted into the atmosphere.

Acid soil : A soil with a pH values <6.6; usually applied to surface layer or root zone, but may be used to characterize any horizon.

Acidity, active : The activity of hydrogen ion in the aqueous phase of a soil; measured and expressed as a pH value.

Acidophile : The microorganisms active in acidic soils with optimum pH range 2.0 to 3.0.

Actinomycete : Also known as 'ray fungi'; this aerobic G+ve bacterial group usually produces characteristic mycelium-like structures and reproduces asexually by spores.

Activated sludge : Solid sediments that has been aerated and subjected to bacterial action.

Active organic matter : A portion of soil organic matter that is relatively easily matabolized by microorganisms and cycles with a half-life in the soil of a few days to a few years.

Adenosine triphosphate (ATP) : Common energy-donating molecule in biochemical reactions; an important compound in transfer of phosphate groups.

Adsorption : The attraction of ions or compounds to the surface of a solid soil colloid adsorbing ions and molecules.

Adsorption complex : The group of organic and inorganic substances in soil capable of adsorbing ions and molecules.

Aeration, soil : The process by which air in the soil is replaced by air from the atmosphere. In a well-aerated soil, the soil air is similar in composition to the atmosphere above the soil. Poorly aerated soils usually contain more carbon dioxide and correspondingly less oxygen than the atmosphere above the soil.

Aerobic : 1) Having molecular oxygen as a part of the environment; 2) Growing only in the presence of molecular oxygen, as aerobic organisms; 3) Occurring only in the presence of molecular oxygen (said of certain chemical or biochemical processes, such as aerobic decomposition).

Aerotolerant anaerobes : Microbes that grow under both aerobic and anaerobic conditions, but do not shift from one mode of metabolism to another as the conditions change. They obtain energy exclusively by fermentation.

Aflatoxin : A metabolic product of *Aspergillus* species, especially *A. flavus* and *A. parasiticus*, possessing carcinogenic properties; a toxin produced in fungal infected cereals and most legumes.

Agar : Complex polysaccharide derived from certain marine red alga that is a gelling agent for solid or semisolid microbiological media. Agar consists of about 70% agarose and 30% agaropectin. Agar can be melted at temperature above 100ºC; gelling temperature is 40-50ºC.

Agarose : Nonsulfated linear polymer consisting of alternating residues of D-galactose and 3,6-anhydro-L-galactose. Agarose is extracted from seaweed, and agarose gels are often used as the resolving medium in electrophoresis.

Aggregate, soil : Many soil particles held in a single mass or cluster such as a clod, crumb, block or prism.

Air porosity : The proportion of the bulk volume of soil that is filled with air at any given time or under a given condition, such as a specified moisture potential; usually the large pores.

Akinete : A non-motile over-sized thick-walled spore formed singly within a cell; a survival structure in cyanobacteria and algae.

Alga (plural, algae) : Phototrophic eukaryotic microorganism. Algae could be unicellular or multicellular. Blue-green algae are not true algae; they belong to a group of bacteria called 'cyanobacteria'.

Aliphatic- An organic compound in which main carbon structure is a straight chain

Alkali soil : A soil that contains sufficient alkali (sodium) to interfere with the growth of most crop plants.

Alkaline soil : Any soil that has pH >7.3; usually applied to surface layer or root zone but may be used to characterize any horizon or a sample thereof.

Alkalophile : Organism that grows best under alkaline conditions (up to a pH of 10.5).

Allelopathy : The process by which one plant may affect other plants by biologically active chemicals introduced into the soil either directly by leaching or exudation from the source plant, or as a result of the decay of plant residues. The effects, though usually negative, may also be positive.

Allochthonous flora : Organisms that are not indigenous to the soil but enter soil by precipitation, diseased tissues, manure and sewage. They may persist for some time but do not contribute in a significant way to ecologically significant transformations or interactions.

Allophane : An alumino-silicate mineral that has an amorphous or poorly crystalline structure and is commonly found in soils developed from volcanic ash.

Alluvial soil : A soil developing from recently deposited alluvium and exhibiting essentially no horizon development or modification of the recently deposited materials.

Aluminosilicates : Compounds containing aluminum, silicon and oxygen as main constituents e.g. mocrocline $KAlSi_3O_8$.

Amensalism : Harmful association in which one species is inhibited/repressed by toxic compound or antibiotic production by other.

Aminization : Enzymatic conversion of proteins to amines and amino acids through hydrolytic reactions so the process leads to proteolysis.

Amino acids : Nitrogen-containing organic acids that couple together to form proteins. Each acid molecule contains one or more amino groups (NH_2) and at least one carboxyl group (COOH). In addition, some amino acids contain sulphur.

Ammonia volatilization : Gaseous loss of NH_3 from soil; accelerated under dry conditions when temperature is high, soil pH alkaline and greater air movement above the soil.

Ammonification : Biochemical process whereby ammoniacal nitrogen is released from nitrogen-containing organic material.

Ammonium fixation : The entrapment of ammonium ions by the mineral or organic fractions of the soil in forms those are insoluble in water and at least temporarily non-exchangeable.

Amorphous material : Non-crystalline constituents of soils.

Anabolism : Metabolic processes involved in the synthesis of cell constituents from simpler molecules. An anabolic process usually requires energy.

Anaerobic : Absence of molecular oxygen; growing in the absence of molecular oxygen, such as anaerobic bacteria; occurring in the absence of molecular oxygen, as a biochemical process.

Anaerobic respiration : Metabolic process whereby electrons are transferred from an organic, or in some cases inorganic compounds, to an inorganic acceptor molecule other than oxygen. The most common acceptors are nitrate, sulfate and carbonate.

Anamorph : Asexual stage of fungal reproduction in which cells are formed by the process of mitosis.

Anhydrobiosis : Life history strategy to tolerate desiccation in which organisms lose water, initiating lower physiological activity; upon rehydration the organism becomes active again.

Anion exchange capacity : The sum total of exchangeable anions that a soil can adsorb, expressed as centimoles of charge per kilogram (cmol kg^{-1}) of soil (or of other adsorbing material such as clay).

Anion : Negatively charged ion; during electrolysis it is attracted to the positively charged anode.

Annelid : Phylum of segmented aquatic and terrestrial worms. Each segment has its own excretory and sensory elements, with specialized functions (e.g. reproduction) associated with specific segments.

Anoxic : An environment without oxygen

Anoxygenic photosynthesis : Type of photosynthesis in green and purple bacteria in which oxygen is not produced.

Antagonism : Negative interaction in which either partner may suffer or one is benefited at the expense of other.

Antagonist : Biological agent that reduces the number or disease-producing activities of a pathogen.

Antheridium : Male gametangium found in the phylum Oomycota (Kingdom Stramenopila) and phylum Ascomycota (Kingdom Fungi).

Anthropogenic : A term referring to human activities.

Antibiosis : Inhibition or lysis of an organism mediated by metabolic products of an antagonist; these products include lytic agents, enzymes, volatile compounds, and other toxic substances.

Antibiotic : Secondary metabolic substance produced by certain organisms, which virtually ceases the active growth of other organisms due to inhibitory or biocidal effect.

Antibody : Protein that is produced by animals in response to the presence of an antigen and that can combine specifically with that antigen.

Antigen : Substance that can incite the production of a specific antibody and that can combine with that antibody.

Antiseptic : Agent that kills or inhibits microbial growth but is not harmful to human tissue.

Apatite : A naturally occurring complex calcium phosphate that is the original source of most of the phosphate fertilizers. Formulas such as $[3Ca_3(PO_4)_2]$; CaF_2 illustrate the complex compounds that make up apatite.

Apothecium : Open ascoma of fungi in the phylum Ascomycota.

Appresoria : The fan shaped complex of narrow branches of fungal hyphae which penetrate fine root cells and produce infection in the roots.

APS : Adenosine 5-phosphosulphate, the activated sulfate involved in reduction of sulphates.

Arbuscular mycorrhiza (AM) : Mycorrhizal type that forms highly branched arbuscules within root cortical cells.

Arbuscule : Special "tree-shaped" structure formed within root cortical cells by arbuscular mycorrhizal fungi.

Arbuscules : The specialized branched hyphae developed intracellularly within the cortical cells and serving as nutrient exchange sites between mycorrhizae and host.

Archaea : Evolutionarily distinct group (domain) of prokaryotes consisting of the methanogens, most extreme halophiles and hyperthermophiles, and *Thermoplasma*.

Archaebacteria : Older term for the Archaea.

Arid, climate : Climate in regions leading to sufficient moisture for crop production without irrigation. In cool regions annual precipitation is <25 cm and in tropical regions as high as 50 cm. Natural vegetation are desert shrubs.

Aromatic : Organic compounds which contain a benzene ring, or a ring with similar chemical characteristics.

Arthropod : Invertebrate with jointed body and limbs (includes insects, arachnids and crustaceans).

Ascoma (plural, ascomata) : Fungal fruiting body that contains ascospores; also termed an ascocarp.

Ascospore : Spores resulting from karyogamy and meiosis that are formed within an ascus; sexual spore of the Ascomycota.

Ascus (plural, asci) : Sac-like cell of the sexual state formed by fungi in the phylum Ascomycota containing ascospores.

Ascus : Well-differentiated fruiting bodies in Ascomycotina which bear sexual spores (ascospores) internally.

Aseptic technique : Manipulating sterile instruments or culture media in such a way as to maintain sterility.

Assimilation : The incorporation of inorganic or organic substances into the cell constituents.

Assimilatory nitrate reduction : The conversion of nitrate to reduced forms of nitrogen for use in the synthesis of amino acids and proteins.

Associative dinitrogen fixation : Close interaction between a free-living diazotrophic organism and a higher plant that results in an enhanced rate of dinitrogen fixation.

Associative symbiosis : Close but relatively casual interaction between two dissimilar organisms or biological systems. The association may be mutually beneficial but is not required for accomplishment of a particular function.

ATP : Adenosine triphosphate; the major carrier of phosphate and growth energy, composed of adenine and three phosphate groups.

Autochthonous organisms : The indigenous microbial population growing slowly in soil containing no easily oxidizable substrate; thought to subsist on the more resistant soil organic matter and little effected by the addition of fresh organic materials. Contrast with zymogenous organisms.

Autolysis : Spontaneous lysis.

Autoradiography : Detecting radioactivity in a sample, such as a cell or gel, by placing it in contact with a photographic film.

Autotroph : An organism capable of utilizing carbon dioxide or carbonates as the sole source of carbon.

Autotrophic nitrification : Oxidation of ammonium to nitrate through the combined action of two chemoautotrophic organisms, one forming nitrite from ammonium and the other oxidizing nitrite to nitrate.

Available nutrient : The portion of any element or compound in the soil that can be readily absorbed and assimilated by growing plants. ("Available" should not be confused with "exchangeable").

Available water : The portion of water in a soil that can be readily absorbed by plant roots. The amount of water released between the field capacity and the permanent wilting point.

Axenic : Literally "without strangers"; a system in which all biological populations are defined, such as a pure culture.

B

B horizon : A soil horizon usually beneath the A horizon that is characterized by one or more of the following; 1) a concentration of silicate clays, iron and aluminum oxides and humus, alone or in combination; 2) a blocky or prismatic structure, and 3) coatings of iron and aluminum oxides that give darker, stronger or redder color.

Bacteriophage : A virus that infects bacteria.

Bacteroid : Irregularly shaped cell form that certain bacteria can assume under some conditions, e.g. *Rhizobium* and *Bradyrhizobium* in root nodules.

Baecyte : Young dwarf daughter cell of cyanobacteria which do not have sufficient time to grow to their full size before next division.

Bar : A unit of pressure equal to one million dynes per square centimeter (10^6 dynes cm^{-2}); it is approximately equal to the pressure of a standard atmosphere.

Basidium : Well differentiated specialized club-shaped hyphal structures borne externally which are usually the sites of nuclear fission and meiosis in Basidiomycotina

Basidophile : The microorganisms active at alkaline pH (>8.0)

Benthic : Aquatic organisms that grow on or inhabit the bottoms of lakes and oceans.

Binary fission : Division of a cell into two by an apparently simple division of nucleus and cytoplasm.

Biodegradable : Subject to degradation by biochemical processes.

Biodynamic farming : A new farming approach for more efficient agriculture and forest management systems with thrust on organic recycling rather than showing complete dependence on inorganic farming thus viewing the whole farm as a living system.

Biofertilizer : Artificially multiplied microbial inoculants helpful to improve plant growth through supplementation of fertilizer requirement of plants or production of growth promoting substances.

Biogas : A byproduct of anaerobic decomposition of organic matter rich in methane.

Biological denitrification : see Denitrification.

Biological nitrogen fixation : Biochemical conversion of inert nitrogen (N_2) molecule into combined nitrogen by either reduction to ammonia (NH_3) or oxidation to nitrate (NO_3^-); usually occurs at ordinary temperatures and pressures;

Biomass : The total mass (dry weight) of living organisms of a specified type (e.g. microbial biomass) in a given environment (e.g. in a cubic meter of soil).

Bioremediation : The decontamination or restoration of polluted or degraded soils by means of enhancing the biochemical degradation or other activities of soil organisms.

Biotechnology : Any technique that uses living organism or a part thereof to make or modify a product, to improve plant or animal or to develop microorganisms for specific purposes.

Biotechnology, soil : Manipulation of soil organisms and their metabolic processes in such a way that leads to the optimization of crop productivity as well as yields maximum resource utilization or protects the environment for better human welfare.

Biotroph : A parasitic organism which obtains its nutrient supply only from a living host tissue regardless of whether or not it can be artificially cultured.

Biotrophy : A type of symbiotrophy wherein derivation of organic compound is from a living cell of symbiotic host.

Buffering capacity : The ability of a soil to resist changes in pH; commonly determined by presence of clay, humus and other colloidal materials.

C

C horizon : A mineral horizon generally beneath the solum that is relatively unaffected by biological activity and pedogenesis and is lacking properties diagnostic of A to B horizon. It may or may not be like the material from which A and B have formed.

Capillary water : The water held in the soil micropores or capillary pores by cohesive forces rather than gravitational forces, usually with a tension >60 cm of water; available for plant growth.

Carbon cycle : The sequence of transformations whereby carbon dioxide is fixed in living organisms by photosynthesis or by chemosynthesis, liberated by respiration and by the death and decomposition of the fixing organism, used by heterotrophic species, and ultimately returned to its original state.

Carbon : nitrogen ratio : The ratio of the weight of organic carbon (C) to the weight of total nitrogen (N) in a soil or in organic material.

Casing soil : A thin layer of soil, peat or sphagnum, added to beds for the induction of mushrooms.

Casts, earthworm : Rounded water-stable soil aggregates that have passed through the gut of an earthworm.

Cat clays : Wet clay soils high in reduced forms of sulphur that, upon being drained, become extremely acidic because of the oxidation of the sulphur compounds and the formation of sulfuric acid; usually found in tidal marshes.

Cation : A positively charged ion; during electrolysis it is attracted to the negatively charged cathode.

Cation exchange : The interchange between a cation in solution and another cation on the surface of any surface-active material such as clay or organic matter.

Cation exchange capacity : The sum total of exchangeable cations that a soil can adsorb. Sometimes it is also called 'total-exchange capacity', 'base-exchange capacity', or 'cation-adsorption capacity' and is expressed in centimoles of charge per kilogram (cmol kg^{-1}) of soil (or of other adsorbing material such as clay).

Cellulose : Water insoluble, semi-crystalline homopolysaccharide of glucose molecules with β $(1 \rightarrow 4)$ linkage; molecular weight varying from 2×10^5 to 2×10^6.

Chelate : (Greek, claw) A type of chemical compound in which a metallic ion is firmly combined with an organic molecule by means of multiple chemical bonds.

Chemoautotroph : The autotrophs which use diverse energy sources (organic or inorganic) and oxidize these reduced inorganic molecules to derive both energy and electrons; CO_2 being their carbon source.

Chemodenitrification : The non-biological (non-enzymatic) conversion of nitrate to gaseous nitrogen form of oxidized nitrogen, principally occurs under well-aerobic conditions through nitrite instability and dismutation to other oxides of nitrogen and to N_2.

Chemotroph : Organisms that obtain energy form the oxidation of chemical compounds.

Chitin : Second most abundant organic compound in nature; it has long chain polysaccharide with amino sugar, N acetyglucosamine in β $(1 \rightarrow 4)$ linkage, as basic unit and empirical formula of $(C_6H_9O_4.NHCOCH_3)_n$.

Chlamydospore : A thick-walled multi-nucleate asexual spore developed from hyphae but not on basidia of conidiophores; produced by many parasitic and mycorrhizal fungi.

Chlorosis : A condition in plants relating to the failure of chlorophyll (the green coloring matter) to develop, chlorotic leaves range from light green through yellow to almost white.

Clay : A smallest soil separate mostly colloidal consisting of particles <0.002 mm in equivalent diameter.

Clay mineral : Naturally occurring inorganic material (usually crystalline) found in soils and other earthy deposits, the particles being of clay size, that is <0.002 mm in diameter.

Claypan : A dense, compact, slowly permeable layer in the subsoil having much higher clay content than the overlying material, from which it is separated by a sharply defined boundary. Claypans are usually hard when dry and plastic and sticky when wet. See also hardpan.

Clod : A compact, coherent mass of soil produced artificially, usually by such human activities as ploughing and digging especially when these operations are performed on soils that are either too wet or too dry for normally tillage operations.

Coenocyte : A multi-nucleate cell or protoplast in which nuclear divisions have not been followed by cytoplasmic cleavage.

Colloid soil : (Greek, glue-like) Organic and inorganic matter with very small particle size and a correspondingly large surface area per unit of mass.

Commensalism : Asymbiotic association in which one species derives benefit from other while other associate apparently remains unaffected e.g. cellulolytic or proteolytic microbes degrade complex macro-molecules to simple monomeric units which are utilized by secondary heterotrophs.

Competition : Negative or harmful interaction in which species compete for the same limiting factor like nutrition, oxygen, space etc. so both suffer e.g. various heterotrophs competing nutrients or space in soil.

Compost : Organic residue or a mixture of organic residues and soil that have been piled, moistened and allowed to undergo biological decomposition. Mineral fertilizers are sometimes added; often called artificial manure or synthetic manure, if produced primarily from plant residues.

Composting : Controlled bioprocessing of fresh organic matter under moist and aerobic conditions carried out by successive microbial populations combining both mesophilic and thermophilic activities leading to the production of CO_2, water, minerals and stabilized organic matter.

Conjugation : The transfer of DNA from one organism to another via contact through plasmids

Conidium : A thin-walled secondary asexual spore borne terminally on a specialized hypha termed a conidiophore.

Copiotroph : Organisms which grow only under conditions of adequate nutrients and carbon supply.

Coralloid : Ectomycorrhizae when appearing as multi-forked root.

Course texture : Texture exhibited by sand, sandy loam and loamy sands, except very fine sandy loams.

Crop rotation : A planned sequence of crop growing in a regularly recurring succession on the same area of land, as contrasted to continuous culture of one crop or growing different crops in haphazard manner.

Cross protection : Protective physical barrier of feeder roots against pathogenic invasion.

Crystal : A homogeneous inorganic substance of definite chemical composition bounded by plane surfaces that form definite angels with each other, thus giving the substance a regular geometrical form.

Cyanobacteria : Photosynthetic prokaryotes containing chlorophyll a and phycobiliprotein pigments and capable of fixing atmospheric nitrogen; exist in unicellular, colonial or filamentous form and some even show simple parechymatous organization.

Cytokinin : N^6-substituted derivatives of adenine that are growth regulators in plants.

D

Deamination : The removal of an amino group (-NH_2), freeing ammonia.

Decomposition : Chemical breakdown of a compound (e. g. a mineral or organic compound) into simpler compound, often accomplished with the aid of microorganisms.

Denitrification : The biological reduction of nitrate or nitrite to molecular nitrogen or to oxides of nitrogen; this process is also termed enzymatic denitrification and biological denitrification.

Detritivore : An organism that subsists on detritus.

Detritus : Debris from dead plants and animals.

Diatomaceous earth : A geologic deposit of fine, grayish, siliceous material composed chiefly or wholly of the remains of diatoms; it may occur as a powder or as porous, rigid material.

Diatoms : Algae having siliceous cell walls that persist as a skeleton after death; any of the microscopic unicellular or colonial algae constituting the class Bacillariaceae; occur abundantly in fresh and salt water and their remains are widely distributed in soils.

Diazotroph : An organism or an association of organisms that can utilize dinitrogen (N_2) for growth.

Diffusion : The movement of atoms in a gaseous mixture or of ions in a solution, primarily as a result of their own random motion.

Disperse : 1) To break up compound particles, such as aggregates, into the individual component particles; 2) To distribute or suspend fine particles, such as clay, in or throughout a dispersion medium, such as water.

Dissimilatory nitrate reduction : The use of nitrate by organisms as an alternate electron acceptor in the absence of O_2 and causing reduction of nitrate to ammonium.

Diversity, microbial : The versatility in microorganisms within and between species/ microbial ecosystem.

DNA : Deoxyribonucleic acid, a type of cell nucleic acid carrying genetic information and containing adenine, guanine, cytosine, thymine, and 2-deoxyribose, linked by phosphodiester bonds.

DNA probe : A short segment of DNA or RNA (often <500 base) that have a reported group (radioactive or fluorescent or enzyme labelled) attached and can bind with high specificity to complementary sequence of target DNA.

DRMO : Deleterious rhizosphere microorganisms which may either be true or weak pathogens.

E

E horizon : Horizon characterized by maximum illuviation (washing out) of silicate clays and iron and aluminum oxides, commonly occur above the B horizon and below A horizon.

Earthworm : Animals of the Lumbricidae family that burrow into and live in soil; mix plant residues into the soil and improve soil aeration.

Ectendomycorrhiza : A type of septate mycorrhizae belonging to class Ascomycetes or Basidiomycetes and sharing some qualities both from ectomycorrhizae and endomycorrhizae;

Ectomycorrhiza : A symbiotic fungus-root association in which fungal hypha forms a compact mantle on the roots; mycelial strands penetrate inward between cortical cells and outward into the surrounding soil.

Ectorhizosphere : Rhizosphere region outside the root region, rich in plant and bacterial mucigel, root hair, plant mucilages, sloughed off root cap cells, etc.

E_h : Redox potential; a measure of the tendency of a substance to accept or donate electrons, dependent on both oxygen content and pH, which is expressed in volts or millivolts.

Enchytraeids : Smallest oligochaetes measuring <1 mm to several centimeters.

Endoenzyme : Intracellular enzymes primarily produced inside a cell.

Endomycorrhiza : A symbiotic fungus-root association with extensive intercellular and intracellular penetration of the root by the fungus; mycelial strands also extend outward into the surrounding soil.

Endophyte : An organism growing within a plant; the association may be parasitic or symbiotic.

Endorhizosphere : The root-epidermis-cortex zone colonized by pathogens and non-pathogens.

Epiphytes : The resident organisms on plant surfaces.

Erosion : Wearing away of land surface by running water, wind and ice, including gravitational creep and detachment and movement of soil rock by water, wind, ice, micro-orthropods etc.

Eu-edaphic : More specialized deep soil inhabiting micro-arthropods lacking pigment and eyes and have soft bodies with short appendages.

Eukaryote : An organism with a membrane bound nucleus within which the genomes are stored as chromosomes, e.g. fungi, protozoa, plants, etc.

Eutrophic : Having concentration of nutrients optimal (or nearly so) for plant or animal growth.

Eutrophication : A process of aging of water bodies caused due to the excessive algal growth; induced by runoff of high carbon and nutrient concentration, leading to oxygen depletion and creation of anaerobic conditions.

Evapotranspiration : The combined loss of water from a given area during a specified period of time by evaporation from the soil surface and by transpiration from plants.

Exoenzyme : Extracellular enzyme produced inside the cell but excreted through cell wall.

Exudates : The compounds of lower molecular weight leaking from root cells, without involving metabolic-mediation, into either intercellular spaces and then to soil via cell conjunction or directly through epidermal cell wall into soil.

F

Facultative organisms : Organisms capable of both aerobic and anaerobic metabolism.

Fanua : The animal life of a region or ecosystem.

Fermentation : Energy-yielding metabolism involving oxidation-reduction reactions in which an organic substrate or organic compounds derived from the substrate serve as the primary electron donor and the terminal electron acceptor; the energy (ATP) is produced by substrate level phosphorylation.

Fertilizer : Any organic or inorganic material of natural or synthetic origin added to a soil to supply certain elements essential for the growth of plants.

Fertility, soil : The quality of a soil that enables it to provide essential chemical elements in quantities and proportions required for optimum plant growth.

Fick's first law : Diffusion of gases is related to gas movement and soil characteristics like concentration of gases in soil and soil depth.

Field capacity : The amount of water present in soil when the downward movement of gravitational water (infiltration) is virtually ceased.

Fine texture : Texture having large quantities of fine fractions especially clay and silt.

First-order reaction : A reaction in which the rate of transformation of the substrate is proportional to the substrate concentration.

Flocculate, soil : To aggregate or clump together individual tiny soil particles, especially fine clay, into small clumps or flocculates; opposite of deflocculate or disperse.

Flush : The appearance of mushroom on the surface of the bed.

Fulvic acid : A mixture of organic substances remaining in solution upon acidification of a dilute alkali extract from the soil.

Fungi : Simple eukaryotes that lack a photosynthetic pigment; the individual cells have a nucleus surrounded by a membrane, and they may be linked together in long filaments called hyphae, which may grow together to form a visible body.

Fungistasis : Inability of non-dormant spore, sclerotia or propagules to germinate even under favourable conditions of pH, temperature and moisture in soil.

G

Gametangium : A differentiated structure in which gametes are produced, or the contents of which may function as a gamete.

Gamete : A sexually reproductive cell or unisexual protoplasmic body incapable of giving rise to another individual without conjugation with another gamete, and the joint production of a zygote.

Gene splicing : The cutting of a gene out of the DNA sequence of one organism and splicing it into the DNA of another.

Genetic engineering : Manipulating the genetic structure of an organism by transfer of genes from one organism to another.

Geosmins : Volatile terpene derivatives produced by some actinomycetes which give soil its earthy smell especially soon after the rainfall.

Gibberellin : Polycylic terpenoids phytohormone.

Gibbisite $(AlOH)_3$: An aluminum trihydroxide mineral common in highly weathered soils e.g. oxisols

Gleization : A soil-forming process under poor drainage conditions, resulting in the reduction of iron and other elements and causing gray colouration and soil mottling.

Gleying : A process involving changes in chemical speciation that result in changes in solubility; characterized by the reduction and subsequent partial solubilization and leaching of iron.

Gravitational potential : That portion of the total soil water potential due to the difference in elevation of the reference pool of pure water and that of the soil water. Since the soil water elevation is usually chosen to be higher than that of the reference pool, the gravitational potential is usually positive.

Gravitational water : Soil water drained through soil by gravitational forces after heavy rains and irrigation.

Green manure : Plant material incorporated with the soil while green, or soon after maturity, for improving the soil.

Greenhouse effect : The entrapment of heat by upper atmosphere gases such as carbon dioxide, water vapour and methane (just as glass traps heat for a greenhouse); increases in the quantities of these gases in the atmosphere will likely result in global warming that may have serious consequences for mankind.

Global warming : The rise in atmospheric temperature due to the excessive release of atmospheric CO_2, methane and other gases.

Grex : Fungal cells on starvation aggregate into a characteristic multicellular mass.

Growth : Orderly increase in all the components of living cell including multiplication.

Growth rate, microbial : The change in the amount of microbial biomass or number per unit time; follows sigmoid growth curve.

H

Halophyte : A plant that requires or tolerates a saline (high salt) environment.

Hardpan : A hardened soil layer, in lower A or B horizon, caused by cementation of soil particles with organic matter or with materials such as silica, sesquioxides, or calcium carbonate. The hardness does not change appreciably with changes in moisture content and pieces of the hard layer do not slake in water.

Harting net : The intercellular complex network of hyphae formed within the intercellular spaces between the cortical cells of a host root found in ectomycorrhizae.

Heavy metals : Metals with particle densities >5.0 Mg m^{-3} (>6 g cm^{-3}).

Heavy soil : A soil with high content of fine separates, particularly clay, or one with a high draw-bar pull, hence difficult to cultivate.

Hematite (Fe_2O_3) **:** Red iron oxide mineral imparting red colour to many soils.

Hemicellulose : Water soluble polysaccharides (polymers of hexoses, pentoses and uronic acids)

Hemi-edaphic : Micro-arthropods characterized by well developed eyes, long appendages and chitinous body with marked pigmentation present on soil surface or in litter layer.

Herbicide : A chemical that kills plants or inhibits their growth; intended for weed control.

Herbivore : A plant-eating animal.

Heterocyst : A large well differentiated specialized thick-walled cell occurring after every 10 to 15 cells in filaments of cyanobacteria; such cells separate intervening hormogonia and often are involved in nitrogen fixation.

Heterotroph : Organism that obtains energy, growth and cell carbon from organic materials.

Hormogonium : A segment or random fragment of filaments, marked off by heterocysts in a cyanobacterial filament, capable of detachment and development into a new filament.

Humic acid : A mixture of variable or indefinite composition of dark organic substances in soil humus, precipitated upon acidification of a dilute alkali extract from soil.

Humic substances : A series of complex, relatively high molecular weight, brown to black coloured humified fraction of organic matter formed by secondary synthesis reactions, that make up 60 to 80% of soil organic matter; generally quite resistant to microbial attack.

Humid climate : Climate in the regions where moisture when distributed normally throughout year should not limit crop production. In cool climate annual precipitation may be as little as 25 cm, in hot climate 150 cm or more.

Humification : The processes involved in the decomposition of organic matter leading to the formation of humus.

Humin : The organic fraction of soil humus that is not dispersible by weak alkali or pyrophosphates.

Humus : The amorphous dark-colored insoluble highly heterogeneous fraction of soil organic matter formed during the decomposition of organic residues; fairly stable lignoprotein complex or amino acid-lignin complex and containing humic and fulvic acids and other poorly defined or unknown substances relatively resistant to decomposition.

Hydraulic conductivity : An expression of the readiness with which a liquid such as water flows through a solid such as soil in response to a given potential gradient.

Hydroponics : Plant production systems that use nutrient solutions and no solid medium to grow plants.

Hydroxyapatite : A member of apatite group of minerals rich in hydroxyl groups; a nearly insoluble calcium phosphate.

Hygrophiles : The water loving microbes showing activities at soil moisture tension >7.1 MPa (pF 4.85).

Hygroscopic water : The water absorbed by dry soil from an atmosphere of high relative humidity or the water remaining in soil after its air-drying.

Hyperbolic reaction : One having a rate constant for which the plot of product formed versus time yields a curve approaching some maximum value, best described by a hyperbolic equation.

Hyphae : Branched or unbranched filaments that from the vegetative form, as in filamentous fungi, algae, and some bacteria.

I

Immobilization : The biochemical conversion of an element from the inorganic to the organic form by microorganisms or plants.

Inoculation : The process of introducing pure or mixed cultures of microorganisms into natural or artificial culture media.

Insecticide : A chemical that kills insects.

K

K selected species : Organisms adapted to live under more stable habitat than 'r' selected species; have greater competitive ability and operate at the levels close to the carrying capacity of the environment.

Karyogamy : The union of gamete nuclei to form a zygote nucleus.

k_c : The amount of microbial C evolved as CO_2.

K_m : The Michaelis constant, the substrate concentration of half the maximum velocity of an enzyme reaction.

L

Labile : A substance that is readily transformed by microorganisms or is readily available for uptake by plants.

Labile, soil organic matter : The organic matter liable to changes, remains unstable and is readily degraded to microbial products (comprises of 10-20% of soil organic matter); important reserviour of plant nutrients.

Langmuir equation : The hyperbolic equation commonly used in physical chemistry to describe adsorption phenomena.

Laterization : A soil-forming process in the humid tropics leading to the formation of soils with a low silica : sesquioxide ratio in clay fractions, low clay activity, low content of most primary minerals and soluble constituents, a high degree of aggregate stability, and usually red in color.

Leaching : Removal of materials in solution from the soil by percolating waters.

Leghemoglobin : An iron-heme protein controlling the oxygen content in legume nodules.

Lignin : The most resistant complex organic part of plant residues entering soil carbon pool; molecular weight varies from 10^4 to 10^6 and is a complex non-uniform polymer of aromatic ring with a basic building unit of phenyl-propane (C_6-C_3) type structure (coniferyl, coumaryl or sinapyl alcohol or guaiacyl prone monomers). It gives strength to plants and is most resistant to microbial attack.

Lime : In strict chemical terms, calcium oxide; a material containing the carbonates, oxides and/or hydroxides of calcium and/or magnesium used to neutralize soil acidity.

Lithotroph : see autotroph

Loam : The textural-class name for soil having a moderate amount of sand, silt, and clay; loam soils contain 7-27% clay, 28-50% silt and 23-52% sand.

Loamy : Intermediate in texture; properties between fine-textured and coarse textured soils. Includes all textural classes with the words loam or loamy as a part of the class name, such as clay loam or loamy sand.

Lysates : The compounds released by autolysis from older epidermal cells or through digestion of cell by bacteria.

Lysimeter : A device for measuring percolation (leaching) and evapotranspiration losses from a column of soil under controlled conditions.

Lysozyme : Enzyme that hydrolyzes peptidoglycans and thus degrades bacterial cell walls.

M

Macrofauna : Soil dwelling metazoan of size >2 to 20 mm.

Macronutrient : A chemical element necessary in large amounts (usually 50 mg kg^{-1} in plants) for the growth of plants; includes C, H, O, N, P, K, Ca, Mg, and S.

Macropore : Soil pores of >100 µm neck size; mostly occur between peds or aggregates; show free air and water movement.

Mantle : A well-developed compact fungal hyphal sheaths, usually 20-100 µm thick, formed around the roots between epidermal and cortical cells, apparently replacing middle lemella.

Matric potential : The portion of total soil water potential formed due to the attractive forces between water and soil solid phases as represented through adsorption and capillarity; it will always be negative.

Mesofauna : The soil-dwelling metazoans, of size <0.2 to 2 mm, large enough to cause disturbance in soil pores when moving through soil; includes nematodes, mites, collembola, enchytracieds, ants, insect larvae, etc.

Mesophile : The microorganism that grows at optimum temperature of 10-20°C.

Mesopore : Soil pores of 30-100 µm neck size; air and water movement medium.

Methaemoglobinaemia : 'Blue baby' syndrome in animals induced due to the accumulation of nitrate nitrogen in water to toxic level; causes defective carriage of O_2 in blood.

Methane (CH_4) **:** An odorless, colorless gas commonly produced under anaerobic conditions; when released to the upper atmosphere methane contributes to global warming.

Michaelis-Menten equation : see Hyperbolic reaction.

Microbial biomass : The living part of soil organic matter excluding plant roots and larger soil animals; represents total weight of living microbial cells in soil.

Microbiota : The soil biota having microscopic dimensions; includes soil microflora, protozoa, and microscopic metazoa.

Microfauna : Soil fauna of <200 µm size; individuals are too small to be clearly distinguished without the use of a microscope e.g. nematodes and protozoa; these are generally capable of moving through existing soil pores.

Microflora : That part of the plant population which consists of individuals too small to be clearly distinguished without the use of a microscope e.g. actinomycetes, algae, bacteria, fungi, etc.

Micronutrient : A chemical element necessary in only extremely small amounts (<50 mg kg^{-1} in the plant) for the growth of plants; examples are B, Cl, Cu, Fe, Mn, and Zn.

Micropore : Soil pore of <30 µm neck size; primarily confined within structural peds or aggregates (intrapeds); air and water movement is slow in micropores.

Mineral soil : A soil consisting predominantly of minerals and having its properties determined predominantly by mineral matter. Usually contains <20% organic matter, but may contain an organic surface layer up to 30 cm thick.

Mineralization : The biochemical conversion of organic nitrogen to mineral (inorganic) form; gross mineralization is the total amount converted, and net mineralization is the gross mineralization minus the growth demand or immobilization by decomposer organisms, usually measured in laboratory incubations.

Minor pathogen : A deleterious microorganism, basically saprophytes but can become weak parasites to juvenile tissues when aggressively colonizing the root surface, produce no visual symptoms and affect plants by their metabolites without parasitizing plant tissues.

Mollisol : A soil that has a dark-coloured high organic matter, highly base saturated surface horizon, usually associated with grasslands.

Monod equation : The hyperbolic equation commonly used to describe microbial growth rate.

Mor humus : Humic soil formed under waterlogged/low aeration conditions with slow organic matter breakdown leading to the formation of peat soil; highly acidic conditions generated inhibit the incorporation of organic matter into inorganic soil fraction so soil surfaces show large buildup/accumulation of organic matter on essentially inorganic soil; a type of forest humus layer of unincorporated organic material, distinct from the mineral soil, unless the latter has been blackened by washing in organic matter.

Mucilages : Poorly defined hydrophilic polyglucuronate and polygalacuronate substances which may be i) originating from root caps and epidermal cells and secreted by Golgi bodies, ii) hydrolysates of polysaccharides of primary cells between epidermal cells and sloughed off root cap cells; or (iii) produced by bacterial degradation of outer primary cell wall of old dead epidermal cells.

Muck : Highly decomposed organic material in which the original plant parts are not recongnizable; contains more mineral matter and is usually darker in color than peat.

Mulch : Any material such as straw, sawdust, leaves, plastic film, and loose soil that is spread upon the surface of the soil and plant roots from the effects of raindrops, soil crusting freezing, evaporation, etc.

Mull humus : A humus-rich layer of forested soil consisting of mixed organic and mineral matter. A mull blends into the upper mineral layers without an abrupt change in soil characteristics; humified organic matter is well incorporated in a mineral soil during decomposition process; formed under moist but well-aerated and neutral or mildly acid conditions.

Mutation : A change in the amount of structure of DNA in chromosomes which produces a discrete heritable difference compared with the "wild type".

Mutualism : A symbiotic positive interaction between species where both the partners are benefited e.g. mycorrhizae-plant interaction or legume-*Rhizobium* symbiosis.

Mycelium : The collective term for fungus filaments or hyphae; the assimilative phase of a body of hyphae.

Mycetismus : Poisoning caused by mushrooms.

Mycorrhizae : Beneficial symbiotic associations between plant roots and fungal mycelia.

Mycorrhizosphere : The altered rhizosphere as a result of mycorrhizal presence; the quality, composition and activity of rhizosphere microorganisms is changed due to the higher input of organic carbon into soil in mycorrhizal plants.

Mycotoxin : Toxins produced by certain moulds which grow on other food products.

Myxamoeba : The naked cell formed when a zoospore sheds its flagella and becomes ameboid.

N

***N*-Acetylglucosamine** and ***N*-Acetylmuramic acid** : Sugar derivatives in the peptidoglycan layer of bacterial cell walls.

NADP : The phosphorylated form of nicotine adenine dinucleotide (NAD); formed when the coenzyme serves as an electron donor in oxidation-reduction reactions.

Necrosis : Death associated with discolouration and dehydration of all or parts of plant organs, such as leaves.

Necrotrophy : A type of symbiotrophy wherein derivation of organic compound is from killed cells of a symbiotic host.

Nematodes : Very small worm abundant in many soils and important because some of them attack and destroy plant roots.

Net primary production (NPP) : The rate of organic carbon production by autotrophs in excess of respiration.

Neutralism : i) Depicts lack of interaction between two organisms in the same ecosystem; ii) two species behave entirely independently, so no specie is benefited or harmed.

Neutrophile : The microorganisms which grow in neutral soil conditions and are sensitive to acid or alkaline conditions.

Ng : nanogram, 10^{-9} grams.

Nitrate reductase : A membrane bound enzyme containing molybdenum and iron, besides labile sulphur groups; involved in reduction of nitrate to nitrite.

Nitric oxide reductase : A membrane bound enzyme involved in the reduction of nitric oxide to nitrous oxide; the enzyme is inhibited by the presence of nitrate.

Nitrification : The biochemical oxidation of ammonium to nitrite and nitrate, predominantly by autotrophic bacteria.

Nitrite reductase *:* A membrane bound enzyme which acts at second stage of denitrification in conversion of nitrite to nitric oxide.

Nitrogen assimilation : The incorporation of nitrogen into organic cell substances by living organisms.

Nitrogen cycle : The sequence of chemical and biological changes undergone by nitrogen as it moves from the atmosphere into water, soil, and living organisms, and upon death of these organisms (plants and animals) nitrogen is recycled through a part or all of the entire process.

Nitrous oxide reductase : A membrane bound unique enzyme in denitrification which is inhibited by acetylene (C_2H_2) and sulphide (HS^-) and is involved in conversion of nitrous oxide to molecular nitrogen.

Nodule, root : Swollen tumor-like structures found on plant roots; often used in reference to those in which symbiotic microorganisms live. Nodules consist of central bacteroid zone surrounded by layers of cortical cells; serves as a protective structure for bacterial nitrogen fixation enzyme, nitrogenase, against oxygen.

Non-humic substances : The portion of soil organic matter comprised of relatively low molecular weight organic substances; mostly identifiable biomolceules.

Non-humified substances : Decomposing residues, excretions and by-products of decomposers and decomposing biomass; these substances have definite/recognizable structural, molecular and other characteristics.

Nucleic acid : A large complex organic macromolecule composed of purine and pyrimidine bases, sugars, and phosphoric; may be combined with proteins as nucleoproteins.

O

O horizon : Organic horizon of mineral soils.

Oligotroph : The organisms growing in nutritionally poor environment.

Order of reaction : The value of exponential used to describe the reaction; reaction function of the concentration of one or more substrates being degraded.

Organic fertilizer : By-product from the processing of animal or plant substances that contain sufficient plant nutrients to be of value as fertilizers.

Organic soil : A soil saturated with water for prolonged periods unless artificially drained and containing at least 18% organic carbon (by weight) with mineral fraction >60% clay; or containing 12% organic carbon if mineral fraction has no clay; or between 12-18% if clay content of mineral fraction is between 0-60%. These are never saturated with water for more than a few days and generally have more than 20% organic carbon. Histisols develop on these organic soil materials.

Organotroph : Organisms obtaining growth energy and cell carbon from organic materials.

Osmotic potential : That portion of soil water potential developed due to the presence of solutes in soil water; the value is generally negative.

Osmotic pressure : Pressure exerted in living bodies as a result of unequal concentrations of salts on both sides of a cell wall or membrane. Water moves from the area of lower salt concentration through membrane to the area of higher salt concentration and, therefore, exerts additional pressure on the side with higher salt concentration.

Osmotrophy : A type of saprotrophy wherein assimilation of organic compounds occurs in a dissolved form.

Oxidation : The loss of electrons by a substance; therefore a gain in positive valence charge, and in some cases, the chemical combination with oxygen gas.

Oxygen diffusion rate : A measure of soil aeration which largely determines the rate at which O_2 can be replenished when used by respiring biomass in soil or removed out of the soil pores by water.

P

PAPS : 3-phosphoadenosine 5-phosphosulfate.

Parasitism : An association in which one specie derives nutrients for growth by parasitizing another live specie e.g. bacteriophage virus parasitizing bacteria; *Gliocladium* parasitizing *Fursrium.*

Pathogen, plant : A deleterious microorganism which penetrates a plant, disrupts its phloem and causes major disturbance/disease.

PCR : Polymerase chain reaction; a method to amplify a specific DNA sequence with the help of a thermostable DNA polymeraze enzyme and specific primer; complementary to target DNA.

P_e : Redox intensity; the tendency to be oxidized or accept electrons, P_e = -log [e^-]; also $P_e = E_h/0.059$

Peat : Unconsolidated soil material consisting largely of undecomposed or only slightly decomposed organic matter accumulated under conditions of excessive moisture.

Peat soil : An organic soil containing more than 50% organic matter.

Pectic substances : Polysaccharides consisting of galactouranan molecules interspersed with a much smaller number of rhaminose molecules and small chains of galacturonan and some other sugars.

Pectin methyl esterases : The enzymes that remove small branches off the pectin chains, alters their solubility and softens the molecules by the action of chain splitting pectinases (polygalacturonases, pectin lyases or transeliminases, etc.).

Ped : A unit of soil structure such as an aggregate, crumb, prism, block, or granule, formed by natural processes (in contrast to a clod, which is formed artificially).

Peptidoglycan : The rigid acetylgucosamine, acetylmuramic acid and amino acid layer of bacterial cell walls.

Percolation, soil water : The downward movement of water through soil; especially the downward flow of water in saturated or nearly saturated soil at hydraulic gradients of the order of 1.0 or less.

Permanent wilting point : The pF at which plants can no longer recover from wilt; usually occurs at near -1.5 MPa.

Permeability, soil : The ease with which gases, liquids or plant roots penetrate or pass through a bulk mass of soil or a layer of soil.

Permease : An enzyme that increases the rate of transfer of a substance across a membrane.

Pesticide : Any agent or chemical or a mixture of chemicals/substances intended to prevent, destroy, repel or irritate one or more pest populations causing economic loss in crop production.

pF : Logarithm of the soil water tension expressed in centimeter height of a column of water; used to measure soil water content.

PGPR : Plant growth promoting rhizosphere microorganisms that have the ability to colonize a rhizosphere aggressively, benefit the plant and inhibit minor pathogens.

pH, soil : The negative logarithm of the hydrogen ion activity (concentration) of a soil. The degree of acidity (or alkalinity) of a soil as determined by means of a glass or other suitable electrode or indicator at a specified moisture content or soil/water ratio, and expressed in terms of pH scale.

Phagotroph : A type of saprotroph wherein assimilation of organic compounds occurs in particulate form.

Phosphatase : The enzymes involved in dephosphorylation of all known organic phosphates of plant origin.

Phosphate fixation : Loss of phosphorus solubility due to adsorption or precipitation reactions in soil.

Photoautotroph : An autotroph that is capable of converting the energy of sunlight into the energy of chemical bonds.

Phototroph : An organism that obtains growth energy from light by means of photosynthetic processes.

Phycocyanin : Blue accessory pigments composed of arginine-asparagine polymers, found in cyanobacteria masking green chlorophyll.

Phycoerythrin : Red accessory pigment composed of arginine-asparagine polymers.

Phylloplane : The leaf surface adhered or anchored by microorganisms.

Phyllosphere : The collective surfaces of aboveground parts of live plants incited by microorganisms.

Phytin : Mostly inositol phosphate; most stable form of organic phosphorus in soil, synthesized both by microorganisms and plants.

Phytoalexin : Antimicrobial secondary metabolite induced and accumulated in plants once challenged by any foreign agent or spores or incompatible microbe; these are produced by healthy cells adjacent to affected one.

Phytohormone : A substance that triggers or initiates biochemical processes leading to the organic formation or influencing other aspects of growth positively; effective in a very low concentration and induces changes in the chemical composition within the plant.

Phytotoxin : Any metabolite primary or secondary, mostly of low molecular weight, which accumulates in toxic concentrations in a system; less commonly transported within plants and is sometimes active in greater concentration than growth regulators.

Pi : Inorganic phosphate.

Plankton : The aquatic organisms those are free-floating in bodies of water.

Plasmodium : A body of naked, multinucleated protoplasm exhibiting amoeboid motion.

Podzolization : A soil-forming process occurring in temperate to temperate-humid climates under coniferous or mixed coniferous and deciduous forest and characterized by a highly leached, whitish gray A_2 horizon.

Polyphosphate : Strong negatively charged poly-anions, the form of phosphorus stored in mycorrhizas; important in cation/anion balance of the fungi.

Pore size distribution : The volume of various sizes of pores in a soil; expressed as percentages of the bulk volume (soil plus pore space).

PPi : Inorganic polyphosphate.

Predation : One specie ingests or kills the other specie for food e.g. protozoa feeding bacteria; slime mould and myxobacteria feeding bacteria in soil.

Primary producer : The organisms which use energy from sun, carbon from atmosphere and water to make organic molecules and cell organelles

Primary production : The rate at which energy is stored by photosynthetic and chemosynthetic activities of producer organisms in the form of organic substances.

Priming effect : The peak of microbial activity during organic matter decomposition, that often stimulates the breakdown of even the original resistant and stable soil humus.

Productivity, soil : The capacity of a soil for producing a specified plant or sequence of plants under a specified system of management. Productivity emphasizes the capacity of soil to produce crops and is expressed in terms of yields.

Profile, soil : A vertical section of the soil through all its horizons and extending into the parent material.

Prokaryote : An organism whose genomes are not contained in a nucleus, e.g. bacteria.

Protein : Any of a group of nitrogen-containing organic compounds formed by the polymerization of a large number of amino acid molecules and that, upon hydrolysis, yields these amino acids. They are essential parts of living matter and are one of the essential food substances of animals.

Protocooperation : The positive interaction between two species but without the co-operation being obligatory for their existence or for their performance of some reaction. Two organisms grow independently under obligatory conditions e.g. nutritional protocooperation between *Proteus vulgaris* and *Bacillus polymyxa* for nicotinic acid and biotin, respectively, excreted by other partner on a medium lacking both the vitamins.

Protozoa : One-celled eukaryotic organisms such as amoeba.

Pseudoplasmodium : see grex

Psychrophile : The cold loving microorganisms growing in temperature range of 0-30ºC with optimum of 10-20ºC.

Puff ball : A type of mushroom edible in young state but sends puff of spores on ripening e.g. *Lycoperdon, Calvatia,* etc.

Q

Q_{10} : The expression to indicate that the rate of a chemical reaction is a direct function of temperature and microbial activity. In most soils, microbial activities (Q_{10}) increase two fold for each 10ºC rise between 0 to 35ºC.

R

R selected species : Organisms which show intrinsic growth rate so have characteristic ability to quickly exploit and exhaust ephemeral resource of habitat they colonize.

Reaction kinetics : Deals with the kinetic study of mechanisms reaction rates, by which reactions occur, and reactant concentration (e.g. substrates, inhibitors, activators and enzymes).

Redox potential : The electrical potential (measured in volts or millivolts) of a system due to the tendency of substances in it; thus to give up or acquires electrons.

Redox intensity (P_e) : The tendency of electrons to flow between compounds and is dependent on both oxygen content and pH; (synonym, redox potential); mathematically $P_e = -\log [e^-]$.

Reduction : The gain of electrons, and therefore the loss of positive valence charge by a substance. In some cases, a loss of oxygen or a gain of hydrogen is also involved.

Regolith : The unconsolidated mantle of weathered rock and soil material on the earth's surface; loose earth materials above solid rock.

Residence time : The time required to decompose an amount of substrate equal to the initial amount.

Resilience, soil : The capacity of soil microbial biomass to maintain its structure, diversity and function; term closely related to the concept of conservation of biodiversity.

Restriction enzymes : Ability to cleave double-stranded DNA at a defined nucleotide sequence.

Rhizobia : Bacteria capable of living symbiotically with higher plants, usually in nodules on the roots of legumes, from which they receive their energy, and capable of converting atmospheric nitrogen to combined organic forms; hence the term symbiotic nitrogen-fixing bacteria.

Rhizoid : Filamentous anchoring organs of some fungi lacking nuclei.

Rhizomorph : Well-differentiated multi-hyphal organs, important for solute transport over longer distances.

Rhizoplane : The zone of attachment of microorganisms to external root surfaces of live plants.

Rhizosphere : A zone of microbial activity within and around the roots rich in organic substrates or soil region immediately surrounding plant roots in which microorganisms are affected by live roots.

RNA : Ribonucleic acid, a linear polymer in which ribose residues are linked by 3'5'phosphodiester linkages to nitrogenous bases such as adenine, guanine, uracil, or cytosine.

Runoff : The portion of the precipitation on an area that is discharged from the area through stream channels; so it is lost without entering the soil is called surface runoff. The portion which enters the soil before reaching the stream is called groundwater runoff or seepage flow from groundwater. In soil science, runoff usually refers to the water lost by surface flow. In geology and hydraulics, runoff usually includes both surface and sub-surface flow.

S

Salinization : The process of accumulation of salts in soil.

Saprotrophy : A situation when derivation of organic compound is from a non-living environment.

Savanna (savannah) : A grassland with scattered trees, either as individuals or clumps; often a transitional type between true grassland and forest.

Secondary mineral : A mineral resulting from the decomposition of a primary mineral or from the re-precipitation of the products of decomposition of a primary mineral.

Secretions : The compounds of low and high molecular weights, actively pumped out of the roots as a result of various metabolic processes.

Selerotia : Perennating mass of interwoven hyphae or an organized structure of certain plant pathogens that differ from spore in germination.

Septate : Separated by cross walls.

Sewage effluent : The liquid part of sewage or wastewater, it is usually treated to remove some portion of the dissolved organic compounds and nutrients present from the original sewage.

Sewage sludge : Settled sewage solids combined with varying amounts of water and dissolved materials, removed from sewage by screening, sedimentation, chemical precipitation, or bacterial digestion.

Sheath : see mantle

Siderophore : A low molecular weight ferrichrome, a complex nitrogen-containing hydroxamate produced by organisms that have strong chelating affinity for iron.

Slow fraction (of soil organic matter**)** : The portion of soil organic matter that can be metabolized with great difficulty by the microorganisms in soil and therefore has a slow turnover rate with a half-life in soil ranging from a few years to a few decades. Often this fraction is the product of some previous decomposition.

Soil : 1) A dynamic natural body composed of mineral and organic materials and living forms in which plants grow; 2) The collection of natural bodies occupying parts of the earth's surface that support plants and that have properties due to the integrated effect of climate and living matter acting upon parent material, as conditioned by relief, over periods of time.

Soil aggregates : Specific microsites in soil with porosity lower than the average value of soil.

Soil biota : A complex entity of diverse organisms whose population always remains in a dynamic equilibrium influencing each other as per the soil fertility status.

Soil conditioner : Any material added to a soil for the purpose of improving its physical condition.

Soil conservation : A combination of all management and land-use methods that safeguard the soil against depletion or deterioration caused by nature or by anthropogenic activities.

Soil enzyme : The sum total of all the enzymes possessed by soil influencing the biochemical processes in it.

Soil fabric : Spatial arrangement of solid particles and associated pores or voids

Soil microbiology : The study of microorganisms living in soil which complete whole or some part of their life cycle in soil.

Soil organic matter : The organic fraction of the soil that includes plant and animal residues, soil organisms, microbial metabolites, and humified constituents; at various stages of decomposition, cells and tissues of soil organisms, and substances synthesized by the soil population; commonly determined as the amount of organic material contained in a soil sample passed through a 2 mm sieve.

Soil profile : A vertical section of soil from the surface through all its horizons including C horizon.

Soil solution : The aqueous liquid phase of the soil and its solutes consisting of ions dissociated from the surfaces of the soil particles and of other soluble materials.

Soil structure : The combination or arrangement of primary soil particles into secondary particles, units or peds. These secondary units may be, but usually are not, arranged in the profile in such a manner as to give a distinctive characteristic pattern. The secondary units are characterized and classified on the basis of size, shape and degree of distinctness into classes, types, and grades, respectively.

Soil water potential (total) : A measure of the difference between the free energy state of soil water and that of pure water. Technically it is defined as the amount of work that must be done per unit quantity of pure water in order to transport reversibly and isothermically an infinitesimal quantity of water from a pool of pure water, at a specified elevation and at atmospheric pressure to the soil water (at the point under consideration). This total potential consists of the following potentials.

Spawn : A pure culture of mushroom fungus grown on an organic rich medium and used for seeding purpose in mushroom cultivation.

Specific growth rate : Intrinsic rate of increase in growth.

Spermosphere : The region around a germinating seed that has greatest potential for microbial effects to have lasting influence on plants.

Spodosol : An order in the U.S. Department of Agriculture Soil Taxonomy in which a mineral illuvial soil horizon accumulates amorphous aluminum and organic carbon, and may or may not accumulate iron; occurs under leaching conditions in forests.

Sporangium : Asexual spore bearing structure which contains kidney-shaped or oval or pear-shaped sporangiophores; may be terminal or intercallary.

Sporocarp : A multicelled body or aggregates of spores.

Stable soil organic matter : Resistant compounds in decomposing residues, microbial products and those formed as a result of interaction of organic compounds with each other and with the mineral components of soil especially clay.

Stalk : Characteristic straight or angular hyphal tips or bulbous spensors of fungal spores.

Starch : The major food reserve of plants; actually admixture of amylose (20%) and amylopectin (80%); both are long chain polymers having repeated units of glucose.

Substrate : A compound on which an enzyme (or organism) acts.

Symbiosis : The living together in intimate association of two dissimilar organisms, the collaboration being not necessarily the mutually beneficial.

Symbiotrophy : The organisms derive food or energy directly from another organism by virtue of permanent intimate contact.

Synergism : A particular type of mutual association where combined activity exceeds the sum of the individual activities.

Systemic : The capacity of pathogens or chemicals to spread throughout plants, rather than remaining localized.

T

Tg : Terragram, 10^{12} grams or one million metric tons.

Thallus : The assimilative part of algae, lichens and fungi.

Therapeutic index : The ratio between the minimum effective dose for the host and the minimum effective dose to kill the intended microorganism.

Thermophile : A heat loving microorganism that can tolerate temperature range of 45-75^{0}C with optimum range of 55-65^{0}C.

Tillage : Mechanical manipulation of soil for modifying soil conditions to suit crop production.

Tolypophagy : Hyphal disintegration within a cell during infection (also known as ptyophagy).

Transduction : The transfer of DNA or a gene or genes from a donor cell or bacterium to a recipient cell or bacterium through virus vectors (bacteriophages).

Transformation, DNA : The transfer of naked DNA from one bacterium to another.

Trichome : Filament of cyanobacteria excluding sheath.

V

Vermiculite : A 2 :1-type silicate clay usually formed from mica that has a high net negative charge stemming mostly from extensive isomorphous substitution of aluminum for silicon in the tetrahedral sheet.

Vesicles : Thin walled spherical to ovate irregularly lobed hyphal expansions of mycorrhizae between host cells, usually filled with lipids and serving as temporary storage organs.

Vesicular : 1) Unconnected voids with smooth walls; 2) Spherical structures formed inside root cortical cells by vesicular, arbuscular mycorrthizal fungi.

Vesicular arbuscular mycorrhiza (VAM) : Obligate symbionts mainly the members of fungal class phycomycetes subclass zygomycetes family endogonaceae; A common endomycorrhizal association characterized by the development of two types of fungal structures i) within roots cells, small structures known as arbuscles; and ii) between root cells, storage organs known as vesicles; host range includes many agricultural and horticultural crops;

Virus : Non-cellular infectious entities; a nucleoprotein entity which can replicate within living cells of the host; passes through bacterium-retaining filters

W

Water potential, soil : A measure of the potential energy of water at a point in a system relative to the potential energy of unbound pure water; mathematical description of water potential is 'ψ'.

Water-stable aggregate : A soil aggregate stable to the action of water such as falling drops, or agitation as in wet-sieving analysis.

Weathering : All physical and chemical changes produced in rocks, at or near the earth's surface, by atmospheric agents.

Wilting point : The point when the forces of adsorption and retention of water in soil become too large that plants can no longer extract water to satisfy their needs; about 15 bars

Wood-Werkman phenomenon : The uptake and incorporation into organic substances of CO_2 by heterotrophs; tracer data show that heterotrophs, although obtaining most of their cell carbon from organic matter, do obtain some cell carbon from CO_2.

X

Xanthoxins : A terpenoid growth inhibitor synthesized from mevalonate.

Xerophiles : The microorganisms showing activities in low soil moisture tension range 30 to 48 MPa (pF 5.48- 5.68).

Y

Yeasts : The unicellular septate ascomycetous fungi which lack mycelial fruiting bodies; the asci are naked, formed by the differentiation of yeast cells or hyphal segments, with no protective sterile tissues associated with them.

Z

Zeolites : Natural three dimensional fine-grained silicate minerals composed of alkali and earth metal crystals that have ability to separate gas molecules on the basis of size and shape; used as molecular sieve in industrial processes.

Zoospore : A motile or swarm spore.

Zero-order reaction : A reaction in which the rate of transformation of a substrate is independent of changes in substrate concentration.

Zygospore : A thick-walled resting spore formed by conjugation of gametes or, in the Zygomycetes, by fusion of similar gametangia.

Zymogenous, organisms : So-called opportunist organisms found in soils in large numbers immediately following addition of readily decomposable organic materials; subsisting on easily decomposable organics; responding quickly to freshly added organic materials.

— *** —

Index

A

Abiogenesis 9
Abiotic stress 317, 322
Abscisic acid 316, 356, 359
Acetylene reduction 14, 226
Acid phosphatase 311
Acid rains 89, 205, 256
Acid soil 191, 203, 237, 287, 343
Acidification 89, 189, 250, 265, 274, 277, 312
Acidophiles 87, 88
Actinomycetes 2, 14, 29, 33, 34, 35, 36, 37, 38, 70, 72, 75, 79, 80, 88, 102, 103, 108, 129, 133, 137, 139, 178, 179, 210, 238, 249, 250, 262, 265, 267, 283, 287, 291, 321, 339, 343, 358, 362, 373, 374, 381, 383, 400, 402
Actinorhizal symbiosis 214
Adenosine triphosphate (atp) 84, 111
Aerobic decomposition 125, 372, 380
Aeroponic culture 323, 327
Aflatoxins 367, 368, 369
Aggregate 33, 45, 71, 72, 73, 75, 76, 81, 85, 88, 93, 94, 95, 119, 127, 148, 149, 191, 199, 204, 208, 217, 306, 311, 316, 341
Agrobacterium 60, 199, 250, 323, 338, 344, 356, 357, 405, 446, 454, 455, 458, 463
Akinete 41
Algae 13, 14, 24, 25, 28, 29, 33, 34, 39, 43, 46, 47, 51, 52, 54, 55, 56, 58, 59, 61, 71, 75, 76, 83, 92, 102, 123, 154, 207, 267, 275, 279, 286, 330, 332, 333, 339, 348, 390, 391, 421, 452
Alkalinity 87, 143, 238, 273, 439
Allophane 169
Alternate nitrogenase 226
Aluminium 93, 246, 252, 258
Aluminosilicates 286
Amensalism 331
Aminization 178, 179
Ammonia volatilization 181, 207
Ammonical-nitrogen 236

Ammonification 13, 89, 92, 141, 178, 179, 190, 273, 401
Ammonium fixation 150, 180
Ammonium loss 207
Ammonium oxidizers 188, 190
Amylopectin 127, 128
Amylose 127, 128
Anabolism 25
Anaerobic digestion 388, 390, 391, 393
Anaerobiosis 10, 75, 85
Annual carbon input 122
Antagonism 314, 327, 331, 332, 332, 361, 457
Anthropogenic 150, 284, 288
Antibiosis 14, 18, 314, 331, 446
Antibiotic markers 105
Antibiotic resistance 365, 366, 462
Antibiotic resistance markers 462
Antibiotics 38, 102, 114, 314, 343, 355, 361, 362, 364, 366, 370, 398, 422
Antibody 105, 106, 458
Antigen 106
Antiseptic 12
Aquaculture 371, 382, 391, 392
Arbuscules 297, 300, 307, 314
Arbutoid mycorrhizae 296, 304
Archaebacteria 4, 5
Arrhenius equation 90
Arsenate 286, 287
Arsenic (As) 286
Arsenic solubility 286
Arsenite 286, 287
Arthropods 57, 60, 61, 67, 139, 443
Arylsulphates 268
Ascomycota 43, 48, 50, 301, 304
Ascospore 48
Ascus 48, 49
Assimilation 25, 41, 179, 183, 184, 185, 227, 228, 229, 230, 251, 260, 267, 268, 291, 334, 343, 381, 387, 445
Assimilation of fixed nitrogen 228
Assimilatory nitrate reduction 194
Assimilatory pathway 194, 229, 267
Associative symbiosis 14, 15, 218, 219
Autotrophic nitrification 190, 201
Autotrophs 25, 122, 188, 190, 192, 249, 334
Auxin 356, 357, 455, 456
Available nutrient 138, 414
Available potassium 282, 282
Axenic 101
Azolla 9, 42, 216, 420, 421, 428, 429, 452
Azolla inoculants 428
Azospirillum 12, 15, 96, 199, 200, 210, 212, 213, 218, 219, 229, 238, 308, 336, 340, 344, 357, 358, 379, 420, 421, 422, 431, 451, 457
Azotobacter 12, 14, 15, 18, 76, 83, 208, 210, 211, 212, 213, 217, 218, 219, 225, 226, 228, 233, 237, 281, 291, 335, 338, 340, 344, 348, 357, 358, 420, 421, 422, 429, 430, 431, 432, 433, 448, 450, 451, 452, 454, 457, 460

B

B exotoxin 439
Bacterial biomass 33, 340
Bacteriophages 29
Bacteroid 214, 219, 220, 222, 223, 237, 240, 424
Baecytes 39
Basidiocarp 303
Basidium 49
Basidophiles 87, 88
Behaviour of humic substances 148
Beijerinkia 76, 211, 340, 348, 416, 420, 429
Bgb flakes 427
Bgb inoculation 427
Bioaccumulation 402, 404
Bioavailability of heavy metal 288
Biochemical oxidation 35, 188
Biochemical reduction 281
Biocontrol agents 336, 388, 442, 446, 447, 450, 457
Bioconversions 17
Biodegradation 17, 402, 407, 411, 413, 414, 415, 416
Biodynamic farming 17
Bioenergy production 371, 388
Biofertilizers 3, 16, 253, 376, 419, 420, 421, 422, 423, 425, 431, 448, 449, 450, 452

Biofuel production 454
Biogas 13, 388, 389, 390, 391, 393, 394
Biogas digesters 389, 390
Biogas plants 389
Biogas production technology 389
Biogenesis 9
Biogeochemical cycling 3, 23, 26, 95, 294
Biogeochemical transformations 121
Bioleaching 292, 293
Bioleaching technique 292
Biolog 106, 107
Biological control of pathogens 445
Biological evolution 4, 6
Biological indicators 380
Biological nitrogen fixation 12, 14, 89, 204, 208, 238, 239, 240, 241, 348, 352, 402, 422, 423, 426, 457
Biological processes 68, 73, 85, 90, 148, 175, 288, 375
Biological productivity 121
Biological transformation 88, 122
Bioluminescence 460, 461, 462
Biomagnification 402, 404
Biomass carbon 109, 111, 126
Biopesticides 417, 419, 434, 435, 436, 448, 449, 450, 452
Bioprocessing engineering 460
Bioremediation 17, 413, 414, 416, 454, 460
Biosafety assessment 461
Biotechnology 16, 17, 21, 67, 97, 119, 171, 293, 326, 352, 370, 394, 416, 451, 452, 453, 454, 457, 458, 463, 464
Biotransformation 454
Boron 292, 308, 328
Browning reaction 145

C

C:N ratio 25, 111, 137, 139, 140, 143, 151, 185, 186
Calcium 8, 33, 235, 246, 249, 253, 283, 291, 308, 312, 318, 424
Capacity factor 247
Carbohydrates 122, 123, 126, 127, 134, 141, 182, 213, 236, 263, 267, 305, 306, 307, 309, 315, 320, 341, 359, 370, 387
Carbon cycle 5, 121, 122, 123, 169, 170, 172
Carbon source 26, 41, 51, 152, 192, 334, 342
Carnivores 28, 51, 60
Carrier-based inoculants 424, 425
Casing soil 386
Cat-clays 265
Cation exchange 94, 127, 143, 148, 382
Cation exchange capacity 382
Cellular slime moulds 45
Cellulolytic culture 374
Cellulose 36, 38, 42, 46, 49, 52, 125, 126, 127, 128, 129, 131, 135, 137, 141, 144, 148, 166, 168, 169, 305, 307, 324, 330, 367, 373, 387, 391, 425, 433
Celluloytic fungi 376
Characters of rhizobial strains 425
Chelates 250, 252, 278, 279, 289, 460
Chelation of iron 277, 311
Chemical energy 93
Chemical evolution 3, 4
Chemoautotrophs 25, 26, 93, 189
Chemodenitrification 197, 198
Chemoheterotrophs 26, 334, 373
Chemolithotrophs 264, 292
Chemostat 156, 157, 168
Chitin 36, 38, 42, 46, 112, 126, 132, 133, 134, 173, 177, 347, 401
Chloroform-fumigation-extraction method 110, 111
Chlorophyll 25, 38, 40, 42, 53, 57, 122, 173, 236, 305, 321, 368, 428
Classification of biofertilizers 420
Clay 3, 4, 33, 62, 69, 71, 72, 74, 76, 77, 80, 93, 94, 95, 105, 113, 127, 133, 138, 139, 142, 143, 146, 147, 149, 150, 152, 165, 169, 174, 175, 180, 182, 203, 204, 207, 245, 246, 248, 252, 259, 283, 311, 344, 345, 382, 392, 404, 405, 409, 414, 415, 425
Clay aggregates 142
Clay fixation 180
Clay minerals 3, 30, 72, 93, 94, 143, 146, 147, 180, 246, 283
Clay soil 182, 382
CO_2 evolution 18, 84, 110, 139, 164
Cobalt 289, 291, 292, 293

Collembola 60
Co-metabolism 412, 414
Commensalism 331, 332
Competition 19, 73, 83, 140, 182, 193, 237, 314, 323, 329, 331, 332, 333, 336, 343, 361, 446, 448
Competition for substrate 446
Complexation/chelation of metals 288
Composition of organic matter 95, 142
Compost 13, 37, 265, 373, 374, 375, 376, 377, 378, 379, 380, 381, 382, 383, 385, 386, 387, 392, 420, 422, 433, 434
Compost heap 373, 374, 378, 381
Compost inoculants 420, 433
Composting 372, 373, 375, 377, 378, 379, 380, 382, 383, 393, 394, 434
Composting process 373, 375, 376, 381
Composting systems 377
Confocal microscopy 100, 120
Conidia 36, 42, 49, 50, 51, 445
Conjugation 39, 366, 410, 415, 454, 455
Consequences of denitrification 205
Copiotrophs 51
Copper 278, 279, 289, 292, 293, 312, 316, 396, 427
Copper deficiency 279
Copper precipitation 279
Copper toxicity 279
Corrosion of iron 278
Crop productivity 15, 21, 67, 119, 171, 173, 249, 370, 395, 421, 422, 450, 454
Cross inoculation 223, 224, 297
Cross inoculation groups 223, 224
Cross protection 313, 448
Crown gall 344, 357, 446, 455, 456
Cryptic growth 168
Cyanobacteria 5, 6, 14, 25, 34, 38, 39, 40, 41, 42, 54, 55, 70, 123, 210, 212, 215, 216, 217, 226, 249, 250, 279, 291, 330, 340, 420, 421, 422, 426, 427, 428
Cysteine desulphahydrase 260
Cytochromes 267, 274
Cytokinins 220, 311, 356, 357, 358, 380

D

Dark field illuminations 100
Deamination 95, 179, 410
Decarboxylation 179
Decay rate constant 168
Decomposition 24, 35, 49, 51, 57, 59, 61, 62, 70, 80, 84, 89, 93, 94, 98, 110, 121, 124, 125, 126, 127, 128, 129, 131, 135, 136, 137, 138, 139, 140, 141, 142, 150, 151, 153, 158, 159, 160, 161, 162, 164, 165, 166, 168, 169, 170, 175, 177, 182, 186, 188, 192, 194, 198, 202, 203, 228, 255, 273, 276, 280, 281, 283, 284, 309, 311, 312, 317, 332, 337, 359, 360, 371, 372, 373, 374, 375, 376, 378, 380, 388, 393, 399, 401, 405, 406, 408, 409, 412, 413, 428
Decomposition of a pollutant 409
Decomposition of organic matter 12, 24, 35, 51, 89, 135, 136, 157, 255, 360, 380, 388
Decomposition process 89, 135, 137, 139, 140, 142, 164, 371, 372, 374, 375, 376, 380, 405
Degradative plasmids 414
Dehydrogenase activity 115
Deleterious Rhizosphere Microorganisms (DRMO) 344
Demethylation 144, 410
D-endotoxin 439, 440
Denitrification 92, 114, 141, 159, 161, 175, 191, 194, 195, 196, 197, 198, 199, 200, 201, 202, 203, 204, 205, 239, 240, 340, 352, 370, 401
Denitrifying bacteria 194, 340, 401
Density centrifugation method 324
Desulfomaculatum 212
Desulfovibrio 212, 267, 268, 278, 279, 332, 340, 370
Desulphuration 454
Detoxification of pesticides in soil 405
Detritus 60, 61, 63, 70
Diazotrophy 210
Dimethyl mercury 284, 285
Dinitrogen reductase 225, 232

Dissimilatory SO_4 reduction 267
Diversity 2, 11, 15, 17, 19, 20, 25, 30, 39, 52, 61, 68, 69, 70, 72, 74, 80, 88, 91, 92, 99, 106, 107, 115, 118, 291, 297, 317, 319, 321, 335, 336, 338, 339, 400, 443, 462
DNA hybridization assays 116
DNA probes 115, 116
DNA recombinant technology 457
Dynamics of organic matter 153

E

Earthworms 56, 62, 63, 65, 68, 77, 89, 129, 189, 380, 381, 404
Ecological balance 400, 461
Ecological imbalance 399
Ecological succession 29, 401
Ecosystem restoration 399
Ectendomycorrhizae 296, 304, 420
Ectomycorrhiza 14, 307, 319
Ectorhizosphere 336
E_h 84, 85, 87, 278
Electron microscopy 100
Elemental sulphur 258, 259, 262, 265, 268
Elicitors 350, 351
Emtomopathogenic fungi 443
Enchytraeids 65, 95
Endomycorrhiza 108, 297, 307, 452, 457
Endophyte 14, 441
Endorhizosphere 336, 338
Endospore 39
Endotoxin production 440
Energy flow 1
Energy source 5, 12, 25, 110, 131, 161, 181, 192, 198, 199, 213, 252, 263, 388, 455
Environmental consequences of so4 269
Enzyme activity 65, 114, 118, 236
Epiphytes 347
Ergot alkaloids 369
Ericoid mycorrhizae 296, 304
Erosion 7, 19, 57, 64, 152, 206, 246, 311, 317, 382, 392, 406
Erosion losses 64, 206
Ethylene 224, 225, 226, 311, 316, 344, 345, 356, 358, 359, 370, 398
Eu-edaphic 60
Eukaryotes 5, 30, 31, 42, 52, 100, 364
Eutrophication 42, 190, 253
Exoenzyme 113
Extraction of metals 292
Exudates 24, 80, 92, 95, 199, 200, 203, 212, 213, 235, 276, 281, 305, 307, 308, 309, 314, 320, 341, 342, 345, 346, 347, 352, 460

F

Facultative organisms 360
Facultative organotrophs 264
Farm yard manure 138, 425
Fate of a pesticide in soil 408
Fate of pesticides 402, 403
Fatty acid profile 107
Fermentation 9, 10, 13, 47, 88, 10, 374, 378, 391, 436, 441, 448, 449, 460, 461
Fermentation techniques 436
Ferrichrome 277
Ferrodoxin 228, 235, 237, 267
Ferrolysis 275
Fertilizer 175, 180, 204, 205, 206, 207, 211, 230, 243, 252, 265, 311, 319, 321, 375, 382, 388, 390, 396, 420, 421, 422, 427, 428, 429, 431
Field capacity 77, 78, 79, 139
Field level constraints 450
First order reaction 159, 160, 163, 168, 187
Flagellate fungi 45, 47
Flavodoxin 226, 232, 267
Flavonoid 220, 230, 234, 235
Fluorescent stains 103
Forms of organic phosphorus 248
Frankia 14, 38, 214, 215, 229, 238, 241, 420, 421
Free living aerobes 210
Free living anaerobes 212
Free living facultative aerobes 211
Free living microaerophiles 212
Free living phototrophs 216
Fulvic acid 145, 147, 148, 249, 259, 278
Fumigants 397, 400, 401, 402
Fungal biomass 109, 306, 308, 311
Fungal sheath 304, 312

Fungi imperfecti 48, 50
Fungicides 237, 279, 320, 396, 397, 399, 400, 401, 402, 404, 406, 412, 417, 448, 458
Fungistasis 346, 351
Fungistatic properties 211
Fungivores 61
Fusarium toxins 369

G

G exotoxins 439
Gaseous composition 74
Gaseous diffusion rates 81
Gaseous losses 203, 256
Gel electrophoresis 118
Generation time 103, 154, 189, 213
Genetic diversity 39
Genetic engineering 15, 16, 30, 240, 326, 454, 455, 457, 459, 463, 464
Genetic modification of plants 455
Genetically engineered microorganism 454
Genetically modified microorganisms 16, 461
Genetics of pesticide 414
Genome tracking 461
Germ theory 9, 10
Gibberellins 211, 218, 237, 311, 358
Gleization 277
Gleying 277
Global warming 122, 169
Glomus 47, 250, 298, 321, 326, 432, 433, 451, 452
Glutamine synthase 183, 184, 227, 229, 231, 445
Glutamine synthetase 183, 184, 227, 229, 233
Gnotobiotic 342
Gogat 183, 184, 185, 227, 229
Granulosis 441, 443
Gravitational potential 78
Gravitational water 77
Green manure 216, 224, 379, 421, 428, 429
Green sulphur bacteria 25, 218, 263
Greenhouse effect 122, 139
Growth curve 154
Growth efficiency 168, 185, 342
Growth hormones 218, 310, 316, 334, 380, 455, 456
Growth yield 83, 155, 157, 185
Gypsum 8, 256, 258, 385

H

Hartig net 303, 304, 306, 307, 308
Haustoria 300, 304
Heavy metals 3, 143, 288, 293, 315, 316, 398
Hematite 277
Hemicellulose 38, 125, 126, 127, 128, 129, 131, 141, 166, 168, 387, 433
Hemi-edaphic 60
Herbicide 404, 445, 452, 458, 459
Heterocysts 40, 216, 217
Heterotrophic nitrification 190, 262
Heterotrophic sulphur oxidizers 262, 265
Heterotrophs 89, 94, 95, 190, 194, 204, 212, 268, 273, 278, 283, 289, 331, 334, 340, 348, 410
Higher fungi 46, 48
Hormogonia 39
Host genotype 237
Host specificity 219, 224, 230, 301, 320
Human pathogens 19
Humic acid 145, 146, 147, 148, 294, 421
Humic substances 141, 143, 145, 148, 169, 170, 171, 172, 382
Humification 64, 142, 374, 376
Humus 3, 8, 19, 29, 71, 76, 95, 97, 125, 126, 132, 139, 141, 142, 143, 144, 147, 148, 149, 150, 152, 153, 172, 175, 180, 186, 187, 245, 248, 252, 259, 276, 287, 292, 306, 322, 372, 399, 400, 404, 409
Humus formation 142, 399, 400
Hydrogen sulphide 205, 262, 278, 370
Hydroponic system 323
Hydroximates 277
Hydroxyapatite 246, 283
Hygrophiles 79
Hygroscopic water 77
Hyperauxy 357
Hyperparasitism 446
Hyperplasia 43
Hypertrophy 43, 357

I

Ice formation 459
Illite 72, 180, 282, 283
Immobilization 135, 136, 137, 140, 177, 178, 179, 181, 182, 183, 185, 186, 202, 205, 239, 245, 249, 251, 252, 261, 262, 372
Immunofluorescence technique 105
Immunofluorscence 100
Indole acetic acid 211, 235, 311, 316, 348, 357, 455
Industrial effluents 388
Infection process 298, 299, 303, 456
Inoculant 224, 424, 425, 427, 430, 449, 450
Inoculation methods 323
Inoculum potential 153, 320, 326, 446
Inorganic mercury 284
Inorganic phosphorus 244, 245, 246, 247, 251
Inorganic sulphur 12, 258, 259, 260, 262
Inositol hexaphosphate 249
Insecticide 400, 401, 407, 435, 438, 441, 451, 458
Integrated nutrient management 419
Intensity factor 247
Iron 3, 5, 13, 76, 98, 190, 196, 222, 225, 226, 235, 236, 246, 250, 252, 258, 262, 263, 264, 269, 273, 274, 275, 276, 277, 278, 279, 282, 284, 286, 290, 291, 292, 311, 312, 316, 318, 343, 353, 426
Iron oxidation 276, 277, 282
Iron oxidizing bacteria 5
Iron reduction 250, 277
Iron solubilization 274
Iron transformation 274

J

Jelly fungi 49

K

K selected organisms 28
K^+ fixation 282
Kaolinite 72, 94
K_c 111
K_m 162, 163, 164, 196
K_n 111, 187

L

Labile phosphorus 247, 247
Landfill application 392
Law of growth 155
Leaching 147, 149, 151, 180, 190, 200, 206, 255, 256, 257, 270, 277, 279, 292, 293, 294, 375, 383, 422, 460
Leaching losses 190, 200, 206, 422
Leaching rates 206
Lead 10, 71, 89, 122, 194, 215, 218, 245, 291, 293, 305, 315, 333, 342, 366, 369, 372, 375, 401, 406, 461
Lectin 220
Leghaemoglobin 210, 214, 222, 223, 235, 236, 237, 292
Lichens 42, 51, 55, 56, 215, 216, 270, 330, 348
Light intensity 319, 320
Lignin 38, 49, 125, 126, 127, 128, 129, 130, 131, 132, 137, 138, 139, 141, 144, 145, 148, 161, 165, 166, 168, 169, 186, 198, 305, 307, 332, 374, 376, 387, 428, 433, 434
Lignin theory 144
Lineweaver-burk double reciprocal plot 163, 164
Lipids 42, 91, 125, 134, 300, 325, 350, 364
Lithotrophs 25
Loam 77, 203, 204
Lower fungi 43, 46, 47, 48
Lysates 342
Lysozyme 116

M

Macro-aggregates 71, 72, 73, 252, 316
Macro-arthropods 61
Macrofauna 2, 56, 61, 67
Maintenance rates 168
Manganese deficiency 282
Manganese oxidizers 281, 286, 312

Manganese oxidizing bacteria 282
Manganese solubility 280
Manganese toxicity 282
Mantle 3, 295, 303, 304, 305, 306, 307, 308, 312, 313, 304
Mass production 325, 426, 430, 430, 431, 432, 441, 443
Mathematical models 165
Matric potential 78
Mean turnover 124, 159, 160, 166
Measurement of soil denitrification 200
Mechanism of N_2 fixation 227
Mechanisms of solubilization 250
Mercaptans 133, 256, 270, 373
Mesofauna 2, 23, 56, 61, 73
Mesophiles 79, 90, 376
Metabolic diversity 99, 106
Metal leaching 460
Metallic mercury 284
Metallic ores 274
Methaemoglobinaemia 190
Methane 5, 13, 62, 132, 139, 181, 205, 226, 290, 332, 372, 388, 390, 391, 394
Methanogenesis 85, 391
Methanogenic activity 391
Method of production 424, 426, 428
Methyl-arsenate 287
Methylation of mercury 284
Mica 4, 56, 72, 282, 283
Michaelis-menten kinetics 161
Microaerophile 211, 212
Micro-aggregates 71, 72, 73, 316
Micro-arthropod 60, 61, 399
Microbial activities 10, 12, 24, 63, 83, 85, 90, 92, 94, 113, 139, 151, 218, 248, 273, 372
Microbial biomass 2, 19, 24, 25, 30, 69, 70, 74, 92, 108, 109, 110, 111, 112, 115, 118, 119, 120, 124, 125, 133, 135, 136, 139, 153, 154, 164, 165, 166, 167, 168, 169, 171, 185, 186, 187, 249, 251, 259, 273, 279, 293, 306, 309, 321, 322, 326, 342, 371, 375, 382, 392, 405
Microbial biomass carbon 109, 111
Microbial biovolume 108
Microbial decomposition 84, 89, 126, 135, 139, 175, 177, 252, 373, 393, 409
Microbial diversity 17, 20, 72, 91, 106, 107, 115, 400
Microbial growth 80, 90, 96, 97, 102, 110, 112, 114, 120, 136, 141, 153, 154, 161, 168, 183, 185, 202, 262, 273, 290, 327, 335, 339, 342, 376, 400, 409, 413, 425
Microbial growth rate 154, 335
Microbial herbicides 445
Microbial inoculants 16, 265, 420, 421, 449, 450, 451
Microbial interactions 74, 329
Microbial mucilages 342
Microbial oxidation 89, 98, 278, 289, 294
Microbial processes 18, 80, 85, 276, 279, 293
Microbial products 71, 132, 137, 167, 168, 355, 356
Microbial reduction 255, 256, 285
Microbial respiration 78, 83, 86, 87, 110, 120, 194, 204, 342, 401
Microbial sensitivity 87, 90
Microbial solubilization of potassium 282
Microbial synthesis 127, 140, 252
Microbiological methods 101
Microfauna 2, 23, 56, 57, 60, 73, 76
Microflora 2, 23, 24, 60, 64, 65, 66, 73, 75, 76, 78, 80, 92, 101, 103, 104, 116, 126, 142, 237, 281, 334, 335, 336, 339, 341, 342, 343, 346, 348, 350, 352, 360, 373, 376, 399, 400, 417, 458
Micronutrient 149, 284, 288
Microscopy 9, 99, 100, 102, 103, 104, 111, 115, 120, 461
Mineral elements 273, 290
Mineral extraction 454
Mineral nitrogen 89, 140, 174, 178, 179, 180, 219, 236
Mineralization 25, 65, 95, 118, 119, 135, 175, 177, 178, 179, 181, 182, 183, 185, 186, 187, 190, 191, 192, 205, 206, 239, 245, 247, 251, 252, 255, 259, 260, 261, 262, 273, 279, 288, 292, 293, 309, 311, 312, 406, 409

Mineralization rate 406
Minerals 3, 7, 18, 19, 30, 72, 93, 94, 96, 97, 143, 146, 147, 148, 152, 180, 236, 244, 245, 246, 258, 267, 269, 274, 277, 279, 282, 283, 289, 290, 291, 293, 294, 307, 387
Minimum inhibitory concentration 362, 363
Modeling 96
Models in decomposition 168
Modified nutrient film technique 323
Molecular biology 15, 16
Molecular methods 115
Molybdenum 15, 194, 196, 225, 226, 227, 291, 293, 426
Molybdenum oxidation 291
Molybdo-protein nitrogenase 225
Monitoring of genetically modified micro-organisms 461
Monoclonal antibody technology 106, 458
Monomethyl mercury 285
Monotropoid mycorrhizae 296
Montmorillonite 72, 94, 282, 283
Mor humus 143
Most probable number method 102
Mould 387, 388
Mucigel 73, 152, 336, 339, 341, 342
Mucilages 152, 281, 317, 336, 342, 366
Mull humus 139
Mushroom cultivation 385
Mushrooms 49, 51, 301, 366, 383, 384, 385, 386, 387, 388, 394
Mutation 4, 366, 439, 448, 449
Mutualism 330
Mutualistic symbiosis 296
Mycorrhiza 152, 295, 305, 307, 308, 310, 317, 320, 321, 324, 327, 328, 421, 448, 450
Mycorrhizal development 85, 319, 320
Mycorrhizal inoculation 321
Mycorrhizal role 318
Mycorrhizosphere 305, 306, 309, 311, 312, 316, 340
Mycotoxicoses 366, 367
Mycotoxins 367, 368

N

Necrogenic toxins 360
Nematicides 396, 398, 402
Nematodes 51, 56, 59, 60, 65, 67, 75, 78, 80, 309, 312, 313, 314, 321, 333, 351, 374, 396, 402, 451
Net primary productivity 123
Neutralism 331, 332
Neutrophiles 88, 89
Nif gene 233, 457, 464
Nif regulation 233
Nitrate 35, 89, 114, 136, 141, 159, 174, 175, 176, 181, 183, 184, 185, 188, 189, 190, 191, 192, 193, 194, 195, 196, 197, 198, 199, 201, 202, 204, 205, 206, 208, 235, 239, 267, 278, 311, 372, 373, 385, 401, 420, 450
Nitrate assimilation 184
Nitrate leaching 190, 206
Nitrate oxidizers 188
Nitrate reductase 184, 195, 196, 202
Nitrate-nitrogen 190, 235
Nitric oxide reductase 195, 196
Nitrification 12, 13, 141, 150, 159, 175, 180, 188, 189, 190, 191, 192, 193, 201, 205, 206, 212, 238, 239, 241, 340, 423, 462
Nitrifiers 35, 80, 188, 189, 191, 192, 193, 241, 374
Nitrifying bacteria 12, 71, 175, 191, 373, 401, 402
Nitrite 88, 141, 183, 184, 188, 189, 190, 191, 192, 193, 194, 195, 196, 197, 198, 201, 202, 205, 273, 333, 401, 410
Nitrite reductase 184, 195, 196
Nitrobacter 26, 154, 188, 189, 191, 193, 250, 401
Nitrogen 173, 174, 175, 176, 177, 178, 179, 180, 181, 182, 183, 184, 185, 186
Nitrogen assimilation 185, 291, 445
Nitrogen cycle 174, 175, 176, 194, 212
Nitrogen cycling processes 92
Nitrogen fixation 11, 12, 14, 15, 42, 89, 175, 204, 208, 211, 212, 213, 214, 215, 217, 218, 223, 224, 225, 227, 228, 229, 230,

231, 233, 235, 236, 237, 238, 239, 240, 241, 279, 282, 291, 352, 401, 402, 422, 423, 424, 426, 428, 430, 431
Nitrogen fixation genes 231, 233
Nitrogen gas 176, 195, 197
Nitrogen metabolism 233
Nitrogen oxidation process 262
Nitrogen oxides 175, 176, 198
Nitrogen sources 173, 436
Nitrogenase 199, 208, 210, 211, 212, 213, 215, 216, 217, 222, 223, 224, 225, 226, 227, 228, 229, 230, 232, 233, 234, 235, 236, 237, 238, 311, 424
Nitrogenase activity 199, 212, 213, 215, 224, 225, 229, 230, 235, 236
Nitrogenase enzyme 208, 212, 217, 225, 226, 227, 228, 229, 233, 236, 274
Nitrogenase structure 240
Nitrosoamines 190
Nitrosomonas 26, 80, 88, 154, 188, 189, 193, 250, 401
Nod genes 220, 230, 233, 234, 235
Nodulation 15, 213, 215, 219, 220, 224, 230, 233, 234, 235, 236, 237, 311, 321, 369, 423, 425, 457, 459
Nodule development 219, 220, 222, 223
Nodule formation 219, 221, 234, 236
Non-humified substances 141
Non-labile phosphorus 247
Non-target effect 399, 400, 402
Ntr genes 233
Nuclear polyhydrosis virus 29
Nucleic acid 112, 116, 134, 177, 249, 355, 364
Nutrient cycling 59, 93, 153, 279, 309, 311, 312, 340, 399
Nutrient flow 93, 323
Nutrient mobilization 309, 310
Nutrient sink 93
Nutrient uptake 51, 136, 153, 168, 310, 312, 320, 321, 322, 326, 340, 343, 350, 383, 421, 432, 433

O

Obligate symbiosis 330
Oligochaetes 62, 65
Oligotrophs 71
Opines 455, 456
Order of reaction 158
Organic acid production 47, 252, 284, 312, 314
Organic carbon 5, 51, 67, 85, 86, 95, 122, 123, 125, 126, 135, 151, 167, 168, 170, 172, 198, 204, 311, 317, 341, 342, 408
Organic chelates 252
Organic fertilizer 265
Organic manure 13, 390, 421, 428
Organic matter turnover 3, 15, 26, 75, 121, 151, 165
Organic phosphorus 247, 248, 249, 250, 251, 252, 307, 311
Organic recycling 17
Organic residues 29, 57, 84, 139, 141, 164, 176, 185, 212, 261, 283, 310, 329, 337, 434
Organic soil 7, 150
Organic sulphur 257, 259, 261, 270
Organic waste recycling 371, 393, 394
Organic wastes 138, 372, 373, 381, 382, 383, 391, 393, 451
Organo-metallic complexes 288
Organotrophs 26, 210, 212, 213, 228, 264
Orthrophosphate 244, 311
Osmotic potential 72, 78
Osmotic pressure 74
Osmotrophy 334
Oxidation-reduction processes 277, 280, 289
Oxygen diffusion rate 81
Ozone 3, 5, 7, 204, 205

P

Parasites 11, 28, 29, 43, 60, 333, 335, 345, 442
Parasitism 51, 296, 331, 333, 335, 352, 446
Patulin 368, 369
PCR 116, 117, 118, 461, 462
Peat soil 279
Pectin methyl esterases 130
Ped formation 149

Pedon 7, 8
Peptidoglycan 38, 177, 364, 365
Persistence 29, 402, 404, 405, 406, 407, 435, 461
Persistence value 405, 406
Pesticide 395, 399, 400, 401, 402, 403, 404, 405, 406, 407, 408, 409, 412
Pesticide application 192, 320, 395, 400, 402
Pesticide chemicals 395
Pesticide movement in soil 404
Pesticide persistence 406
Pesticides 15, 19, 33, 143, 149, 256, 284, 287, 320, 332, 351, 388, 395, 396, 399, 400, 401, 402, 403, 404, 405, 406, 407, 408, 409, 410, 412, 413, 414, 415, 416, 417, 427, 427, 434, 435, 437, 452, 455, 458, 462
Pf value 77
Phases of organic matter decomposition 374
Phosphatase 95, 105, 115, 248, 251, 253, 311, 312, 401
Phosphatase activity 115, 312
Phosphate fixation 153, 246, 247
Phosphate translocation 307
Phosphorus 3, 4, 15, 55, 172, 236, 243, 244, 245, 246, 247, 248, 249, 265, 270, 296, 307, 308, 309, 310, 311, 312, 314, 316, 317, 319, 321, 327, 330, 340, 343, 371, 372, 375, 382, 390, 420, 427, 431, 432, 433
Phosphorus availability 246, 247, 251
Phosphorus cycle 244
Phosphorus deficiency 319
Phosphorus fixation 246, 247, 252
Phosphorus mineralization 251
Phosphorus reserviours 245
Phosphorus solubilizers 249, 375
Phosphorus sources 243
Photoautotrophs 25, 93
Photobiont 55
Photoheterotrophs 26
Photosynthetic bacteria 25, 28, 68, 86, 199, 455
Phototrophic purple non-sulphur bacteria 217
Phycobiont 55
Phycocyanin 38
Phycoerythrin 38
Phyllosphere 14, 212, 213, 340, 347, 348, 350, 353, 400
Phyllosphere organotrophs 212
Phytates 248
Phytin 248
Phytoalexins 313, 349, 350, 351
Phytohormone 355
Phytopathogens 49, 347, 364
Phytotoxins 313, 341, 355, 360
Plant diseases 10, 38, 328, 360, 364, 422, 430, 446, 452, 458
Plant pathogens 11, 29, 47, 51, 66, 312, 339, 342, 345, 346, 351, 360, 448, 452
Plant roots 13, 15, 23, 24, 59, 60, 61, 66, 69, 75, 77, 81, 83, 89, 92, 97, 101, 108, 168, 178, 182, 183, 193, 199, 218, 237, 244, 247, 248, 249, 295, 298, 306, 311, 320, 327, 328, 336, 339, 340, 342, 346, 351, 432
Plant-soil-mycorrhizae relationship 296
Plasmids 230, 233, 234, 235, 290, 366, 414, 416, 440, 454, 458
Plasmodium 43, 45
Podzol 89, 151, 278
Podzolization 277
Pollutants 15, 17, 399, 400, 404, 413, 414, 415
Pollution 20, 56, 169, 270, 293, 348, 382, 388, 392, 400, 402, 416, 420, 434, 435, 450, 454, 460
Pollution treatment 454
Polymeraze Chain Reaction (PCR) 116
Polyphenol mechanism 144
Polyphenols 138, 144, 276, 278
Polyphosphate 39, 251
Pore space 15, 72, 83
Potassium 3, 180, 243, 282, 283, 294, 385, 390
Potassium solubilization 282
Precipitation 5, 70, 89, 97, 145, 149, 151, 246, 274, 276, 279, 281, 288, 318
Predation 58, 94, 314, 331, 333, 446

Primary colonizers 42, 55, 56, 92, 137
Primary consumers 28
Primary minerals 93, 258, 283
Primary producers 28, 41, 55, 122, 125, 126
Primary production 28, 122, 123, 124, 135, 151
Primary productivity 1, 74, 123, 135, 150, 455
Priming effect 136
Production constraints 448
Profile development 149
Prokaryotes 5, 6, 30, 31, 100, 208, 227, 457
Proteins 28, 42, 65, 91, 100, 105, 112, 113, 125, 126, 132, 133, 134, 144, 147, 176, 177, 178, 182, 183, 184, 194, 220, 223, 225, 229, 233, 234, 255, 256, 258, 278, 289, 434, 438, 439, 440, 451
Proteolysis 175, 178
Protocooperation 330, 331, 332
Protozoa 13, 24, 28, 33, 34, 43, 56, 57, 331, 333, 339, 340, 374, 383, 402, 414
Pseudomonas 28, 33, 85, 128, 129, 130, 154, 178, 194, 198, 199, 213, 219, 248, 249, 250, 338, 339, 340, 343, 344, 346, 348, 356, 357, 358, 359, 420, 431, 432, 440, 441, 446, 454, 457, 458, 459, 460, 462
Psychrophiles 91
Puff balls 49
Purple and green sulphur bacteria 263
Purple sulphur bacteria 217
Putrefaction 10, 194

Q

Q_{10} 90
Quality standards 450
Quantification of pesticide degradation 412
Quantity factor 247

R

Rate of disappearance of substrate 162
Reaction kinetics 158, 159, 160, 161, 169, 187, 201
Reaction rates 161
Recalcitrants 17, 392
Reclamation of metals 274
Reductive deamination 179
Regolith 69
Regulation of gene function 233
Regulatory gene 233, 235
Residence time 159, 166, 311, 404, 413
Residual toxicity 404
Residue decomposition rate 169
Resistance development 407, 434
Resistance markers 366, 462
Resistance of pesticides 402
Resistant soil humates 141
Respiratory sulphate reduction 267, 268
Restriction enzymes 118
Rhizobacteria 199, 311, 344
Rhizobium 12, 76, 129, 199, 200, 210, 213, 215, 216, 219, 220, 222, 223, 224, 228, 230, 233, 234, 235, 237, 239, 250, 284, 287, 292, 308, 309, 326, 330, 331, 332, 335, 343, 356, 357, 358, 369, 401, 420, 421, 422, 423, 424, 425, 430, 432, 433, 448, 450, 451, 454, 457
Rhizobium specificity 424
Rhizodepositions 339, 342
Rhizomorph 303, 315
Rhizoplane 298, 310, 337, 338
Rhizosphere 13, 24, 26, 29, 199, 204, 206, 211, 213, 218, 219, 237, 277, 281, 282, 298, 305, 306, 310, 311, 312, 314, 317, 321, 322, 323, 325, 326, 332, 335, 336, 337, 338, 339, 340, 341, 342, 343, 344, 345, 346, 347, 350, 351, 352, 353, 382, 429, 432, 440, 446
Rhizosphere effect 337, 339, 346, 347
Rhizosphere microflora 92, 336, 342, 350, 352, 360, 458
Rhizosphere organotrophs 213
Rock phosphate 15, 265, 379
Root colonization 215, 313, 319, 320, 321, 326, 338, 345
Root exudates 24, 95, 124, 126, 200, 203, 213, 220, 235, 276, 305, 307, 308, 309, 320, 337, 342, 345, 346, 347, 400, 460
Root nodules 11, 12, 214, 215, 219, 223, 236, 237, 284

Root organ culture 323
Root pathogen 313, 433

S

Salinity 70, 238, 316, 433
Salinization 143
Saprophytes 37, 61, 135, 334, 335, 343, 344, 345, 448
Saprotrophy 51, 334
Sclerotia 48, 303, 321, 346
Secondary consumers 28, 125, 175
Secondary minerals 93, 245
Sediments 119, 172, 174, 212, 244, 256, 262, 267, 280, 284, 286, 288, 415
Selenates 287
Selenium 172, 287, 290, 293
Serine sulphahydrase 260
Sesquioxides 8, 145, 248
Sewage sludge 12, 17, 19, 138, 284, 288, 378, 383, 388, 392
Sheath 14, 38, 41, 218, 276, 282, 295, 303, 304, 312, 316, 317
Siderophores 277, 318, 343, 351
Slime moulds 43, 45
Slugs 56, 57, 65, 95
Smelting 292
Snails 56, 57, 65, 427
SO_2 pollution 270
Sodium 104, 111, 114, 116, 246, 283, 284, 286
Soil acidification 89
Soil acidity 89, 197, 203, 237, 337
Soil aeration 61, 81, 84, 90, 191, 192, 316
Soil aggregate formation 71, 75, 95, 148
Soil aggregation 51, 97
Soil animals 23, 60, 65, 68, 85, 86, 89, 108, 129, 135, 138, 158, 208, 212, 329, 399
Soil biochemistry 119
Soil biological diversity 88
Soil biota 23, 24, 67, 69, 70, 75, 80, 81, 92, 93, 104, 122, 329, 333, 399, 462
Soil biotechnology 21, 67, 97, 119, 171, 370, 453, 454, 457, 458
Soil carbon pool 122, 130
Soil conditioners 372
Soil conservation 310, 316
Soil depth 32, 71, 81, 125, 141, 250
Soil development 7, 23, 70, 150, 454
Soil ecology 3, 67, 97, 171, 326, 370, 463
Soil ecosystem 1, 15, 24, 66, 78, 81, 86, 87, 90, 92, 95, 96, 106, 107, 110, 114, 121, 133, 143, 169, 262, 288, 330, 362, 396, 400, 413, 454, 461
Soil environment 5, 68, 69, 79, 90, 101, 118, 140, 157, 190, 191, 262, 286, 333, 336, 416, 420, 455, 460, 461
Soil enzymes 15, 96, 112, 113, 115, 119, 120, 253, 321
Soil ethylene 359
Soil fabric 71
Soil fauna 24, 56, 61, 65, 68, 70, 78, 92, 103, 321, 414
Soil fertility 8, 12, 14, 15, 19, 21, 64, 66, 68, 92, 97, 98, 149, 152, 172, 253, 271, 279, 339, 371, 380, 421, 422, 431, 451, 454
Soil formation 7, 15, 18, 41, 42, 55, 57, 62, 93, 150, 151, 170, 258, 274, 277, 278, 289
Soil forming processes 71
Soil fungi 14, 42, 47, 51, 66, 67, 80, 82, 85, 135, 154, 186, 192, 351, 400
Soil health 2, 17, 18, 19, 20, 330, 433, 449
Soil horizon 8, 277
Soil humus 122, 126, 132, 142, 153, 175, 306, 409
Soil invertebrates 56
Soil microbial activity 126
Soil microbial diversity 107
Soil microbiology 1, 2, 8, 13, 15, 16, 20, 21, 65, 66, 68, 96, 97, 98, 99, 115, 119, 120, 169, 171, 253, 270, 293, 327, 353, 370, 452
Soil mineral fraction 93
Soil minerals 93, 94, 96, 97, 148, 267, 282
Soil moisture 63, 78, 79, 80, 82, 83, 92, 101, 191, 203, 263, 392
Soil molluscs 65
Soil nitrogen 175, 177, 181, 182, 205, 224, 238
Soil nutrients 51, 93, 96, 419, 430

Soil organic carbon 23, 95, 126
Soil organic matter 13, 17, 26, 27, 57, 62, 63, 73, 74, 75, 76, 94, 95, 98, 108, 110, 121, 122, 124, 126, 132, 133, 136, 137, 138, 139, 140, 141, 145, 146, 149, 150, 151, 152, 153, 164, 165, 166, 167, 168, 169, 172, 174, 175, 177, 181, 182, 185, 186, 201, 245, 249, 250, 251, 252, 292, 342, 353, 408, 409, 446, 450
Soil organic matter decomposition 62, 153
Soil organic matter turnover 26, 75, 121, 165
Soil organic nitrogen 175, 304
Soil oxygen 81
Soil permeability 206, 247
Soil ph 33, 85, 87, 88, 89, 189, 191, 207, 208, 247, 248, 262, 265, 269, 292, 312, 343, 347, 426
Soil pores 29, 60, 72, 81, 83, 202, 462
Soil productivity 18, 19, 62, 149, 189
Soil profile 7, 8, 58, 70, 89, 149, 180, 206, 408, 462
Soil quality 16, 17, 19, 381, 423
Soil reaction 70, 87, 88, 150, 191, 201, 203, 237
Soil respiration 24, 65, 83, 84, 109, 110, 402, 414
Soil respiration rates 84
Soil sickness 13, 19
Soil solution 17, 78, 82, 87, 93, 159, 180, 199, 207, 243, 244, 247, 248, 249, 251, 262, 268, 276, 280, 283, 286, 288, 310, 318, 408, 409
Soil structure 3, 15, 38, 74, 76, 80, 103, 149, 170, 202, 247, 309, 345, 382
Soil temperature 33, 89, 90, 92, 101, 192, 197, 201, 203, 207, 251, 319, 344, 345
Soil water potential 78
Solar radiation 92, 93
Sordia 56
Spawn 385, 386
Spawn run 385
Spawning rate 386
Specific growth rate 155
Spermosphere 334, 335, 370
Sporangium 45, 46, 47, 313
Sporocarp 45, 259, 298, 301, 303
Stability of humus 142, 150
Stable soil organic matter 137
Stachybotrys toxins 369
Stages in biogas production 390
Starch 38, 40, 127, 128, 135, 237, 314
Stem nodules 15, 214, 224, 241
Succession of microorganisms 136
Sugar fungi 47, 136
Suicidal genes 462
Sulphatases 95, 260
Sulphate 89, 116, 35, 185, 256, 257, 258, 259, 260, 261, 263, 267, 268, 269, 270, 277, 279, 332, 370, 385, 427, 428
Sulphate adenyltransferase 267
Sulphate esters 257, 259, 261
Sulphate reducing bacteria 267, 268, 277
Sulphate reduction 267, 268, 269
Sulphate reduction pathway 268
Sulphite 258, 261, 267
Sulphur 3, 5, 7, 12, 13, 15, 25, 35, 55, 87, 89, 132, 148, 172, 196, 199, 200, 212, 217, 218, 239, 244, 253, 255, 256, 257, 258, 259, 260, 261, 262, 263, 264, 265, 267, 268, 269, 270, 271, 273, 279, 282, 283, 287, 289, 291, 292, 293, 294, 308, 321, 332, 397
Sulphur assimilation pathways 260
Sulphur cycle 244, 256, 257, 270, 271
Sulphur mineralization 259, 260, 261
Sulphur oxidation 35, 262, 263, 264, 265, 271
Sulphur oxide 256
Sulphur oxidizers 262, 265, 278
Sulphur oxidizing bacteria 13, 212, 269
Suphuric acid 265
Symbiont 15
Symbiont 200, 212, 219, 303, 306, 307, 309, 428
Symbiosis 14, 15, 51, 213, 214, 215, 218, 219, 220, 224, 229, 230, 239, 241, 296, 306, 309, 312, 327, 330
Symbiotic nitrogen fixation 14, 35, 175, 240
Symbiotrophy 334

Synergism 65, 331, 332
Synergistic 309, 321, 322

T

Termite mounds 62
Termites 56, 57, 61, 62, 439
Terrestrial ecosystems 21, 25, 68, 98
Tertiary consumers 28
Tetrathionate 258
Therapeutic index 362
Thermophiles 91, 376
Thermophilic fungi 102, 129, 373
Thiobacillus 15, 26, 88, 94, 194, 199, 200, 250, 262, 263, 265, 266, 269, 460
Thiosulphate 199, 258, 261, 262, 264, 267
Thylakoids 40, 53
Ti plasmid 455, 456, 463
Toxin production 348
Transduction 366, 454
Transformation kinetics 15, 193
Transformation of copper 278
Transformation of manganese 280
Transformation of mercury in soil 285
Transgenic insecticidal plants 440
Translocation 78, 238, 248, 282, 292, 307
Trichomes 38, 347
Trichothecene toxins 369
Turnover time 124, 125, 138, 159, 160, 164, 165, 166, 167, 307

U

Uranium 100, 292

V

Vam spore staining 325
Vanadium 226, 235, 237
Vanadium 291
Vermicomposting 380, 382
Vermiculite 72, 94, 180, 282, 283, 323, 324, 425
Vesicles 214, 215, 217, 297, 298, 300, 304
Vesicular arbuscular mycorrhizae 296, 297, 431, 432
Viruses 4, 24, 29, 30, 75, 237, 334, 383, 387, 434, 436, 441, 442, 443
Volatile sulphur compounds 260, 270
Volatilization 175, 177, 181, 205, 207, 208, 270, 404, 405, 406, 408, 412

W

Water activity 74, 90
Water fern 9, 42, 420, 428
Water holding capacity 83, 143, 182, 191, 203, 206, 382
Water moulds 45, 46
Water movement 62, 72, 152
Water percolation 62, 149, 316
Water pollution 293, 400
Water stress 78, 79, 80, 191, 315, 321
Wet-sieving and decantation procedure 324
White rot fungi 131
Wilting point 77, 191
Worm cast 62

X

Xanthoxins 359
Xenobiotic 35, 404, 413, 460, 462
Xerophiles 79

Y

Yeasts 42, 43, 48, 49, 51, 58, 76, 79, 102, 284, 285

Z

Zero order reaction 163, 187
Zinc 283, 284, 289, 292, 293, 310, 312, 321, 398, 433
Zygospore 298
Zymogens 26, 136

— *** —